GEOLOGY
and the
ENVIRONMENT

The Brooks/Cole Earth Science and Astronomy Series

Internet and Media Products in the Brooks/Cole Earth Science Series:

Blue Skies, College Edition, CD-ROM, *Samson*

Brooks/Cole's Earth Science Resource Center: http://www.brookscole.com/geo

Earth Online: An Internet Guide for Earth Science, *Ritter*

In-Terra-Active 2.0: Physical Geology Interactive Student CD-ROM, *Brown and Dunning*

The Brooks/Cole Geolink: A Geology and Oceanography Presentation Tool

The Brooks/Cole Astrolink: An Astronomy Presentation Tool

The Geology Workbook for the Web, *Blackerby*

Earth Science

Earth Science, *Dutch, Monroe, Moran*

Earth Online: An Internet Guide for Earth Science, *Ritter*

Earth Science Today, *Murphy and Nance*

Meteorology

Meteorology Today, 6th, *Ahrens*

Essentials of Meteorology, 2nd, *Ahrens*

Meteorology Today for Scientists and Engineers, *Stull*

Astronomy

Horizons: Exploring the Universe, 6th, *Seeds*

Foundations of Astronomy, 6th, *Seeds*

Universe: Origins and Evolution, *Snow and Brownsberger*

Astronomy: The Cosmic Journey, 5th, *Hartmann and Impey*

Moons and Planets, and Introduction to Planetary Science, 4th, *Hartmann*

Astronomy on the Web, *Kurland*

Current Perspectives in Astronomy and Physics, *Pompea*

The Universe Revealed, *Impey and Hartmann*

Great Ideas for Teaching Astronomy, *Pompea*

Astronomy: The Solar System and Beyond, 2nd, *Seeds*

Physical Geology

Physical Geology: Exploring the Earth, 3rd, *Monroe and Wicander*

Essentials of Geology, 2nd, *Wicander and Monroe*

Historical Geology

Historical Geology: Evolution of Earth and Life Through Time, 3rd, *Wicander and Monroe*

Combined Physical and Historical Geology

The Changing Earth, 2nd, *Monroe and Wicander*

Environmental Geology

Geology and the Environment, 3rd, *Pipkin and Trent*

Environmental Science

Living in the Environment: Principles, Connections, and Solutions, 12th, *Miller*

Environmental Science: Working with the Earth, 8th, *Miller*

Sustaining the Earth: An Integrated Approach, 5th, *Miller*

Current Issues Reader for Geology

Current Perspectives in Geology, 2000 edition, *McKinney, McHugh, Meadows*

Oceanography

Essentials of Oceanography, 2nd, *Garrison*

Oceanography: An Invitation to Marine Science, 3rd, *Garrison*

Oceanography: An Introduction, 5th, *Ingmanson and Wallace*

Introduction to Ocean Science, *Segar*

Marine Life and the Sea, *Milne*

third | edition

GEOLOGY
and the
ENVIRONMENT

BERNARD W. PIPKIN
University of Southern California

D. D. TRENT
Citrus College

BROOKS/COLE
™
THOMSON LEARNING

Australia • Canada • Mexico • Singapore • Spain • United Kingdom • United States

BROOKS/COLE

THOMSON LEARNING

Sponsoring Editor: *Nina Horne*
Project Development Editor: *Marie Carigma-Sambilay*
Marketing: *Rachel Alvelais/Carla Martin-Falcone/Mandie Houghan*
Editorial Assistant: *John-Paul Ramin*
Production Editor: *Tessa Avila*
Production Service: *Helen Walden*
Manuscript Editor: *Carole Bailey Reagle*
Permissions Editor: *Mary Kay Hancharick*

Interior Design: *Terri Wright*
Cover Design: *Irene Morris*
Cover Photo: *Christopher Talbot Frank/Panoramic Images*
Interior Illustration: *Precision Graphics*
Print Buyer: *Vena Dyer*
Typesetting: *Parkwood Composition Service*
Printing and Binding: *Transcontinental Printing, Inc.*

For more information about this or any other Brooks/Cole product, contact:
BROOKS/COLE
511 Forest Lodge Road
Pacific Grove, CA 93950 USA
www.brookscole.com
1-800-423-0563 (Thomson Learning Academic Resource Center)

Printed in Canada

10 9 8 7 6 5 4 3 2 1

Library of Congress Cataloging-in-Publication Data

Pipkin, Bernard W.
  Geology and the environment / Bernard W. Pipkin, D.D. Trent.--3rd ed.
    p. cm.
  Includes bibliographical references and index.
  ISBN 0-534-51383-2
  1. Environmental geology.   I. Trent, D. D.   II. Title.

QE38 .P48 2000
550--dc21

DR. BERNARD W. PIPKIN is Professor Emeritus at the University of Southern California. He received his doctorate from the University of Arizona and is a licensed geologist and certified engineering geologist in the state of California. After graduation he worked for the U.S. Army Corps of Engineers on a variety of military and civil projects, from large dams to pioneer roads in rough terrain for microwave sites. He has been a consulting engineering geologist as well as a university teacher since 1965.

Dr. Pipkin is past president of the National Association of Geology Teachers and is a Fellow in the Geological Society of America. He hosted the PBS 30-part program *Oceanus* that won a local Emmy for Best Educational Television Series. He recently shared the Clare Holdredge Award (1995) with Richard Proctor from the Association of Engineering Geologists for their book *Engineering Geology Practice in Southern California*.

Dr. Pipkin is a private pilot with a flight instructor's rating; he took many of the aerial photos in the book. He and his wife Faye have three grown children and live south of Los Angeles in Palos Verdes.

DR. D. D. "DEE" TRENT has been working at, or teaching, geology since 1955. After graduating from college, he was employed by an oil company where his geological skills were sharpened with projects in Utah, Arizona, California, and Alaska. When the company decided to send him to Libya, he decided it was time to become a college geology teacher.

He has taught 28 years at Citrus Community College in Glendora, California, and along the way has worked with the National Park Service, picked up a Ph.D. at the University of Arizona, done field research on glaciers in Alaska and California, visited numerous mines in the United States and Germany, mapped deserts in California and Arizona, and served as an adjunct faculty member at the University of Southern California, where he taught Field Geology. When not involved with geology, he's either skiing or playing the banjo in a dixieland band.

Dr. Trent and his wife Pat have two grown children and they live in Claremont, California.

brief | contents

contents

chapter one

Humans, Geology, and the Environment 1

chapter two

Getting Around in Geology 22

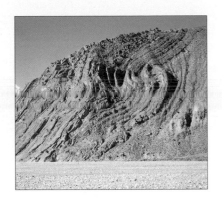

chapter
three

Plate Tectonics   56

chapter
four

Earthquakes and Human Activities   80

chapter five

Volcanoes   132

chapter six

Soils, Weathering, and Erosion   170

chapter seven

Mass Wasting and Subsidence   203

chapter eight

Fresh-Water Resources   241

chapter nine

Surface Hydrologic Hazards   274

chapter ten

The Coastal Ocean Environment   310

chapter eleven

Glaciation and Long-Term Climate Change   347

chapter twelve

Arid Lands and Desertification   390

chapter thirteen

Mineral Resources and Society   403

chapter fourteen

Energy Resources   450

chapter fifteen

Waste Management and Geology   503

Our Changing Planet

Environmental geology is the study of the relationship between humans and their geologic environment. An underlying assumption is that this relationship is interactive. Not only do naturally occurring geologic phenomena affect the lives of people each day, but human activities affect geological processes, sometimes with tragic consequences. Renowned historian Will Durant noted that our physical environment "exists by geological consent, subject to change without notice." We may regard this as the historian's assurance that during our lifetimes, we will probably be subject to an earthquake, a flood, a landslide, volcanic activity, or some other significant geologic hazard. No one is truly educated until she or he knows about our geological environment and these hazards.

Environmental geology is dynamic; that is, it is the geology of "now" and not some distant past. While this edition was being written, for example, the Izmit, Turkey, earthquake occurred, the largest and most deadly earthquake in Turkey in 60 years. It claimed 30,000 lives and caused some $10 billion in economic losses. Not too long after that an Amtrak passenger train traveling through the California desert was derailed by a very strong earthquake, the Hector Mine earthquake. This earthquake resulted in only a few injuries and no deaths because it occurred in an area where there are few people. Unprecedented emissions of sulfur dioxide gas at Hawaii's Kilauea Volcano recently forced short-term closure of Volcanoes National Park. Visibility was so reduced in the park that there was real concern for public safety and for the health of people with respiratory ailments. An earthquake was felt in several counties in Mississippi—serving to remind us that the most widely felt and perhaps the strongest earthquake in U.S. history was centered near New Madrid, Missouri, not too far north of Mississippi. Hurricane Mitch in 1998 resulted in mudflows in Central America that killed more than 10,000 people, and then in 1999 Hurricane Floyd produced torrential rains in North Carolina that killed millions of farm animals. The human misery and economic impact of these events is difficult to evaluate, but they illustrate that virtually no area is free of geologic hazards.

In a society where science and technology are interwoven with economics and political action, an understanding of the sciences is increasingly important. The National Science Foundation, the National Center for Earth Science Education, and several other prestigious earth-science organizations are promoting earth-science literacy in the United States for students from elementary school through college. The hope is that through education, today's students will become better stewards of our planet than their predecessors have been. How long can civilization sustain itself if it continues to harvest the earth's natural resources at the current rate? How much responsibility should we as individuals take for maintaining the earth that we share, and how much can we accomplish as individuals? Earth-science literacy opens our minds to such questions and allows us to give them better answers. It helps us to appreciate the earth's beauty, but also to recognize its limitations.

One thing is certain: the earth can sustain just so much life, and it is now being pushed to the limit. The fragility of earth was described very eloquently by *Apollo 14* astronaut Edgar Mitchell:

> It is so incredibly impressive when you look back at our planet from out there in space and you realize so forcibly that it's a closed system—that we don't have unlimited resources, that there's only so much air and so much water. You get out there in space and you say to yourself "That's home. That's the only home we have, and the only one we're going to have for a long time." We had better take care of it. We don't get a second chance.

About This Book

The authors, longtime friends and former classmates, have concentrated on different aspects of geology. One of us (Pipkin) has specialized in the study of geologic hazards and their mitigation—more specifically, in the safe siting of engineering works ranging from dams and tunnels to single-family dwellings. The other (Trent) has focused on geologic fieldwork, exploration for natural resources, and glacier studies. As colleagues in the teaching profession, it seemed natural for us to combine our past experiences and mutual interests in teaching and environmental protection in a textbook.

Geology and the Environment, Third Edition, is intended to fulfill the needs of a one-semester college course for students with little or no science background. It examines geologic principles, processes, and phenomena and relates them to human activities. What you learn in this course will be useful to you throughout your life. It will serve you as you form opinions about environmental issues, select a home site, or evaluate real property in a business venture. At the very least, when you understand how the earth "works," any fears you have about such things as earthquakes, hurricanes, and volcanic eruptions should be lessened.

This book presents geology that can be applied to improving the human endeavor. Chapter 1 presents an overview of the underlying cause of many of our present environmental problems. The stress of overpopulation on the environment has resulted in water and air pollution, land degradation, occupation of lands subject to geological hazards, and uncontrolled extraction of resources. The impact of overpopulation and the relatively new awareness of the environment are reasons that colleges are adding "environmental" courses such as environmental geology to their curriculums.

Because this book is intended for students who are not (yet!) geology majors, Chapters 2 and 3 offer some basic information useful for "getting around in geology." Chapter 2 presents information that is needed for the book's subsequent examinations of geological hazards and processes, mineral resources and so on. The emphasis is that the earth is an interacting "system." Students will see that one cannot divorce atmospheric and oceanic processes from processes that occur on the solid earth surface, nor moving tectonic plates (Chapter 3) from volcanic and seismic activity.

Following these groundwork chapters, the focus shifts to various types of geologic processes and hazards. Earthquakes and volcanoes are two of the most spectacular expressions of the energy contained *within* the earth, and Chapters 4 and 5 examine how this energy is transferred to the earth's surface—where we live—and how we can learn to recognize the signals that "something's up" and work to minimize the damage when the inevitable occurs.

The next two chapters focus on geologic processes that occur on or very near the surface of the earth. Much of the soil that we build our homes on and grow our food in today is bedrock of the far distant past that has been chemically and physically altered through weathering processes (Chapter 6). Present-day rocks and soils continue to undergo these processes, in some cases assisted by human activities such as construction, mining, and extraction of water from the earth. The results—mass wasting (landslides) and ground subsidence and collapse—cause deaths and millions of dollars of damage each year. These ground-failure problems are examined in detail in Chapter 7.

Earth is sometimes described as the water planet, and the presence or absence of water is essential in defining the geological and biological environments in which we live. Chapter 8 deals with fresh-water resources, with the emphasis on underground water and how we can protect and conserve this most essential resource. Streams and rivers are also vital water and esthetic resources. They sustain life along their banks,

and some of them are important means of transportation. They also pose hazards to people who live along their banks, however, most notably floods. These hazards are the subject of Chapter 9.

In Chapter 10 we look at the effects of the largest bodies of water on earth, the oceans, on life. Coastal geologic processes, beaches, and hurricanes—the most life-threatening of all natural hazards—are examined. Because oceans play an important role in global climate, the phenomenon known as El Niño is discussed.

Glaciers and glaciation and their impacts on humans are the subject of Chapter 11. This chapter logically includes a discussion of long-term climate change, which leads naturally to the consideration of the potential for global warming that is so much on people's minds today. The theme of climate extremes continues in Chapter 12, which deals with arid environments and the consequences of human-induced desertification.

Geology and the Environment, Third Edition, closes with a look at mineral resources—their origins, limits, extraction, and processing—and at the potential health hazards to humans from mining and smelting activities (Chapter 13). All sources of energy that can be utilized by humans, their geology and limitations, are considered in Chapter 14. The geology of disposing of waste materials generated by affluent, energy-rich societies is discussed in the final chapter.

DISTINCTIVE FEATURES OF THE BOOK

At most schools, the course for which this book is used is a component of the general education curriculum. General education broadens and enriches students' lives and minds beyond the specialization of their major interest. To this end, **Galleries** of photos at the end of each chapter illustrate many geologic wonders and some oddities. For example, at the end of Chapter 5 (Volcanoes), there are photographs of often-visited scenic volcanic features with explanations of how they formed. The Galleries are intended to stimulate students' curiosity and appreciation for natural geologic wonders.

Case Studies in each chapter highlight the relevance of the text discussion. These cover a broad spectrum of subjects and geographic areas, but many of them focus on the causes and aftereffects of bad environmental and geological planning. Also within each chapter are several provocative **Consider This** questions. They require students to apply the information just presented in the text and, thus, reinforce the learning. The questions will stimulate classroom discussion as well.

At the end of each chapter is a list of **Key Terms** introduced in the chapter, a **Summary** of the chapter in outline form, **Study Questions** geared to test understanding of the chapter's key concepts, and a list of books, articles, and Web sites **For Further Information.**

THE THIRD EDITION

- New to *Geology and the Environment,* Third Edition, are the chapter openers. The intriguing text and photos will capture students' attention and draw them into the chapter contents. All data have been updated where possible and appropriate.
- Many fresh case studies have been added.
- Because recent natural disasters are always newsworthy and instructive, earthquakes in Turkey and Taiwan and Hurricane Floyd are discussed in detail.
- The popular Galleries are still at the end of the chapters, offering some new glimpses of the unusual.
- Subsidence and mass wasting are now covered in one chapter (Chapter 7).
- Water material has been reorganized into chapters on water resources (Chapter 8) and hydrologic hazards (Chapter 9).
- A generous sampling of World Wide Web sites appears at the end of each chapter.
- We continue to employ the systems approach in this edition–the idea that all of the earth's reservoirs (atmosphere, hydrosphere, solid earth, biosphere, and extraterrestrial), and the processes acting within them are interconnected.
- Although we cite the various environmental laws in the text where they are applicable, a table of the most important environmental legislation appears in Chapter 1 for easy reference.
- Of course, the excellent quality and quantity of illustrations and figures have been maintained. Since our First Edition, adoptors have praised the fit of the art and photos to the text discussions.

Supplements

BROOKS/COLE THOMSON LEARNING EARTH SCIENCE RESOURCE CENTER

This full-service Web site offers an array of teaching and learning resources. It contains online testing, tutorials, discussion forums, maps, and many more features to make your course an interactive experience.

INFOTRAC® COLLEGE EDITION

Packaged free with each copy of the text, InfoTrac is an online library offering the full text of articles from hundreds of scholarly and popular publications. It is updated daily, and adopters and their students receive unlimited access for one academic term (four months).

EARTH SYSTEMS TODAY™ CD

This comprehensive CD can be found on the back inside cover of your book. Directly related to the text, the CD covers both geology and earth-science concepts. It allows students to manipulate live data sets, perform interactive exercises, review concepts, and quiz themselves.

STUDY GUIDE

The study guide is designed to help students master the course, with key concepts, terms, and review questions.

Acknowledgments

We gratefully acknowledge the thoughtful and helpful reviews by the following individuals, some of whom have also reviewed for earlier editions:

Thomas B. Anderson, *Sonoma State University*
Robert Boutilier, *Bridgewater State College*
Kathleen M. Bower, *Eastern Illinois University*
Christopher Cirmo, *State University of New York, Cortland*
Rachael Craig, *Kent State University*
Charles DeMets, *University of Wisconsin, Madison*
John Field, *Western Washington University*
Tark S. Hamilton, *Consultant*
Gilbert Hanson, *State University of New York, Stony Brook*
Douglas W. Haywick, *University of South Alabama*
David D. Jackson, *UCLA*
Hobart King, *Mansfield University*
Robert Kuhlman, *Montgomery County Community College*
Kenneth A. LaSota, *Robert Morris College*
Berry Lyons, *University of Alabama*
Harmon Maher, *University of Nebraska, Omaha*
William Mode, *University of Wisconsin, Oshkosh*
Robert Sanford, *University of Southern Maine*
Steven Schafersman, *Miami University*
Hongbing Sun, *Rider University*
Terry Swanson, *University of Washington*
Glenn D. Thackray, *Idaho State University*

We would be remiss if we did not acknowledge the contributions of other reviewers and those who have contributed published and unpublished materials and photographs as this book has evolved to its Third

Edition. These people have truly helped us build this book. The list is long, and our thanks are many to:

James L. Baer, *Brigham Young University*
William B. N. Berry, *University of California, Berkeley*
Lynn A. Brant, *University of Northern Iowa*
Don W. Byerly, *University of Tennessee, Knoxville*
Susan M. Cashman, *Humboldt State University*
Robert D. Cody, *Iowa State University*
Mark W. Evans, *Emory University*
Robert B. Furlong, *Wayne State University*
Bryan Gregor, *Wright State University*
Roger D. Hoggan, *Ricks College*
Alan C. Hurt, *San Bernardino Valley College*
Michael Lyle, *Tidewater Community College*
Lawrence Lundgren, *University of Rochester*
Garry McKenzie, *Ohio State University*
Robert Meade, *California State University, Los Angeles*
Marie Morisawa, *State University of New York, Binghamton*
William J. Neal, *Grand Valley State University*
Michael J. Nelson, *University of Alabama, Birmingham*
June A. Oberdorfer, *San Jose State University*
Charles Rovey, *Southwest Missouri State University*
Susan C. Slaymaker, *California State University, Sacramento*
Frederick M. Soster, *DePauw University*
Douglas J. Lathwell, *Cornell University*
Jack A. Muncy, *Tennessee Valley Authority*
T. K. Buntzen, *Alaska Department of Natural Resources*
Chuck Meyers, *U.S. Dept. of the Interior, Surface Mining Reclamation and Enforcement*
Ed Nuhfer, *University of Colorado, Denver*
Peter W. Weigand, *California State University, Northridge*
Joe Snowden, *University of Southeastern Missouri*
James L. Schrack, *Arco, Anaconda, Montana*
Marcie Kerner, *Arco, Anaconda, Montana*
Herbert G. Adams, *California State University, Northridge*
Lisa DuBois, *San Diego State University*
Edward B. Evenson, *Lehigh University*
Raymond W. Grant, *Mesa Community College*
Roger D. Hoggan, *Ricks College*
Steve Kenaga, *Grand Valley State University*

Barbara Hill, *Onondaga Community College*
Rita Leafgren, *University of Northern Colorado*
Joan Licari, *Cerritos College*
Larry Mayer, *Miami University*
David McConnell, *University of Akron*
James Neiheisel, *George Mason University*
Geoffrey Seltzer, *Syracuse University*
Albert J. Robb, III, *Mobil Exploration and Producing U.S., Inc., Liberal, Kansas*
Joan Van Velsor, *California Department of Transportation*
Lydia K. Fox, *University of the Pacific*
Bill Kane, *University of the Pacific*
Randall Jibson, *U.S.G.S.*
John E. Gray, *U.S.G.S.*
Edwin Harp, *U.S.G.S.*
Lynn Highland, *U.S.G.S.*
Robert Schuster, *U.S.G.S.*
Pam Irvine, *California Division of Mines and Geology*
Siang Tan, *California Division of Mines and Geology*
John Gamble and Ed Belcher, *Wellington, New Zealand*
Rick Giardino, *Texas A&M University*
Kenneth Ashton, *West Virginia Geological Survey*
Peter Martini, *University of Guelph*
Ward Chesworth, *University of Guelph*
Marion M. Gallant, *Colorado Department of Public Health and Environment*
Simon Young, *Montserrat Volcano Observatory*

Special thanks are due our colleagues Professors Leon Teng and James Dolan of the University of Southern California. They kept us current on the 1999 earthquakes in Taiwan and Turkey and provided the spectacular photographs of those events. In addition, Carole Bailey Reagle, who has edited the manuscripts of all three editions, is responsible for any "silk purse" created from our "sow's ear." Finally, we thank John-Paul Ramin of Brooks/Cole for his encouragement and guidance in our preparation of the manuscript.

BERNARD W. PIPKIN
D. D. TRENT

GEOLOGY
and the
ENVIRONMENT

chapter one

Humans, Geology, and the Environment

ASSOCIATED PRESS, 25 AUGUST 1999 "Turkey's Prime Minister has promised stricter building rules to prevent the shoddy construction blamed for the thousands of deaths from last week's massive earthquake. Eight days after the powerful 7.4-magnitude temblor reduced much of western Turkey to rubble, an estimated 200,000 survivors remained camped out in parks and on vacant lots."

The cause of the earthquake was a sudden rupture along northern Turkey's Anatolia fault. The severe ground motion that was generated collapsed hundreds of buildings, many of which had been built below earthquake-resistant standards. Chapter 4 provides more information about this event. Interestingly, the Anatolia fault and California's infamous San Andreas fault have some striking geographic similarities; see ◆▶ Figure 1. Let's hope that is where the likeness ends. ■

◆▶ FIGURE 1 Comparison of two earthquake-producing faults, the North Anatolian fault in Turkey and the San Andreas fault in California. The faults have the same sense of motion, are about the same age and length, and are capable of similar-magnitude earthquakes. The earthquake sequences shown on the North Anatolian fault represent events that occurred, one after the other from east to west, along the fault (see Chapter 4).

◆▶ FIGURE 2 Total collapse of a building during the earthquake of August 17, 1999, near I'zmit, Turkey. This apartment was built on top of the North Anatolia fault, which moved two meters at this location. Building inhabitants are not likely to survive such structural failure (see Chapter 4).

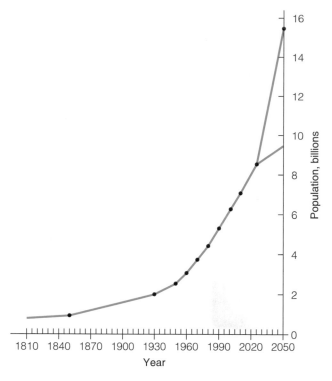

◆ FIGURE 1.1 World population projections to the year 2050 based on minimum and maximum birthrate estimates. Each tick on the time scale is 5 years.

As we began this work in 1999 we realized that few authors would be able to claim a "millennium edition." We realized also that the very concept of the millennium year 2000 ("Y2K") is based on our use of the Christian calendar. For Muslims, it is the year 1377; Jews view it as the calendar year 5759; and Hindus say it is the year 5101. Keep in mind that at the first millennium, the year 1000 of the Christian calendar (Y1K), there were no printing presses, calendars were rare, almost all information was disseminated by word of mouth, and few people had much to celebrate. The science of geology had not developed, and people didn't even realize they *had* a "geological environment."

How things have changed in a millennium! It is estimated that a quarter-billion people lived on earth 2,000 years ago. It took 15 centuries from year A.D. 1 to double the population to a half-billion people. The next doubling took only two to three centuries, as the population passed one billion in about 1830. That population doubled in only one century, to two billion in 1930, and the next doubling, to four billion people, took only 44 years (1977). In 1999 world population was six billion people, and the doubling to eight billion was expected to occur by about the year 2020 (◆ Figure 1.1). This growth pattern should be a wake-up call for citizens concerned about the earth's capacity to sustain the human population.

What Is Environmental Geology?

Geology (Greek *geo,* "earth"; *logos,* "knowledge of") is the study of the earth, and **environmental geology** is the application of geological principles to the effects of humans on their physical environment. Environmental geology is the geology of "now," as opposed to what happened five million years ago, and seeks ways for the

earth's inhabitants to interact safely with their geological surroundings. There is much evidence that the environmental problems of today have been developing for a long time, not the least of which is the explosive world-population growth since the early 1900s. In 1999 the U.S. population reached 274 million people, many of whom had experienced the effects of seven of the country's ten most costly natural disasters in history in the preceding decade. Two of these, the 1994 Northridge, California, earthquake (Chapter 4) and Florida's Hurricane Andrew in 1992 (Chapter 10), together cost at least $45 billion. They were the most expensive disasters to date in North America (◆ Figure 1.2).

The financial cost of natural disasters in the United States rose to about a billion dollars per week during the 1990s, a cost that many experts say will continue to rise (◆ Figure 1.3). Phenomena such as El Niño and global warming amplify the impact of flooding and hurricanes, and inflation diminishes the value of the dollar, but those things do not adequately explain the huge increase in U.S. costs for natural disasters in the 1990s. As ◆ Figure 1.4 shows, there has been no obvious increase in the number or severity of strong earthquakes that can account for the increased costs. It appears that nature is not becoming more violent; rather, that we are becoming more vulnerable because of population and land-occupation trends in our society. That is, we are occupying more lands that are in harm's way.

(a)

(b)

◆ **FIGURE 1.2** (a) One aftermath of the 1994 Northridge, California, earthquake in Studio City was a luxury car with its hind wheels on the street and its front wheels in the garage, which tumbled down a steep slope. (b) Homestead, Florida, after passage of Hurricane Andrew's 350 km (220 mi) per hour winds.

◆ **FIGURE 1.3** The cost of natural disasters in the United States in five-year increments, 1952–1996. The graph would appear slightly different if different five-year periods were used, but the sharp increase in the 1990s is obvious.

It was succinctly stated that "With an improved understanding, we can create public awareness that most natural disasters are not random acts, but rather the direct and predictable consequences of inappropriate land use . . . (G. van der Vink; see the "For Further Information" list at the end of the chapter). This could be rephrased, "Earthquakes don't kill people; buildings do." As populations grow, more buildings can be expected to topple on more people during earthquakes.

Some of life's choices have geological ramifications. For example, where one decides to live may depend upon the stability of a hillside, a neighborhood's low potential for flooding, or a community's distance from an active earthquake fault. A person's ability to make a sound decision in such a case would be enhanced by knowledge of environmental geology.

Environmental geology is largely concerned with surface geologic processes, processes that not only make human life possible but that also may threaten life. Some of these processes are soil formation and erosion, water supply and floods, landslides, and natural and human-made pollution. It applies the science of geology to problems arising from the complex interactions of water, air, soil, solid earth, and life. As such, it encompasses all the traditional branches of geological study such as mineralogy, petrology (rocks), geomorphology (landforms), hydrology (water), geophysics, and geochemistry. ◆ Figure 1.5 illustrates environmental geology's relationship to the traditional branches of geologic study and practice.

THE ENVIRONMENTAL GEOLOGIST

Environmental geology is a science, and like other scientists, the environmental geologist formulates hypotheses and then tests them. He or she uses observation, mapping and sampling of geologic materials, measurements and sometimes experiments to collect data. In all sciences the goal is to obtain results that can be reproduced by other scientists. The next step of the process is data interpretation, a totally different activity

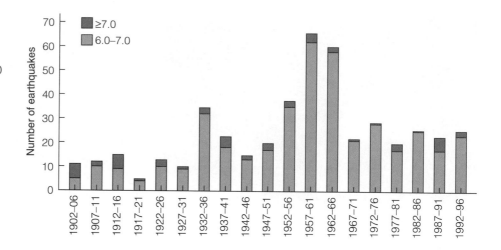

◇▶ FIGURE 1.4 Earthquakes in the United States with Richter magnitudes of 6.0 and above, 1902–1996. Earthquakes of greater than Richter-magnitude 7.0 are indicated. Magnitudes of 6.0 and above are designated as "damaging" earthquakes; those of 7.0 or more, "major" ones.

that occurs in the brain of the investigator. From this interpretation, we may arrive at conclusions. This is much the same process that a medical doctor follows with a patient, and the geologist, just as the physician, does not know for sure that a conclusion is correct until "the patient is opened up." Similarly, conclusions are not cast in concrete; they may have to be changed when additional data become available. These data may come from new excavations, drill holes, roadcuts, or geophysical investigations.

Oil and mining companies were the biggest employers of geologists from the turn of the last century

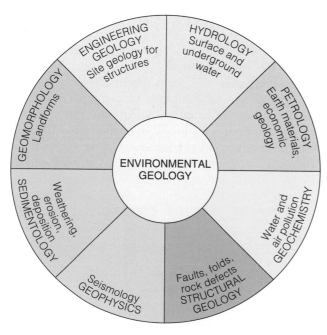

◇▶ FIGURE 1.5 The relationship between environmental geology and other geological sciences. Environmental geology incorporates many skills and specialties of the broad discipline of geology.

until the late 1950s. As cities grew, however, the need to protect public health and welfare created a large demand for **engineering geologists,** specialists who work with engineers on development and construction projects. Many older, uncontrolled projects were becoming public-health hazards, but they lay undiscovered. Then, with the discoveries of buried toxic waste at Love Canal in 1978, of leaking oil and gasoline tanks all over the country, and of agricultural chemicals in rivers and lakes, the specialty we now call *environmental* geology evolved. Its work is protecting existing environments and correcting past transgressions. Many environmental geologists work in areas of high geologic-hazard potential, such as sites subject to earthquakes or landslides (◇▶ Figure 1.6). Their efforts have significantly reduced injuries and economic losses. In the past much of this work was strictly within the realm of the engineering geologist, but today the distinction between "engineering" and "environmental" geology is blurred if it exists at all. (In fact, the old *Bulletin of the Association of Engineering Geologists* bears a new name, *The Bulletin of Environmental and Engineering Geosciences.*)

The work of many environmental geologists today is in identifying adverse geologic conditions and potential problems in areas being considered for development. Others work in resource planning. The demand for resources increases as cities grow. This requires advance planning for local sources of building materials, adequate good-quality drinking water, and determining optimal uses for given parcels of land. Many cities' growth problems are reducible to engineering or geological ones.

TOOLS OF THE TRADE

Multipurpose maps of areas being considered for development are essential to effective planning, and they

(a)

(b)

◆▶ **FIGURE 1.6** (a) Geologists use tilt-up drilling rigs on hillsides and in close quarters such as people's backyards. At this residence a landslide is encroaching upon the swimming pool. (b) Geologists carefully log the sidewalls of a trench across the San Andreas fault near Wrightwood, California, to find and date the youngest rupture (an 1857 earthquake) and then older breaks. The data allow earthquake recurrence intervals for the fault to be determined.

must be available long before the bulldozers arrive. Maps (◆▶ Figure 1.7) can be generated that show areas of unstable ground, locations of active earthquake faults, areas with high flood potential, the depth to underground water, and possible mineral resources. Such maps can also indicate areas that are unsuitable for residential or commercial development but which could be used for parks and recreation. The generation of a multipurpose map usually begins with a geologic map that shows the distribution and spatial relationships of rocks below and at the surface of the earth and the area's drainage patterns. Geologic mapping is a fundamental part of most earth-science studies. It is used in such fields as applied-geology investigations; water-resource planning; and land-use planning for waste-disposal sites, housing developments, and trans-

portation. Geologic maps can be adapted to the special requirements of a given project or long-term plan.

Nature is a good teacher, and many significant advances in environmental geology have resulted from studying "case histories," serious attempts to describe and analyze geologic events that have occurred. Several strong earthquakes that caused extensive damage initiated reevaluations of what had been considered state-of-the-art earthquake-resistant design of freeways, high-rise buildings, parking structures, apartments, and even single-family dwellings. They were the earthquakes in San Francisco (1989); Northridge, California (1994); Kobe, Japan (1995); and Izmit, Turkey (1999). Although none of these was a "great" one (>8.0 on the Richter magnitude scale), the Kobe earthquake cost $100 billion, the world record for monetary losses from a natural disaster as of the year 2000. (Hurricane Andrew cost an estimated $20

(a)

(b)

◆ **FIGURE 1.7** Hazard maps of the San Francisco Bay region. (a) Maximum predicted ground shaking due to strong earthquake motion. Ground shaking is a function of the geologic foundation at the site, such as soft bay muds in the Marina district (violent shaking) or hard bedrock on Nob Hill (weak shaking). (b) Risk of earthquake-induced landslides along the San Andreas fault west of San Francisco Bay.

billion, and the Northridge earthquake an estimated $25 billion, huge sums but much less than Kobe's. It follows that research and design improvements that cut damage losses are well worth the expenditure. The eruption of Mount Pinatubo in the Philippines is another example. It was considered a dormant volcano when the U.S. Air Force built the immense Clark Air Force Base nearby. Its 1991 eruption forced abandonment of the air base (◆ Figure 1.8; see Chapter 5).

Other advances in environmental geology are developed in response to problems that become apparent less catastrophically. U.S. laws now require underground gasoline-storage tanks to be examined for

◆ **FIGURE 1.8** Volcanic ash rises several kilometers above Mount Pinatubo, the Philippine Islands, June 1992.

leaks. If leaks are found, the contaminated soil must be "cleaned" and new leak-proof tanks must be installed. New technologies for sanitizing hydrocarbon-contaminated underground water and soils have been developed in response to the laws (see Chapter 8). The deadline for replacing identified leaking tanks—which became known as "yank-a-tank" in the environmental field—was the end of 1998. Service stations that were not compliant by then were no longer permitted to store or dispense gasoline.

Solid-waste disposal is an ongoing major problem. It requires identifying geologically acceptable disposal sites, including sites judged to be safe for sequestering hazardous substances (see Chapter 15).

Extracting water from beneath the land surface for human use may seem harmless enough, but in Venice, Italy, this is causing the land beneath the city to subside. This sinking causes periodic flooding, with serious consequences for the residents and their largest source of revenue, tourism. Similarly, subsidence due to extraction of underground water has altered the flow of surface water and storm drains over large areas of the U.S. Gulf Coast and the Great Valley of California.

At the core of these and many other environmental problems is the burgeoning world population. Every one of these persons wishes to share in the good life, but this growing population exists on a planet with limited air, land, water, and mineral resources.

Too Many People

The concept of **carrying capacity**—the idea that a given amount of land or other critical resource can support a limited number of people or animals—has intrigued scientists, economists, and philosophers for centuries. Thomas Malthus postulated in 1812 that populations grow geometrically, whereas food supplies increase only linearly, which leads to a gap between demand (population) and supply (food) that eventually results in famine. Human life is inextricably linked to resource utilization, environmental quality, and economic well being. Thus, a good starting point for any environmental study is to learn how populations grow, a study referred to as **population dynamics.**

World population increased dramatically in the twentieth century. In 1900, the population of the earth was estimated at 1.6 billion people. By 2000 this figure had quadrupled, to 6 billion. Stated another way, it took about 10,000 generations for humankind to reach the first two billion in population, but only one generation to add the most recent two billion. Pogo, a popular comic strip character of the 1960s, once said, "Yep, son, we have met the enemy, and he is us," and this applies to the global population dilemma. Much of the recent dramatic increase in population can be explained by a reduction in the death rate due to overall better health globally.

◆ **FIGURE 1.9** Linear versus exponential growth. (a) The principle. (b) The growth of a savings account at 10 percent interest per year. Although the text example uses the dollar as the currency unit, the unit could be pounds, yen, or rupees—and the principle is the same for population growth. Note that the doubling time for exponential growth at a 10 percent annual rate is about 7 years.

EXPONENTIAL GROWTH AND THE POPULATION BOMB

Growth and increases are familiar processes to all of us; we see them in our national debt, taxes, rents, food costs, and so on. Increases that occur in nearly equal increments are described as *linear*. Examples include your electric bill variation with usage of electricity and growth of the grass in your lawn between weekly cuttings. Some other things grow by increments that become larger with time, and this kind of growth is described as *geometric* or *exponential*. Examples include the expansion of weeds in a neglected garden and historic increases in the national debt. The growth of a savings account also is exponential, assuming no withdrawals. A thousand dollars invested at 10 percent annual interest rate earns $100 the first year and becomes $1,100. During the second year $1,100 at 10 percent earns $110, and the value of the account grows to $1,210, and so on in each succeeding year. Thus, interest is earned on interest (compounded) so that at the end of ten years, for instance, the account balance is not just $1,000 *plus* 10 percent interest *times* 10 years, or $2,000, but $2,593.74.

It may seem incongruous that to our well-being and way of life the most important increases are not interest rates or the national debt, but the explosive growth of world population and natural-resource consumption. The graphs of ◆ Figure 1.9 contrast linear (straight) and exponential (**J**-shaped) growth curves. Although a population growth rate of 2 percent per year or a similar increase in the consumption of a

(a)

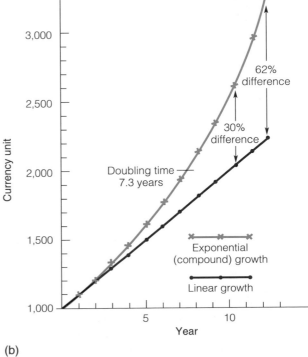

(b)

⊙ TABLE 1.1 **How Populations Grow**	
GROWTH RATE, %	DOUBLING TIME, YEARS*
1.0	70.0
2.0	35.0
3.0	23.3
4.0	17.5
5.0	14.0
6.0	11.7
7.0	10.0

*Calculated by using the formula 70 ÷ growth rate (%), which yields a close approximation up to a growth rate of 10%.

⊙ TABLE 1.2 **World Population and Growth Rate, 1950–2000**		
YEAR	POPULATION, BILLIONS	GROWTH RATE, %*
1950	2.52	
1955	2.75	1.77
1960	3.03	1.95
1965	3.34	1.99
1970	3.77	1.90
1975	4.08	1.84
1980	4.45	1.81
1985	4.85	1.75
1990	5.30	1.70
1995	5.76	1.68
2000	6.26	1.56

*Average annual rate for the previous five-year period. SOURCES: Estimates in United Nations, *Demographics Yearbooks* for 1985 and 1990, and in *World Resources, 1994–95* (New York: Oxford University Press).

particular natural resource may seem trivial, when we realize that this growth is exponential, our assessment changes. Valid concerns for the future of the human condition arise. As world population increases exponentially, natural resources are rapidly being depleted; even if geological exploration can double or triple known oil or mineral reserves, it will make little difference in the useful lifetime of those resources (see Case Study 1.1).

Population growth rate is determined by subtracting a population's death rate from its birthrate, with death and birth rates expressed as deaths and live births per 1,000 people per year, respectively. For example, a birthrate of 20 *minus* a death rate of 10 would yield a growth rate of 10 per 1,000 persons, or 1.0 percent. World growth rates were low in 1900 (about 0.8%), and they have fluctuated with time. They reached a high of about 2.06 percent during the 1960s and then have slowly declined to the year 2000. At first glance you might think that a growth rate of 1 percent would yield a population **doubling time** of 100 years (since 1% per year × 100 years = 100%). This is not the case, however. Populations grow like savings accounts with compounded interest. Just as you can earn interest on interest, the people that are added produce more people, resulting in a shorter doubling time. A very close approximation of doubling time can be obtained by the "rule of 70"; that is,

Population doubling time = 70 ÷ growth rate (%)

Thus, the doubling time for a 1 percent growth rate would be 70 years; that for a 2 percent growth rate would be 35 years, and so forth. ⊙ Table 1.1 provides some examples. The global population growth and the average annual growth rate for 1950 to 2000 are shown in ⊙ Table 1.2. Although some progress has been made in slowing the growth rate, the number of people

added each year continues to grow larger. The world population grew by almost a billion people in the decade 1990–2000, the largest increment ever for a single decade.

U.S. POPULATION GROWTH AND IMMIGRATION

In 1999 an estimated 274 million people were living in the United States. They generated 15.9 live births per thousand people and a death rate of 9.0 per 1,000, for a growth rate of 6.9 per 1,000, or 0.69 percent. At first glance these figures appear to be real progress toward *zero population growth,* a condition that exists when, on average, as many people die as are born each year. However, even at this low rate the population would double in 100 years, without accounting for immigration—which adds significantly to the U.S. population, particularly in Texas, Florida, and California. From the 1920s through the 1960s about 200,000 immigrants came to the United States each year. Since then this number has grown dramatically for a variety of political and economic reasons. By 1993 there were 810,000 legal immigrants *plus* an estimated 200,000 illegal ones, for a total of about a million new persons—five times the annual immigration of 30 years earlier. From the year 2000 through the end of the next century, it is estimated that 90 percent of the population growth in the United States will be immigrants and their progeny.

Case Study | 1.1

A Billion Big Macs, Please!

In New Delhi, India, there stands a two-story digital clock with numbers that change upward every few minutes, just like the signs at some MacDonald's franchises that report the number of hamburgers sold to date. One day in August 1999 the clock read 998,499,510 at a momentary pause in its upward tick at the rate of 33 a minute, 2,000 an hour, 48,000 a day. The numbers represent the people living in India. Experts estimate that the population surpassed 1 billion during that last week in August 1999. The earth now has two "billionaires," China and India, and they're both still counting (◉ Table 1).

The United Nations Conference on Population convened in Cairo, Egypt, late in 1994. After extensive debate the four main blocs at the conference—Muslim, Roman Catholic, emerging nations, and developed Western nations—adopted a carefully worded consensus "program." The thrust of the program is to hold down population growth by family planning and by giving women the power to make decisions about their own lives. The program states that women have the right to decide "freely and *responsibly*" (author's emphasis) how many children they have and when they have them. The program is not binding and is probably being interpreted differently in different countries. The Vatican and many Muslim nations were unhappy with the section on reproductive rights because they believe those rights belong to couples, rather than to individuals, and because they object to abortion.

According to a United Nations spokesperson, if family planning succeeds, the world population will rise to 7.27 billion by the year 2015. If it fails, the population will grow to 7.92 billion by 2015 and explode to 12.5 billion by 2050.

Some of us will be able to monitor these numbers. It appears that population momentum, even with the most successful birth-control efforts, will generate 10 billion people by the year 2050. This dramatic growth is driven by increases in longevity and by the continuing growth in the number of women of child-bearing age.

◯ TABLE 1	Ten Most Populous Countries in the World, and Projections to 2050				
RANK	1998		2050		
	COUNTRY	POPULATION, MILLIONS	COUNTRY	POPULATION, MILLIONS	RANK CHANGE
1	China	1,255	India	1,533	+1
2	India	976	China	1,517	−1
3	United States	274	Pakistan	357	+4
4	Indonesia	207	United States	348	−1
5	Brazil	165	Nigeria	339	+5
6	Russia	148	Indonesia	318	−2
7	Pakistan	147	Brazil	243	−2
8	Japan	126	Bangladesh	218	+1
9	Bangladesh	124	Ethiopia	213	+11
10	Nigeria	122	Iran	170	+4

SOURCE: United Nations, *World Population Prospects: The 1996 Revision* (New York: United Nations).

WORLD POPULATION GROWTH, RESOURCES, AND THE ENVIRONMENT

In 1900 the world population growth rate was 1 percent. It climbed to a historical high of 2.2 percent in 1964, and then gradually dropped to 1.4 percent in 1997—a good sign. Some experts expect this growth rate to remain fairly steady for at least two decades and then to drop. Other experts expect education and birth-control measures to lower the growth rate steadily. Whatever the reality, in 1999 80 million people were added to a world population of 6 billion people. The most significant world trend is that death rates are falling in poor and rich countries alike, while birthrates remain high in most poor countries and low in most rich ones. Exceptions are the generally higher death rates of Africa and the high birthrates of the rich oil-producing countries.

Due to varying estimates of *growth rate momentum,* world population projections for 2050 range from a low of 9–10 billion to a high of 15 billion people. China provides a good example of population momentum. Utilizing education and incentives, the country cut its growth rate dramatically in a few decades to 1.1 percent from more than 2.0 percent. With a population in excess of 1.3 billion, however, 11–13 million people are added every year even with a low growth rate! Keep in mind that the population growth curve is the typical **J-curve** for exponential growth of anything—slow growth at first followed by compounding and very rapid growth (see Figure 1.1).

The tripling of the population in the twentieth century has placed extreme demands on soil, water, forest, energy, and mineral resources. As a population increases, marginal lands are tilled because of the need for more food, and thus, poor agricultural practices become exaggerated. In addition, between 1940 and 1990, atmospheric carbon dioxide (CO_2) levels increased 13 percent, which may have implications for global warming, and the protective stratospheric ozone layer decreased 2 percent worldwide (more than that over Antarctica; see Chapter 11). None of these problems is insurmountable, but human ingenuity will certainly be taxed to overcome the challenges they present.

Overpopulation Scenario One: The Sky Is Falling

This scenario is fairly simple. The earth has too many people and finite resources for sustaining them. The carrying capacity of the earth has either been reached or is fast approaching, and the basic resources for life—soil, water, energy, and clean air—are being taxed to their limit. Students of population dynamics, economics, and environmental science generally agree that the

earth will be hard pressed to sustain the 9–10 billion people projected for 2050, based upon the *lowest* of the projected world growth rates. If nothing is done to stabilize population, future population reduction will occur by increased fatalities from natural disasters, mass starvation, epidemics, and ecological degradation.

Overpopulation Scenario Two: The Gaia Hypothesis

The name of Gaia, the Greek goddess of the earth, has been applied by James Lovelock and U.S. biologist Lynn Margulis to their hypothesis that the earth is a superorganism whose environment is controlled by the plants and animals that inhabit it, rather than vice-versa. According to this hypothesis, the planet is "self-adjusting," so to speak. The super-organism regulates its environment with a complex system of mechanisms and buffers, just as an animal adjusts to varying temperatures, chemistry, and other environmental factors. For example, carbon dioxide, CO_2, a gas that is required for photosynthesis and therefore all of life, also acts as an insulating blanket in that it holds the sun's heat close to the earth. Terrestrial and marine plants act as regulators that "pump out" excess CO_2 from the atmosphere when great amounts are added by volcanic activity, which, if it were to accumulate, would smother humans and other animals.

According to the Gaia hypothesis, the last ice age ended abruptly when atmospheric CO_2 doubled due to a sudden failure of the "pumps," not by the slow processes of geochemistry as invoked by conventional scientific theory.

The Gaia hypothesis is appealing to some people because it seems to offer an easy solution for our environmental problems. However, toward the end of the twenty-first century the amount of CO_2 in the atmosphere will be doubled, and even Lovelock admits that "it is near certain that the new state will be less favorable for humans than the one we enjoy now."

Toward a Sustainable Society

A **sustainable society** is one that satisfies its needs without jeopardizing the needs of future generations. The current generation *should* strive to pass along its legacy of food, fuel, clean air and water, and mineral resources for material needs to the generations of the twenty-first century. We know what needs to be done to accomplish this, but knowing what to do and implementing a long-range plan for carrying it out are entirely different things. A promising development in industrialized nations is the growing respect for basing an economy on steady-state conditions, rather than on continual growth. The conceptual ideal is sometimes referred to as the *spaceship economy,* since all resources must be conserved, shared, and recycled in space travel. This, together with population control, could lead in time to a sustainable society (see Case Study 1.2).

Do efforts toward a sustainable society threaten "the American way of life"? Alma Paty of the American Mining Congress says *yes.* With only five percent of the world's population, the United States has been consuming more than a third of its raw materials. Our high standard of living has been supporting two cars, a nice big home in the "burbs," beryllium–copper golf clubs, personal computers, and odor-killing cat litter. We have gizmos galore, and they all require the use of mined raw materials. Unfortunately, most of us seem to have the mentality that cars and microwaves come from factories, not from mines, and that production and consumption can continue at their present levels forever. The truth, Paty points out, is that resources are limited, and sustainable development requires a life-style of "sophisticated modesty" characterized by more use of public transportation, smaller houses, and fewer gadgets and adult toys.

ENERGY AND GLOBAL WARMING

After population growth, global warming is potentially the most serious long-term environmental problem facing humankind. It is closely associated with the increase in atmospheric carbon dioxide produced by the burning of coal, oil, and natural gas ($\blacklozenge$ Figure 1.10). Other greenhouse gases, such as methane, nitrous oxide, ozone, and chlorofluorocarbons (CFCs), are also

◆ **FIGURE 1.10** Estimated atmospheric carbon dioxide changes from 1700 to 1980 based on data from air bubbles trapped in Antarctic ice cores and on atmospheric measurements.

increasing in the atmosphere, and this, too, is closely associated with population growth. It is estimated that these gases may contribute to about half of the global warming projected for the twenty-first century.

Carbon-dioxide concentrations in the atmosphere today (as measured atop Hawaii's Mauna Loa volcano; see Chapter 5) are at their highest levels of the past 160,000 years, as indicated by CO_2 in air bubbles in Antarctic deep ice cores (see Chapter 11). If the goal is to stabilize climate, calculations indicate that the CO_2 emissions must be lowered from their 1990 level of 22 billion tons per year to 8 billion tons per year. In developed countries this would require an 85 percent per capita reduction in fossil-fuel consumption, which is unrealistic in the short term. As a start, a 5.2 percent average reduction has been agreed to by 38 industrialized nations, and there is some doubt that even this can be accomplished (Case Study 1.3). Industrial nations consume energy at a prodigious rate—2–4 times the global per capita mean and 10–20 times that of Africa and Australasia, which includes Polynesia, New Zealand, Australia, the Philippines, and their dependencies ($\blacklozenge$ Figure 1.11). In the early 1990s half a billion motor vehicles were registered in the world, and this figure is expected to double by 2025—each consuming an average of two gallons of gasoline per day. Thus, other means of powering those vehicles must be

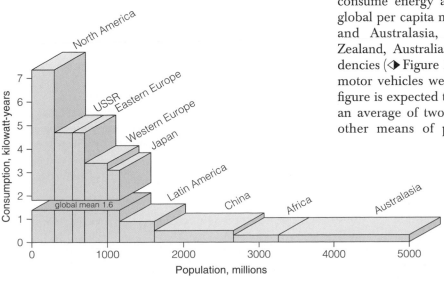

◆ **FIGURE 1.11** Per capita energy consumption in kilowatt-years (kwy) and population by geographic or political region for 1987. Note that the industrialized nations consumed 2–4 times the global per capita mean and 10–20 times that of such areas as Australasia and Africa.

Case Study | 1.2

Earth Day 1995—Assessing 25 Years of "Thinking Green"

In April 1995 the United States celebrated the 25th anniversary of the first Earth Day with much fanfare. Looking back at the achievements of those years, all that fuss seems to have been warranted.

During the 1960s a relatively small group of environmentalists, conservationists, and scientists, all widely regarded at the time as silly do-gooders and "tree huggers," began raising the public awareness that there is such a thing as "the environment" and that it needs protection. One of them was an outspoken biologist named Barry Commoner. In his 1971 book, *The Closing Circle,* Commoner advanced four laws of ecology that have had great impact on our thinking:

1. Everything is connected to everything else.
2. Everything must go somewhere.
3. Nature knows best.
4. There is no such thing as a free lunch.

Today, most of us appreciate Commoner's first law, the concept that all elements of an ecosystem are related and that disrupting any of them may adversely impact the entire system. The second law stresses that humans do not destroy matter but simply alter it into forms that may be detrimental to health and the environment. Consider toxic waste and greenhouse gases, for example. Number three, "Nature knows best," seems simple enough, but it presents an enormous educational problem because it conflicts with the normal human tendency to attempt to control nature. This dangerous tendency has been reinforced in our Judeo-Christian culture by the biblical

> Be fruitful and multiply and replenish the earth, and subdue it; and have dominion over the fish of the sea, the fowl of the air, and over every living thing that moveth upon the earth (Genesis 1:28).

Too many times we have not "listened" to nature, even after carrying out exhaustive scientific investigations, with disastrous results, as illustrated by many examples detailed in this book. We are learning that we cannot force nature to absorb our wastes and poisons beyond its capacity to do so without damaging the ecosphere. The last law puts the teeth in the first three; it assures us that we will pay for any ill-gotten

benefits. The cost of polluting streams, oceans, or soil or of overpopulating unfit lands is eventual remediation expense, resource depletion, health problems, natural disasters, global warming, or the like.

Before 1970 little serious thought was given to environmental education. Today, however, one would be hard pressed to find a college, university, law school, or graduate program that does not offer courses dealing with environmental issues. "Green" thinking is now ingrained in policies, practices, and activities of industry, government, and the public. Great progress has been made. Some successes of the 25 years since the first Earth Day are these:

- Hydrocarbon emissions from automobiles have decreased by almost half.
- The number of large cities violating clean-air standards has dropped from 40 to 9.
- The use of unleaded gasoline has reduced lead emissions by 98 percent.
- Industrial toxic-waste spills have declined 43 percent.
- The California gray whale, the bald eagle, and the American alligator have been removed from the endangered species list.
- Mandatory recycling programs have been implemented in 6,600 U.S. cities. The amount of waste that is recycled increased from 17 percent in 1990 to 22 percent in 1995.
- 100 million acres have been set aside as wilderness areas.
- Since 1979 the country has invoked energy savings valued at almost five times all the new sources of energy combined. Some people call these savings "negawatts."
- The number of environmental groups on college and university campuses grew from 50 in 1989 to more than 2,000 in 1995.

Underlying these achievements is a large body of federal and state laws enacted since the 1960s to clean up and protect the environment. Some of them are listed in ● Table 1.

On the negative side we still have tons of highly radioactive spent nuclear fuel (see Chapter 14) and stockpiles of nuclear weapons awaiting the development of safe disposal

⦿ TABLE 1	Some Important U.S. Environmental Laws, 1963–1994	

1963	Clean Air Act	1978	Energy Policy and Conservation Act
1964	Wilderness Act—sets aside lands permanently for public use, restricts resource development activities	1980	Comprehensive Environmental Response, Compensation, and Liability Act (CERCLA)*—established Superfund to clean up most severely polluted sites Low-Level Radioactive Waste Act
1970	National Environmental Policy Act (NEPA)* Environmental Protection Agency (EPA) established	1982	Nuclear Waste Policy Act*
1971	Vermont Bottle Bill—"Recycle or pay."	1984	Resource Conservation and Recovery Act amendment
1972	Water Pollution Control Act Marine Mammal Protection Act Pesticide Act Coastal Zone Management Act	1986	Safe Drinking Water Act amendment Superfund amendment
1973	Endangered Species Act* Oregon Bottle Bill	1987	Montreal Protocol—bans chlorofluorocarbons (CFCs) Marine Plastic Pollution Control Act
1974	Safe Drinking Water Act*—protects drinking water, establishes EPA water-quality standards	1988	Alternative Motor Fuels Act
1976	Resource Conservation and Recovery Act (RCRA)*—controls hazardous and toxic wastes	1990	National Environmental Education Act Clean Air Act amendment
1977	Clean Air Act amendment Ocean Dumping Act amendment	1994	California Desert Protection Act—protects environmentally sensitive desert areas of Western U.S.

*Landmark legislation.
SOURCES: Various.

facilities. The government estimates this task may cost future generations $300 billion. There were 108 million cars in the United States on the first Earth Day. Today there are more than 200 million, and the number of miles traveled has nearly doubled. Although per capita energy use has decreased, total energy use is up significantly (see Chapter 14). The period from March to December in 1994 was the warmest on record in the United States, and sea level rose about three millimeters on many stable coasts. If that is shown to be the start of a trend, it could be strong evidence for global warming (see Chapter 11).

In the mid 1990s the policies and practices of the Environmental Protection Agency have received severe criticism by members of Congress and some sectors of the public. Complying with some EPA requirements has proven to be so expensive that some industries are unable to compete in world markets, thus costing jobs for U.S. workers. Many legislators want the states to take over the policing role, thereby reducing federal interference with business. Some studies have shown that states with the best environmental protection have the healthiest economies, which may lead other states to conclude that it would be good business to "clean up their acts." Americans have become "lite green" say some institute and university researchers. "Lite" green or "dark" green, the record of the first 25 years of Earth Days is pretty good.

Case Study | 1.3

The Kyoto Protocol—A Lot of Hot Air?

At the 1997 Climate Treaty Meeting in Kyoto, Japan, 159 participating nations signed a historic document called *the Kyoto Protocol*. It is the first international agreement setting legally binding limits on emissions of greenhouse gases. Although industrialized nations comprise only 20 percent of the world's population, they have released into the atmosphere 90 percent of all global carbon emissions from human sources to date. The largest polluters are required to make the largest sacrifices, and the 38 industrialized nations participating agreed to reduce their greenhouse-gas emissions by an average of 5.2 percent below their 1990 levels between 2008 and 2012. Specifically, the European Union countries are to reduce their emissions by 8 percent below those levels, the United States is to reduce its emissions by 7 percent, and Japan by 6 percent. The six targeted greenhouse gases are carbon dioxide, methane, nitrous oxide, hydrofluorocarbons, perfluorocarbons, and sulphur hexafluoride. The Kyoto Protocol is only a beginning; deeper cuts will be needed in the phase to begin after 2012. Some other details of the treaty are that:

- Developing countries are not required to make any cuts unless they choose to.
- Countries with gas-absorbing "sinks," such as extensive forests and jungles that absorb CO_2, can receive credit toward their emission-reduction goals.
- Industrialized nations may trade transfers of energy-efficient technology to poorer nations for credit toward their own emissions-reduction requirements.

- Emissions trading and methods of enforcement are to be discussed at future meetings.

The Protocol's "flip" side has been called the "7% solution that isn't" by energy companies (understandable) and some eminent economists. The economic impact upon industrialized nations is the most serious drawback. Meeting the Kyoto target could cause as much as a two percent drop in the U.S. gross domestic product (GDP), or about $227 billion. It could devastate some industries, coal mining for instance, and the economic impact would not be spread evenly among the population. In contrast to particular other environmental problems, the climate problem is not one that can be "fixed" by a single legislative act. According to M.I.T. economics professor Richard Schmalensee, a better long-term solution that would not impose such enormous economic burdens would be to:

- conduct serious research in low-carbon energy technologies,
- involve developing nations, such as China, for example, now, before they become large enough consumers and less willing to cut emissions, and
- build an international framework for dealing with climate change.

Although meeting the Kyoto target would not be a solution to global warming, it would be a first small step.

developed—perhaps fuel cells; liquid or gaseous fuels derived from coal, natural gas, or biomass (alcohol); or even solar energy (see Chapter 14).

As developing nations raise their standards of living, their energy consumption will also increase, and competition between the developed and the developing nations for energy supplies will increase. This competition will surely increase as the world's coal and oil reserves decrease. In the decade 1982–1992 energy consumption increased more than 50 percent in Asia, 40 percent in Africa, and 30 percent in Latin America, compared to increases of 8 percent or less in the United States and Europe. Per capita oil consumption actually decreased 16 percent during the 1980s in the United States. The developing nations were using 30 percent

of the world's energy in the early 1990s, a usage that was predicted to double by 2015. It is in the best interest of the industrialized nations to undertake research and development of alternative energy sources and to expand their current methods of reducing energy usage.

The only options for large-scale power production sufficient to drive economic growth and human requirements in the near future are solar-energy conversion and nuclear energy, including both fission and fusion. Solar-derived electricity at present costs one-third less than electricity from a nuclear power plant and it lacks the hazards associated with radioactive substances. On the negative side, with present technology, solar-energy conversion requires vast areas of land and an effective storage system (see Chapter 14).

⊙ TABLE 1.3	Estimated World Soil Degradation, 1945–1990		
REGION	DEGRADED LAND AREA AS A PERCENTAGE OF VEGETATED LAND		
	Total	*Light Erosion**	*Moderate to Extreme Erosion***
World	17	7	10
Asia	20	7	13
South America	14	6	8
Europe	23	6	17
Africa	22	8	14
North America (U.S., Canada)	5	1	4
Central America, Mexico	25	1	24

*Light: crop yields reduced less than 10%.
**Moderate: crop yields reduced 10%–50%. Severe: crop yields reduced more than 50%. Extreme: no crop growth possible. About 9 million hectares worldwide exhibit extreme erosion, less than 0.5 percent of all degraded lands.
SOURCE: Various sources compiled by World Resources Institute and the United Nations, World Resources 1992–93 (New York: Oxford University Press, 1992).

THE LAND

The amount of cultivatable soil limits the number of humans the earth can support, because soil determines grain production and therefore food supplies. Arable soils worldwide suffered slight to severe degradation between 1945 and 1990 due to poor agricultural practices, natural erosion, and erosion accelerated by deforestation (⊙ Table 1.3). The world's farmers lost an estimated 500 million tons of topsoil in the 1970s and 80s, an amount equal to the tillable area of India and France combined. Waterlogging and salt-contamination of soils are reducing the productivity of at least a fourth of the world's cropland, and human-generated smog (ozone) and acid rain also are taking their toll on crops. It is the lack of soil, sometimes combined with drought, that is the cause of famine, not soil infertility. (Soils are discussed in Chapter 6.)

WATER—A LIMITING RESOURCE

Mark Twain's famous quote, "Whiskey is for drinking; water is for fighting over," was prophetic. Little water from the Nile River makes it to the Mediterranean; the Ganges River is a trickle in the dry season, and the Colorado River is a joke when it crosses the international border with Mexico. With the populations of Egypt and her neighbors projected to double between 2000 and 2050, and as the inevitable growth of India and Bangladesh occurs, there is bound to be competition for the use of water on the Nile and the Ganges. Two-thirds of the world's irrigated lands lie in Asia, where populations and water needs also are greatest. The Yellow River ran dry in 1972 for the first time in China's 3,000-year history, and water tables fell on every continent in the 1990s. Water withdrawals exceed water recharge in the southern Great Plains and the entire southwestern United States (see Chapter 8).

Seventy percent of water that is diverted from rivers or pumped from underground is used for irrigating crops upon which entire cultures depend. It is said that India is on a temporary "free ride," expanding irrigation at the expense of stored underground water. Underground water needs to be managed, as it is not a limitless resource. When this fact strikes home, India's grain production could fall as much as 25 percent. For a country that adds 18 million people every year, the consequences could be disastrous.

Consider This... ?

Few people doubt the existence of ozone holes in our atmosphere and the increased flux of ultraviolet radiation to the earth's surface due to ozone depletion by CFCs in the stratosphere. Global warming and its cause, however, are subjects of heated (no joke intended) debate. The earth *is* getting warmer, and the big question is whether the cause is anthropogenic (humans' production of greenhouse gases), or just part of a natural climatic rhythm of warming and cooling. Think hard about the long-term impact of global warming on the hydrosphere, atmosphere, biosphere, and on humans, and then answer the question, Is it wise to wait, perhaps several decades, to know for certain if the warming is natural or anthropogenic before reducing greenhouse-gas emissions? ■

Gallery

Overpopulation Problems

As emphasized in this chapter, overpopulation is causing environmental damage, societal problems, and human suffering. In some countries population growth is utterly out of control, and too little is being done to educate the world's people on the long-term dangers of this uncontrolled growth. In other countries—China, for instance—measures and incentives are in place for curbing the growth and achieving a sustainable population. This photo gallery highlights the dilemma.

◆ FIGURE 1.12 Coney Island beach in the 1930s, when the U.S. population was a fraction of what it is today. Imagine this beach on a summer weekend today.

◆ FIGURE 1.13 Important Hindu pilgrimage site on the Ganges River, Vārānasi (Benares), State of Uttar Pradesh, India. The waters of the Ganges are used for irrigation, drinking, bathing, and religious rituals. India's population grew from 853 million in 1990 to one billion in 1999 (see Case Study 1.1).

(a)

(b)

(c)

◁▶ **FIGURE 1.14** Deforestation near the city of Manaus in Amazonia, Brazil's largest state. (a) False-color satellite photo of the confluence of the Rio Negro (*black*) and the sediment-laden Amazon River (*light green*). Healthy rain forest is shown in red, and deforested areas appear in light colors. Manaus is on the point of land that extends into the Rio Negro. The waters of the Rio Negro are recognizable for more than 80 kilometers (48 mi) downstream in the Amazon. (b) The Rio Negro and the natural Amazon rain forest at Manaus. (c) Adjacent rain forests have been removed as the population of the area has grown. Manaus's population is now more than a million people, and the city is rapidly expanding into areas that were once sparsely populated.

◁▶ **FIGURE 1.15** A model Chinese family in front of a billboard encouraging the one-child family; Beijing. Although the People's Republic of China is pushing population control, its population of more than a billion people makes continued growth inevitable.

FORESTS

Deforestation continues to be one of the hottest environmental issues of the day. Although most deforestation is occurring in the tropics and remains largely unmeasured, the Food and Agricultural Organization (FAO) of the United Nations reported in 1991 that 17 million hectares (42.5 million acres) of tropical rain forest were being deforested per year, up 50 percent from the early 1980s rate. Deforestation leads to loss of habitat and decreased biodiversity, contributes to climate change by adding CO_2 to the atmosphere, and often results in soil degradation. ◆ Figure 1.16 shows late 1980s estimates of annual rates of deforestation in Asia, Latin America, and parts of Africa. Deforestation estimates do not include forests that are logged out or clear-cut and then replanted or allowed to rejuvenate.

One reason for clearing forests is to convert land to agricultural use. More than half of the tropical-forest lands destroyed each year are in Brazil. It should be noted, however, that Brazil plans to set aside 1.6 million square kilometers (618,000 mi^2), about 20 percent of its land area, as public reserves and parks. The Indonesian government, in a full-page advertisement in a U.S. magazine in 1989, explained that the aspirations of its 170 million people were the same as those of citizens in developed nations, and that in order to better their lives, 20 percent of Indonesian forests were to be converted to plantations of teak, rubber, rice, and coffee. A poor African farmer, on the other hand, has little choice but to clear land to grow crops just for subsistence. How a country uses its land is a complex political and sociological issue, and there are no simple ways to reconcile the conflicting human needs that the issue involves.

Logging is another cause of deforestation, but it has been demonstrated that selective logging in tropical forests is sustainable if the logged lands are allowed to rejuvenate. In 1988 Thailand was inundated with 40 inches of rain in five days. Thousands of logs left on hillsides to dry floated off in mud and water that engulfed entire villages, killing 350 people. The Thai government banned logging, and was caught between the companies that wanted compensation for lost business and the landless poor who were given farms in the previously logged areas.

A third cause of deforestation is the demand for wood fuel and forest products. Many of the forests in

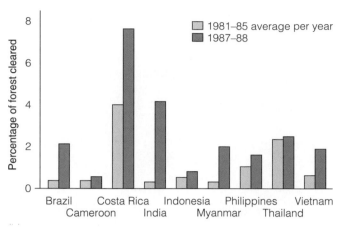

◆ **FIGURE 1.16** Average annual percentage of forest cleared in nine tropical countries, 1981–1985 and 1987–1988. Although Costa Rica lost the largest percentage of its forests in these periods, it deforested the smallest total area of the nine countries in these years.

India and the Sahel of West Africa have been decimated for use as domestic fuel and in light industry.

Deforestation is second only to the burning of fossil fuels as a source of atmospheric CO_2. Forests store 450 billion metric tons of carbon. When cleared, they can no longer sequester the carbon, and it is released rapidly to the atmosphere when trees are burned, or slowly if they are left to decay.

OTHER RESOURCES

Our throw-away society, though certainly a convenient one, has become a recycling society as natural resources dwindle and waste-disposal sites become scarce and expensive to operate. Most of what we used to discard after one use—paper, plastic, glass, aluminum, and even steel—is now recycled. Trash disposal has become so expensive, at least in urban settings where most trash is generated, that recycling is close to being cost-effective. Unless alternative energy sources are maximally utilized, oil reserves will be exhausted in your lifetime or in your children's. United States coal reserves are estimated at a 400-year reserve, but such pollution considerations as smog, acid rain, and global warming will continue to limit coal's use. World reserves of metallic and nonmetallic resources, many of which are in abundant supply at present, can certainly be extended by recycling. The geological and environmental consequences of mining and energy consumption and the problems associated with waste disposal are considered in Chapters 13 through 15.

Summary

ENVIRONMENTAL GEOLOGY

DEFINED
The interaction of humans and the geological environment.

UTILIZES
Most of the traditional geologic specialties such as petrology, structural geology, sedimentology, economic geology, engineering geology, geophysics, and geochemistry.

AREAS OF INTEREST
Geologic hazards such as volcanism, earthquakes, and flooding; mineral and water resources; and land-use planning—should I build my house on this hillside?

GOALS
To predict or anticipate geologic problems by collecting data in the field and analyzing it in the laboratory. Data may be portrayed on special-use maps and utilized by planners and public officials.

CASE HISTORIES
Nature is a good teacher, and geologists learn much by studying case histories of geologic events.

POPULATION DYNAMICS

CARRYING CAPACITY
Defined for our purposes as the number of people the earth can support. This number is uncertain but lies somewhere between the 1995 population of 5.76 billion and the 10–15 billion projected for the year 2050.

POPULATION GROWTH RATE
Defined as live births per 1,000 persons (that is, the birthrate) minus the death rate per 1,000. A birthrate of 20 *minus* a death rate of ten *equals* a growth rate of 10 per 1,000 persons, or 1%.

DOUBLING TIME
The number of years in which a population is doubled. It can be projected closely by dividing 70 by the percentage birthrate. The 1995–96 world growth rate of 1.68 percent yields a doubling time of 40 years.

OVERPOPULATION SCENARIOS

UNCONTROLLED GROWTH, WORST-CASE SCENARIO
The earth becomes uninhabitable as resources dwindle, sea-level rises due to global warming and inundates coastal cities, and hazardous ultraviolet radiation strikes the earth's surface due to depletion of the ozone layer.

GAIA HYPOTHESIS
Life is an important regulator of the earth environment. If left alone, natural systems are self-regulating; for example, as carbon dioxide builds up in the atmosphere, plant growth and the oceans will remove it, thereby reducing global warming. Human intervention into natural systems (by deforestation, for example) has rendered a basic premise of this hypothesis untenable.

SUSTAINABLE SOCIETY

DEFINED
Society's current needs are satisfied without jeopardizing future generations. To accomplish this, population control is needed.

WATER IMPLICATIONS
Water is a finite resource and needs to be managed. Simply put, no water, no food.

ENERGY IMPLICATIONS
Increasing energy demands dictate that we seek alternatives to burning the earth's reserves of limited fossil fuels. Alternatives include solar, wind, geothermal, and nuclear energy.

LAND IMPLICATIONS
Soil erosion and the occupation of marginal lands subject to geologic hazards are problems that must be addressed.

FORESTS IMPLICATIONS
Deforestation is the second-greatest contributor to the problem of human-induced carbon dioxide accumulating in the atmosphere. It results in loss of habitat for humans and animals and contributes to desertification.

OTHER RESOURCES IMPLICATIONS
We must invest our efforts in recycling and conserving all resources.

◆ Key Terms

carrying capacity	environmental geology	population dynamics	sustainable society
doubling time	J-curve	population growth rate	zero population growth
engineering geology			

◆ Study Questions

1. Physical geology is the study of earth materials, the processes that act upon them, and the resulting products. How does environmental geology differ from physical geology? In order to answer this question you might compare textbooks in the two subject areas.
2. Distinguish between engineering geology and environmental geology.
3. Which natural resources are most important in determining the carrying capacity of the earth?
4. The United States has a current growth rate of 0.7 percent. At this rate how much time is required for a population to double? Discuss the impact of a doubling of your community's population in that length of time.
5. Of what value are case histories in solving environmental geological problems?
6. What is the impact of overpopulation on energy use and global warming? How does this bode for the well-being of future generations? In what parts of the world will the largest percentage increases in energy consumption probably occur in the next decade, and why?
7. What are some of the adverse consequences of clearing tropical rain forests?

◆ For Further Information

BOOKS AND PERIODICALS

Brown, L., and others. 1990. *The state of the world, 1990.* New York: W. W. Norton and Co.

Blue Planet Group. 1991. *Blue planet: Now or never,* case study 9734. Ottawa, Ontario, KIG 5J4, Canada: Blue Planet Group.

Brown, L., and others. 1999. *The state of the world, 1999.* New York: W. W. Norton and Co.

Cohen, J. E. 1999. A global garden, *The ZPG reporter* 31, no. 1 (February).

Ehrlich, Paul R., and Anne Ehrlich. 1970. *Population, resources, and environment: Issues in human ecology.* New York: W. H. Freeman and Co.

_____. 1987. *Earth.* New York: Franklin Watts.

Gore, Al. 1992. *Earth in balance: Ecology and the human spirit.* Boston: Houghton Mifflin Co.

Joseph, Lawrence E. 1990. *Gaia: The growth of an idea.* New York: St. Martin's Press.

Knickerbocker, Brad. 1995. Earth Day at 25: A U.S. environmental report card. *Christian Science monitor,* April 18.

Lovelock, James. 1988. *The ages of Gaia: A biography of the living earth.* New York: W. W. Norton and Co.

Mileti, Dennis. 1999. *Disasters by design: A reassessment of natural hazards in the United States.* Washington, D.C.: Joseph Henry Press/National Academy Press.

Parfit, Michael. 1998. *Living with natural hazards,* Washington, D.C.: *National geographic,* July.

Paty, Alma Hale. 1993. Sustainable development: End of the American dream? *Geotimes,* July: 6.

Silver, Cheryl S., and Ruth S. DeFries. 1990. *One earth one future: Our changing global environment.* Washington, D.C.: National Academy of Sciences, National Academy Press.

United Nations. 1980. *World population trends and policies, 1979 monitoring report, Volume 1: Population trends.* New York: United Nations.

Van der Vink, G., and others. 1998. Why the United States is becoming more vulnerable to natural disasters. *EOS* 79, no. 44 (November 3).

World Resources Institute and the United Nations Development and Environment Programs. 1992–93. *Toward sustainable development: A guide to the global environment.* New York: Oxford University Press.

World Resources Institute and the United Nations Development and Environment Programs. 1994–95. *People and the environment: Resource consumption, population growth, and women.* New York: Oxford University Press.

◆ World Wide Web Activities

U.S. Census Bureau POPClocks
http://www.census.gov/main/www/popclock.html
This site shows the current United States and world population figures. Press **Reload** on your computer for the latest figures. Click on the American flag to see more details about U.S. population increases, including birth and death statistics.

1. How many people are added to the U.S. population every second?
2. How many people have been added to the world population since October 1999, when the six billionth human was born?

EERI videoclip of the Kobe, Japan Earthquake, January 17, 1995
http://www.eeri.org/Publications/Video/Kobe.html
View a 3-minute, 34-second videoclip from the Earthquake Engineering Research Institute of the earthquake

that killed about 5,000 people and was the most costly of any natural disaster on record.

1. What was the magnitude of this quake, and what specific effects of the quake were responsible for so many deaths?

U.S.G.S. Page on Various Changes, 1980–1997, in the High Plains Aquifer
http://www-ne.cr.usgs.gov/highplains/hp97_web_report/ fs-124-99.htm

This Web page has a report and map explaining the changes in the High Plains Aquifer under eight states east of the Rocky Mountains, which is the supplying watershed for this resource. Click on the map to see a larger, more readable version.

1. What human efforts have slowed depletion of the aquifer in recent years?

 # Earth Systems Today CD Correlations

CHAPTER CONCEPT: The Rock Cycle and Rock Families
This module provides an interactive rock cycle diagram that defines the three rock families and illustrates the processes of rock formation.

SECTION MODULE: Main Activity: Rocks and the Rock Cycle
Use the interactive functions of the right screen to see videos of rock formation for the three families and specific examples of the processes that link one family to another. Link to the **sedimentary rock laboratory** to answer the following question.

QUESTION: *What specific processes act on weathered remnants of preexisting rock and create sedimentary rock? Where do these processes operate?*

CHAPTER CONCEPT: As Human Populations Expand, More of Us Are 'In Harm's Way'
Human populations tend to be concentrated in coastal regions, in low-lying regions, and in low latitudes. Especially heavy concentrations of population can be found in India, eastern China, western Europe, and southern Mexico.

SECTION MODULE: Earth Today
This module will link you to live data regarding natural disasters and unusual natural phenomena. Keeping in mind the population patterns outlined above, look over the global map to see whether any of those specified areas are experiencing current activity.

QUESTION: *What specific natural disasters threatening to humans are currently being reported, and where are they? Are any events focused on major population areas?*

chapter two

Getting Around in Geology

◁▷ **FIGURE 1** A view in the vicinity of the *Mars Pathfinder* landing site. The "marscape" seen is a portion of Ares Vallis (Mars Valley), with the rover Sojourner backed up to the rock named Yogi. Sojourner has placed an alpha proton X-ray spectrometer instrument against Yogi in order to determine its elemental composition. Yogi is about 6.5 meters from the *Mars Pathfinder* lander and stands about 1 meter (3 ft) high. Sojourner is 30 centimeters (1 ft) tall. On the lower left is the ramp that Sojourner used to exit the lander. On the horizon, about 150 meters (500 ft) distant, is a rock known as the Couch.

Barnacle Bill, Yogi Bear, and Scooby Doo–Planetary Rock Stars

About five million years ago an interstellar cloud of gas and dust began to collapse, probably under the force of its own gravity. As it shrank it started spinning, which caused the cloud to flatten into a disc shape. Eventually, the compression accompanying the collapse created temperatures and pressures in the center of the cloud sufficient to initiate nuclear reactions. Hydrogen atoms fused, forming helium and releasing enormous energy. Our primordial sun began to shine. Around the sun a cooler disc, the solar nebula, contained the elemental leftovers of the solar implosion that eventually would form the planets. Low-boiling-point compounds such as methane, ammonia, and various ices condensed in the cold far reaches of the nebula to form the outer planets, the very large *Jovian planets.* Closer to the sun, metallic and rocky substances with high melting points condensed and collided to form the four dense inner planets, the *terrestrial planets.* Of that group, the third and fourth planets from the sun are most interesting to geologists.

What better ways to celebrate Independence Day 1997 than to receive transmissions from NASA's *Mars Pathfinder* showing a Martian landscape once ravaged by floods and now a

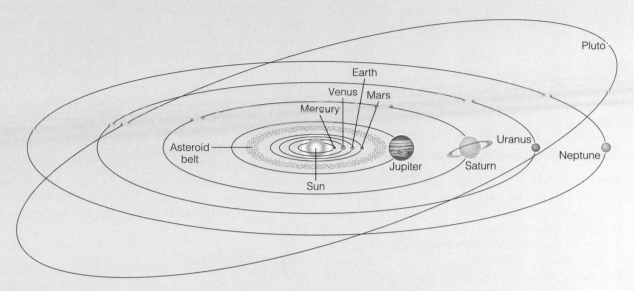

◇ **FIGURE 2** Diagram of the solar system showing the relative sizes and positions of the planets. Note Pluto's discordant orbit. It causes astronomers to debate whether Pluto should be considered a planet.

surface of wind-blown soil. Aha! Mars did indeed have an atmosphere and liquid water at one time. *Mars Pathfinder* carried the Jet Propulsion Laboratory's (JPL) remote rover, Sojourner, named for the nineteenth-century civil rights activist and evangelist Sojourner Truth. The funny-looking little ten-inch high rover carried an X-ray spectrometer and a miniature lab for analyzing specimens of Martian rocks and soil on the spot. *Mars Pathfinder* was the media event of the summer. Its Web site took 265 million hits, and toy stores' stocks of Mattel's lander and rover models sold out quickly.

During Sojourner's first week on Mars the little rover collected data from soil and a football-size rock affectionately dubbed "Barnacle Bill." Other rocks named Yogi and Scooby Doo made headlines. The naming of rocks goes back to the 1970s *Viking* missions. Investigators found that naming items in the transmitted photographs made it easier for them to remember the geography of the sample field.

Most rocks on Earth, the Moon, and Mars are silicates, composed of silica (SiO_2) and aluminum. The amount of SiO_2 in a rock that was once molten

is a key indicator of how "distilled" the melt that cooled to form the rock was. Low-silica rocks (<50%), such as the black lava basalt found around the Hawaiian Islands, come from "primitive" melts, melts that have not lost any of their original high-temperature, low-silica minerals. This is what researchers expected to find on Mars, because Mars has the largest volcanoes in the solar system. Barnacle Bill, however, contained moderate amounts of SiO_2, similar to rocks found in mountainous regions where plate margins converge, like the Andes (see Chapter 3). The rock andesite is named for this region and is close to the composition of Barnacle Bill. It appears that both basalt and andesite are found on Mars, even though Mars has had no plate tectonic motion.

A Martian day, known as a "sol," is 24 hours and 37 minutes long. *Mars Pathfinder* was designed to transmit for about 30 sols before going dead. It blinked out on 27 September 1997, almost two months longer-lived than expected. NASA and JPL were tickled Martian pink. ■

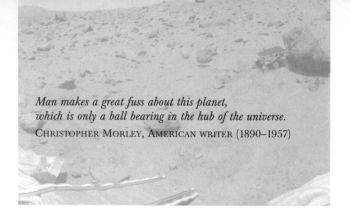

Man makes a great fuss about this planet,
which is only a ball bearing in the hub of the universe.
CHRISTOPHER MORLEY, AMERICAN WRITER (1890–1957)

Geologic materials such as rocks, valuable mineral deposits, and soils are an integral part of the human geologic environment. Rocks form the geologic foundations and provide building material for many of our structures. In addition, rocks are involved in many surface earth processes that are hazardous to humans, such as landslides and rockfalls. In this chapter we will discuss briefly how the planet earth formed and how its rocky crust and atmosphere arrived at their present condition. Environmental geologists are often asked about future geologic events, and in order to better understand the future a geologist must learn as much as possible about the past. An important concept in geology known as the Law of Uniformitarianism is that all geologic processes active today have been active in the past, only at differing rates. The law can be paraphrased as "the present is the key to the past." For example, by studying the geologically recent past, we have learned that North America was subject to multiple glacial advances in the past 1.6 million years, a period of time known as the Ice Ages. As a result, scientists monitor and computer-model weather and climate in order to anticipate long-term trends and perhaps even the next ice age (see Chapter 11). In this regard, we might also say "the present is the key to the future."

The Earth System

If you were to ask the earth, if the earth could speak, "How is your geology today," or, "How is your bios-

◁▶ **FIGURE 2.1** View down Lone Pine Canyon toward Valhalla, Sequoia National Park, California. Granites and country rocks of the Sierra Nevada were sculpted into broad valleys and pinnacles by glaciers that existed in the region less than 20,000 years ago. Bare rock and thin soils testify to the erosive power of ice.

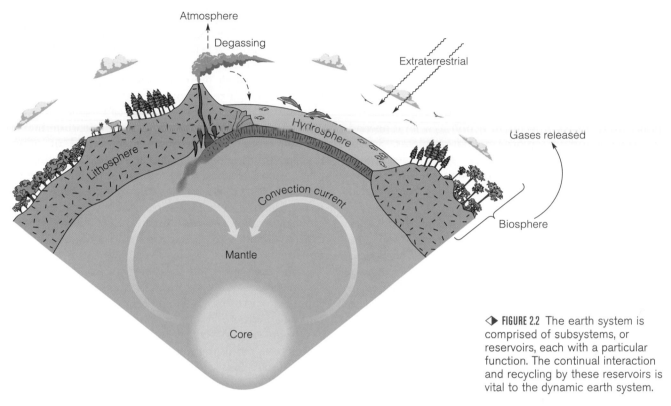

◁▶ FIGURE 2.2 The earth system is comprised of subsystems, or reservoirs, each with a particular function. The continual interaction and recycling by these reservoirs is vital to the dynamic earth system.

phere," the reply in all likelihood would be, "How's my *what?*" Just as the elements of human physiological systems are connected and interdependent, so on earth there are linkages and interactions between the rocky surface and the atmosphere and oceans, and between these systems and all plant and animal life. Our interest extends from the ecosphere, where life evolves, to the reactions between the crust and its fluid and gaseous envelopes, and even to the earth's core, where magnetic fields that protect most organisms from lethal radiation are generated. An example of a little-known connection is the one that exists between dust storms in the Gobi and Takla Makan deserts in China and the North Pacific Ocean. Airborne silt from those deserts is carried all the way to the North Pacific, where it increases the nutrient content and fertility of this otherwise barren oceanic realm. The periodic weather phenomenon known as *El Niño,* which heavily impacts the west coasts of South and North America, has its origins in ocean–atmosphere interactions called the *Southern Oscillation* in the western Pacific thousands of miles away. We now know that ENSO—the acronym for *El Niño–Southern Oscillation*—is a global weather event whose effects are not limited to floods along the west coasts of the Americas.

A system is defined as a set of concepts or parts that work together to perform a particular function. A system may comprise many subsystems. Transportation is an easily understood system, whether it be by plane, train, or car. Drive the freeways in a large city

and see how they are designed as a system with subsystems such as interchanges and on- and off-ramps. All must work, or the system will start to break down. The five interconnected *reservoirs,* or subsystems, of the complete earth system are

- the *solid earth,* which includes soil, the rocks beneath our feet, the *lithosphere* (the earth's rocky outer shell), and the very hot, mushy interior
- the *hydrosphere,* all water on and in the earth, including oceans, lakes, rivers, clouds, underground water, and glaciers
- the *biosphere,* all living and dead organic components of the planet. The *ecosphere* is a subsystem of the biosphere that includes the biosphere and its interactions with the other reservoirs.
- the *atmosphere,* the gaseous envelope surrounding the earth
- the *extraterrestrial,* which includes such things as energy from the sun, meteors, tides, back-radiation of energy from the earth to space, and the ozone layer.

◁▶ Figure 2.2 portrays the earth system.

Most of earth's systems are *closed systems;* that is, there are no additions to or subtractions from the system. Energy is an exception; it is part of an *open system.* Energy from the sun drives many earth systems, and some of earth's energy is back-radiated to space.

The solid-earth sciences attempt to understand the past, present, and future behavior of the whole earth system; that is, they take a planetary approach to the study of global change by studying the present and the past. The spheres of the earth system are all linked, and separate disciplines such as oceanography, geology, hydrology, and atmospheric physics have simply developed in order to organize information and award academic degrees. Over the past century the sciences have become increasingly segmented and specialized, going in the exact opposite direction of what is needed for solving today's environmental problems. Strong disciplinary studies are needed, but we also need a systems approach for understanding how the earth works and how a sustainable environment can be provided for future generations.

The surface of the lithosphere is a major boundary in the earth system, as this is where energy from the sun is transferred to the planet and its inhabitants. It is also at this boundary that we find landslides, volcanic activity, earthquakes, severe weather, and long-term global weather changes that impact humans. "The whole of humankind," wrote noted historian Norman Pounds, "has been in the shadow of an ice age." When Columbus was a teenager, the earth fell into a period of severe cold called the "Little Ice Age," which lasted 400 years to the mid 1800s. The earth is a dynamic system, so there is nothing new about global change. What *is* new is how *fast* change seems to be taking place today. Temperatures changed at a rate of about .05° C per century during the Little Ice Age as best as we can determine from written records and ice cores. The rate of temperature change today is about ten times faster, even in the most conservative models. We must learn how much of this increase is anthropogenic (due to human activities). This is the first time humans have had the luxury of being able to look into the future and predict ten years in advance what the climate will be like—and humans today have the intellect, technology, and obligation to mitigate or at least to slow adverse anthropogenic changes.

The reservoirs, except for energy, represent closed systems in which the mass movements of material that we call *geologic cycles* (rock and hydrologic cycles) and *biogeochemical cycles* (cycling of life-forming elements such as carbon and nitrogen) occur. The most readily understood of these cycles is probably the hydrologic cycle, by which water on land and at the sea surface is evaporated to the atmosphere and transported as clouds, which then redistribute the moisture around the earth (see Figure 8.2). It is a short-term cycle. The rock cycle, on the other hand, is a long-term cycle involving slow changes, some of them taking millions of years (see Figure 2.12). Less obvious is the recycling of the elements hydrogen, oxygen, carbon, nitrogen, and potassium, the main players in biogeochemical cycles. The atmosphere and hydrosphere get carbon, nitro-

gen, and oxygen by interacting with the solid earth and oceans, by weathering, and by volcanic action, and the cycle is completed when animals and plants give up these elements to the atmosphere, lithosphere, and hydrosphere by their life processes or after death by decomposition. Current knowledge of the carbon cycle is woefully inadequate, as scientists cannot account for about 25 percent of sequestered carbon (carbon that is isolated from the environment by burial or other means). The assumption is that it is buried beneath sediment in the oceans and in soil on land. Knowledge of the carbon cycle is of major importance in predicting future global climate and, specifically, the amount of global warming that can be caused by anthropogenic carbon dioxide (CO_2) and methane (CH_4), so-called greenhouse gases.

The time scale and areal extent of selected earth-system processes are summarized in ◆ Figure 2.3. We cannot do anything about such processes as mountain building and petroleum generation, but we can reduce death and injury from geologic hazards and the amount of atmospheric pollution, soil erosion, and other human-caused environmental problems.

Earth Materials

ELEMENTS, ATOMS, AND ATOMIC STRUCTURE

Elements are substances that cannot be changed into other substances by normal chemical methods. They are composed of infinitesimal particles called **atoms.** In 1870, Lothar Meyer published a table of the 57 then-known elements arranged in the order of their atomic weights. Meyer left blank spaces in the table wherever elements of particular weights were not known. About the same time Dmitry Mendeleyev developed a similar table, but his table was based upon the similar chemical properties of the particular known elements. Where an element's placement based on weight did not group it with elements of similar properties, Mendeleyev did not hesitate to suggest that its weight had been measured incorrectly. He also predicted the properties of the "missing" elements, and very shortly his predictions proved correct with the discovery of the elements gallium, scandium, and germanium (Appendix 1).

The weight of an atom of an element is contained almost entirely in its **nucleus,** which contains protons (atomic weight = 1, electrical charge = 1^+) and neutrons (atomic weight = 1, electrical charge = 0). An element's **atomic number** is the number of protons in its nucleus, and this number is unique to that element. The weight of an atom—its **atomic mass**—is the sum of

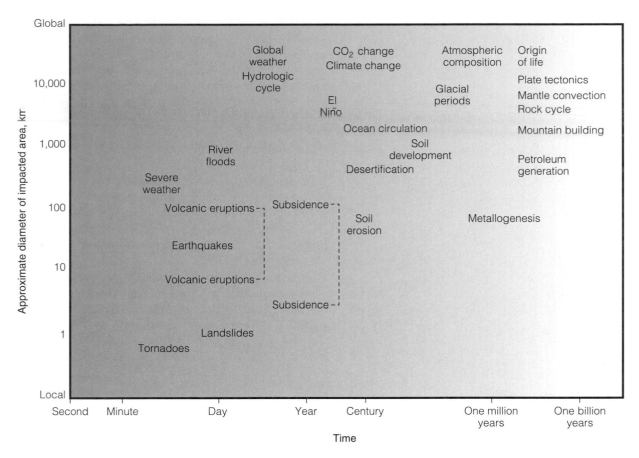

◆ FIGURE 2.3 Approximate space and time scales of earth-system processes. The space scale indicates the approximate diameter of the region impacted by the earth-system process.

its nuclear protons and neutrons. For example, $^4_2$He denotes the element helium, which has 2 protons (denoted by the subscript) and two neutrons, for an atomic mass of 4 (superscript). Electrical neutrality of the atom is provided by balancing the positive proton charges by an equal number of negatively charged electrons orbiting in shells around the nucleus (◆ Figure 2.4). **Ions** are atoms that are positively or negatively charged owing to a loss or gain of electrons (e^-) in the outer electron shell. Thus, sodium (Na) may lose an electron in its outer shell to become sodium ion (Na$^+$, a *cation*), and chlorine (Cl) may gain an electron to become chloride ion (Cl$^-$, an *anion*). *Valence* refers to ionic charge, and sodium and chlorine are said to have valences of 1$^+$ and 1$^-$, respectively. Elemental sodium is an unstable metallic solid, and chlorine is a poisonous gas. When ionized, they may combine in an orderly fashion to form the mineral halite (NaCl), common table salt, a substance necessary for human existence (◆ Figure 2.5).

Isotopes are forms (or species) of an element that have different atomic masses. The element uranium, for instance, always has 92 protons in its nucleus, but

varying numbers of neutrons define its isotopes. For example, Uranium-238 (^{238}U) is the common naturally occurring isotope of uranium. It weighs 238 atomic mass units and contains 92 protons and 146 (238 *minus* 92) neutrons. Uranium-235 is the rare isotope (0.7% of all uranium) that has 143 neutrons. Similarly, ^{12}C is the common isotope of carbon, but ^{13}C and ^{14}C also exist. A modern periodic table of the elements that shows their atomic number and mass appears in Appendix 1.

MINERALS

The earth's crust is composed of rocks, and rocks are made up of one or more minerals. Although most people seldom think about minerals that do not have nutritional importance or great beauty and value such as gold and silver, the mineral kingdom encompasses a broad spectrum of chemical compositions. **Minerals** are defined as *naturally occurring, inorganic, crystalline* substances, each with a narrow range of *chemical compositions* and characteristic *physical properties*. Thus, neither artificial gems such as zirconia nor organic deposits such as coal and oil are minerals by definition. The

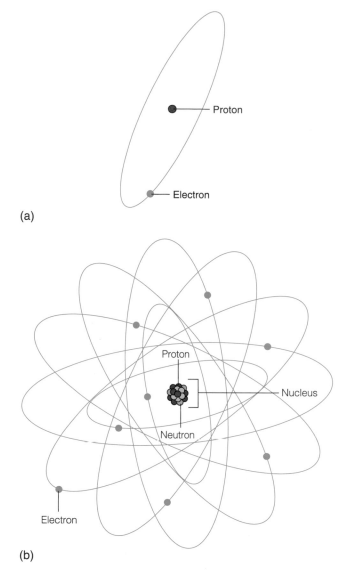

(a)

(b)

◆ **FIGURE 2.4** Structures of (a) a hydrogen atom, in which an electron orbits a single proton, and (b) an oxygen atom, in which 8 electrons orbit a dense nucleus containing 8 protons and 8 neutrons.

crystal form of a mineral reflects an orderly atomic structure, which in turn determines the crystal shape. Solids that have random or noncrystalline atomic structures—glass and opal, for example—are described as **amorphous.** Minerals composed of a single element, such as copper and carbon (as in graphite and diamonds), are known as *native elements*. Most minerals, however, are composed of combinations of two or more elements. Although more than 3,000 minerals have been named and described, only about 20 make up the bulk of the earth's crustal rocks. The chemistry of minerals is the basis of their classification, and the most common minerals are primarily composed of the elements oxygen (O), silicon (Si), aluminum (Al), and iron (Fe). This is not surprising, because oxygen and silicon compose 75 percent of crustal rocks by weight. A memory aid for these elements in order is "OSAF."

Geologists classify minerals into groups that share similar negatively charged ions (anions) or ion groups (radicals). Oxygen (O^{-2}) may combine with iron to form hematite (Fe_2O_3), a member of the *oxide group* of minerals. The *sulfide minerals* typically are combinations of a metal with sulfur. Examples include galena (lead sulfide, PbS), chalcopyrite (copper-iron sulfide, $CuFeS_2$), and pyrite (iron sulfide, FeS_2), the "fool's gold" of inexperienced prospectors. *Carbonate minerals* contain the negatively charged $(CO_3)^{-2}$ ion. Calcite ($CaCO_3$), the principal mineral of limestone and marble, is an example. ◉ Table 2.1 summarizes some of the mineral groups most important to geologists.

The most common and important minerals are the *silicates,* which are composed of combinations of oxygen and silicon with or without metallic elements. The basic building block of a silicate mineral is the silica tetrahedron (◆ Figure 2.6a). It consists of one silicon atom surrounded by four oxygen atoms at the corners of a four-faced tetrahedron. The manner in which the

◆ **FIGURE 2.5** Two models of the atomic structure of the mineral halite (NaCl), common table salt. (a) A packing model that shows the location and relative sizes of the sodium and chloride ions. (b) A ball-and-stick model that shows the mineral's cubic crystal structure.

(a)

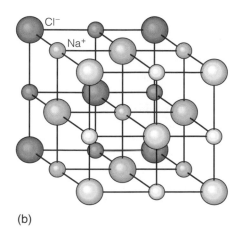

(b)

TABLE 2.1 Some Important Mineral Groups*

MINERAL GROUP	NEGATIVELY CHARGED ION OR ION GROUP	EXAMPLES	COMPOSITION
Carbonate	$(CO_3)^{-2}$	Calcite	$CaCO_3$
		Dolomite	$CaMg(CO_3)_2$
Halide	CL^{-1}, F^{-1}	Halite	$NaCl$
Hydroxide	$(OH)^{-1}$	Limonite	$FeO(OH) \cdot H_2O$
Native element	—	Gold	Au
		Diamond	C
Oxide	O^{-2}	Hematite	Fe_2O_3
Silicate	$(SiO_4)^{-4}$	Quartz	SiO_2
		Olivine	$(Mg,Fe)_2SiO_4$
Sulfate	$(SO_4)^{-2}$	Gypsum	$CaSO_4 \cdot 2H_2O$
Sulfide	S^{-2}	Galena	PbS

*A table of the elements and their symbols appears in Appendix 1.

tetrahedra are packed or arranged in the mineral structure is the basis for classifying silicates. Some of the ways the tetrahedra may be arranged are in layers or sheets, as in the clay minerals and mica; in long chains, as in the amphibole and pyroxene groups; and in three-dimensional networks, as in quartz and feldspar (◆ Figure 2.6b, Case Study 2.1).

ROCK-FORMING SILICATES

Most rocks are aggregates of minerals, and feldspar is the most abundant of the silicate minerals in the earth's crust (◆ Figure 2.7). Feldspar is actually a mineral *group* that ranges in chemical composition from potassium-rich orthoclase found in granite (◆ Figure 2.8) to calcium-rich plagioclase found in basalt and gabbro. The ferromagnesian silicates are rich in iron and magnesium in addition to silicon and oxygen. Most of them are dark-colored, either black, brown, or green (◆ Figure 2.9). Important among these are hornblende (amphibole), augite (pyroxene), biotite mica, and olivine, which is believed to be the most abundant mineral just below the outer crust of the earth (◉ Table 2.2).

◆ **FIGURE 2.6** (a) Model of the silica tetrahedron, showing the unsatisfied negative charge at each oxygen that allows it to form (b) chains, sheets, and networks.

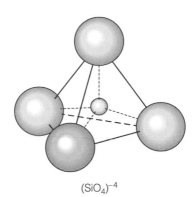

$(SiO_4)^{-4}$

	Formula of negatively charged ion group	Example
Isolated tetrahedra	$(SiO_4)^{-4}$	Olivine
Continuous chains of tetrahedra	$(SiO_3)^{-2}$	Pyroxene group (augite)
	$(Si_4O_{11})^{-6}$	Amphibole group (hornblende)
Continuous sheets	$(Si_4O_{10})^{-4}$	Mica(muscovite)
Three-dimensional networks	Too complex to be shown by a simple two-dimensional drawing $(SiO_2)^0$ $(Si_3AlO_8)^{-1}$ $(Si_2Al_2O_8)^{-2}$	Quartz Orthoclase feldspar Plagioclase feldspar

(a)

(b)

◆ FIGURE 2.7 Some common rock-forming silicate minerals: (a) feldspar, (b) quartz, (c) muscovite, and (d) hornblende.

(a)

(b)

(c)

(d)

MINERAL IDENTIFICATION

Serious students of mineralogy are able to identify several hundred minerals without destructive testing. They are able to do this because minerals have distinctive physical properties, most of which are easily determined and associated with particular mineral species (Appendix 2). With practice, one can build up a mental catalog of minerals, just like building a vocabulary in a foreign language. The most useful physical properties are hardness, cleavage, crystal form, and to lesser degrees, color (which is variable) and luster. Mineralogists use the **Mohs hardness scale,** developed by Friedrich Mohs in 1812 (◉ Table 2.3). Mohs assigned a hardness value (H) of 10 to diamond and of 1 to talc, a very soft mineral. Diamond scratches everything; quartz ($H = 7$) and feldspar ($H = 6$) can scratch glass ($H = 5\frac{1}{2}–6$) and calcite ($H = 3$), and your fingernail can scratch gypsum ($H = 2$).

Cleavage refers to the characteristic way particular minerals split along definite planes as determined by their crystal structure. Mica has perfect cleavage in 1 direction; this is called *basal* cleavage, because it is parallel to the basal plane of the crystal structure. Feldspars split in 2 directions; halite, which has *cubic* structure, in 3 directions; and so on, as shown in ◆ Figure 2.10. Minerals that have perfect cleavage can be split readily by a tap with a rock hammer or even peeled apart, as in the case of mica. Some minerals do not cleave but have distinctive **fracture** patterns that

◆ FIGURE 2.8 This granite is an aggregate of the minerals quartz, biotite, and feldspar.

Granite (rock)

Quartz (mineral)

Biotite (mineral)

Feldspar (mineral)

◆ **FIGURE 2.9** Some common ferromagnesian silicates:
(a) olivine; (b) augite, a pyroxene-group mineral;
(c) hornblende, an amphibole-group mineral; and
(d) biotite mica.

help one to identify them. The common crystal forms shown in ◆ Figure 2.11, can be useful in identifying a particular mineral. Color tends to vary within a given mineral species, so it is not a reliable identifying property. **Luster,** how a mineral reflects light, is useful,

however. We recognize metallic and nonmetallic lusters, the latter being divided into types such as glassy, oily, greasy, and earthy. (Case Study 2.2 discusses some gem minerals.)

ROCKS

Rocks are defined as consolidated or poorly consolidated aggregates of one or more minerals, glass, or solidified organic matter (such as coal) that cover a significant part of the earth's crust. There are three classes of rocks, based upon their origin: igneous, sedimentary, and metamorphic. **Igneous** (Latin *ignis,* "fire") rock is crystallized from molten or partly molten material. **Sedimentary rocks** include both *lithified* (that is, turned to stone) fragments of preexisting rock and rocks that were formed from chemical or biological action. **Metamorphic** rocks are those that have been changed, essentially in the solid state, by heat, fluids, and/or pressure within the earth.

The **rock cycle** is one of many natural cycles on earth. The illustration of it in ◆ Figure 2.12 shows the interactions of energy, earth materials, and geologic processes that form and destroy rocks and minerals. The rock cycle is essentially a closed system; "what goes around comes around," so to speak. In a simple cycle there might be a sequence of formation, destruction, and alteration of rocks by earth processes. For example, an igneous rock may be eroded to form sediment that subsequently becomes a sedimentary rock,

○ TABLE 2.2 **Some Common Rock-Forming Minerals**

MINERAL	ABUNDANCE IN CRUST, %	ROCK IN WHICH FOUND
Plagioclase*	39	Igneous rocks mostly
Quartz	12	Detrital sedimentary rocks, granites
Orthoclase**	12	Granites, detrital sedimentary rocks
Pyroxenes	11	Dark-colored igneous rocks
Micas	5	All rock types as accessory minerals
Amphiboles	5	Granites and other igneous rocks
Clay minerals	5	Shales, slates, decomposed granites
Olivine	3	Iron-rich igneous rocks, basalt
Others	11	Rock salt, gypsum, limestone, etc.

*A series of six minerals within the plagioclase group from albite ($NaAlSi_3O_8$) to anorthite ($CaAl_2Si_2O_8$).
**Feldspar group of minerals.

○ TABLE 2.3 **Mohs Hardness Scale**

HARDNESS	MINERAL	COMMON EXAMPLE
1	Talc	Pencil lead 1–2
2	Gypsum	
		Fingernail 2½
3	Calcite	Copper penny 3
		Brass
4	Fluorite	Iron
5	Apatite	Tooth enamel
		Knife blade
		Glass 5½–6
6	Orthoclase	
		Steel file 6½
7	Quartz	
8	Topaz	
9	Corundum	Sapphire, ruby
10	Diamond	Synthetic diamond

Case Study

Case Study | 2.1

Minerals, Cancer, and OSHA—Fact and Fiction

When an international health organization designated quartz as "probably" carcinogenic (cancer-causing) to humans, the U.S. Occupational Safety and Health Agency (OSHA) immediately went into action. It issued a mandate requiring products containing more than one tenth of one percent (0.1%) of quartz or "free silica" (SiO_2) to display warning signs. Because quartz is chemically inert and because it is the most common mineral species, questions arose as to whether this requirement extended to sandpaper, beach sand, and unpaved roads that may produce clouds of fine-grained quartz dust. Questions—and more—arose when some Delaware truck drivers were "ticketed" for not displaying the required "quartz on board" signs during transport of crushed rock to construction sites. Such well-meaning OSHA actions have been perceived as ridiculous. They illustrate what happens when a government authority establishes health regulations without understanding the substances involved. In this case OSHA did not understand the where, how, and chemistry of the mineral—its mineralogy. Unless one has had a long-term craving to inhale beach sand or crushed rock, there are no health hazards in the examples cited. Nonetheless, because exposure to very large quantities of any substance, even table salt or vitamins, can be harmful to health, it is certain that long-term workplace exposure to dusts from some minerals can pose health problems.

The World Health Organization's Group 1 classification of "known" carcinogenic minerals includes asbestos minerals (there are six), erionite (a zeolite), and minerals containing arsenic, chromium, and nickel. Its Group 2 of "probable" carcinogenic minerals includes all radioactive minerals and minerals containing lead, beryllium, and silica. The classification is based upon suspected relationships between specific diseases and workplace exposure to the identified substances and on laboratory experiments with animals.

A major human health concern is the relationship between lung cancer and inhaling the fine fibers of asbestos minerals. The long, fibrous crystal forms of asbestos belonging to the amphibole group (see hornblende structure, Fig-

ure 2.6) are considered the most carcinogenic. Of these, blue asbestos, crocidolite (pronounced **krō-sid'-əl-īt**; ◈ Figure 1), is thought to be most dangerous and is believed to cause mesothelioma, a relatively rare cancer of the lining of the heart and lungs. Although crocidolite constitutes only 5 percent of all industrial asbestos, its morbidity statistics are grim indeed. For example, a 1990 survey of 33 men who, in 1953, worked in a factory where crocidolite was utilized in manufacturing cigarette filters found that 19 of them had died of asbestos-related diseases.

Common white asbestos, chrysotile (pronounced **kris'-ə-tīl**; ◈ Figure 2), is a sheet-structure mineral similar to mica. It appears under high magnification as long fibers, which are in reality rolled sheets, much like long rolls of gift-wrapping paper. Chrysotile constitutes 95 percent of all industrial asbestos and has never been demonstrated to cause cancer at the levels found in most schools and public

◈ FIGURE 1 Deadly crocidolite, "blue asbestos," magnified about 100 times. The mineral appears in its characteristic blue and green colors when viewed through a petrographic microscope. It was used in only a small percentage of commercial products.

which may then be metamorphosed by heat and pressure to become a metamorphic rock, and then melted to become an igneous rock.

Throughout this book we emphasize those rocks that, because of their composition or structure, are involved in geologic events that endanger human life or well-being or that are important as resources.

Consider This... ? Your school district is told it must remove all asbestos insulation from its older buildings at considerable cost to the taxpayers, and at the expense of teachers' salaries and educational materials. What questions should be asked of the government agency requiring this action, and what position will you take if you learn that the asbestos is the long-chain, fibrous type, rather than the sheet-structure, mica type (Case Study 2.1)? ■

◇▶ FIGURE 2 Chrysotile asbestos (×100) has many uses in homes, offices, and public buildings. Although it is the target of removal in thousands of structures, at low levels it poses no documented risk. Note the difference between chrysotile and crocidolite's crystal structures.

buildings. A study conducted in Thetford, Quebec, Canada, evaluated persons who had been exposed to high concentrations of white asbestos in their workplace and found no deaths attributable to it. Many of the diagnosed chrysotile-related lung problems have been traced back to improper handling of the material and inadequate respiratory protection during asbestos removal procedures. Experimental studies at Virginia Tech have shown that respirable-sized 1-nannometer-diameter chrysotile fibers dissolve in about nine months under conditions expected in lung tissue. This very short lifetime is difficult to reconcile with the general observations that asbestos-related diseases appear in asbestos workers only after many years on the job.

Inhaling finely divided airborne asbestos fibers in large quantities is definitely a health hazard. Under the Toxic Substance Control Act of 1972, the EPA mandated in 1986 that *all* asbestos minerals be treated as identical health hazards, regardless of their mineralogy, and that they be removed. By the mid 1990s the cost of removal had amounted to hundreds of millions of dollars, and the total cumulative cost for remediation of rental and commercial buildings, litigation,

and enforcement was estimated at $100 billion. The EPA requirements were sweeping. They applied to buildings containing the relatively benign white asbestos and to those where airborne concentrations were so low that they could not be measured. The fear of asbestos and the EPA regulations are based on the idea that there is no safe "threshold" concentration; that even one fiber can kill. No doctor or public health official would agree with this. A New York school district's administration was besieged by irate parents when it admitted that it could not *guarantee* that *all* asbestos had been removed from its buildings. The district had spent millions of dollars to remove it.

Industry and government can ill afford to squander capital and tax dollars on poorly conceived hazard-mitigation requirements based upon the "no threshold" theory of health risk; that is, that one molecule of pesticide or one asbestos fiber can cause health problems. Many government regulations regarding "toxic substances" address risks to humans that are no greater than those incurred by routinely drinking several cups of coffee each day or eating a peanut butter sandwich for lunch. The "no threshold" criterion forces "carcinogen" classification upon otherwise benign minerals and many other useful substances and needlessly increases public anxiety. (Remember when synthetic sweeteners and cellular telephones were labeled carcinogenic?) The costs of removal and liability protection in such cases are horrendous, and a fiscal crisis in the environmental field is acknowledged. Liability extends to building owners, real estate agents who sell the buildings, banks that lend money for property purchases, and purchasers of land upon which any material classified as hazardous is found.

A person's chances of being struck by a lightning bolt are about 35 per 1 million lifetimes, and the risk of a non-smoker's dying from asbestos exposure is about 1 in 100,000—about a third of the chance of being struck by lightning. Incidentally, the risk of dying from cigarette-related diseases is about 1 in 5 for smokers. Certainly, some substances in the environment pose dangers, and the key to risk reduction and longevity is awareness. Just as a reasonable person would not play golf during a violent thunderstorm because of the risk of being struck by lightning on the course, one should not handle hazardous substances without wearing protective gear.

Igneous Rocks

Igneous rocks are classified according to their texture and their mineral composition. A rock's texture is a function of the size and shape of its mineral grains, and for igneous rocks this is determined by how fast or slowly a melted mass cools. If **magma**—molten rock within the earth—cools slowly, large crystals develop, and a rock with coarse-grained, **phaneritic texture** such as granite is formed (see Figure 2.8). The resulting large rock mass formed within the earth is known as a **batholith.** Greater than 100 square kilometers (39

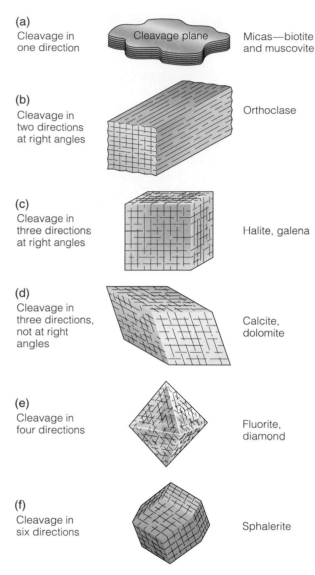

(a)
Cleavage in
one direction

Cleavage plane

Micas—biotite
and muscovite

(b)
Cleavage in
two directions
at right angles

Orthoclase

(c)
Cleavage in
three directions
at right angles

Halite, galena

(d)
Cleavage in
three directions,
not at right
angles

Calcite,
dolomite

(e)
Cleavage in
four directions

Fluorite,
diamond

(f)
Cleavage in
six directions

Sphalerite

◆ **FIGURE 2.10** Types of cleavage and typical minerals in which they occur.

mi^2) in area by definition, batholiths are mostly granitic in composition and show evidence of having invaded and pushed aside the *country rock* into which they intruded. Also, many batholiths have formed in the cores of mountain ranges during mountain-building episodes. Batholiths are a type of igneous mass referred to as *plutons,* and rocks of batholiths are described as **plutonic** (after Pluto, the Greek god of the underworld), because they were formed at great depth. The Sierra Nevada of California is an example of an uplifted and eroded mountain range with an exposed core composed of many plutons (◆ Figure 2.13).

Most igneous rocks—whether on Earth, Moon, or Mars—are composed of silicate mineral-containing silicon, oxygen, and aluminum. Silicon combined with oxygen is the compound *silica,* SiO$_2$. The percentage of silica in an igneous rock is a measure of how "distilled," or fractionated, its original magma was. The higher the percentage of silica in an igneous rock, the more steps of fractionation it has gone through. ◆ Figure 2.14 illustrates the process. Minerals that are rich in magnesium and iron and low in silica (minerals such as pyroxene, amphibole, and olivine) will crystallize and settle out of the magma first (◆ Figure 2.14b). The resulting rocks, relatively rich in magnesium and iron and poor in silica, are thus designated **mafic** (*ma-* for magnesium, *-fic* for iron). The remaining magma has less magnesium and iron and a greater percentage of silica. When the next batch of minerals settles out, perhaps with quartz crystals (◆ Figure 2.14c), rocks of *intermediate* composition are formed. During the fractionation process the magma becomes richer and richer in silica and more impoverished in magnesium and iron. This late-stage magma is largely silica, alumina, and other elements that combine at lower temperatures. The resulting rocks are rich in feldspar and silica, and are thus described as **felsic** (*fel-* for feldspar, *-sic* for silica). "Primitive" magmas (<50% SiO$_2$) produce mafic gabbro and its volcanic equivalent, basalt, the rock common to Hawaii volcanoes. More "evolved" mag-

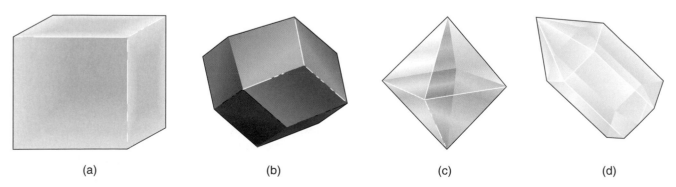

(a)

(b)

(c)

(d)

◆ **FIGURE 2.11** Mineral crystal shapes: (a) cube (halite), (b) 12-sided dodecahedron (garnet), (c) 8-sided octahedron (diamond, fluorite), and (d) 6-sided hexagonal prism (quartz).

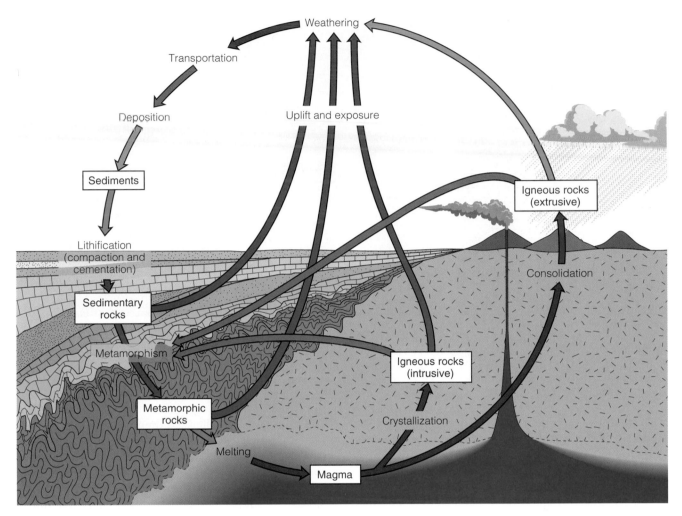

◆ **FIGURE 2.12** The rock cycle. The three rock types are interrelated by internal and external processes involving the atmosphere, ocean, biosphere, crust, and upper mantle.

◆ **FIGURE 2.13** The Sierra Nevada batholith and Mount Whitney, looking west from Owens Valley; east-central California.

mas (>50% SiO_2) produce rocks of intermediate composition, such as diorite and its volcanic equivalent, andesite. Highly evolved magmas produce felsic rocks such as granite and its volcanic equivalent, rhyolite. Mafic rocks are black or dark-colored, and felsic rocks are whitish and pinkish. Intermediate igneous rocks vary in color between these extremes according to their percentage of silica.

Lava is molten material at the earth's surface produced by volcanic activity. Lava cools rapidly, resulting in restricted crystal growth and a fine-grained, or **aphanitic,** texture. An example of a rock with aphanitic texture is rhyolite, which has about the same composition as granite but a much finer texture. The grain size of an igneous rock can tell us if it is an **intrusive** (cooled within the earth) or an **extrusive** (cooled at the surface of the earth) rock. The classification by texture and composition of igneous rocks is shown in ◆ Figure 2.15. Note that for each composition there are pairs of rocks that are distinguished by

(a)

(b)

(c)

◆ **FIGURE 2.14** The relative proportions of colored candies change as particular colors are preferentially selected. This is analogous to the process of fractionation of magma as particular minerals are removed by settling and the proportion of the other components increases.

their texture. Granite and rhyolite, diorite and andesite, and gabbro and basalt are the most common pairs found in nature. The end member shown in Figure 2.15, peridotite, is rarely found because it originates deep within the earth. Three important and relatively common glassy (composed of amorphous or finely crystalline SiO_2) volcanic rocks are obsidian, which looks like black glass; pumice, which has the composition of glass but is not really glassy-looking; and tuff, consolidated volcanic ash or cinders (◆ Figure 2.16). Tuff has a **pyroclastic** (literally "fire-broken") texture, resulting from fragmentation during violent volcanic eruptions.

Sedimentary Rocks

Sediment is particulate matter derived from the physical or chemical weathering of materials of the earth's crust and by certain organic processes. It may be transported and redeposited by streams, glaciers, wind, or waves. Sedimentary rock is sediment that has become **lithified**—turned to stone—by pressure from deep burial, by cementation, or by both of these processes. **Detrital sedimentary rocks** are those that are composed of *detritus,* fragments of preexisting rocks and minerals. Their texture, which makes this composition quite obvious, is described as **clastic.** Clastic sedimentary rocks are classified according to their grain size. Thus sand-size sediment lithifies to sandstone, clay to shale, and gravel to conglomerate (◆ Figures 2.17 and

2.18 a, b, and c). Other sediments result from chemical and biological activity. **Chemical sedimentary rocks,** which may be clastic or nonclastic, include chemically precipitated limestone, rock salt, and gypsum. **Biogenic sedimentary rocks** are produced directly by biological activity, such as coal (lithified plant debris), some limestone and chalk ($CaCO_3$, shell material), and chert (siliceous shells) (◆ Figure 2.18 d and e). ◉ Table 2.4 presents the classification of sedimentary rocks.

A distinguishing characteristic of most sedimentary rock is bedding, or **stratification** into layers. Because shale is very thinly stratified, or *laminated,* it splits into thin sheets. Some sedimentary rocks, such as sandstone and limestone, may occur in beds that are several feet thick. Other structures in these rocks give hints as to how the rock formed. *Cross-bedding 2*—stratification that is inclined at an angle to the main stratification—indicates the influence of wind or water currents. Thick

Consider This... ? Forensic geologists are specialists that are sometimes called upon to help solve crimes. They can, for example, compare the mineralogy of soil on a suspect's shoes with the soil at the scene of the crime. Also, they use geologic clues to reconstruct events that happened millions of years ago. For instance, huge round scars (craters) on the surface of the earth can indicate ancient meteorite impacts. What are some of the "clues" in sedimentary rocks that tell us about the ancient environments in which they formed? ■

Rhyolite   Andesite   Basalt

Granite   Diorite   Gabbro   Peridotite

◀▶ **FIGURE 2.15** Classification of common igneous rocks by their mineralogy and texture. Relative proportions of mineral components are shown for each rock type. Fine-grained *aphanitic* texture results from rapid cooling of molten material. Coarse-grained *phaneritic* texture results from slow cooling.

cross-beds and frosted quartz grains in sandstone indicate an ancient desert sand-dune environment (◀▶ Figure 2.19). Some shale and claystone show polygonal *mud cracks* on bedding planes, similar to those found on the surface of modern dry lakes; these indicate desiccation in a subaerial environment (◀▶ Figure 2.20). *Ripple marks,* low, parallel ridges in deposits of fine sand and silt, may be asymmetrical or symmetrical (◀▶ Figure 2.21). Asymmetrical ripples indicate a unidirectional current; symmetrical ones, the back-and-forth motion produced by waves in shallow water. Only in sedimentary rock do we find abundant fossils, the remains or traces of life.

Metamorphic Rocks

Rocks that have been changed from preexisting rocks by heat, pressure, or chemical processes are classified as metamorphic rocks. The process of metamorphism results in new structures, textures, and minerals. **Foliation** (Latin *folium,* "leaf") is the flattening and layering of minerals by nonuniform stresses. Foliated metamorphic rocks are classified by the development of this structure. Slate, schist, and gneiss are foliated, for example, and they are identified by the thickness or crudeness of their foliation (◀▶ Figure 2.22). Metamorphic rocks may also form by **recrystallization.** This

(a)

(b)

(c)

◆ **FIGURE 2.16** Extrusive igneous rocks: (a) glassy obsidian; (b) pumice, which is formed from gas-charged magmas; and (c) tuff. The lavas that form obsidian and pumice cool so rapidly that crystals do not grow to any size. The reference scale in each photo is 5 centimeters long; 5 centimeters is about 2 inches.

occurs when a rock is heated and strained by uniform stresses so that larger, more perfect grains result or new minerals form. In this manner a limestone may recrystallize to marble or a quartz sandstone to quartzite, one of the most resistant of all rocks (◆ Figure 2.23). ◉ Table 2.5 summarizes the characteristics of common metamorphic rocks.

ROCK DEFECTS

Some rocks have structures that geologists view as "defects"; that is, surfaces along which landslides or rockfalls may occur. Almost any planar structure, such as a bedding plane in sedimentary rock or a foliation plane in metamorphic rock, holds potential for rock slides or falls. The orientation of a plane in space, such as a stratification plane, a fault, or a joint, may be defined by its **dip** and **strike** (Appendix 3). **Joints** are rock fractures without displacement and they occur in all rock types. They are commonly found in parallel sets spaced several feet apart (◆ Figure 2.24). **Faults** are also fractures in crustal rocks, but they differ from joints in that some movement or displacement has occurred along the fault surface. Faults also are found in all rock types and are potential surfaces of "failure." We will examine the relationships of rock defects to geologic hazards when we discuss landslides, subsidence, and earthquakes.

Geologic Time

Interest in extremely long periods of time sets geology and astronomy apart from other sciences. Geologists think in terms of billions of years for the age of the earth and its oldest rocks, numbers that, like the national debt, are not easily comprehended. Time is such

Consider This... ? Rock climbing is a popular sport in many parts of the world. A favorite spot in the U.S. West is Yosemite Valley, a picturesque area carved by glaciers into a uniform granite body known as the Cathedral Peak Granite (see Figure 2.33). Why do you suppose climbers would prefer to climb near-vertical cliffs of granite, as opposed to similar cliffs composed of schist, gneiss, or shale? ■

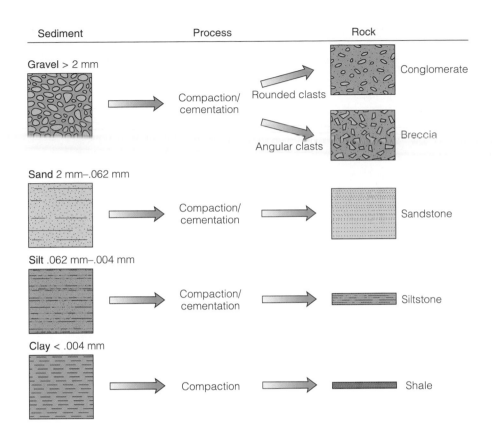

◆▶ FIGURE 2.17 How sediments are transformed into clastic sedimentary rocks. The sequence of lithification is *deposition* to *compaction* to *cementation* to hard rock.

◆▶ FIGURE 2.18 Common *clastic* sedimentary rocks: (a) shale, (b) sandstone, (c) conglomerate. *Biogenic* sedimentary rocks: (d) limestone composed entirely of shell materials (called *coquina*), (e) coal.

(a)   (b)   (c)

(d)   (e)

○ TABLE 2.4   Classification of Sedimentary Rocks

DETRITAL SEDIMENTARY ROCKS (CLASTIC TEXTURE)		
Sediment	*Description*	*Rock Name*
Gravel (>2.0 mm)	Rounded rock fragments	Conglomerate
	Angular rock fragments	Breccia
Sand (0.062–2.0 mm)	Quartz predominant	Quartz sandstone
	>25% feldspars	Arkose
Silt (0.004–0.062 mm)	Quartz predominant, gritty feel	Siltstone
Clay, mud (<0.004 mm)	Laminated, splits into thin sheets	Shale
	Thick beds, blocky	Mudstone

CHEMICAL SEDIMENTARY ROCKS		
Texture	*Composition*	*Rock Name*
Clastic	Calcite ($CaCO_3$)	Limestone
	Dolomite [$CaMg(CO_3)_2$]	Dolostone
Crystalline	Halite (NaCl)	Rock salt
	Gypsum ($CaSO_4 \cdot 2H_2O$)	Rock gypsum

BIOGENIC SEDIMENTARY ROCKS		
Texture	*Composition*	*Rock Name*
Clastic	Shell calcite, skeletons, broken shells	Limestone, coquina
	Microscopic shells ($CaCO_3$)	Chalk
Nonclastic (altered)	Microscopic shells (SiO_2), recrystallized silica	Chert
	Consolidated plant remains (largely carbon)	Coal

◆ **FIGURE 2.19** Wind-blown sandstone exhibiting large cross-beds; Zion National Park, Utah.

◆ **FIGURE 2.20** Mud cracks in an old clay mine; Ione, California.

(a)

◆ FIGURE 2.21 Ripple marks found in sedimentary rocks. (a) Asymmetrical ripple marks are common in streambeds and are also found on bedding planes in sedimentary rocks. (b) Symmetrical ripple marks are due to oscillating water motion; that is, wave action.

(b)

(c)

◆ FIGURE 2.22 Metamorphic rocks: (a) coarsely foliated gneiss, (b) mica schist showing wavy or crinkly foliation, (c) finely foliated slate showing slaty cleavage.

ancient events relevant to environmental geology? Ongoing research is being directed toward determining recurrence intervals for large earthquakes and volcanic eruptions. It is postulated that if we know the average time interval between catastrophic events over the past several thousand years, we can predict more reliably when the next such events will occur.

Geologists use two different means of dating rocks and geologic events. In the field we normally use **relative age dating,** which establishes the sequence of geologic events. For example, we can establish that "these" rock strata are oldest, or that "this" igneous rock is younger than that rock. In the laboratory, **absolute age dating** can be carried out. This yields an age in years before the present for an event, rock, or artifact.

Relative age is determined by applying geologic laws based upon similar reasoning. For instance, the Law of Superposition tells us that in a stack of undeformed sedimentary rocks, the stratum (layer) at the top is the youngest. The Law of Cross-cutting Relationships tells us that a fault is younger than the youngest rocks it displaces or cuts. Similarly, we know that a pluton is younger than the rocks it intrudes (◆ Figure 2.25).

Using these laws, geologists arranged a great thickness of sedimentary rocks and their contained fossils representing an immense span of geologic time. The geologic age of a particular sequence of rock was then determined by applying the Law of Fossil Succession, the observed chronologic sequence of life-forms through geologic time. This allows fossiliferous rocks from two widely separated areas to be correlated by matching key fossils or groups of fossils found in the rocks of the two areas (◆ Figure 2.25). Using such indicator fossils and radioactive dating methods, geologists have developed the geologic time scale to chronicle the documented events of earth history (◆ Figure 2.26). Note that the scale is divided into units of time

Case Study | 2.2

A Girl's Best Friend and Other Gems

Rarity and beauty are what make precious and semi-precious gems so desirable. Those shown in ◆ Figure 1 are but a sample of several hundred valued gem minerals and their varieties. The largest uncut diamond in the world, the Cullinan diamond (◆ Figure 1, part a), was recovered in 1905 from the Premier Mine at Kimberley, Transvaal, South Africa; the huge, colorless stone weighed 3,106 carats (1.3 pounds). The smooth face of the Cullinan replica shown in the photograph replicates the stone's natural cleavage surface, which indicates that it was not the entire stone. Unfortunately, the remainder of the Cullinan lies undiscovered in a deep volcanic pipe—which is where all diamonds originate. The Transvaal government bought the diamond for $1.6 million and gave it uncut to King Edward VII of England. After months of study, an Amsterdam diamond cutter's decisive blow yielded nine large stones and about 100 smaller ones, all of them flawless. All are now in the British crown jewels collection.

Diamonds are the high-pressure form of carbon and are the hardest natural substance known. Curiously, graphite, the high-temperature, low-pressure form of carbon, is among the softest minerals known.

Diamonds are found mostly in Africa, Australia, Russia, India, and South America (◉ Table 1). Near Yellowknife, Northwest Territories in Canada, the Ekati diamond mine opened in the late 1990s with great expectations. Diamond finds in the United States are shown on the map of ◆ Figure 2. Most are solitary finds, and their sources are unknown. At Murfreesboro, Arkansas, however, thousands of diamonds have been recovered from an eroded diamond "pipe," one

(a)   (b)

◆ **FIGURE 1** (a) Exact replica of the Cullinan diamond with the two largest stones cut from it, the Great Star of Africa (530.2 carats) and the Second Star of Africa (317 carats), which now reside in the British crown jewels collection. (b) Uncut precious opal. (c) Uncut emerald.

(c)

○ **TABLE 1**	**Major World Diamond Producers, 1997***
COUNTRY	**VALUE, $ MILLIONS**
Botswana	$1,461
Russia	1,210
Angola	1,035
South Africa	971
Congo	715
Namibia	410
Australia	288

*Just seven countries account for 90% of all rough-diamond production.
SOURCE: Larry Gordon, *Los Angeles Times*, 4 November 1998.

during which rocks were deposited, life evolved, and significant geologic events such as mountain building occurred. Eons are the longest time intervals, followed, respectively, by eras, periods, and epochs. The Phanerozoic ("revealed life") Eon began 570 million years ago with the Cambrian Period, the rocks of which contain the first extensive fossils of organisms with hard skeletons. Because of the significance of the Cambrian Period, the informal term *Precambrian* is widely used to denote the time before it, which ex-tends back to the formation of the earth 4.6 billion years ago. Note that the Precambrian is divided into the Archean and Proterozoic Eons, with the Archean Eon extending back to the formation of the oldest known in-place rocks about 3.9 billion years ago. Precambrian time accounts for 88 percent of geologic time, and the Phanerozoic for a mere 12 percent. The eras of geologic time correspond to the relative complexity of life forms: Paleozoic (oldest life), Mesozoic (middle life), and Cenozoic (most recent life). Environ-

◁▷ **FIGURE 2** Reported diamond finds in glacial drift in the Great Lakes region. Numerals indicate the number of diamonds found at the location; single finds have been reported at the other locations.

weighing 40 carats. For the price of admission, one can dig for diamonds here in the "blue ground" of the eroded pipe. One lucky hunter uncovered a 15-carat stone.

Precious opal (Sanskrit *upālā,* "precious stone"; ◁▷ Figure 1b) occurs in shallow surface diggings, and because it lacks crystalline form, it actually is not a mineral. It is amorphous hydrated silica and gets its "fire" from an orderly arrangement of silica spheres interspersed with water molecules. Black and fire opals are the most desirable, and each stone is priced individually.

Emerald, the most valuable mineral carat-for-carat, is the precious variety of the silicate mineral beryl (◁▷ Figure 1c). It is found in coarse-grained granites called *pegmatites* (see Chapter 13) and in mica schist.

Quartz, plain old SiO_2, one of the most common minerals, has a host of semiprecious varieties such as purple amethyst, banded agate, yellow citrine, and smoky and rose quartz (◁▷ Figure 3). Searching for polishing-quality quartz varieties is the passion of many "rock hounds," and good crystal specimens also are cherished.

◁▷ **FIGURE 3** Some varieties of quartz: (a) colorless crystals, (b) smokey quartz, (c) amethyst, (d) agate, and (e) rose quartz.

mental geologists are most interested in the events of the past few million years, a mere heartbeat in the history of the earth.

Absolute age-dating requires some kind of natural clock. The ticks of the clock may be the annual growth rings of trees or established rates of disintegration of radioactive elements to form other elements. At the turn of the twentieth century, American chemist and physicist Bertram Borden Boltwood (1870–1927) discovered that the ratio of lead to uranium in uranium-bearing rocks increases as the rocks' ages increase. He developed a process for determining the age of ancient geologic events that is unaffected by heat or pressure—**radiometric dating.** The "ticks" of the radioactive clocks are radioactive decay processes—spontaneous disintegrations of the nuclei of heavier elements such as uranium and thorium to lead. A radioactive element may decay to another element, or to an isotope of the same element. This decay occurs at a precise rate that can be determined experimentally. The most common

Metamorphism

(a)

Metamorphism

(b)

◆ **FIGURE 2.23** (a) Limestone recrystallizes to form white marble. (b) Quartzite, the hardest and most durable common rock, is metamorphosed sandstone.

emissions are alpha particles ($^4_2\alpha$), which are helium atoms, and beta particles (β^-), which are nuclear electrons. New *radiogenic* "daughter" elements or isotopes result from this **alpha decay** and **beta decay** (◆ Figure 2.27).

For example, of the three isotopes of carbon, ^{12}C, ^{13}C, and ^{14}C, only ^{14}C is radioactive, and this radioactivity can be used to date events between a few hundred and a few tens-of-thousands of years ago. Carbon-14 is formed continually in the upper atmosphere by neutron bombardment of nitrogen, and it exists in a fixed ratio to the common isotope, ^{12}C. All plants and animals contain radioactive ^{14}C in equilibrium with the atmospheric abundance until they die, at which time ^{14}C begins to decrease in abundance, and along with it, the object's radioactivity. Thus by measuring the radioactivity of an ancient parchment, log, or piece of charcoal and comparing the measurement with the activity of a modern standard, the age of archaeological materials and geological events can be determined (◆ Figure

◯ TABLE 2.5 **Common Metamorphic Rocks**

ROCK	PARENT ROCK	CHARACTERISTICS
Foliated or Layered		
Slate	Shale and mudstone	Splits into thin sheets
Schist	Fine-grained rocks, siltstone, shale, tuff	Mica minerals often crinkled
Gneiss	Coarse-grained rocks	Dark and light layers of aligned minerals
Nonfoliated or Recrystallized		
Marble	Limestone	Interlocking crystals
Quartzite	Sandstone	Interlocking, almost fused quartz grains

◆ **FIGURE 2.24** Joints in basalt flows near Chico, California.

<◆ **FIGURE 2.25** Geologic cross sections illustrating the Laws of Superposition, Cross-cutting Relationships (intrusion and faulting), and Fossil Succession. The limestone beds are correlative because they contain the same fossils. The numerals indicate relative ages, with 1 being the youngest.

2.28). Carbon-14 is formed by the collision of cosmic neutrons with ^{14}N in the atmosphere, and then it decays back to ^{14}N by emitting a nuclear electron (ß$^-$).

$$^{14}_{7}N + neutron \longrightarrow ^{14}_{6}C + proton$$

$$^{14}_{6}C \longrightarrow ^{14}_{7}N + ß^-$$

Both carbon—the radioactive "parent" element—and nitrogen—the radiogenic "daughter"—have 14 atomic mass units. However, one of ^{14}C's neutrons is converted to a proton by the emission of a beta particle, and the carbon changes (or *transmutes*) to nitrogen. This process proceeds at a set rate that can be expressed as a **half-life,** the time required for half of a population of radioactive atoms to decay. For ^{14}C this is about 5,730 years.

Radioactive elements decay exponentially; that is, in two half-lives one fourth of the original number of atoms remain, in three half-lives one eighth remain, and so on (◆ Figure 2.29). So few parent atoms remain after seven or eight half-lives (less than 1 percent) that experimental uncertainty creates limits for the various radiometric dating methods.

Whereas the practical age limit for dating carbon-bearing materials such as wood, paper, and cloth is about 40,000 to 50,000 years, ^{238}U disintegrates to ^{206}Pb and has a half-life of 4.5×10^9 years. Uranium-238 emits alpha particles (helium atoms). Since both alpha and beta disintegrations can be measured with a Geiger counter, we have a means of determining the half-lives of a geological or archaeological sample. ◉ Table 2.6 shows radioactive parents, daughters, and half-lives commonly used in age-dating.

Age of the Earth

Before the advent of radiometric dating, determining the age of the earth was a source of controversy between established religious interpretations and early scientists. Archbishop Ussher (1585–1656), the Archbishop of Armagh and a professor at Trinity College in Dublin, declared that the earth was formed in the year 4004 B.C. Not only did Ussher provide the year but also the day, October 23, and the time, 9:00 A.M. Ussher has many detractors, but Stephen J. Gould of Harvard University, though not proposing acceptance of Ussher's date, was not one of them. Gould pointed out that Ussher's work was good scholarship for its time, because other religious scholars had extrapolated from Greek and Hebrew scriptures that the earth was formed in 5500 B.C. and 3761 B.C., respectively. Ussher based his age determination on the verse in the Bible that says "one day is with the Lord as a thousand years" (2 Peter 3:8). Because the Bible also says that God created heaven and earth in 6 days, Ussher arrived at 4004 years B.C.—the extra four years because he believed Christ's birth year was wrong by that amount of time.

Case Study | 2.3

Minerals, Cats, and the Litter Belt

Many of Lovell, Wyoming's 2,100 inhabitants once depended upon the oil industry for their livelihood. The people here mine bentonite, which is used in oil drilling. Mixed with water, bentonite forms a "mud" that is circulated through the drill pipe and back to the ground surface outside the pipe. It forms a filter cake on the walls of the hole that prevents entry of water and loss of the drilling fluid into porous formations. It also cools the drill bit as the drill churns through hard rock and carries the drilled rock debris up to the surface. Oil prices dropped in the late 1980s, and Lovell all but dried up (Chapter 14).

Bentonite is formed by geochemical alteration of volcanic ash to form clay minerals, the most important of which is *smectite*. One of smectite's unusual properties is that it can absorb water equivalent to many times its original volume. The clay may swell as it absorbs water, forming a gooey mess, or it may crumble into fluffy pellets, depending on the amount of water and the initial size of the clay aggregates. Land areas underlain by bentonite may absorb so much water in the rainy season that they become impassable. Also, bentonitic soils can swell and cause severe damage to structures founded in it or on it (see Chapter 6). It has many uses. Native Americans once used it to remove oil and dirt from buffalo hides. It is used today in the landscaping of golf greens, to filter and clarify wine, to line landfills and ponds to prevent water loss, and in the manufacture of crayons.

Lovell was a potential ghost town without oil drilling. Bentonite to the rescue! It began when kitty litter replaced the sand box in 1987. The new product utilized a variety of clay-type minerals in the litter mix to absorb moisture. An enterprising chemist recognized the potential of bentonite, with its great moisture-absorbing properties, as cat litter. He reasoned that, with the proper grain size, the bentonite would clump when the cat did his duty, so that the soiled clumps could easily be removed from the litter pan. He was right, and the rest, as they say, is history (◇ Figure 1).

Today, almost all cat litter contains bentonite, and the Lovell mine is back in full-swing operation. The plant employs 60 people, and new houses are being built in the community to accommodate them. Bentonites in Wyoming are the result of volcanic eruptions in Cretaceous time, when dinosaurs roamed the land. The swath of mineable bentonites from Lovell to Casper has been dubbed the "Litter Belt" in Wyoming.

◇ FIGURE 1

Ussher's age for the earth was accepted by many as "gospel" for almost 200 years.

By the late 1800s geologists believed that the earth was on the order of 100 million years old. They reached their estimates by dividing the total thickness of sedimentary rocks (tens of kilometers) by an assumed annual rate of deposition (mm/year). Evolutionists such as Charles Darwin thought that geologic time must be almost limitless in order that minute changes in organisms could eventually produce the present diversity of species. Both geologists and evolutionists were embarrassed when the British physicist William Thomson (later Lord Kelvin) demonstrated with elegant mathematics how the earth could be no older than 400 million years, and maybe as young as 20 million years. Thomson based this on the rate of cooling of an initially molten earth and the assumption that the material composing the earth was incapable of creating new heat through time. He did not know about radioactivity, which adds heat to rocks in the crust and mantle.

Bertram Boltwood postulated that older uranium-bearing minerals should carry a higher proportion of lead than younger samples. He analyzed a number of specimens of known relative age, and the absolute ages he came up with ranged from 410 million to 2.2 billion years old. These ages put Lord Kelvin's dates based on cooling rates to rest and ushered in the new radiometric dating technique. By extrapolating backward to the time when no radiogenic lead had been produced

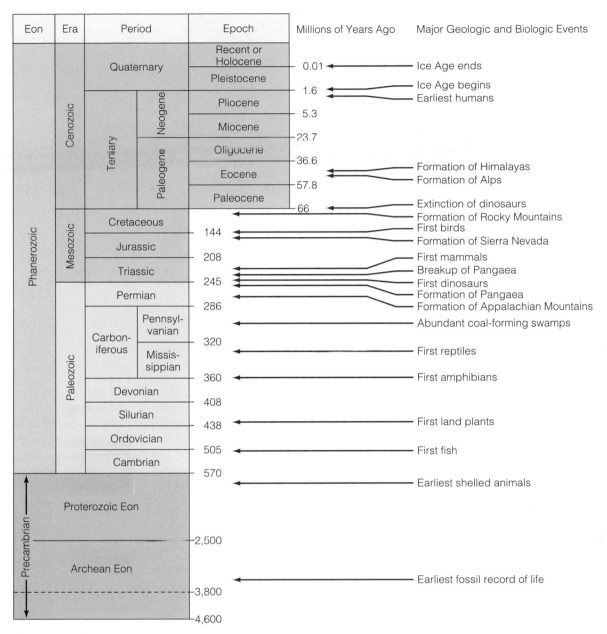

Eon	Era	Period		Epoch	Millions of Years Ago	Major Geologic and Biologic Events
Phanerozoic	Cenozoic	Quaternary		Recent or Holocene	0.01	Ice Age ends
				Pleistocene	1.6	Ice Age begins — Earliest humans
		Tertiary	Neogene	Pliocene	5.3	
				Miocene	23.7	
			Paleogene	Oligocene	36.6	
				Eocene	57.8	Formation of Himalayas — Formation of Alps
				Paleocene	66	Extinction of dinosaurs — Formation of Rocky Mountains
	Mesozoic	Cretaceous			144	First birds — Formation of Sierra Nevada
		Jurassic			208	First mammals
		Triassic			245	Breakup of Pangaea — First dinosaurs — Formation of Pangaea
		Permian			286	Formation of Appalachian Mountains
	Paleozoic	Carboniferous	Pennsylvanian		320	Abundant coal-forming swamps
			Mississippian		360	First reptiles
		Devonian			408	First amphibians
		Silurian			438	First land plants
		Ordovician			505	First fish
		Cambrian			570	Earliest shelled animals
Precambrian		Proterozoic Eon			2,500	
		Archean Eon			3,800	Earliest fossil record of life
					4,600	

◆▶ **FIGURE 2.26** Geologic time scale.

on earth, we arrive at an age of 4.6 billion years for the earth. This corresponds to the dates obtained from meteorites and lunar rocks, which are part of our solar system. The oldest known intact terrestrial rocks are found in the Acasta Gneiss of the Slave geological province of Canada's Northwest Territories. Analyses of lead to uranium ratios on the gneiss's zircon minerals indicate that the rocks are 3.96 billion years old. However, older detrital zircons on the order of 4.0–4.3 billion years old have been found in western Australia, indicating that some stable continental crust was present as early as 4.3 billion years ago (◉ Table 2.7). Suffice it to say that the earth is very old and that there has been

abundant time to produce the features we see today (◆▶ Figure 2.30).

Environmental geology deals mostly with the present and the recent past—the time interval known as the *Holocene Epoch* (10,000 years ago to the present) of the *Quaternary Period*. However, because the Great Ice Age advances of the Pleistocene Epoch of the Quaternary Period (which preceded our own Holocene Epoch) had such a tremendous impact on the landscape of today, we will investigate the probable causes of the "ice ages" in order to speculate a bit about what the future might bring. In addition, Pleistocene glacial deposits are valuable sources of underground water, and they have

◆▶ **FIGURE 2.27** (a) Alpha emission, whereby a heavy nucleus spontaneously emits a helium atom and is reduced 4 atomic mass units and 2 atomic numbers. (b) Emission of a nuclear electron (ß⁻ particle), which changes a neutron to a proton and thereby forms a new element without a change of mass.

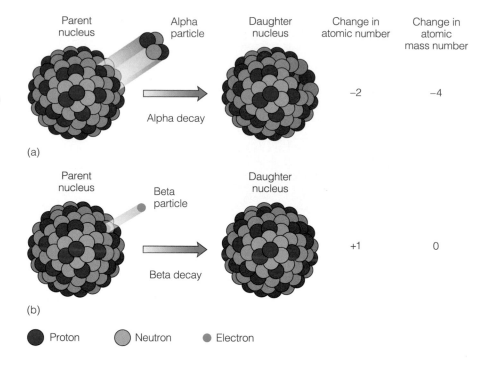

economic value as sources of building materials. We will see that "ice ages" have occurred several times throughout geologic history (see Chapter 11).

What impresses one most about geologic time is how short the period of human life on earth has been. If we could compress the 4.6 billion years since the earth formed into one calendar year, *Homo sapiens* would appear about 30 minutes before midnight on December 31 (see ◆▶ Figure 2.30). The last ice-age glaciers would begin wasting away a bit more than 2 minutes before midnight, and written history would exist for only the last 30 seconds of the year. Perhaps we should keep this calendar in mind when we hear that the dinosaurs were an unsuccessful group of reptiles. After all, they endured 50 times longer than hominids have existed on the planet to date. At the present rate

⊙ **TABLE 2.6** **Isotopes Used in Age-Dating**				
ISOTOPES		**PARENT'S HALF-LIFE, YEARS**	**EFFECTIVE DATING RANGE, YEARS**	**MATERIAL THAT CAN BE DATED**
Parent	*Daughter*			
Uranium-238	Lead-206	4.5 billion	10 million to 4.6 billion	Zircon
				Uraninite
Uranium-235	Lead-207	704 million		
Thorium-232	Lead-208	14 billion		
Rubidium-87	Strontium-87	48.8 billion	10 million to 4.6 billion	Muscovite
				Biotite
				Orthoclase
				Whole metamorphic or igneous rock
Potassium-40	Argon-40	1.3 billion	100,000 to 4.6 billion	Glauconite Hornblende
				Muscovite Whole volcanic rock
				Biotite
Carbon-14	Nitrogen-14	5,730 years	less than 100,000	Shell, bones, charcoal

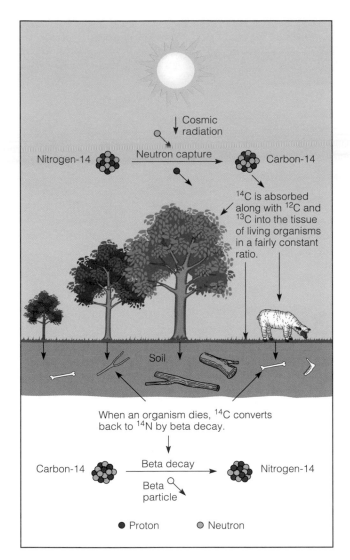

◁▷ **FIGURE 2.29** Decay of a radioactive parent element with time. Each time unit is one half-life. Note that after two half-lives one fourth of the parent element remains, and that after three half-lives one eighth remains.

◁▷ **FIGURE 2.28** Carbon-14 is formed from nitrogen-14 by neutron capture and subsequent proton emission. Carbon-14 decays back to nitrogen-14 by emission of a nuclear electron (ß⁻).

of population growth, there is some doubt that humankind as we know it will be able to survive anywhere near that long.

The earth we see today is the product of earth processes acting on earth materials over the immensity of geologic time. The response of the materials to the processes has been the formation of mountain ranges, plateaus, and the multitude of other physical features found on the planet. Perhaps the greatest of these processes has been the formation and movement of platelike segments of the earth's surface, which influence the distribution of earthquakes and volcanic eruptions, global geography, mineral deposits, mountain ranges, and even climate. These *plate tectonic* processes are the subject of the next chapter.

⊙ TABLE 2.7	**Earth's Oldest Known Materials**	
MATERIAL	**LOCATION**	**AGE, BILLIONS OF YEARS**
Crust	Zircon minerals in rocks of western Australia	4.0–4.3
Rock	Zircon minerals in the Acasta Gneiss, N. W. Terr., Canada	3.96
Sedimentary rock	Isua Greenstone Belt, Greenland	3.7–3.75
Fossils	Algae and bacteria	3.5

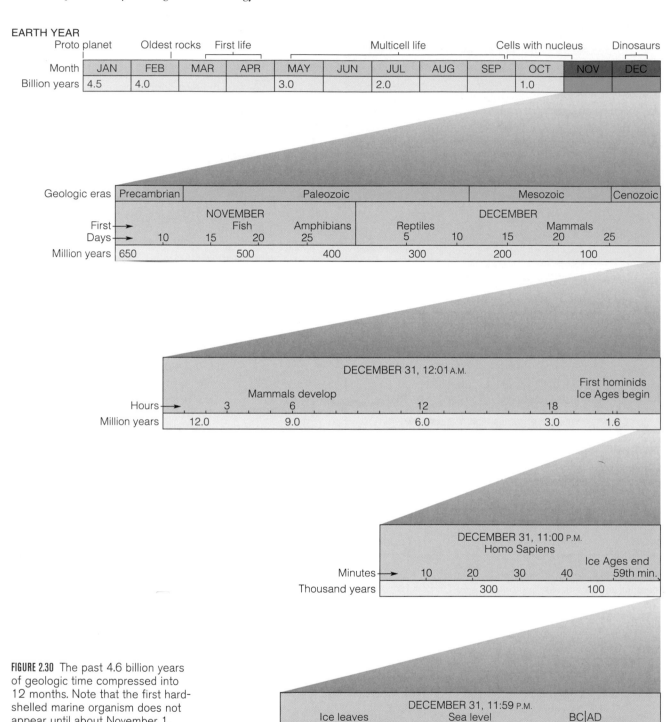

FIGURE 2.30 The past 4.6 billion years of geologic time compressed into 12 months. Note that the first hard-shelled marine organism does not appear until about November 1, and that humans as we know them have been around only since the last hour before the New Year.

Gallery

A Rock Collection

Specific rocks have had historical, aesthetic, and practical significance to humans; Plymouth Rock, the Rock of Gibraltar, the biblical Rock of Ages, and the Rosetta Stone are examples. Various geologic agents of erosion have shaped the rocks of the Grand Canyon, Bryce Canyon, Yosemite, and other national parks into areas that we admire for their great beauty. Early humans lived in rock caves and threw rocks to defend themselves against their enemies and

(a)

◆ FIGURE 2.31 Ayers Rock, the world's largest sandstone monolith, is one of Australia's leading tourist attractions. An erosional remnant of a previous highland, it rises abruptly above the surrounding flat land. In geological terminology it is called an *inselberg* (German "island mountain") because of its resemblance to an island rising from the sea. Aborigines (native Australians) call Ayers Rock *Uluru.* In 1958 Ayers Rock and the rocks of the Olgas nearby were incorporated into Uluru National Park. (a) Ayers Rock rises 348 meters (1,141 ft) above the surrounding plain, and is scaled by 80 percent of the visitors to the park. (b) The Olgas, a circular grouping of more than 30 red domes, is a subdued but still spectacular sandstone inselberg in Uluru National Park. Its highest point, Mt. Olga, rises 460 meters (1,500 ft) above the plain. The Olgas' Aboriginal name is *katajuta,* "many heads." Ayers Rock is in the background.

(b)

to kill animals for food. Over time humans came to build crude rock shelters, and then rock fences, and eventually beautiful castles and abbeys. Today, rock and rock products are important mineral resources, used mostly as building materials. The beauty and usefulness of rocks are celebrated in Figures 2.31 to 2.34.

◀▶ **FIGURE 2.32** Sixty-foot-high representations of the heads of four presidents were carved into Mount Rushmore's Harney Peak Granite in southwestern South Dakota between 1927 and 1941. The great work was possible because the Precambrian granite is relatively homogeneous (massive) and not riddled with defects. Why would such a project not have been considered if Mount Rushmore were composed of slate, shale, or schist? Can you identify the presidents honored here?

◀▶ **FIGURE 2.33** President Theodore Roosevelt (*left*) and naturalist John Muir at Yosemite Valley in 1903. The rock on which they were photographed is Cathedral Peak Granite, one of the many granite intrusions that make up the Sierra Nevada batholith. Note the rounded exfoliation domes in the background (see Chapter 6).

◀▶ **FIGURE 2.34** A skillfully assembled sandstone wall in Yorkshire, England. Yorkshire is the largest county (shire) in England, and literally thousands of kilometers of these walls are found there. This commonly occurring sandstone breaks readily into platy and blocky pieces, which makes it perfect for wall building.

Summary

SOLAR SYSTEM

DESCRIPTION
Formed 5 billion years ago from an interstellar cloud (nebula).

PROBABLE CAUSE
The nebula collapsed as a result of a gravitational attraction. The nebula started to spin and condense. The inner part of the cloud became the hot protosun and it was surrounded by a cooler, disc-shaped cloud, the solar nebula, which formed the large outer, Jovian planets. Rocky material closest to the sun formed the inner, terrestrial planets. Mars rocks are similar to Earth rocks.

EARTH SYSTEM
The earth system has five interconnecting subsystems, or reservoirs, sometimes called "spheres." They are the solid earth (includes lithosphere and soil), the atmosphere, the hydrosphere, the biosphere, and the extraterrestrial. Most are closed systems, such as the rock cycle and the hydrologic cycle, but energy is the exception. Continual interaction and recycling by these reservoirs is vital to the dynamic earth system.

EARTH MATERIALS

CRUST
Outermost rocky layer of earth composed of rocks that are aggregates of minerals.

MINERALS
Naturally occurring inorganic substances with a definite set of physical properties and a narrow range of chemical compositions.

CLASSIFICATION BY
Chemistry of anions (negative ions or radicals), such as oxides, sulfides, carbonates, etc. Silicates are the most common and important mineral group and are composed of silica tetrahedra $(SiO_4)^{-4}$ units that are joined to form chains, sheets, and networks.

IDENTIFICATION
Physical properties such as hardness, cleavage, crystal structure, fracture pattern, and luster.

ROCKS
Consolidated or poorly consolidated aggregates of one or more minerals or organic matter.

CLASSES
Igneous—rock formed by crystallization of molten or partially molten material.
Sedimentary—layered rock resulting from consolidation and lithification of sediment.
Metamorphic—preexisting rock that has been changed by heat, pressure, or chemically active fluids.

CLASSIFICATION BY
Texture and composition.

STRUCTURES
Many planar rock structures may be viewed as defects along which landslides, rockfalls, or other potentially hazardous events may occur. Stratification in sedimentary rocks and foliation in metamorphic rocks are common planar features. Faults and joints occur in all classes of rock.

ROCK CYCLE
A sequence of events by which rocks are formed, altered, destroyed, and reformed as a result of internal and external earth processes.

GEOLOGIC TIME

RELATIVE TIME
The sequential order of geologic events established by using basic geologic principles or laws.

GEOLOGIC TIME SCALE
The division of geologic history into eons, eras, periods, and epochs. The earliest 88 percent of geologic time is known informally as *Precambrian* time. Precambrian rocks are generally not fossiliferous. Rocks of the Cambrian Period and younger contain good fossil records of shelled and skeletonized organisms. Environmental geology is concerned mainly with geologic events of the present epoch, the Holocene, and the one that preceded it, the Pleistocene.

ABSOLUTE AGE
Absolute dating methods use some natural "clock." For long periods of time, such clocks are known rates of radioactive decay. A parent radioactive isotope decays to a daughter isotope at a rate that can be determined experimentally. This established rate, expressed as a half-life, enables us to determine the age of a sample of material.

AGE OF THE EARTH
Comparing present U/Pb ratios with those in iron meteorites, we can extrapolate backward and obtain an earth age of 4.6 billion years old.

ROCKS AND CRUST
The oldest known in-place rocks are 3.96 billion years old. Dating of minerals derived from older crust yields an age of 4.0–4.3 billion years for stable continental crust.

◆ Key Terms

absolute age dating	cleavage	ion	phaneritic texture
alpha decay	detrital sedimentary rocks	isotope	plutonic
amorphous	dip	joint	pyroclastic
aphanitic texture	element	lava	radiometric dating
atom	extrusive	lithified	recrystallization
atomic mass	fault	luster	relative age dating
atomic number	felsic	mafic	rock
batholith	foliation	magma	rock cycle
beta decay	fracture	metamorphic rocks	sedimentary rocks
biogenic sedimentary rocks	half-life	mineral	stratification
chemical sedimentary rocks	igneous rocks	Mohs hardness scale	strike
clastic	intrusive	nucleus	

◆ Study Questions

1. What physical and chemical factors are the bases of the rock-classification system? What are the three classes of rocks, and how does each form? Identify one characteristic of each class that usually makes it readily distinguishable from the other rock classes.
2. How and where do batholiths form? What type of rock most commonly forms in batholiths?
3. What are some of the planar surfaces in rocks that may be weak and thus lead to various types of slope failure?
4. How may structures in sedimentary rocks be used to reconstruct past environments?
5. How may absolute age-dating techniques be used to the betterment of human existence?
6. The earth is 4.6 billion years old, and humans have been on earth for only a few hundred thousand years—but a blink of an eye in the great length of geologic time. What changes of a global nature have humans invoked on the earth's natural systems (water, air, ice, the solid earth, and biology) in this short length of time? Which impacts are reversible, and which ones cannot be reversed or mitigated?
7. What is the most common mineral species? Name several rocks in which it is a prominent constituent.
8. What is cleavage? How can it serve as an aid in identifying minerals?
9. The most common intrusive igneous rock is composed mostly of (1) the most common mineral and (2) a mineral of the most common mineral group. Name the rock and its constituent minerals.
10. What mineral is found in both limestone and marble?

◆ For Further Information

BOOKS AND PERIODICALS

Albritton, C. C. 1984. Geologic time. *Journal of geological education* 32, no. 1: 29–47.

Bowring, S. A.; I. S. Williams; and W. Compston. 1989. 3.96 Ga gneisses from the Slave province, Northwest Territories, Canada. *Geology* 17: 971–975.

Brown, V. M., and J. A. Harrell. 1991. Megascopic classification of rocks. *Journal of geological education* 39: 379.

Dietrich, R. V., and Brian Skinner. 1979. *Rocks and rock minerals.* New York: John Wiley and Sons.

Edgett, Ken. 1997. Mars Pathfinder, Sojourner, and a cast of rock stars captivate earthlings. *Earth in space* 10, no. 3: 1–16.

Eicher, D. L. 1976. *Geologic time,* 2d ed. Englewood Cliffs, N.J.: Prentice-Hall, Inc.

Gunter, Mickey Eugene. 1994. Asbestos as a metaphor for teaching risk perception. *Journal of geological education* 42: 17.

Harvey, Carolyn, and Mark Rollinson. 1987. *Asbestos in the schools.* New York: Praeger.

Hurley, Patrick. 1959. *How old is the earth?* New York: Anchor Books.

Libby, W. F. 1955. *Radiocarbon dating.* Chicago: University of Chicago Press.

Mackenzie, Fred T. G., and Judith A. Mackenzie. 1995. *Our changing planet: An introduction to earth system science and global environmental change.* Englewood Cliffs, N.J.: Prentice-Hall, Inc.

National Research Council, National Academy of Sciences. 1993. *Solid-earth sciences and society.* Washington, D.C.: National Research Council Commission on Geosciences, Environment, and Resources, National Academy of Sciences Press.

Newcott, William R. 1998. Return to Mars. *National geographic* 194, no. 2: 2–29.

Nuhfer, E. B.; R. J. Proctor; and Paul H. Moser. 1993. *The cit-*

izen's guide to geologic hazards. Arvada, Colo.: American Institute of Professional Geologists.

Parker-Pope, T. 1997. Cat litter breathes new life into region of bentonite mines. *The Wall Street journal,* 1 April.

Skinner, H. Catherine W., and Malcom Ross. 1994. Geology and health. *Geotimes* (Washington, D.C.: American Geological Institute), January: 11–12.

_____. 1994. Minerals and cancer. *Geotimes,* January: 13–15.

Snow, T. P. 1993. *Essentials of the dynamic universe.* St. Paul, Minn.: West Publishing Company.

Wicander, Reed, and James Monroe. 1999. *Essentials of geology.* Pacific Grove, California: Brooks/Cole.

◆ World Wide Web Activities

NASA *Mars Pathfinder* On-site Photos of the Martian Surface
http://nssdc.gsfc.nasa.gov/planetary/image/marspath_81205_map.gif
This panorama photo taken by the *Mars Pathfinder* camera shows Barnacle Bill, Yogi, and Scooby Doo, along with other features of the Martian surface, from the *Mars Pathfinder* landau to the Martian horizon. Sojourner appears in the picture also. Note that The Wedge and Flat Top both appear to have angular shapes associated on Earth with wind abrasion.

Narration of Diamond Search and Discovery in North America
http://www.ingersoll-rand.com/compair/aprmay99/fpg_01.htm
"Diamonds on Ice" uses text and images to tell the exciting story of the first major diamond mining operation in North America, from discovery (1980) through development, production, and marketing of the first diamonds in 1999. The text highlights the technology, from roads of ice to computer modeling, that was used to bring this remote arctic site into production.

1. Canadian exploration geologist Chuck Fipke found minerals associated with diamonds in surface deposits, and traced them to the primary diamond pipe. What aspect of these minerals did he use to figure out which was 'pay dirt'?

Detailed On-line Version of the Geologic Time Scale
http://www.ucmp.berkeley.edu/help/timeform.html
This site posts a version of the Geologic Time Scale with eons, eras, periods, and epochs spelled out and age-bracketed. Clicking on a time division will take you to an overview of that division, plus buttons to access information about the division's stratigraphy, ancient life, tectonics, and localities where rocks of that time are exposed. Links to museum displays and collections are included.

1. What are the three main hypotheses regarding the cause of the Permian mass extinction, the largest extinction event in Earth history?

Earth Systems Today CD Correlations

CHAPTER CONCEPT: In the Igneous Rock Family, Crystal Size and Cooling History Are Related
At the bottom of the Rock Cycle diagram you may press on one of the rock family representatives to enter a rock laboratory, which demonstrates the formation of representative rock types.

SECTION MODULE: Main Activity: Rocks and the Rock Cycle; Igneous Rock Laboratory
Here you will see igneous rocks form as crystals grow from a melt. Note how the rate of cooling will affect the final texture of the rock.

QUESTION: *What is the difference in cooling history between a phaneritic rock and a porphyry, and how is this manifested in the final rock?*

CHAPTER CONCEPT: Measuring Absolute Time with Radioactive Elements
You will see in this module an animation of radioactive decay that visually emphasizes that the process is nonlinear; over time the production of daughter slows as the amount of parent decreases.

SECTION MODULE: Main Activity: Geologic Time; Absolute Time
Watch the animation, and examine the table of commonly used radiogenic systems.

QUESTION: *$^{14}C/^{14}N$ dating will generally not yield a meaningful date, and thus cannot be used, on materials older than about 60,000 years. About how many half-lives is this for the $^{14}C/^{14}N$ reaction? At 60,000 years, what fraction of parent is present? Can you conclude from this WHY the scheme won't work for older materials?*

chapter | three

Plate Tectonics

◁▷ FIGURE 1 Dead Sea rift zone

"And the Walls Came Tumbling Down"

The prominent lineation down the center of this satellite photo is the Dead Sea rift zone, an active tectonic plate margin (◁▷ Figure 1). It is the Holy Land of biblical times and an area of political and religious conflict. Included in the view are the lands of Jordan, Israel, Syria, and part of Saudi Arabia and the Jordan River, the Mediterranean Sea, the Gulf of Aqaba, and the Dead Sea. The Dead Sea's shoreline is the lowest point on the earth's surface, averaging 396 meters (1300 ft) below sea level. This valley was part of the sea 65 million years ago, and its desiccation may be the source of salt deposits in the region. The Dead Sea is fed by the Jordan River, but its level has dropped 19 meters (63 ft) in the past hundred years due to development along the river.

The holy book tells us that the Lord appointed Joshua to lead the Israelites to Canaan, the land of milk and honey. That "Promised Land" was in the vicinity of present-day Jericho. Jericho is at least 8,000 years old, the oldest almost continuously occupied city yet discovered by archaeologists. Karnak, Giza, and the tomb of Tutankamen are all young in comparison. The scriptures also tell us (in Joshua 6) that Joshua's orders were to march his army around the walled city for six days, blowing seven ram's-horn trumpets. On the seventh day, upon their blowing the trumpets, the walls of the city would fall flat, and the city could then be destroyed.

Jericho (see map) is in a rift valley formed by lateral slip on faults between the African plate to the west and the Arabian plate to the east (◁▷ Figure 2). Although both plates are moving

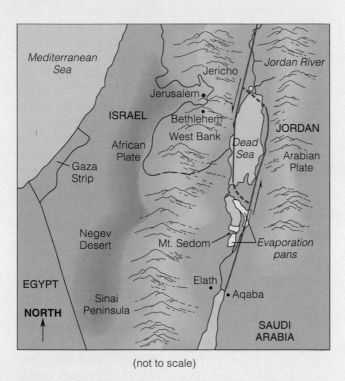

◆ **FIGURE 2** The Dead Sea, Jordan River, and biblical villages and lands surrounding the sea.

(a)   (b)

◆ **FIGURE 3** (a) The Dead Sea transform fault, a special variety of strike-slip fault, separates two northward-moving plates, the slower-moving African plate (left) and the faster-moving Arabian plate. (b) Overlapping faults between the West Bank and Jordan outline a pull-apart basin in which the Dead Sea resides. **D** indicates displacement on the fault.

northward, the African plate is moving more slowly, so the Arabian plate's *relative* motion is to the north, and the African plate's is to the south (◆ Figure 3). Motion of this type on faults results in a "pull-apart" basin, or *graben,* as shown in the figure. This explains how the floor of the valley came to be 400 meters (1,300 ft) below sea level.

The ruins of Jericho included warehouses that contained grain. It was not the Israeli *modus operandi* to attack and destroy after a grain harvest. They usually attacked just before grain was

harvested so that a siege period would be effective and they could later harvest the grain for themselves.

Similar plate boundaries at other places in the world have produced destructive earthquakes. Examples include California and Turkey. An alternative explanation for the Israelites' success at Jericho, and perhaps a more plausible one, is that the walls were felled by an earthquake caused by the Dead Sea faults so that the earth shook and the mud-brick "walls came tumbling down." ■

During the 1960s our concepts of the architecture of the earth's crust and how the earth "works" were completely revolutionized by the theory of **plate tectonics** (Greek *tektonikos,* "builder"). The theory maintains that the lithosphere, the outer rocky shell of the earth, consists of a number of rigid plates, 70–150 kilometers (40–90 mi) thick, that slide under, past, or away from one another. Continents, although parts of the moving plates, are but passive passengers. Where plates interact with each other, earthquakes and volcanic activity occur, mountain ranges may form, and internal heat is released to the earth's surface. Plate tectonics integrates two older explanations for the distribution of the earth's physical and geophysical features—the theories of continental drift and sea-floor spreading.

Continental Drift

What ultimately was to become the theory of plate tectonics began in 1910 when Alfred Wegener (1880–1930), a German meteorologist, presented a paper to the Frankfurt Geological Society in which he expressed his opinion that the continents have moved great horizontal distances in the past 100 million years. He called his hypothesis *die Verschiebung der Kontinente,* meaning "continental displacement," which English-speaking scientists translated to **continental drift,** the name that stuck. Wegener called the late Paleozoic supercontinent—a large landmass consisting of all the continents we know today—**Pangaea** (Greek "all land"; ◆ Figure 3.1). He theorized that Pangaea split apart during Mesozoic time and that the continents then moved steadily in diverse directions to their present positions. Wegener explained that as the continents drifted westward (*Westwanderung*), they plowed up mountain ranges such as the Andes and the North American Coast Ranges. Australia became geographically isolated, and the Atlantic Ocean became larger at the expense of a diminishing Pacific Ocean.

Strong evidence in favor of the hypothesis, according to Wegener and many Southern Hemisphere geolo-

gists, is the distribution of late Paleozoic (Permian Period) glacial deposits and plant fossils known collectively as the *Gondwana Succession.* These distinctive fossil land plants (◆ Figure 3.2) and their associated glacial deposits are found only in South America, Africa, India, Australia, and Antarctica (◆ Figure 3.3). The only explanation for this, it was argued, is that these continents were once connected in the southern part of Pangaea. Some years earlier this hypothetical former landmass with the distinct Permian glaciation pattern had been named **Gondwanaland,** or simply Gondwana. (The northern landmass had been named **Laurasia.**)

Additional paleontological evidence supporting continental drift is supplied by the amphibian and reptile fossils currently found on widely separated Gondwana continents. In particular are the fossil remains of the Permian fresh water reptile *Mesosaurus* found in Brazil and South Africa. It is inconceivable that these freshwater creatures swam or island-hopped across the salty Atlantic Ocean to an identical freshwater habitat on the opposite side. It seems much more reasonable that *Mesosaurus* inhabited large Permian lakes and/or rivers spanning parts of both continents when they were joined, and that the rocks containing their fossils were separated when Pangaea split apart.

Wegener believed that the ultimate proof of drifting was the absolute motion of Greenland relative to Europe that had been found. He relied upon surveys made between 1823 and 1870—later proved in error—indicating that Greenland had moved 420 meters (¼ mi) to the west in 47 years (an astounding 8.9 m or 29 ft per year). Nonetheless, persuasive arguments against drifting were founded on the lack of a known mechanism for driving the system and on the rather simple logic that if the rocks beneath the oceans were suffi-

Late Carboniferous Period (250 million years ago)

◆ **FIGURE 3.1** The earth of 250 million years ago according to Alfred Wegener. The supercontinent Pangaea began to break up about 100 million years ago, and the individual continents, roughly outlined as we know them today, have since then "drifted" to their present positions. The stippled areas represent widespread shallow seas over the continents in which sediments and fossils accumulated.

(a)

(b)

◀▶ FIGURE 3.2 Plant fossils of the Gondwana Succession. These fossils are found only in Gondwana; that is, the present-day continents of South America, Africa, India, Australia, and Antarctica. Shown are *Glossopteris* leaves from the Upper Permian (a) Dunedoo Formation and (b) Illawarra Coal Measures in Australia.

ciently rigid to cause mountain ranges to crumple up along the forward edges of the moving continents, then how could they *move*. Wegener spent his entire life defending his hypothesis in scientific papers and five editions of his classic book *The Origin of Continents and Oceans*. He died on an expedition to Greenland, where, it is said, his theory was born 30 years earlier when he saw a giant iceberg float away from a calving glacier. Unfortunately, he was never able to reconcile the forces necessary to drive continental drift, and his hypothesis gained relatively few adherents until 30 years after his death. Nevertheless, there persisted the nagging correlations of fossils, mountain ranges, and ages of rocks and geologic provinces across the Atlantic and Indian Oceans. Wegener deserves our admiration for his dogged determination, scientific insight, and ability to synthesize seemingly disparate data. To evaluate Wegener's attractive but controver-

sial concept requires some knowledge of the structure and composition of the earth's interior. We examine this subject next.

STRUCTURE OF THE EARTH

The interior of the earth is differentiated into distinct layers separated by boundaries called **discontinuities;** see ◀▶ Figure 3.4. The model of the earth's interior in the figure is not speculation. Our "X-ray" vision is provided by earthquake (seismic) waves, which give us clues to the physical and chemical nature of the earth's interior. Although seismic waves cannot tell us exactly what composes the earth hundreds of kilometers below its surface, their velocities and travel paths give us information about the physical properties (density, elasticity, etc.) of the materials, from which we can make deductions about their composition. The **lithosphere** is the solid, rigid part of the earth comprising the crust and the upper mantle. It extends 100–150 kilometers (60–90 mi) below the surface of the continents and is 70–80 kilometers (40–50 mi) thick beneath the oceans. What we regard as the plates in plate tectonic theory are composed of lithosphere. The crust, or upper part of the lithosphere, consists of relatively light (granitic) continental crust and somewhat denser (basaltic) oceanic crust. Continental crust has a density of about 2.7 grams per cubic centimeter (gm/cm^3) and stands topographically higher than the denser oceanic (basalt) crust with a density of 3.0 gm/cm^3. This "floating" of lighter (less dense) lithospheric blocks on denser asthenosphere is known as **isostasy.** At the base of the crust is the **Mohorovičić discontinuity** (usually called simply "M" or *Moho*), which is indicated by a drastic change in the behavior of seismic waves due to a change in the chemical or physical nature of the material. The thickness of the crust and depth to the "M" varies, but it is 20–90 kilometers (12–56 mi) beneath the continents and 5–10 kilometers (3–6 mi) beneath the sea floor. The crust is thickest under mountain ranges and thinnest under ocean basins (◀▶ Figure 3.5).

At the base of the lithosphere is a zone known as the **asthenosphere** (Greek, "weak rock"). It is about 250 kilometers (150 mi) thick, extending from the base of the lithosphere to a depth of about 350 kilometers (210 mi). In the upper part of the asthenosphere is a zone of low seismic velocities, indicating that the mantle here is close to its melting point and may actually contain melted portions. It is a zone of mobility that detaches the overlying rigid lithosphere from the plastic lower mantle (Figure 3.5). The distinction then between the lithosphere and the asthenosphere is one of physical state and not chemistry; the lithosphere is rigid, and the asthenosphere is plastic because of the higher temperatures at that depth.

(a)

(b)

Glaciated area
Arrows indicate the direction of glacial movement based on striations preserved in bedrock.

◀▶ **FIGURE 3.3** (a) The directions of glacial grooves preserved in bedrock on continents as they are now positioned show the direction of glacier movement during Permian (Gondwana Succession) time. For a large glacier or a number of glaciers to produce the directions of the observed grooves would otherwise require a source area in the Indian and South Atlantic Oceans. (b) With Gondwana continents reunited and an ice sheet over a south pole in present-day South Africa, the directions of glacier motion are resolved.
 (c) Gondwana glacial grooves at Hallet's Cove, Australia. Formed in Permian time, these grooves are more than 200 million years old. The width of the area shown in the photo is approximately 2 meters.

(c)

What drives plate motion is still being debated, but it is most certainly a combination of gravity acting on the lithosphere and thermal convection currents in the asthenosphere that drag the overlying plates along with them as though on a conveyor belt. Thermal convection is analogous to warm air rising and cool air sinking in closed loops in our atmosphere. We can observe a similar convective phenomenon by carefully pouring cold milk into hot coffee or by gently heating tomato

soup or oatmeal on the kitchen stove. The heated liquid rises in the center of the pan and moves laterally to sink as it cools. Convection serves as a first approximation of the driving force for plate tectonics, completed by gravity-induced "push" at mid-ocean ridges and slab "pull" at trenches (◀▶ Figure 3.6).

To complete the picture of the earth's interior, we find a very dense molten outer core and a solid inner core, both believed to be composed of a nickel–iron

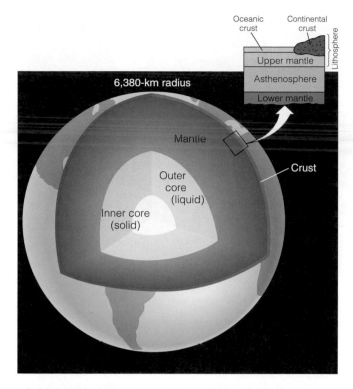

◆ **FIGURE 3.4** The internal structure of the earth. The lower part of the figure shows the layers that are defined by contrasts in density, as deduced from observations of seismic waves. The upper part of the figure illustrates the relative thickness of the crust in continental areas as compared with the sea floor. The lithosphere is comprised of the rigid tectonic plates moving on a plastic asthenosphere.

alloy. This composition is suggested by theory and by analogy to nickel–iron meteorites that strike the earth, which may represent chunks of a disintegrated planetary body.

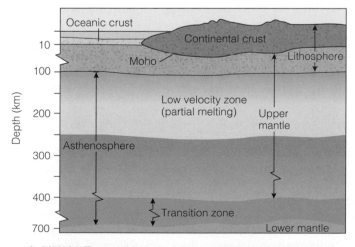

◆ **FIGURE 3.5** The earth's outer layers are divided into *crust* and *mantle* based on rock type, and into *lithosphere* and *asthenosphere* based on rigidity.

Sea-Floor Spreading

U.S. geologists Harry Hess and Robert Dietz independently concluded in 1960 that new sea floor is being created at mid-ocean ridges by volcanic activity. Inasmuch as the earth is not growing larger, they said, an equal amount of oceanic crust is probably being lost at trenches (Figure 3.6). They envisioned the driving force to be convection currents in the mantle caused by heat differences within the earth and gravity, an explanation that had been proposed 15 years earlier by British geologist Arthur Holmes. According to this theory, convection carries the rigid new crust away from the mid-ocean ridges like a conveyor belt and drives it into the mantle at trenches. This new theory, called **sea-floor spreading** by Dietz, explained the distribution of mid-ocean ridges and trenches quite nicely.

Plate Tectonics: The Global Theory

Plate tectonics is a unifying concept in that it provides a coherent model of the outer part of the earth and how it works. According to the theory, the outer part of the earth is composed of seven large and a number of small interacting plates. There are only three ways these plates can interact and, therefore, three types of boundaries: divergent, convergent, and transform. Each of the three boundaries is produced by different stresses:

- At **divergent boundaries** there is *tension;* stresses acting in opposite directions away from bodies tend to pull them apart.
- At **convergent boundaries** there is *compression;* stresses converge in the same plane on bodies and tend to shorten or compress them.
- At **transform boundaries** *shear* forces acting in opposite directions in different planes cause adjacent parts of a body to slide relative to each other in a direction parallel to a fracture or plane of contact (◆ Figure 3.7).

Where plates move apart, openings are created and molten material fills them, forming new oceanic lithosphere. These divergent boundaries, also known as **spreading centers,** occur at mid-ocean ridges. Here we find high heat flow, usually mild volcanic activity, and shallow (< 40 km or 25 mi deep) earthquakes.

Where plates come together, or collide, at convergent boundaries, crust is destroyed. One form of convergence occurs where an oceanic plate is driven beneath another oceanic plate or a continental plate; this is called **subduction.** Where two continental plates

◆ **FIGURE 3.6** The mechanism of plate motion. Under *ridge push* a plate slides down a slope of a high-standing mid-ocean ridge toward a deep trench, where it re-enters the earth aided by *slab pull*. Both push and pull are gravity-induced. *Mantle drag* between convection currents in the plastic asthenosphere and the base of the rigid lithosphere may also contribute to plate motion.

collide, folded mountain ranges may form. The Himalayas originated this way when the sediments forming them were caught in a "vise" between the Indian plate and the Eurasian plate. There we find strong earthquakes but no true subduction, because the lighter continental crust cannot sink into denser mantle. Where there is subduction, the descending plates may be identified by a zone of earthquake **foci** (points within the earth where earthquakes originate) inclined at an angle of about 60 degrees to the earth's surface. This zone was discovered by Japanese seismologist Kiyoo Wadati in 1927 and is called the **Benioff zone,** in honor of Hugo Benioff who mapped the zones worldwide. Only at subduction zones do we find earthquake foci greater than 400 kilometers (250 mi) deep. This fact is cited as evidence that one rigid plate is sinking beneath another, thereby creating earthquakes at great depth. Subduction is most commonly associated with volcanic island arcs, such as the Japanese and Aleutian Islands, and the most destructive earthquakes and explosive volcanoes occur at these boundaries. (Earthquakes and volcanoes are the subjects of Chapters 4 and 5.)

At transform boundaries, plates slide past one another along faults, and crust is neither created nor destroyed. Also known as **transform faults,** we find that portions of mid-ocean ridges and trenches are offset or that the fault connects a ridge and a trench. At the San Andreas fault, an example of a transform boundary, we find only earthquakes and no volcanic activity directly related to the boundary (◆ Figure 3.8). Regardless of the type of boundary, earthquakes outline the major plates (◆ Figure 3.9).

Plate boundaries may also be described by what happens to the crust at that plate margin. Where plates diverge at a spreading center, new ocean crust is created, and we have a *constructive* plate boundary. Where an oceanic plate sinks and disappears at a subduction zone, the plate boundary is described as *destructive*. Finally, at transform boundaries crust is neither created nor destroyed, and we have a *conservative* boundary. Because these designations describe how the margin is modified, they also serve as a memory aid for what processes are taking place there.

PROOF OF PLATE TECTONICS

The essential evidence for sea-floor spreading and plate tectonics is found by studying a totally unrelated phenomenon, the magnetic forces within the earth. This study determines the direction and strength of the earth's ancient "fossil" magnetic fields—its **paleomagnetism**—by establishing the **polarity** of magnetic minerals contained in ancient rocks, preferably rocks of known age. The earth is essentially a magnetic dipole, like a bar magnet, that has a positive (north) pole and negative (south) pole. *Polarity* refers to the orientation of a magnet's positive or negative pole that is exhibited when the earth's north magnetic pole attracts the north-seeking end of a compass needle. Studies in the mid 1960s indicated that the earth's magnetic field has reversed periodically throughout geologic time; that is, the north magnetic pole became the south magnetic pole, and vice versa. Rocks that formed when the earth's magnetic field was the same as today's field

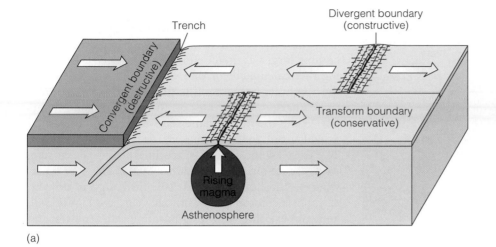

Trench

Divergent boundary
(constructive)

Convergent boundary
(destructive)

Transform boundary
(conservative)

Rising
magma

Asthenosphere

(a)

◆▶ FIGURE 3.7 (a) Divergent, convergent, and transform plate boundaries. Each type of boundary is characterized by a particular type of plate motion and stress. (b) Stresses found at plate boundaries:
(1) *Compression.* Two forces acting in opposite directions in the same plane results in shortening of crust by folding and faulting.
(2) *Tension.* Extension, or pulling the crust apart, causes faulting. (Rifting is shown.)
(3) *Shear.* Two forces acting in opposite directions on different planes causes faulting and displacement along closely spaced planes.

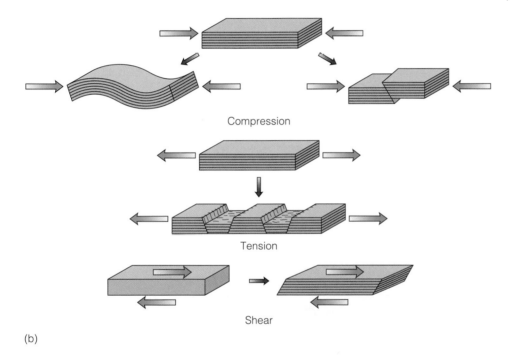

Compression

Tension

Shear

(b)

yield a strong magnetic signal, or *positive anomaly,* whereas those that formed when the field was the opposite of today's field (reversed) yield a weak signal, a *negative anomaly.* An **anomaly** is a departure from what is expected or from the mean. These anomalous signals can be mapped one sample at a time in a laboratory or on a large scale with an airborne or shipboard instrument called a *magnetometer.* We now know that during the past five million years the earth's magnetic poles have flip-flopped many times. The long periods of normal and reversed polarity are called *magnetic epochs,* and within the epochs there were shorter reversals, called *magnetic events* (◆▶ Figure 3.10).

Oceanographers and geologists had long been puzzled by the striped magnetic-anomaly patterns mapped in sea-floor rocks that parallel mid-ocean ridges (◆▶ Figure 3.11). Based on U.S. geologists' findings on the timing of magnetic reversals, British researchers reasoned that new sea floor created at mid-ocean ridges should preserve the magnetic field of the time it was formed and that, if sea-floor spreading had occurred, the pattern should change from normal to reversed as one travels away from the ridge. Repeated traverses by ships over several spreading centers were made, and just such a parallel, symmetric pattern of normally and reversely magnetized rocks was confirmed

◀▶ **FIGURE 3.8** The San Andreas fault is a transform fault, one of the plate boundary types, between the North American and Pacific plates.

(Figure 3.11). In fact, at least 171 reversals over the past 76 million years are now known. The rate at which the sea floor is moving away from a ridge axis may be calculated by measuring the distance from the ridge to a magnetic anomaly of known age, and then simply dividing the distance by the age of the anomaly. The rate at which the ocean is opening up, becoming larger, is then determined as twice the spreading rate calculated for one side of the ridge, because the spreading movement is presumed to be symmetrical. Plates move at rates on the order of centimeters per year (about the same rate as human fingernail growth), not meters per year as Wegener believed. The East Pacific Rise is spreading at the fastest rate, with a rate up to 18 centimeters (7 in) per year, whereas the northern Mid-Atlantic Ridge is slowest, at about 2 centimeters (¾ in) per year (see Figure 3.9). Techniques using very-long-baseline radio interferometry (VLBI), satellite laser ranging (SLR), and global positioning system (GPS) are now providing direct measurements that confirm the longer-term magnetic measurements. As these techniques' use expands, it will be possible to monitor surface deformation directly. This will contribute to our understanding of strain buildup that leads to earthquakes as well as other earth processes.

◀▶ **FIGURE 3.9** The earth's major tectonic plates and types of plate boundaries. Regardless of the type of plate boundary, each one is delineated by earthquake epicenters (not shown). The arrows indicate the present direction of plate motion; the numbers, the rate of spreading in centimeters per year. The rate of spreading is determined by dividing the distance to a known magnetic anomaly in kilometers ($\times 10^5$ = cm) by the age of that anomaly in years (= cm/y).

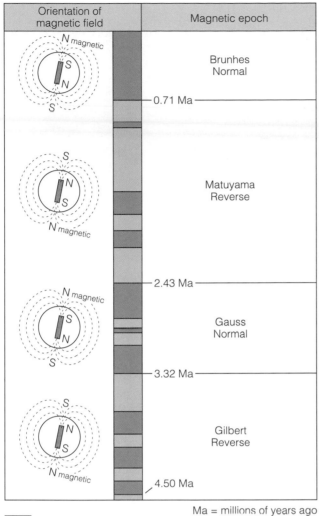

Orientation of magnetic field	Magnetic epoch
	Brunhes Normal
	—0.71 Ma—
	Matuyama Reverse
	—2.43 Ma—
	Gauss Normal
	—3.32 Ma—
	Gilbert Reverse
	—4.50 Ma—

Ma = millions of years ago

█ Normal

░ Reverse

◆▶ **FIGURE 3.10** The earth's magnetic field is much like the field that would be generated by a bar magnet inclined at 11° from the earth's rotational axis. Reversals of the field have occurred periodically, leaving "fossil" magnetism in rocks that can be dated. Each epoch is named after an important student of earth magnetism.

HOW PLATE TECTONICS WORKS

As can be seen from the model of plate-tectonic movement in ◆▶ Figure 3.12, plate boundaries determine the distribution and type of volcanic and earthquake activity. This relatively new global theory also explains other things that have puzzled scientists in the past. For instance, in the mid twentieth century many geologists believed that ocean basins and continents were stable, semipermanent features of the earth's crust. This belief

(a)

(b)

↑ = Normal magnetic polarity

↓ = Reversed magnetic polarity

←—→ = Direction of plate movement

mya = Millions of years ago

◆▶ **FIGURE 3.11** (a) Symmetric pattern of magnetic variation across part of the Mid-Atlantic Ridge southwest of Iceland. Colors indicate sea-floor rocks with matching polarity. (b) The origin of the magnetic anomaly pattern across mid-ocean ridges. Lava erupting along the crest of the ridge records the earth's prevailing magnetic polarity as it cools and solidifies. Sea-floor spreading then carries the rock away from the ridge in both directions as new lava erupts on the ridge axis.

was refuted when ancient sea-floor rocks could not be found. The oldest sea floor known is less than 200 million years old, representing just five percent of geologic time. Plate tectonics resolves this problem by allowing

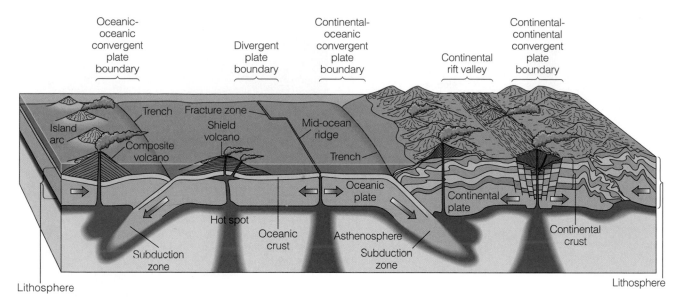

◁▷ **FIGURE 3.12** The plate tectonic model. Divergent and convergent boundaries and their relationships to sea-floor and continental volcanoes are shown. Fracture zones on the sea floor are transform boundaries, and the mid-ocean ridge is offset along these features. Rift-valley formation as a result of divergence under the continent is also shown.

that the rocks of the ocean basins have been recycled back into the asthenosphere many times. Therefore, no rocks or sediments older than the last opening cycle would be expected. In fact, if we multiply the average spreading rate by the length of the mid-ocean ridge system, we find that new oceanic crust is formed at a rate of 2.8 square kilometers (1.1 mi²) per year. Because our oceans cover an area of 310 million square kilometers (120 million mi²), they could have formed in as little as 110 million years, and over the past two billion years, as many as 20 oceans may have been created and destroyed! Land geology also supports at least two and as many as five complete openings and closings of all ocean basins.

The *Wilson cycle* is an evolutionary sequence for oceans named for its developer, J. Tuzo Wilson, a Canadian geophysicist and a major contributor to plate tectonic theory:

EVOLUTIONARY STAGE	EXAMPLE
Embryonic	Rift valleys of East Africa
Youthful	Gulf of California, Red Sea
Mature	Atlantic Ocean (growing larger)
Declining	Pacific Ocean (becoming smaller)
Terminal	Mediterranean Sea (closing, almost extinct)

The embryonic-stage rift valleys of East Africa exemplify down-dropped fault features called **grabens** that result when the crust is being pulled apart, "rifted," by convection in the mantle below the continent (◁▷ Figure 3.13a and b). As rifting progresses, the crust becomes thinner, volcanism occurs, and eventually a new ocean basin forms with a floor of oceanic crust and an oceanic spreading center in the middle. In this manner long, narrow seas, such as the present-day Red Sea (◁▷ Figure 3.13c, and ◁▷ Figure 3.14) and Gulf of California, form. The South Atlantic Ocean, for example, began as a rift in Pangaea between present-day South America and Africa in late Triassic time about 180 million years ago. The North Atlantic Ocean opened later. Thus the entire Atlantic basin has matured to its present size in the last 180 million years (◁▷ Figure 3.13d). It has done so at the expense of the declining-stage Pacific Ocean basin, which is growing smaller as North and South America move westward. The Mediterranean Sea, on the other hand, is slowly being "terminated" as the African and Eurasian plates converge on the basin. Evidence for this closure is found around the Mediterranean basin, where we find strong earthquakes, active volcanoes, and the east–west-trending Atlas and Alps mountain ranges.

At the end of the Paleozoic Era all continental masses were welded into one supercontinent, Pangaea. Pangaea began to break up (rift) in the Triassic and is still doing so today (◁▷ Figure 3.15). A variation of J. Tuzo Wilson's ocean-basins hypothesis is that at least

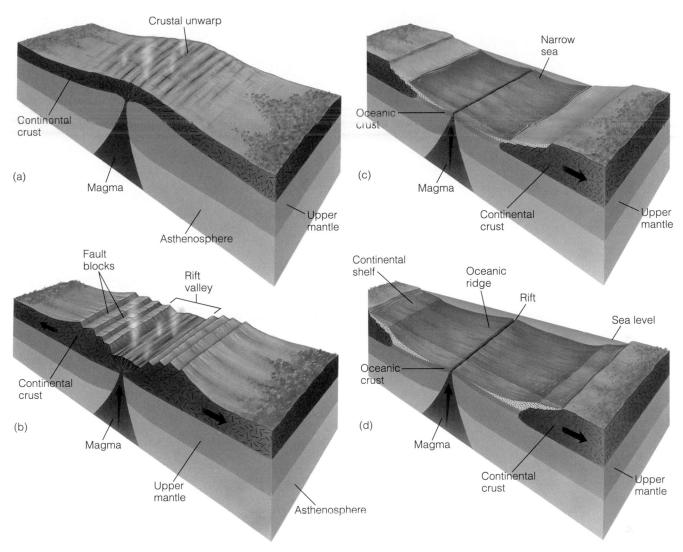

◆ FIGURE 3.13 History of a divergent plate boundary and development of an ocean. (a) Rising magma beneath a continent raises the crust, producing cracks and fractures. (b) As the crust is stretched and thinned, downfaulted rift valleys develop, and lava flows into them. (c) Continued spreading further separates the continental crust, and a narrow sea develops. (d) Spreading continues, forming an oceanic-ridge system, and an ocean basin develops and grows.

one and maybe more supercontinents existed before Pangaea that broke up and reformed in cycles spanning about 500 million years. According to this variation, Pangaea was formed at the end of the Paleozoic Era from fragments of a supercontinent that existed at the end of the Proterozoic Eon. A convincing argument for this *supercontinent cycle* is the regularity of mountain-building episodes caused by continental collisions toward the end of each cycle. The formation of a supercontinent (and mountain building) occurs about every 400–500 million years, followed by episodes of rifting, formation of intercontinent ocean basins, and drifting of fragments that then collide and join to form mountains and another supercontinent.

> **Consider This... ?**   It is said that all knowledge of the natural world is potentially beneficial to humanity, even knowledge about past conditions that have subsequently changed. How might our knowledge of past and present plate positions and subsequent motions be applied to benefit the human race? ■

A dramatic evolution is apparent in Asia. There the Indian plate moved thousands of kilometers northward at about five centimeters (2 in) per year and collided with the Eurasian plate, folding, faulting, and uplifting the sediments of the ancient Tethys Sea to form the Himalaya Mountains. Where

(a)

(b)

(c)

◆ **FIGURE 3.14** (a) Two large rift valleys are forming in East Africa as continental crust stretches along a divergent plate margin. The large lake between the rifts is Lake Victoria. The Red Sea and the Gulf of Aden exemplify more advanced stages of rifting and ocean development. (b) Looking down the Great Rift Valley of East Africa with Lake Magadi, Kenya, in the foreground. Because of high evaporation rates and lack of exterior drainage Lake Magadi is a soda lake rich in sodium carbonate, a salt that whitens mud flats on the lake's shore. A Marabou stork spreads its wings on the opposite shore. (c) Satellite photo of the Sinai Peninsula, looking north. The Red Sea is at the bottom, the Gulf of Suez on the left, and the Gulf of Aqaba on the right. The Red Sea is a youthful ocean basin. Very likely, it resembles the South Atlantic Ocean of 270 million years ago.

continent–continent collisions occur, mountain ranges form and catastrophic earthquakes occur. This is because both colliding plates are composed of low-density granitic crust, and thus, neither can subduct completely beneath the other into the heavier mantle (Case Study 3.1).

Within some ocean basins there are chains of aligned volcanic islands that become progressively

(a)

(b)

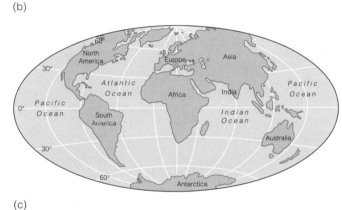

(c)

◀◈ **FIGURE 3.15** Paleogeographic reconstruction of continents' movements over the past 180 million years during (a) the early Triassic Period, (b) late in the Cretaceous Period, and (c) as it is today. Note that the Panthalassa Ocean was the "one" ocean dominating the globe, as Pangaea was the "one" continent. The maps show present-day coastlines; ancient coastlines differed from these. Compare these maps with Figure 3.1, which was originally published in 1915 and revised by Wegener in 1929 just before his death.

older in one direction. The Hawaiian Islands, Line Islands, and Tuamotus are such chains in the Pacific Ocean. Although the Galápagos Islands off the coast of Ecuador—made famous by the writings of Charles Darwin—are only roughly aligned, they are of similar origin. These lines-of-islands are not associated with plate boundaries, but have formed over rising plumes

of lava, or **hot spots,** in the mantle. These rising intraplate plumes penetrate the general convective circulation in the mantle that drives plate motion and are independent of spreading centers. As a tectonic plate moves over one of these plumes, it carries with it the volcano that formed there from the plume, and a new volcano forms on the sea floor in its place. A line of extinct volcanoes forms over time, with their ages increasing with distance from the hot spot. In the case of the Hawaiian chain, each island formed over the stationary hot spot and was then carried northwestward on the moving lithospheric plate (◈ Figure 3.16). The Island of Hawaii is the youngest of the Hawaiian hot-spot volcanoes and is the only one with active volcanism today. A line drawn from the Island of Hawaii to Midway Island, the oldest of the chain, would represent an arrow marking the direction and rate of movement of the Pacific plate in this region. The track of the Hawaiian Islands forms a "lazy *L*" with the Emperor Seamount chain, a line of submerged volcanoes that extends northward from Midway, tracing a different direction of plate movement prior to 40 million years ago. Of greater environmental concern because of population densities are hot spots that occur beneath the continents. Yellowstone National Park, for example, is situated above a midplate hot spot with a potential for destructive volcanic activity. About a third of the more than 120 identified hot spots that have been active within the last ten million years penetrate continental areas (see Figure 3.9).

Much of the sea-floor evidence that supports the plate tectonics theory comes from the Deep-Sea Drilling Project funded by the National Science Foundation. During a series of cruises in the Mediterranean Sea the researchers discovered thick salt deposits beneath the sea floor, indicating that about ten million years ago (Miocene time) the sea was isolated from the Atlantic Ocean and that the sea totally evaporated (◈ Figure 3.17a). Access of Atlantic Ocean water to the Mediterranean through the Strait of Gibraltar was cut off by uplift as the African plate moved northward, and also by a global lowering of sea level at the time. Further drilling yielded sediments indicating that the floors of the two Mediterranean basins once resembled the great dry lakes of present-day deserts in the western United States. Occasionally they were occupied by water when the connection with the Atlantic was periodically reestablished (Figure 3.17a), but generally they were lifeless salt flats. The Mediterranean must have looked as Death Valley does today from a high vantage point, except that major rivers such as the Rhone, Danube, and Nile cascaded down the continental slopes to the flats below.

Case Study | 3.1

Exotic Terranes—A Continental Mosaic

ecause continental plates divide, collide, and slip along faults, it is not surprising that bits and pieces of continental crust existing within the oceans ultimately collide with continents at subduction zones and become stuck there. Because of their low density, they stand high above oceanic crust (due to isostatic equilibrium), and these microplates or microcontinents become plastered, *accreted,* to larger continental plates when they collide with them. The accreted plates are known as **terranes,** defined as fault-bounded blocks of rock with histories quite different from those of adjacent rocks or terranes.* In size they may be several thousand square kilometers or just a few tens of square kilometers, and they may become part of a continent composed of many terranes. Oceanic crust and the sediment resting on it may be scraped up onto the continents. Terranes may consist of almost any type of rock, but each one is fault-bounded, has a paleomagnetic signature indicating a distant origin, and has little geology in common with adjacent terranes or with the continental **craton**—the continent's stable core.

The terrane concept developed in Alaska when geologic mapping for land use planning revealed that the usually predictable pattern of rocks and structures was not valid for any distance. In fact, the rocks the geologists found a few kilometers away were almost always of a "wrong" composition and age. Further studies revealed that Alaska is a collection of microplates (terranes)—tectonic flotsam and jetsam, if you wish—that have mashed

*Terrane, as noted, is a geological term describing the area or surface over which a particular rock or group of rocks is prevalent. *Terrain* is a geographical term referring to the topography or physical features of a tract of land.

◇ **FIGURE 1** Accreted terranes of the west coast of the United States and Canada. The Salina, Wrangellia, Alexander, Stikine, Klamath Mountains, Ruby, and Nixon Fork blocks (dark brown) were probably once parts of other continents and have been displaced long distances. The terranes shown as light-green areas are probably displaced parts of North America, the stable North American craton (darker green). The lighter-brown areas represent rocks that have not traveled great distances from their places of origin.

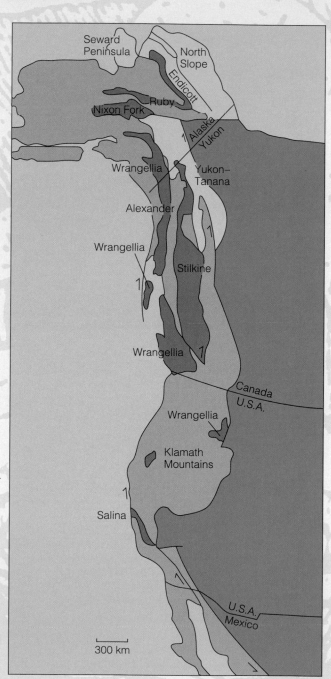

together over the past 160 million years and that are still arriving from the south (◁▶ Figure 1). One block, the Wrangellia terrane, was an island during Triassic time and it has a paleomagnetic signature of rocks that formed 16° from the equator. It is not known whether Wrangellia formed north or south of the equator, because it is not known whether the magnetic field at the time was normal or reversed. In either case, the terrane traveled a long distance to become part of present-day Alaska (◁▶ Figure 2). It now appears that about 25 percent of the western edge of North America, from Alaska to Baja California, formed in this way; that is, by bits and pieces being grafted onto the core of the continent and thus enlarging it. In the distant future, part of California may become an exotic terrane of Alaska.

If you've deduced that terrane accretion onto the continents (also known as *docking*) is a characteristic of active continental margins, you are right (◁▶ Figure 3). The east coast of North America is a rifted (pulled-apart), or *passive*, margin. Material is added to it by river sediment forming flat-lying sedimentary rocks that become part of the continent. Such rocks are not subject to the mountain-building forces of active margins and they accumulate in thick, undisturbed sequences.

◁▶ FIGURE 3 Formation of exotic terranes by a small continental mass and a submarine volcano being scraped off from the subducting plate and onto the continental mass. Many portions of the west coasts of the United States and Canada have grown in this manner.

◁▶ FIGURE 2 An analysis of Wrangellia terrane trajectories for the period 180 Ma to 100 Ma, assuming initial position in the Northern Hemisphere. In this interpretation both Vancouver Island and the Alaskan terranes (see Figure 1) rode with the oceanic plate until they collided (docked) at their present positions on the North American continent about 100 million years ago. The data points (the dots) represent paleomagnetic evidence at the respective locations.

3.8–5.6 Ma
Kauai

Sea level

2.3–3.3 Ma
Oahu

1.3–1.8 Ma
Molakai

0–1.3 Ma
Maui

0–0.7 Ma
Hawaii   Loihi

Plate motion

Hot spot

(a)

▶ **FIGURE 3.16** (a) Cross section of six islands formed and forming over the Hawaiian "hot spot." The Hawaiian Islands are sequentially older in a northwest direction. The new volcano Loihi is forming underwater off the southeast end of the Hawaiian chain. (b) Kinked hotspot trajectories in the Pacific. The kink in each island chain occurred about 40 million years ago due to a change in the Pacific plate's direction of motion.

(b)

Consider This… ?   Remembering that the "earth system" is composed of many different cycles, such as the rock cycle and the supercontinent cycle, predict the general arrangement of oceanic and continental plates in, say, 75 million years. In particular, will the continents be more or less consolidated than they are now? ■

About five million years ago the Strait of Gibraltar reopened, and the Atlantic Ocean flowed into the Mediterranean Sea as a waterfall or cascade of great magnitude (▶ Figure 3.17b). This inflow filled the Mediterranean in a few thousand years. The sediments on top of the five-million-year-old salt-flat deposits contain fossils of marine organisms, indicating rapid filling with normal sea water, and no abnormal salinity, which would have occurred if the filling were slow.

Plate tectonics theory unifies and incorporates the basic concepts of continental drift and sea-floor spreading. In one way or another, plate tectonics explains why nearly all geologic phenomena occur, such

◆ **FIGURE 3.17** Desiccation of the eastern and western Mediterranean basins about ten million years ago (Miocene time), when its connection with the Atlantic Ocean through the Strait of Gibraltar was blocked. (a) Total evaporation took as little as 1,000 years, during which salt was deposited and dry lakes or salt lakes existed in the basins. Some spillover from the Red Sea occurred in the eastern Mediterranean. (b) Two-kilometer-thick deposits of sediments and evaporites (especially gypsum and halite) beneath the floors of these basins indicate that the cycle of filling, evaporation, and refilling was repeated many times (2 km ≈ 6,500 ft or 1¼ mi).

(a)

(b)

as why volcanoes are either explosive or quiet at given locations, why mountain ranges are located where they are, and why large faults such as the San Andreas in California cut across the landscape and occasionally generate damaging earthquakes. The theory has revolutionized our understanding of how the earth works and has provided geologists with a useful model in their searches for new mineral and oil deposits.

biosphere. Sea-floor vents have nurtured new life-forms and contribute to the hydrosphere's changing chemistry. Plate motion produces earthquakes that disrupt the biosphere and hydrosphere, and changing continental connections create different environments for the biosphere, including humans. The fossil record validates this, and the impact of shifting plates on the biosphere has been important in developing the theory.

THE EARTH'S RESERVOIRS AND PLATE TECTONICS

Plate tectonic processes are responsible for producing igneous and metamorphic rocks at plate boundaries and completing the rock cycle (see Figure 2.12). Volcanic action near boundaries releases heat, water, and gases from the earth's interior to the atmosphere and hydrosphere, where they alter the chemistry of these reservoirs. Some of these gases are returned to earth in aqueous form to enhance surface weathering processes that produce soils that are essential to the

Consider This...❓ A stockbroker might speculate on what the market will do the next day, a business person might try to forecast six-month trends, but only an earth scientist will speculate on what the face of our planet will look like a million years from now. Here's your chance! What will be the fate of (1) the Mediterranean Sea, (2) the Atlantic and Pacific Ocean basins, and (3) the west coast of the United States exclusive of Alaska if the present plate motion continues for the next 20 million years? Hazard a guess. ■

Gallery

Visions of How the Earth Works

The evolution of modern plate-tectonic theory started with Alfred Wegener's 1912 hypothesis of Continental Drift (◇ Figure 3.18). Fifty years later the idea of sea-floor spreading was proposed. Although scientists were skeptical at first, this far-reaching concept led to the unifying theory of plate tectonics (◇ Figure 3.19).

◇ **FIGURE 3.18** Alfred Wegener formulated his hypothesis that continents move about on the face of the earth while sitting out a long Arctic winter in Greenland. His ideas were considered so outrageous that more than fifty years passed before scientists realized he was on the right track.

◇ **FIGURE 3.19** Evidence of colliding and sliding tectonic plates. Collision of the Indian and Asian plates produced the great Himalaya Mountains. The rocks of Nepal's Dhaulāgiri I (8,172 m, 26,810 ft), one of the highest peaks in the world, were uplifted and contorted by the continent–continent collision. The steep-sided mountain's south wall is 15,000 feet high.

(a)

(b)

(c)

◆▶ FIGURE 3.20 (a) "Black smokers" of mid-oceanic ridge. Hot brine loaded with minerals leached from basalt emanates from vents on ridges and builds tall "chimneys." (b) Giant tube worms thrive around the "smokers" and live on bacteria in the Galapagos Rift (see Chapter 5). (c) Clams more than 25 centimeters (10 in) long inhabit the warm waters near the vents in the Galapagos Rift.

Summary

CONTINENTAL DRIFT
DESCRIPTION
Alfred Wegener's theory that during the Permian Period there was a single supercontinent, Pangaea, which split apart during Jurassic time into individual continents that have since migrated to their present positions.

CAUSE
The theory provided no explanation for a driving force.

EVIDENCE
1. Similar glacial deposits and plant fossils in Permian rocks (the Gondwana Succession) in Africa, South America, India, Australia, and Antarctica.
2. Similarities between fossils, ages of rocks, and geologic provinces on opposite sides of the Atlantic Ocean.
3. The jigsaw fit of the continents if the Atlantic and Indian Oceans were closed.

SEA-FLOOR SPREADING
DESCRIPTION
Theory independently proposed by both Harry Hess and Robert Dietz that new oceanic crust is created at mid-ocean ridges and moves laterally across the ocean basins to trenches, where it sinks and returns to the mantle.

CAUSE
Convection cells in the mantle generated by thermal and density differences carry the rigid overlying crust (lithosphere) away from the ridges until the crust cools and descends toward the mantle at oceanic trenches.

EVIDENCE
1. Distribution of deep-focus earthquakes at trenches.
2. Lack of ancient sediments on the sea floor.
3. Increasing age of sea-floor sediments away from mid-ocean ridges.

PLATE TECTONICS
DESCRIPTION
Theory that unifies and incorporates the earlier theories of continental drift and sea-floor spreading. According to it, the lithosphere is composed of seven large plates and a number of smaller ones that move about on the surface of the earth. Plates interact in only three ways: they *collide* at convergent boundaries marked by trenches or mountain ranges; they *move apart,* forming new sea floor, at divergent boundaries marked by mid-ocean ridges; or they *slip by one another* along transform boundaries.

CAUSE
Convection currents in the asthenosphere (the plastic layer of the upper mantle) carry the plates as passive passengers from divergent boundaries to convergent and conservative boundaries. Gravity pulls plates down into the mantle at convergent boundaries.

EVIDENCE
1. Land geology, ocean drilling, and geophysical data support past openings and closings of the ocean basins (Wilson cycle).
2. Magnetic anomaly patterns ("stripes") on both sides of mid-ocean ridges.
3. Chains of oceanic volcanoes and seamounts that systematically increase in age with distance from the present locations of active volcanism (hot spots) indicate that plates are moving over fixed plumes of lava coming from deep in the mantle.
4. The geological and geophysical evidence for continental drift and sea-floor spreading also support plate tectonics.

EXPLAINS
1. The thick salt deposits beneath the Mediterranean Sea and its desiccation 10 million years ago.
2. The existence of narrow seas and gulfs such as the Gulf of California and the Red Sea as spreading centers.
3. The location and geology of the Himalaya Mountains.
4. The location of violent earthquakes and volcanoes at convergent margins.

◆ Key Terms

anomaly	convergent boundary	focus (pl. *foci*)	isostasy
asthenosphere	craton	Gondwanaland	Laurasia
Benioff zone	discontinuity	graben	lithosphere
continental drift	divergent boundary	hot spot (plume)	

Mohorovičić discontinuity ("M" or Moho)
paleomagnetism

Pangaea
plate tectonics
polarity

sea-floor spreading
spreading center
subduction

terrane
transform boundary
transform fault

◆ Study Questions

1. What geologic and geographic evidence did Alfred Wegener cite for continental drift?
2. How do we know that the interior of the earth is hot, and why has it not cooled entirely in the last 4.5 billion years?
3. How do density differences cause movement in fluids? In your own words describe heat transfer by convection currents, giving some illustrations from everyday life.
4. What are the major spheres, or shells, of the earth's interior based upon density? What is meant by the term *lithosphere*? Describe the lithosphere and tell how it is separated from the deep mantle.
5. What is a tectonic "plate"? Give an example (exact geographic location) of each type of plate boundary. How fast do plates move?
6. What geophysical observations and measurements support the model of sea-floor spreading and plate tectonics?
7. How can the change in direction of a line of oceanic islands—such as that between the Hawaiian Islands chain and the Emperor Seamount chain—be explained?
8. Draw a cross section of a typical convergent plate boundary (a subduction zone) and label the following: trench, volcanic areas, Benioff zone, lithosphere, asthenosphere.
9. How might the distribution of the Permian Gondwana Succession be explained without invoking plate tectonics (ancient landmass, land bridges, etc.)? Might an alternative explanation be more reasonable than drifting continents and plate tectonics?
10. How have the narrow bodies of the Red Sea and the Gulf of California formed? Study a map, seeking an obvious relationship between the Red Sea and the Dead Sea.

◆ For Further Information

BOOKS AND PERIODICALS

Ballard, Robert. 1983. *Exploring our living planet.* Washington, D.C.: National Geographic Society.

Bonatti, Enrico. 1994. The earth's mantle below the oceans. *Scientific American,* March: 44–51.

Cox, Allan. 1986. *Plate tectonics: How it works.* Palo Alto, Calif.: Blackwell Scientific Publications.

Dalziel, Ian W. D. 1995. Earth before Pangaea. *Scientific American,* January: 58.

Debiche, M. G.; Allan Cox; and D. Engebretson. 1987. *The motion of allochthonous terranes across the north Pacific basin.* Geological Society of America Special Paper 207, 49 pp.

Hallam, A. 1973. *A revolution in the earth sciences.* Oxford, U.K.: Oxford University Press.

Hsu, Ken. 1983. *The Mediterranean was a desert.* Princeton, N.J.: Princeton University Press.

Kearey, P., and F. J. Vine. 1996. *Global tectonics,* 2d ed. Palo Alto, Calif.: Blackwell Scientific.

Larson, Robert. 1995. The mid-Cretaceous superplume episode. *Scientific American,* January: 82–86.

Moores, E. M., ed. 1990. Shaping the earth: tectonics of continents and oceans. *Scientific American readings.* New York: W. H. Freeman and Co.

Murphy, J. Brendan, and R. Damian Nance. 1992. Mountain belts and the supercontinent cycle. *Scientific American,* April: 84–91.

Nance, R. Damian, T. R. Worsley, and J. B. Moody. 1988. The supercontinent cycle. *Scientific American* 259: 172–79.

Schneider, David. 1997. Hot spotting. *Scientific American,* April: 22–23.

Sullivan, Walter. 1974. *Continents in motion: The new earth debate.* New York: McGraw-Hill Book Co.

Tarling, D., and Maureen Tarling. 1970. *Continental drift: A study of the earth's moving surface.* New York: Anchor Books.

Uyeda, Seiya. 1977. *The new view of the earth.* New York: W. H. Freeman.

Wegener, Alfred. 1929, 1966 translation. *The origin of continents and oceans.* New York: Dover Publications.

Wilson, J. Tuzo. 1972. Continents adrift. *Scientific American readings*. San Francisco: W. H. Freeman.

INTERNET SITE
University of Nevada–Reno. Site features several Web pages of tutorial material on plate tectonics, the earth's interior, and earthquakes. Interesting pages that make high-quality images.
http://www.seismo.unr.edu/ftp/pub/louie/class/100/plate-tectonics.html

◆ World Wide Web Activities

Interactive Plate Tectonics for the Phanerozoic
http://www.scotese.com/newpage13.htm
At this Web site, Christopher Scotese has posted six virtual reality animations showing plate motion scenarios for specific segments of geologic time. Especially important are *The Assembly of Pangaea During the Paleozoic,* and *The Breakup of Pangaea (last 200 million years).* Once loaded, the images may be manipulated backward and forward in time by moving your cursor over the maps.
1. View *The Breakup of Pangaea.* Which major modern ocean area was present before Pangaea was disassembled?
2. View *The Collision of India and Asia.* For how long was India a continent?

Animation of the Movement of Southern Hemisphere Continents
http://www.earth.monash.edu.au/ ~ greg/Gond.html
Alfred Wegener was one of the few European geoscientists to study in the Southern Hemisphere, and field work he did there contributed enormously to his theory of continental drift. This Web site shows animated movement of the southern continents between 200 million years ago (Gondwanaland) and the present.

Color Map of Seafloor Ages
http://gdcinfo.agg.nrcan.gc.ca/toc.html?/app/agegrid.html
This global map shows the age of the seafloor using color bands. Examine the map, consult the legend, and then answer these questions:

1. Both the central Atlantic Ocean and eastern Pacific Ocean have active spreading ridges, but the Pacific ridge has a much wider red central portion than the Atlantic ridge. What does this tell us about the rate of seafloor spreading in the Pacific versus the Atlantic?
2. Where would you find the oldest seafloor in the Pacific Ocean?

Details on the Trace of a "Hot Spot" in the Northwestern United States
http://volcano.und.nodak.edu/vwdocs/volc_images/north_america/yellowstone.html
This page offers maps, a satellite image, cross section, and text regarding the hot spot volcanism along the Snake River Plain (Idaho) and Yellowstone National Park (Wyoming).
1. What other volcanic region on the geographic (first) map may be related to the Snake River Plain–Yellowstone hot spot?
2. Why do the volcanic features of the Snake River Plain become younger to the east?

Earth Systems Today CD Correlations

CHAPTER CONCEPT: Plates Can Interact at Three Types of Boundaries
You will use the interactive plate boundary map in the module to study the geographic distribution of the three boundary types, and their relationship to earthquakes and volcanism.

SECTION MODULE: Plate Tectonics: Main Activity, Overview, and Map
Read the **Overview** on the left screen, and then use the **interactive map** on the right screen to locate the three dif-

ferent plate boundary types. Display Holocene volcanoes and recent earthquakes for comparison.

QUESTION: *What is the direction of motion of the Pacific Plate? How can you tell? Describe the distribution of the three plate boundary types around the edge of this large lithosphere segment.*

CHAPTER CONCEPT: Oceans Have an Evolutionary Sequence

You will use the interactive **Divergent Boundary** map in the module to study stages of ocean basin development and actual examples as they exist today.

SECTION MODULE: Plate Tectonics: Introductory Animation, Divergent Boundaries

Read the **Overview** on the left screen, and then use the **interactive map** on the right screen to run the animation sequence of an opening ocean.

QUESTION: *What physical features characterize the mature (stage 3) ocean basin?*

chapter | four

Earthquakes and Human Activities

(a)

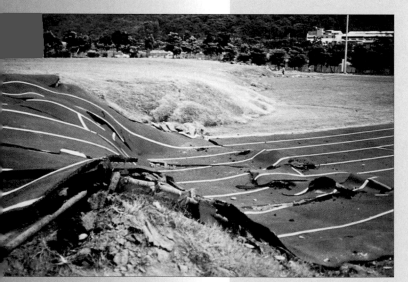

(b)

Tragedy in Taiwan

In the early morning of 21 September 1999, an earthquake struck the central part of Taiwan near the village of Chi-Chi. The largest earthquake in at least a hundred years for the island, it registered a local Richter magnitude of 7.3 and a moment magnitude of 7.6. Sudden movement on the Chelungpu fault caused the earthquake, and the fault ruptured the ground surface over a distance of 85 kilometers (50 miles). The total movement on the fault was 9.8 meters (315 ft), ranking it among the larger slips measured on earthquake-producing faults. ◇▶ Figure 1 shows a running track traversed by the fault at Kuang Fu High School. With the vertical offset of 2.5 meters (8.1 ft), it is obvious that no 100-meter dash records will be broken here (see also Figure 4.53). Severe shaking damage to nearby towns and cities resulted in 2,300 fatalities and 8,700 injuries. A high-rise structure in the capitol, Taipei, 150 kilometers (90 mi) north of the epicenter, collapsed, probably due to poor construction. Seventy closely spaced seismographs provided seismologists one of the best data sets ever recorded for an earthquake.

◇▶ **FIGURE 1** Offset on the track of Kuang Fu High School, Central Taiwan. (a) Looking down the track at a 2.5-meter (8.1-ft) offset. (b) Looking along the Chelungpu fault, across the crumpled track, and the grass escarpment formed when the fault moved.

◇▶ **FIGURE 2** The plate-tectonic setting of Taiwan and rate of relative motion between the Eurasian plate and the Philippine Sea plate. Star indicates the location of the Chi-Chi earthquake epicenter.

Taiwan's plate-tectonic setting is complex. The Philippine Sea plate is subducting into the Ryukyu trench to the north, and the Eurasian plate is doing the same thing into the Manila trench to the south (◆ Figure 2). Thus, Taiwan is in a zone of colliding plates. The relative rate of plate convergence is about 8.2 centimeters (3.2 in) per year—a very rapid rate indeed. In fact, Taiwan had had four earthquakes with magnitudes greater than 7.0 in the century prior to the Chi-Chi event, reflecting the high tectonic stresses in the Taiwan collision zone. ■

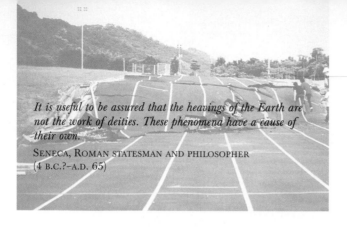

After experiencing an earthquake in Concepción, Chile, in 1835, Charles Darwin noted that "A bad earthquake at once destroys the oldest associations; the world, the very emblem of all that is solid, had moved beneath our feet like crust over a fluid." Darwin's reflections are vivid, and many of us have felt the same way when experiencing strong earthquake motion. What Darwin did not know was the cause of earthquakes and the resulting ground motion.

The Nature of Earthquakes

Earthquakes are the result of abrupt movements on **faults**—fractures in the earth's lithosphere. The types of faults and the earth forces that cause them are shown in ◆ Figure 4.1. The movements occur as the earth's crustal plates slip past, under, and away from one another. Because the **stress** (force per unit area) that produces **strain** (deformation) can be transmitted long distances in rocks, active faults do not necessarily occur exactly on a plate boundary, but they generally occur in the vicinity of one. The mechanism by which stressed rocks store up strain energy along a fault to produce an earthquake was explained by Harold F. Reid after the great San Francisco earthquake of 1906. Reid proposed a mechanism to explain the shaking, which resulted from movement on the San Andreas fault, known as the **elastic rebound theory** (◆ Figure 4.2). According to this theory, when sufficient strain energy has accumulated in rocks, they may rupture rapidly—just as a rubber band breaks when it is stretched too far—and the stored strain energy is released as vibrations that radiate outward in all directions (◆ Figure 4.3).

Most earthquakes are generated by movements on faults within the crust and upper mantle that do not produce ruptures at the ground surface. We can thus recognize a point within the earth where the fault rupture starts, the **focus**, and the **epicenter**, the point on the earth surface directly above the focus (Figure 4.3). Elastic waves—vibrations—move out spherically in all directions from the focus and strike the surface of the earth.

Damaging earthquake foci are generally within a few kilometers of the earth's surface. Deep-focus earthquakes, on the other hand, those whose foci are 300–700 kilometers (190–440 mi) below the surface, do little or no damage. Earthquake foci are not known below 700 kilometers. This indicates that the mantle at that depth behaves *plastically* due to high temperatures and confining pressures, deforming continuously as ductile substances do, rather than storing up strain energy.

The vibration produced by an earthquake is complex, but it can be described as three distinctly different types of waves (◆ Figure 4.4). Primary waves, **P-waves,** and secondary waves, **S-waves,** are generated at the focus and travel through the interior of the earth; thus, they are known as **body waves.** They are designated P- and S-waves because they are the first (*p*rimary) and second (*s*econdary) waves to arrive from distant earthquakes. As these body waves strike the earth's surface, they generate surface waves that are analogous to water ripples on a pond.

P-waves (◆ Figure 4.4a) are **longitudinal waves;** the solids, liquids, and gases through which they travel are alternately compressed and expanded in the same direction the waves move. Their velocity depends upon the resistance to change in volume (in compressibility) and shape of the material through which they travel. P-waves travel about 300 meters (1,000 ft) per second in air, 300–1,000 meters (1,000–3,000 ft) per second in soil, and faster than 5 kilometers (3 mi) per second in solid rock at the surface. P-wave velocity increases with depth, because the materials composing the mantle and core become less compressible with depth. This increase causes the waves' travel paths to be bowed downward as they move through the earth. P-waves speed through the earth with a velocity of about 10 kilometers (6 mi) per second. At the ground surface they have very small amplitudes (motion) and cause little property damage. Although they are physically identical to sound waves, they vibrate at frequencies below what the human ear can detect. The earthquake noise that has been reported could be P-waves of a slightly higher frequency or some other, related vibrations whose frequencies are in the audible range.

S-waves (◆ Figure 4.4b) are **transverse (shear) waves;** they produce ground motion perpendicular to their direction of travel. This causes the rocks through which they travel to be twisted and sheared. These waves have the shape produced when one end of a garden hose or a loosely hanging rope is given a vigorous flip. They can travel only through material that resists shearing; that is, material that resists two forces acting in opposite directions in different planes. Thus, S-waves travel only through solids, because liquids and gases have no shear strength (try piling water in a mound). S-wave ground motion may be largely in the horizontal plane and can

(b)

(d)

(f)

◆▶ FIGURE 4.1 (a) Normal fault geometry and (b) examples in a road cut on I-40 east of Kingman, Arizona. (c) Reverse fault geometry and (d) example in Death Valley, California. (e) Right-lateral strike-slip fault geometry and (f) right-lateral offset of resistant sedimentary rock ridges, Frontier Formation, Wyoming.

(a) Original position

(b) Deformation

(c) Rupture and release of energy

(d) Rocks rebound to original undeformed shape

(e)

◆▶ **FIGURE 4.2** (a–d) The cycle of elastic-strain buildup and release for a right-lateral strike-slip fault according to Reid's elastic rebound theory of earthquakes. At the instant of rupture, (c), energy is released in the form of earthquake waves that radiate out in all directions. (e) Right-lateral offset of a fence by 2.5 meters (8 ft) by displacement on the San Andreas fault in 1906; Marin County, California.

result in considerable property damage. S-wave velocity is approximately 5.2 kilometers (3.2 mi) per second from distant earthquakes; hence, they arrive after the P-waves.

◆▶ **FIGURE 4.3** The focus of an earthquake is the point in the earth where fault rupture begins, and the epicenter is directly above the focus at the ground surface. Seismic-wave energy moves out in all directions from the focus.

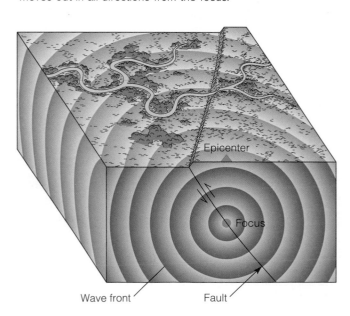

Unexpended P- and S-wave energy bouncing off the earth's surface generates the complex surface waves (◆▶ Figure 4.4c). These waves produce a rolling motion at the ground surface. In fact, slight motion sickness is a common response to surface waves of long duration. The waves are the result of a complex interaction of several wave types, the most important of which are *Love waves,* which exhibit horizontal motion normal to the direction of travel, and *Rayleigh waves,* which exhibit retrograde (opposite to the direction of travel) elliptical motion in a plane perpendicular to the ground surface. Surface waves have amplitudes that are greatest in near-surface unconsolidated layers and are the most destructive of the earthquake waves.

LOCATING THE EPICENTER

A **seismograph** is an instrument designed specifically to detect, measure, and record vibrations in the earth's crust (◆▶ Figure 4.5). It is relatively simple in concept (◆▶ Figure 4.6), but some very sophisticated electronic systems have been developed. Seismic data are recorded onto a **seismogram** (◆▶ Figure 4.7). Because seismographs are extremely sensitive to vibrations of any kind, they are installed in quiet areas such as abandoned oil and water wells, cemeteries, and parks.

P-wave propagation

Dilation   Compression   Dilation   Compression   Unstressed condition

(a)

◆▶ FIGURE 4.4 Ground motion during passage of earthquake waves. (a) P-waves compress and expand the earth. (b) S-waves move in all directions perpendicular to the wave advance, but only the horizontal motion is shown in this diagram. (c) Surface waves create surface undulations that result from a combination of the retrograde elliptical motions of Rayleigh waves (shown) and Love waves, which move from side to side at right angles to the direction of wave propagation.

S-wave propagation

Vertical plane

(b)

Surface-wave propagation

(c)

When an earthquake occurs, the distance to its epicenter can be approximated by computing the difference in P- and S-wave arrival times at various seismograph stations. Although the method actually used by seismologists today is more accurate and determines the depth and location of the quake's focus and its epicenter, what is described here serves to show how seismic-wave arrival times can be used to determine the distance to an epicenter. Because it is known that the two kinds of waves are generated simultaneously at the earthquake focus and that P-waves travel faster than S-waves, it is possible to calculate where the waves started.

Imagine that two trains leave a station at the same time, one traveling at 60 kilometers per hour and the other at 30 km/h, and that the second train passes your house one hour after the first train goes by. If you know their speeds, you can readily calculate your distance from the train station of origin as 60 kilometers. The relationship between distance to an epicenter and arrival times of P- and S-waves is illustrated in ◆ Figure 4.8.

◁▶ **FIGURE 4.5** Seismometer (*left*) and seismogram (*right*) used to demonstrate method of detecting and recording earthquakes. The record on the seismogram is not an earthquake, but represents the seismometer's detection of vibrations from students' footsteps amplified about 2,000 times.

The epicenter will lie somewhere on a circle whose center is the seismograph and whose radius is the distance from the seismograph to the epicenter. The problem is then to determine where on the circumference of the circle the epicenter is. To learn this, more data are needed; specifically, the distances to the epicenter from

◁▶ **FIGURE 4.6** The first earthquake detector, invented about A.D. 130 by Chinese scholar Chang Heng. Balls held in the dragons' mouths were aligned with a pendulum inside the vase. When an earthquake occurred the pendulum swung and pushed balls into the mouths of the frogs aligned with the pendulum's swing. If the frogs facing north and south contained balls, for example, Chang Heng could say that the earthquake epicenter was north or south of the instrument.

two other seismograph stations. Then the intersection of the circles drawn around each of the three stations—a method of map location called *triangulation*—specifies the epicenter of the earthquake (◁▶ Figure 4.9).

EARTHQUAKE MEASUREMENT
Intensity Scales

The reactions of people (geologists included) to an earthquake, or an earthquake prediction, typically range from mild curiosity to outright panic. However, a sampling of the reactions of people who have been subjected to an earthquake can be put to good use. Numerical values can be assigned to the individuals' perceptions of earthquake shaking and local damage, which can then be contoured upon a map. One **intensity scale** developed for measuring these perceptions is the 1931 **Modified Mercalli scale (*MM*).** The scale's values range from *MM* = I (denoting not felt at all) to *MM* = XII (denoting widespread destruction), and they are keyed to specific U.S. architectural and building specifications. (See Appendix 4.) People's perceptions and responses are compiled from returned questionnaires, and then lines of earthquake intensity, called **isoseismals,** are plotted on maps. Isoseismals enclose areas of equal earthquake damage and can indicate areas of weak rock or soil as well as areas of substandard building construction. Such maps have proved useful to planners and building officials in revising building codes and safe construction standards (◁▶ Figure 4.10). It is difficult to compare earthquakes in different regions using this scale because of differences in construction practices, population density, and sociological factors.

Stone Canyon Reservoir

Epicentral distance 14 km, East-West component 372 cm / sec^2

31
00.000   10.000   20.000   30.000   40.000   50.000

◆▶ **FIGURE 4.7** Seismogram of the main shock of the Northridge earthquake, 17 January 1994. The time in minutes and seconds after 4:00 A.M. appears at the bottom. Distance from the epicenter and direction and ground acceleration at the recording site are also indicated. (Acceleration is explained in the next section.)

Richter Magnitude Scale—The Best-Known Scale

The best-known measure of earthquake strength is the **Richter magnitude scale,** which was introduced in 1935 by Charles Richter and Beno Gutenberg at the California Institute of Technology. It is a scale of the energy released by an earthquake and thus, in contrast to the intensity scale, may be used to compare earthquakes

◆▶ **FIGURE 4.8** Generalized graph of distance versus travel time for P- and S-waves. Note that the P-wave has traveled farther from the origin of the earthquake than the S-wave at any given elapsed time. The curvature of both wave paths is due to increases in velocity with depth (distance from epicenter).

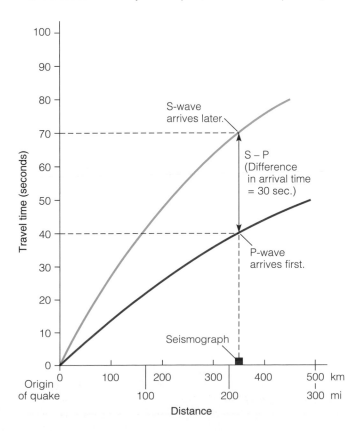

in widely separated geographic areas. It is calculated by measuring the maximum amplitude of the ground motion as shown on the seismogram using a specified seismic wave, usually the surface wave. Next, the seismologist "corrects" the measured amplitude (in microns) to what a "standard" seismograph would record at the station. After an additional correction for distance from the epicenter, the Richter **magnitude** is the common logarithm of that ground motion in microns. For example, a magnitude-4 earthquake is specified as having a corrected ground motion of 10,000 microns ($\log_{10}$ of 10,000 = 4) and thus can be compared to any other earthquake for which the same corrections have been made. It should be noted that because the scale is logarithmic, each whole number represents a ground shaking (at the seismograph site) 10 times greater than the next-lower number. Thus a magnitude-7 produces 10 times greater shaking than a magnitude-6; 100 times that of a magnitude-5; and 1,000 times that of a magnitude-4. Total energy released, on the other hand, varies logarithmically as some exponent of 30. Compared to the energy released by a magnitude-5 earthquake, a $M = 6$ releases 30 times (30^1) more energy; a $M = 7$ releases 900 times (30^2) more; and a $M = 8$ releases 27,000 (30^3) times more energy (◆▶ Figure 4.11a).

The Richter scale is open-ended; that is, theoretically it has no upper limit. However, rocks in nature do have a limited ability to store strain energy without rupturing, and no earthquake has been observed with a Richter magnitude greater than 8.9, yet.

Moment Magnitude—The Most Widely Used Scale

Seismologists have abandoned Richter magnitudes in favor of **moment magnitudes** (M_w or M) for describing earthquakes. The reason is that Richter magnitudes do not accurately portray the energy released by large earthquakes on faults with great rupture lengths. The seismic waves used to determine the Richter magnitude come from only a small part of the fault rupture

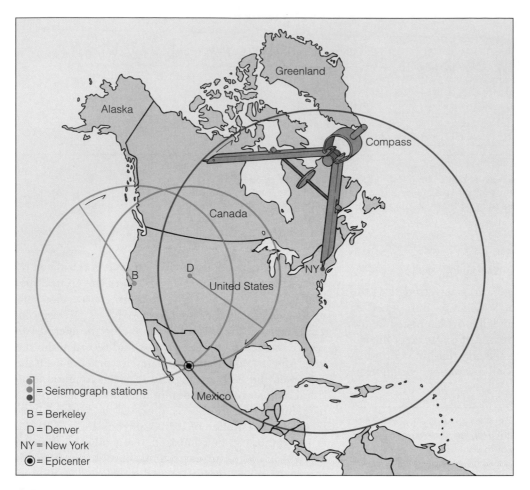

◁▶ **FIGURE 4.9** An earthquake epicenter can be closely approximated by triangulation from three seismic stations.

and, hence, cannot provide an accurate measure of the total seismic energy released by a very large event.

Moment magnitude is derived from seismic moment, M_0 (in dyne centimeters), which is proportional to the average displacement (slip) on the fault *times* the rupture area on the fault surface *times* the rigidity of the faulted rock. The amount of seismic energy (in ergs) released from the ruptured fault surface is linearly related to seismic moment by a simple factor, whereas Richter magnitude is logarithmically related to energy. Because of the linear relationship of seismic moment to energy released, the equivalent energy released by other natural and human-caused phenomena can be conveniently compared to earthquakes' moment magnitudes on a graph (◁▶ Figure 4.11b).

Consider This… ?   You are shopping for a house in earthquake country. How could an isoseismal map of an earthquake whose epicenter was near a house you think you would like to buy be useful? What patterns would you look for when examining the isoseismal map? ■

Moment magnitudes (M_w) are derived from seismic moments (M_0) by the formula: $M_w = (\tfrac{2}{3} \log M_0 - 10.7)$. This table compares the two most commonly used scales for selected significant earthquakes:

EARTHQUAKE	RICHTER MAGNITUDE	MOMENT MAGNITUDE
Chile, 1960	8.3	9.5
Alaska, 1964	8.4	9.2
New Madrid, 1812	8.7 (est.)	8.1 (est.)
Mexico City, 1985	8.1	8.1
San Francisco, 1906	8.3 (est.)	7.7
Loma Prieta, 1989	7.1	7.0
San Fernando, 1971	6.4	6.7
Northridge, 1994	6.4	6.7
Kobe, Japan, 1995	7.2 JMA*	6.9

*The Richter scale is not used in Japan. The official earthquake scale there is the scale developed by the Japanese Meteorological Agency (JMA).

◁▶ **FIGURE 4.10** Isoseismal map of the near-field (close to the epicenter) Modified Mercalli scale intensities of the Northridge earthquake. Intensity-9 values are assigned to locations where there are spectacular partial collapses of modern store buildings, destroyed wood-frame apartment buildings, and collapses of elevated freeways. Intensity-6 values are assigned to locations where there are some broken windows, a few instances of fallen plaster or damaged unreinforced masonry chimneys, and many items fallen from shelves.

At the close of the twentieth century the Chilean earthquake of 1960 ($M_w = 9.5$) still held the record for the greatest seismic moment and energy release ever measured; that is, it had the longest fault rupture and the greatest displacement. The Alaskan earthquake of 1964 had the highest Richter magnitude of the 1900s and theoretically more severe ground motion than the event in Chile.

FAULT CREEP, THE "NONEARTHQUAKE"

Some faults move almost continuously, or in short spurts, and do not produce detectable earthquakes. This type of movement, called **fault creep,** is well known but poorly understood. The Hayward fault, which is part of the San Andreas fault system, is an example of a creeping fault. It is just east of the San Andreas fault and runs through the cities of Hayward and Berkeley, California. (It's on the map in Figure 4.48). Creep along this fault causes displacements of millimeters per year, and is one way that this plate boundary accommodates motion between the Pacific and North American plates. An

alternative "accommodation" of plate motion on the San Andreas system is the storage of the strain energy and periodic release as a large displacement on a fault that generates a damaging earthquake. The obvious benefit of a creeping fault does not come without a price, as the residents of Hollister, California, can testify. The "creeping" Calaveras fault runs through their town and, as in Hayward, it gradually displaces curbs, sidewalks, and even residences (◁▶ Figure 4.12).

A good strategy in areas subject to fault creep is to map the surface trace of the fault precisely so that it can be avoided in future construction. Land adjacent to such known faults can be designated as limited-use areas. It may provide recreation space for local residents, for example.

An interesting lack of geological foresight is seen at the University of California–Berkeley Memorial Stadium. The huge structure was built on the Hayward fault before the hazard of fault creep was recognized, and creep-caused damage to the stadium's drainage system requires periodic maintenance. It is ironic that creep damage occurs at an institution that is highly regarded for seismological research and that installed the first seismometer in the United States.

Consider This... ? "Then the Lord rained upon Sodom and Gomorrha brimstone and fire. . . . And he overthrew those cities, and all the plain about . . ." (Genesis 19: 24–25). Sodom and Gomorrha are believed to have been located near the modern industrial city of Sedom in Israel at the south end of the Dead Sea. Could some geological event have caused the destruction of these towns? If so, what events might have been possible? ■

Seismic Design Considerations

GROUND SHAKING

Movements, particularly rapid back-and-forth displacements, are the real "killers" during an earthquake.

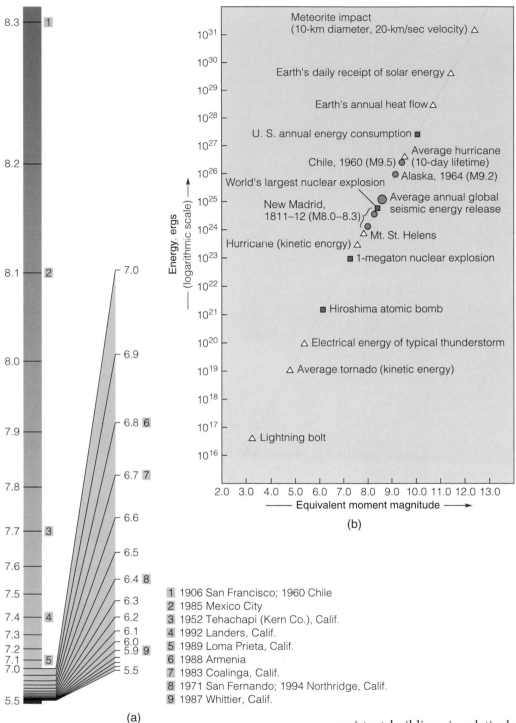

1   1906 San Francisco; 1960 Chile
2   1985 Mexico City
3   1952 Tehachapi (Kern Co.), Calif.
4   1992 Landers, Calif.
5   1989 Loma Prieta, Calif.
6   1988 Armenia
7   1983 Coalinga, Calif.
8   1971 San Fernando; 1994 Northridge, Calif.
9   1987 Whittier, Calif.

(a)

◀▶ **FIGURE 4.11** Earthquake magnitude and energy.
(a) Richter magnitudes of nine selected earthquakes,
1906–1994. (b) Equivalent moment magnitudes of energetic
human-caused events (*squares*), natural events (*triangles*),
and large earthquakes (*circles*). The erg is a unit of energy, or
work, in the metric (cgs) system. To lift a one-pound weight
one foot requires 1.4×10^7 ergs.

Shear and surface waves are the culprits, and the potential for them must be evaluated when establishing design
specifications. The design objective for earthquake-

resistant buildings is relatively straightforward: structures should be designed to withstand the maximum
potential horizontal ground acceleration expected in
the particular region. Engineers call this acceleration
base shear, and it is usually expressed as a percentage of
the *acceleration of gravity* (*g*). On earth, *g* is the acceleration of a falling object in a vacuum (9.8 m/sec^2, or 32
ft/sec^2). In your car it is equivalent to accelerating
from a dead stop through 100 meters in 4.5 seconds.
An acceleration of 1 *g* downward produces weightless-
ness, and a fraction of 1 *g* in the horizontal direction

◆ **FIGURE 4.12** Fault creep along the Hayward fault at Hayward, California. Creep of a few millimeters per year had resulted in about 18 centimeters (7 in) of offset by the time this photo was taken.

can cause buildings to separate from their foundations or to collapse completely. An analogy is to imagine rapidly pulling a carpet on which a person is standing; most assuredly the person will topple.

The effect of high horizontal acceleration on poorly constructed buildings is twofold. Flexible-frame structures may be deformed from cube-shaped to rhomb-shaped, or they may be knocked off their foundations (◆ Figure 4.13). More rigid multistory buildings may suffer "story shift" if floors and walls are not adequately tied together (◆ Figure 4.14). The result is a shifting of floor levels and the collapse of one floor upon another like a stack of pancakes. The terrible death toll (over 25,000 fatalities) in Armenia in December 1988 was largely due to this kind of collapse. There, prefabricated multistory buildings simply came apart because their design could not accommodate the shearing force (◆ Figure 4.15). Such structural failures are not survivable by inhabitants and they clearly illustrate the adage that "earthquakes don't kill people; buildings do."

Damage due to shearing forces can be mitigated by bolting frame houses to their foundation and by *shear walls*. An example of a shear wall is plywood sheeting nailed in place over a wood frame, which makes the structure highly resistant to deformation. Wall framing, usually two-by-fours, should be nailed very securely to a wooden sill that is bolted to the foundation. Diagonal bracing and blocking also provide shear resistance

(◆ Figure 4.16). L-shaped structures may suffer damage where they join, as each wing of the structure vibrates independently. Such damage can be minimized by designing *seismic joints* between the building wings or between adjacent buildings of different heights. These joints are filled with a compressible substance such as felt, plastic, or rubber that will accommodate movement between the structures (Figure 4.16).

Wave period is the time interval between arrivals of successive wave crests, or of equivalent points of waves, and it is expressed as T in seconds. It is an important consideration when assessing a structure's potential for seismic damage, because if a building's natural period of vibration is equal to that of seismic waves, a condition of resonance exists. **Resonance** occurs when a building sways in step with an oscillatory seismic wave. As a structure sways back and forth under resonant conditions, it gets a push in its direction of sway with the passage of each seismic wave. This causes the sway to increase, just as pushing a child's swing at the proper moments makes it go higher with each push. Resonance also may cause a wine glass to shatter when an operatic soprano sings just the right note or frequency.

Low-rise buildings have short natural wave periods (0.05–0.1 seconds), and high-rise buildings have long natural periods (1–2 seconds). Therefore, high-frequency (short-period) waves affect single-family dwellings and low-rise buildings, and long-period (low-frequency) waves affect tall structures. Close to an earthquake epicenter, high-frequency waves dominate, and thus more extensive home and low-rise damage can be expected. With distance from the epicenter, the short-period wave energy is absorbed or dissipated, resulting in the domination of longer-period waves.

LANDSLIDES

Thousands of landslides are triggered by earthquakes in mountainous or hilly terrain; there were an estimated 17,000 during the Northridge earthquake alone (see Case Study 4.1). The 1989 Loma Prieta earthquake caused landsliding in the Santa Cruz Mountains and adjacent parts of the California Coast Ranges. Such

Consider This... ? Over time tectonic stresses strain the rocks along a fault and increase the chance of sudden rupture and a potentially damaging earthquake. For this reason, it is generally believed that a region that has had many small earthquakes (magnitude <4.0) has a low probability for a large earthquake. *Question:* For a magnitude-7.0 earthquake with a 50-year recurrence interval, how many magnitude-4.0 earthquakes would be required to "balance the energy books"? How many small earthquakes a day is this? Is this belief realistic in a seismically active region based upon your calculations? ■

(a)

(b)

(c)

(d)

◆▶ **FIGURE 4.13** (a) Strong horizontal motion may deform a house from a cube to a rhomboid or knock it from its foundation completely. (b) A Coalinga, California, frame house that was deformed by the magnitude-6.3 earthquake in 1983. (c) A cripple-wall consists of short vertical members that connect the floor of the house to the foundation. Cripple-walls are common in older construction. (d) A Watsonville, California, house that was knocked off its cripple-wall base during the Loma Prieta earthquake of 1989. Without exception, cripple-walls bent or folded over to the north relative to their foundations.

◁▶ FIGURE 4.14 (Total vertical collapse as a result of "story shift"; Mexico City, 1985. Such structural failures are not survivable.

◁▶ FIGURE 4.15 (Wreckage of an older three-story building in Leninakan, Armenia, 1988. A collapsed nine-story building is at the right. Membrane (floor) failures in precast-concrete-frame structures were common here; 132 of these buildings "pancaked," leaving little space for occupant survival.

areas are slide prone under the best of conditions, and even a small earthquake will trigger many slope failures. In the greater San Francisco Bay area there was an estimated $10 million damage to homes, utilities, and transportation systems because of landslides and surficial ground failures resulting from the Loma Prieta earthquake.

One of the worst earthquake-triggered tragedies occurred in Peru in 1970. An earthquake-initiated avalanche consisting of a mixture of snow, ice, and rock gave way at an elevation of 6,000 meters (20,000 ft) on the western slopes of Nevado Huascarán. A kilometer wide and 1.5 kilometers long, it accumulated water as it plummeted downhill at more than 160 kilometers per hour (100 mph), jumping a natural barrier, and burying 25,000 residents and the villages of Yungay and Ranrahirca. The inundated areas remain completely covered with mud and boulders, leaving little indication of what lies below (◁▶ Figure 4.17).

GROUND OR FOUNDATION FAILURE

Liquefaction is the sudden loss of strength of water-saturated, sandy soils resulting from shaking during an earthquake. Sometimes called **spontaneous liquefaction,** it can cause large ground cracks to open, lending support to the ancient myth that the earth opens up to swallow people and animals during earthquakes. Shaking can cause saturated sands to consolidate and thus to occupy a smaller volume. If the water is slow in draining from the consolidated material, the overlying soil comes to be supported only by pore water, which has no resistance to shearing. This may cause buildings to settle, earth dams to fail, and sand below the ground surface to blow out through openings at the surface much like small volcanoes (◁▶ Figure 4.18). Liquefac-

tion at shallow depth may result in extensive lateral movement or spreading of the ground, leaving great cracks and openings.

Ground areas most susceptible to liquefaction are those that are underlain at shallow depth—usually less than 30 feet—by layers of water-saturated fine sand. With subsurface geologic data obtained from water wells and foundation borings, liquefaction-susceptibility maps have been prepared for many seismically active areas in the United States.

Similar failures occur in certain clays that lose their strength when they are shaken or remolded. Such clays are called *quick clays* and they are natural aggregations of fine-grained clays and water. They have the peculiar property of turning from a solid (actually a gel-like state) to a liquid when they are agitated by an earthquake, an explosion, or even vibrations from pile driving. They occur in deposits of glacial-marine or glacial-lake origin and are therefore found mostly in northern latitudes, particularly in Scandinavia, Canada, and the New England states. Failure of quick clays underlying Anchorage, Alaska, produced extensive lateral spreading throughout the city in the 1964 earthquake (◁▶ Figure 4.19).

The physics of failure in spontaneous liquefaction and in quick clays is similar. When the earth materials are water-saturated and the earth shakes, the loosely packed sand consolidates or the clay collapses like a house of cards. The pore-water pressure pushing the grains apart becomes greater than the grain-to-grain friction, and the material becomes "quick" or "liquefies" (◁▶ Figure 4.20). The potential for such geologic

Case Study | 4.1

Earthquakes, Landslides, and Disease

The January 1994 Northridge earthquake caused 17,000 landslides over an area of about 10,000 square kilometers (4,000 mi²), according to E. L. Harp and Randall Jibson of the U.S. Geological Survey (◇ Figure 1). The ground failures occurred up to 70 kilometers (43 mi) from the epicenter, with the slides and falls concentrated in the dry, weakly cemented Neogene sedimentary rocks of the Santa Susana Mountains. Most failures were thin, less than 5 meters (16 ft) thick, although failures more than 10 meters (32 ft) thick were not uncommon. The largest landslide involved a volume greater than 200,000 cubic meters. The Santa Susana Mountains are near a plate-tectonic boundary (the San Andreas fault) and are being rapidly uplifted and deeply eroded by running water. The soft Neogene marine and nonmarine sediments there are among the most susceptible in the world to seismically induced ground failures. The toes of some of the landslides advanced as much as 200 meters (650 ft) from their point of origin, a condition that could be disastrous if it occurred in developed areas of the Santa Susana Mountains.

Dynamically induced (caused by earthquake shaking) landslides are not particularly newsworthy, but these slides caused an outbreak of coccidiodomycosis (CM), commonly known as "valley fever." Endemic to the desert Southwest, the disease causes flulike symptoms and, in severe cases, can be fatal. It is caused by the victim inhaling airborne *Coccidioides immitis* spores that reside in the top 20 centimeters of the soil. From January 24 to March 15, 166 people were diagnosed with valley fever symptoms in Ventura County, up from only 53 cases in all of 1993. Most of the cases were reported from the Simi Valley, an area in which only 14 percent of the county's population resides (◇ Figure 2).

Large clouds of dust hung over the Santa Susana Mountains for several days after the earthquake, promoted by the lack of winter rains preceding the quake. During this period, pressure-gradient winds known locally as "Santa Ana"

◇ **FIGURE 1** Thousands of landslides in the Santa Susana Mountains triggered by the Northridge earthquake created a tremendous dust cloud that remained over the mountains for several days.

winds of 10–15 knots (11–17 mph) blew into the Simi Valley, carrying in spore-laden dust from the Santa Susana Mountains to the northeast. Researchers believe that the more "competent" metamorphic rocks of the San Gabriel Mountains northeast of the San Fernando Valley probably account for the lack of cases reported in the epicentral region.

The U.S. Center for Disease Control and Prevention in Atlanta, Georgia, concluded that there is epidemiological, geological, and meteorological evidence to establish a direct association between the Northridge earthquake and the outbreak of CM in the Simi Valley. Earthquake ground motion triggered many landslides, which generated the spore-laden dust that was carried into the Simi Valley. This is the first report of an earthquake-associated outbreak of valley fever, even though many earthquakes have occurred in CM endemic areas.

◇ **FIGURE 2** Histogram of the almost-epidemic outbreak of coccidioidomycosis (valley fever) in Ventura County in early 1994 following the Northridge earthquake. The affected area was downwind from the dust clouds shown in Figure 1.

(a)   (b)   (c)

◆▶ **FIGURE 4.16** Methods of reinforcing structures against base shear. (a) Diagonal cross-members and blocks resist horizontal earthquake motion (shear). (b) Plywood sheeting forms a competent shear wall, and metal "L" braces and bolts tie the structure to the foundation. (c) Seismic joint between two wings of a classroom building; San Bernardino Valley College, California. The joint is the vertical, dark path in the center of the photo. The active San Jacinto fault runs beneath the campus of this community college.

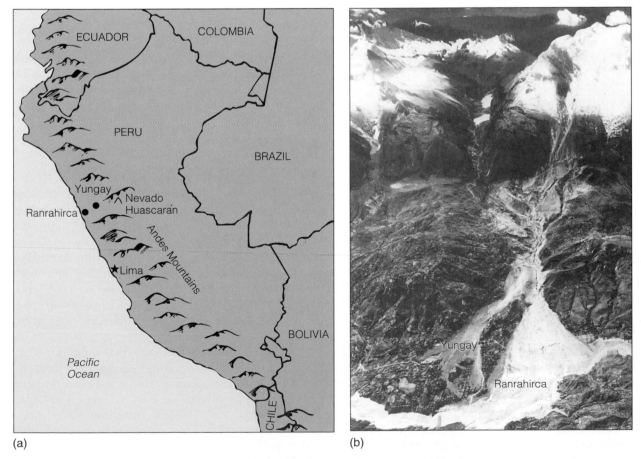

(a)   (b)

◆▶ **FIGURE 4.17** (a) and (b). A 1970 earthquake triggered a rock and snow avalanche down the slopes of Nevado Huascarán in Peru that buried the towns of Yungay and Ranrahirca. The depth of mud at Yungay was as much as 4.6 meters (15 ft).

(a)

(b)

◆▶ **FIGURE 4.18** (a) These apartment buildings tilted as a result of soil liquefaction in Niigata, Japan, in 1964. Many residents of the building in the center exited by walking down the side of the structure. (b) "Sand blows" in a lettuce field near Castroville, California, caused by liquefaction during the Loma Prieta earthquake.

conditions is not easily recognized. In many cases it can be determined only by information gained from bore holes drilled deeper than 30 meters (100 ft). Because of this expense, many site investigations do not include deep drilling, and the condition can be unsuspected until an earthquake occurs.

GROUND RUPTURE AND CHANGES IN GROUND LEVEL

Structures that straddle an active fault may be destroyed by actual ground shifting and the formation of a fault scarp (◆▶ Figure 4.21a). By excavating trenches across fault zones, geologists can usually locate past rupture surfaces that may reactivate. In 1970 the California legislature enacted the Alquist–Priolo Special Study Zones Act, which was renamed the *Earthquake Fault Zones Act* in 1995. It mandates that all known active faults in the state be accurately mapped and zoned for seismic

◀▶ **FIGURE 4.19** "House of cards" collapse of quick clay structure in Turnagain Heights, Anchorage, Alaska. Total destruction occurred within the slide area, which is now called Earthquake Park.

safety. The act provides funds for state and private geologists to locate the youngest fault ruptures within a zone and requires city and county governments to limit land use adjacent to identified faults within their jurisdictions. Ironically, the faults responsible for the 1992 Landers earthquake ($M_w = 7.3$) were designated as Earthquake Fault Zones just prior to the June 28 event.

Changes in ground level as a result of faulting may have an impact, particularly in coastal areas that are uplifted or down-dropped. For instance, during the 1964 Alaskan event, parts of the Gulf of Alaska thrust upward 11 meters (36 ft), exposing vast tracts of former tidelands on island and mainland coasts (◆▶ Figure 4.21b).

FIRES

Fires caused by ruptured gas mains or fallen electric power lines can add considerably to the damage caused by an earthquake. In fact, most damage attributed to the San Francisco earthquake of 1906 and much of that in Kobe, Japan, in 1995 was due to the uncontrolled fires that followed the earthquakes. The Kobe quake hit at breakfast time. In neighborhoods crowded with wooden structures, fires erupted when natural-gas lines broke and falling debris tipped over kerosene stoves. Broken water mains made fire-fighting efforts futile.

Many gas pipelines that cross active faults in California today are above ground in order to facilitate repair in case of fault rupture. Fire is always a threat in urban areas after an earthquake (◆▶ Figure 4.22). One of the principal "do's" for citizens immediately after a quake is to shut off the gas supply to homes and other buildings in order to prevent gas leaks into the structure from damaged lines. This in itself will save many lives (see Figure 4.51).

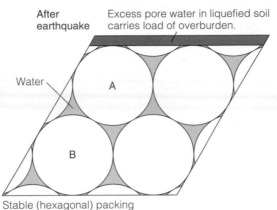

Before earthquake   Water

After earthquake   Excess pore water in liquefied soil carries load of overburden.

Water

A

B

Unstable (cubic) packing

A

B

Stable (hexagonal) packing

◁▶ **FIGURE 4.20** Liquefaction (lateral spreading) due to repacking of spheres (idealized grains of sand) during an earthquake. The earthquake's shaking causes the solids to become packed more efficiently and thus to occupy less volume. A part of the overburden load is supported by water, which has no resistance to lateral motion.

TSUNAMIS

The most myth-ridden hazard associated with earthquakes is **tsunamis** (pronounced "soo-nah´-meez"), or seismic sea waves. *Tsunami* is a Japanese word meaning "great wave in harbor" and it is appropriate, because these waves most commonly wreak death and destruction along shorelines in bays and harbors. The Japanese record of tsunamis goes back 2,000 years, and their power is dramatically displayed in the well-known print by Hokusai (◁▶ Figure 4.23). Tsunamis are impulsively generated waves that are produced most commonly by underwater earthquakes and less commonly by volcanic eruptions, submarine landslides, or meteorite impacts. These impulses give rise to waves that move out from their source in all directions—analogous to those produced when you drop a rock into a pond. Tsunamis in deep water may have velocities exceeding 800 kilometers per hour (480 mph) and they could easily keep pace with a Boeing 747. With wavelengths in excess of

(a)

(b)

◁▶ **FIGURE 4.21** (a) The 1992 Landers earthquake ($M_w = 7.3$) in the Colorado Desert of southern California was felt from Phoenix, Arizona, to Reno, Nevada. Right-lateral offset of 4.27 meters (14 ft) on the Johnson Valley fault created a 2-meter (6.5-ft) vertical scarp due to lateral offset of a ridge. Providing a scale for this photograph, the geologist is 185 cm tall. (b) Fault scarp on Montague Island in Prince William Sound, Alaska. In places, the scarp is 4.8 meters (16 ft) high between the white (shell-covered) uplifted sea floor (right) and the landward area.

◇▶ **FIGURE 4.22** Fire in San Francisco's Marina district after the 1989 earthquake. Liquefaction caused gas mains to break, and fire ensued. The structure at the far right collapsed due to liquefaction and foundation failure.

◇▶ **FIGURE 4.23** "The Breaking Wave off Kanagawa," wood-block color print by Katsushika Hokusai (1760–1849) from the series "Thirty-six Views of Mount Fuji," 1826–33.

160 kilometers (100 mi) and wave heights of only a meter or two in the open ocean, however, tsunamis have such gentle slopes that they go unnoticed (◆ Figure 4.24). As the leading waves move into shallow water they slow to freeway speeds, and the waves behind begin to overtake them—a process called *shoaling* (see Chapter 10). Thus, the energy contained in many waves is condensed into a smaller volume of water, and the waves increase in height and steepness. Eventually, the wave runs ashore as a huge breaking wave, as a wall of white, foamy water, or as a tidelike flood.

Earthquake-induced tsunami (magnitude >7) may rise to heights of more than 30 meters (about 100 ft), like one did in 1993 at Okushiri, Japan, killing 239 people. The time between crests of the incoming waves is about 20 minutes, and the first crest may not be the largest. The *trough* of the initial wave may arrive first, and many lives have been lost by curious persons who wandered into tidal regions exposed as water withdrew. In Lisbon, Portugal, on All Soul's Day in 1755, many people were in church when an earthquake struck. The worshipers ran outside to escape falling debris and fire, many of them joining a group seeking safety on the waterfront. There was a quiet withdrawal of water followed by a huge wave minutes later. Sixty thousand people died. The length of time of strong shaking has been estimated from historical accounts of how many *Ave Marias* and *Pater Nosters* the churchgoers recited during their fearful experience.

Water withdrawal before the first high wave also contributed to fatalities in the 1946 Hawaiian tsunami; many people walked offshore to collect exposed mollusks just before the first wave crest arrived. A good rule of thumb is to head for high ground at once should you witness a sudden recession of water along a beach or coastline.

Tsunamis are most common within the Pacific Ocean, where they are generated mostly by large-scale dip-slip faulting in subduction zones that ring the basin.

Japanese studies have shown that strike-slip faulting seldom produces tsunamis. In the late twentieth century earthquake-generated tsunami around the Pacific Rim damaged Chile (1960), Alaska and the Pacific Northwest (1964), Nicaragua (1992), and New Guinea (1998). The Hawaiian Islands are very tsunami-prone due to multiple tsunami source areas. Between 1850 and 2000 the Islands averaged a damaging tsunami every two years (◆ Figure 4.25). A tsunami's damage potential depends upon the earthquake mechanism, the distance the waves travel to the shoreline, the offshore topography, and configuration of the coastline. In 1998 a community on a low, coastal sandbar in Papua, New Guinea, was obliterated by tsunami waves of up to 15 meters (50 ft) high that killed 2,200 people and left few survivors. Experts were puzzled, because the magnitude-7.1 earthquake was considered too small to generate such huge waves. This led to speculation that the earthquake triggered a submarine landslide or an explosion of methane hydrates (see the Chapter 14 opener) that added strength to the tsunami.

In 1992, twenty-six towns along 250 kilometers (150 mi) of Nicaragua's Pacific coast were struck by a 10-meter-high (33 ft) wave that killed 170 people and left 13,000 homeless (◆ Figure 4.26). Normally, coastal inhabitants run for high ground when they feel rapid (short-period, <1 second) earthquake shaking. Only slight tremors were experienced in this case, however, and people did not suspect danger. The high-frequency seismic waves generated at the earthquake focus were weak and they were barely felt onshore. Most of the earthquake energy was transmitted in much longer waves (long-period, up to 200 seconds) that were not noticed or were ignored. This kind of quake, known as a "slow" or "silent" earthquake, was first described by Hiroo Kanamori of the California Institute of Technology. Kanamori believes these quakes may be ten times as strong as the magnitude determined from short-period seismometer readings. Slow earthquakes of

◆ **FIGURE 4.24** Evolution of a tsunami from deep to shallow water. (a) In deep water, the tsunami wave has a low height, a very long length, and a high velocity. (b) The wave moves out in all directions from an area of uplift. This example is of the 1964 Gulf of Alaska tsunami. (c) Upon arriving in shallow water, the wave's length decreases and its height increases until eventually the wave breaks and runs up onto the land, where it causes damage.

(a)

(b)

(c)

◇▶ **FIGURE 4.25** Tsunami sources and travel paths to the Hawaiian Islands, 1900–1983.

local origin are particularly dangerous in coastal areas, because they produce little precursory ground shaking that serves to warn people that a tsunami may be imminent. Scientists believe that up to ten percent of all tsunamis are caused by this kind of earthquake. Certainly, the earthquakes that produced both the New Guinea and the Nicaragua tsunamis were considered slower than normal.

◇▶ **FIGURE 4.27** U.S. earthquake zones capable of generating tsunami are the Alaska–Aleutian seismic zone and the Cascadia subduction zone. The five Pacific states are subject to tsunami. The insert shows a tsunami detector that rests on the sea floor. It transmits acoustic signals to a surface buoy, which relays signals to a NOAA satellite, and then to ground-based warning centers.

◇▶ **FIGURE 4.26** Damage at El Transisto, Nicaragua, the area most impacted by the September 1992 tsunami. Most of the 200 homes in El Transisto were destroyed, and 16 people, mostly children, lost their lives. The first wave was relatively small, but it provided a warning. The able-bodied ran to higher ground, where they escaped the force of the powerful third wave.

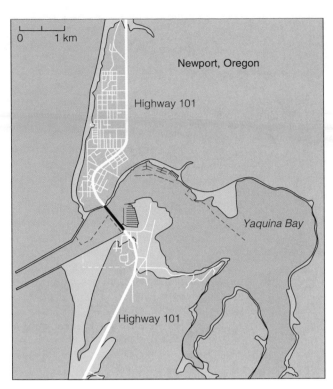

◁▷ **FIGURE 4.28** Potential tsunami inundation areas (green) at Newport, Oregon.

The entire Pacific Rim is vulnerable to tsunami inundation. Geologic evidence indicates that within the past 2,000 years several large earthquakes have generated destructive tsunami on the shores of Washington, Oregon, and Northern California. The risk becomes clearer when we recognize that 90 percent of the deaths in the Alaska earthquake of 1964 were tsunami related, and that the geology of the Cascadia subduction zone is very similar to the Alaska coast (◁▷ Figure 4.27). The best defense against tsunami risk is an early warning system. Useful (life-saving) warning times for the half-million people who live and work in potential tsunami inundation areas of the Pacific Northwest range from a few minutes in the southern part of the Cascadia subduction zone to 20–40 minutes in the northern parts of the zone.

In April 1992, a magnitude-7.1 earthquake generated a small tsunami at Cape Mendocino, California. This event triggered establishment of the National Tsunami Hazard Mitigation Program, involving the five Pacific states and three federal agencies—the U.S. Geological Survey, NOAA, and FEMA. By the late 1990s inundation maps were being prepared for parts of Oregon and Washington, and public-education programs had been implemented (◁▷ Figure 4.28). Most important is the development of a warning system based upon real-time tsunami detection. The current Pacific-wide warning system is based on earthquake foci, and about an hour is required to issue a warning.

It is therefore useful only for coastal areas more than 750 kilometers (465 mi) from the tsunami source, the approximate distance a tsunami could travel in that amount of time. Regional systems warn in about ten minutes, making them useful within 100–750 kilometers (62–465 mi) of the source, and local systems warn in about five minutes, which is useful <100 kilometers (62 mi) from the source. However, not all earthquakes generate tsunamis, and 15 of the 20 warnings issued between 1946 and 2000 were false alarms. This has led to public distrust of early earthquake-based warning systems (Case Study 4.2).

The goal now is to place real-time deep-ocean tsunami detectors on the sea floor that transmit data acoustically to a floating buoy, which relays the information to a satellite that, in turn, transmits it to a ground station (see Figure 4.27). The sensitive instruments can detect tsunami as small as 3 centimeters (1.17 in) high and transmit the data to the ground station in three minutes. For nearby tsunami sources, three minutes from detection to an alert is a life-saving warning time. Six instruments were in operation by 1998, and 38 instruments will eventually be installed and maintained in the Pacific Ocean. Lest you think that only the Pacific Northwest, Hawaii, and Alaska are at risk, tsunami are purported to have struck Santa Barbara, California, in 1812 and Santa Monica, California, in 1930.

Some Major Earthquakes, 1989–1999

Each year in the world, on average, there are at least two earthquakes of magnitude 8.0 or greater and 20 earthquakes in the magnitude 7.0 to 7.9 range. Release of seismic energy in the form of earthquakes has occurred throughout geologic time, and recorded history contains many references to strong earthquakes. More than 3,000 years of seismicity is documented in China, and Strabo's *Geography* mentions an earthquake in 373 B.C. in Greece. Thus, humans have probably always been subject to earthquakes.

This section presents case histories of four late-twentieth-century world-class earthquakes that had significantly different impacts on communities and their inhabitants. They are Loma Prieta, California ($M_w = 6.9$); Northridge, California ($M_w = 6.7$); Kobe, Japan ($M_w = 6.9$); and Izmit, Turkey ($M_w = 7.8$). All of them occurred near dense population centers and resulted in deaths and costly property damage. Some comparisons are useful here. For example, compared to the energy released by the Loma Prieta earthquake:

■ the Kobe earthquake released the same amount, but it did 12 times the dollar-equivalent damage;

Case Study | 4.2

Distant Tsunamis—The Silent Threat

mple warning had been given the residents of Hawaii as the Chilean tsunami raced across the Pacific Ocean at jetliner speeds. A warning had been issued about 6:45 P.M. 22 May 1960 that large waves were expected to reach Hilo on the Island of Hawaii about midnight. Coastal warning sirens had wailed at 8:30 P.M. and continued for half an hour.

At midnight a wave arrived, but it was only a few feet high. Hundreds of people had stayed home, and those that had left assumed the danger was over and returned home. The highest wave struck at 1:04 A.M. Sixty-one people died, and another 282 suffered severe injuries (◁▶ Figure 1). Jerry Eaton of the U.S. Geological Survey, who was monitoring water height at the Wailuku Bridge (◁▶ Figure 2), described the approaching wave as "an ominous noise, a faint rumble like a distant train, that came from the darkness far out in Hilo Bay." A restaurant memorialized the event with a waterline on its window (◁▶ Figure 3).

The city of Hilo, with its bay facing northeast toward the Aleutian trench, is a coastal section with high expected wave heights. This is due to the coastline's orientation and the bay's funnel shape, which concentrates tsunami energy. An engineering study in the aftermath of the 1960 tsunami resulted in some recommendations for mitigating tsunami damage in that city. Parking meters in the runup area were bent flat in the direction of wave travel (◁▶ Figure 4), with the direction varying throughout the affected area. It was recommended that new structures be situated with their narrowest dimension aligned in the direction indicated by nearby bent parking meters. Several waterfront structures had open fronts that allowed water to pass through them, demolishing the solid bearing walls at the rear holding up the roof. Waterfront buildings are now designed with open parking or glassed-in space on ground level, so that much damage from water impact can be avoided as wave runup flows through the open part of the structure.

◁▶ FIGURE 1 Water-level measurements beneath the Wailuku River bridge during the first hours of the tsunami of 23 May 1960.

◆ FIGURE 2 Area inundated by the tsunami. There were 61 fatalities in this small area. Banyan Drive, the horseshoe-shaped street that projects seaward from the electric plant, is the location of the city's resort hotels and restaurants, which were completely swamped.

◆ FIGURE 3 A line painted on the window of a rebuilt Banyan Drive restaurant commemorates the height to which the tsunami waters rose.

◆ FIGURE 4 Parking meters bent by the 1960 tsunami at Hilo, Hawaii. The meters resemble arrows aligned with the wave runup.

○ **TABLE 4.1** Selected Significant* World Earthquakes, in Chronological Order			
Location	**Year**	**Richter Magnitude****	**Impact**
San Francisco	1906	8.3	700 killed, $7 million damage, fire
Messina (Sicily)	1908	7.5	160,000 killed
Tokyo, Japan	1923	8.3	140,000+ killed, fire
Assam, India	1950	8.4	30,000 killed
Chile	1960	M_w 9.5	5,700 killed, 58,000 homes destroyed, tsunami
Alaska	1964	M_w 9.2	131 killed, tsunami
Guatemala	1976	7.5	23,000 killed
T'ang-shan, China	1976	7.9	330,000 killed
Mexico City	1985	8.1	10,000+ reported killed
Armenia	1988	6.8	55,000 killed
Loma Prieta (California)	1989	M_w 6.9	67 killed, $8 billion damage
Northern Iran	1990	7.7	40,000 killed
Maharashtra, India	1993	M_w 6.4	30,000+ killed, widespread collapse
Northridge, California	1994	M_w 6.7	60 killed, $25 billion damage
Neftegorsk, Russia	1995	M_w 7.5	2,000+ killed, widespread collapse
Kobe, Japan	1995	M_w 6.9	5,378 killed, $100 billion damage
Izmit (Kocaeli), Turkey	1999	M_w 7.4	15,370 killed, $10–$20 billion damage
Chi-Chi, Taiwan	1999	M_w 7.6	2,300+ killed
Hector Mine, California	1999	M_w 7.1	No deaths, little damage

*A "significant" earthquake is defined as one that registers a moment magnitude (M_w) of at least 6.5 or a lesser one that causes considerable damage or loss of life. The world averages 60 significant earthquakes per year.
**Estimated prior to 1935. M_w = Moment magnitude.
SOURCES: U.S. Geological Survey: *Earthquakes and Volcanoes;* California Division of Mines and Geology; *Academic American Encyclopedia* (1990, Grolier Electronic Publishing Company).

- the Northridge event released half as much energy, but it did 3 times the dollar damage;
- the Izmit earthquake released 6 times as much energy, and more than 220 times as many people died (○ Table 4.1).

The case histories show that earthquake magnitude is not the best measure of an earthquake's impacts on humans. This may be best illustrated by the Hector Mine earthquake in California on 16 October 1999. It was an $M_w = 7.1$—potentially a damaging, life-threatening earthquake—but it occurred in the Mojave Desert, miles from anywhere. Damage was limited to little more than a derailed Amtrak passenger train.

LOMA PRIETA, 1989

The Loma Prieta earthquake of 17 October 1989, a magnitude-6.7 event, was certainly the most thoroughly televised earthquake to date. It occurred at 5:04 P.M. (PDT), just prior to the third game of the World Series in San Francisco, canceling the game and leaving a hundred million television fans bewildered when their screens went blank. Strong motion lasted only 15 seconds, but the quake caused 67 deaths and more than $8 billion in property damage in the heavily populated San Francisco Bay area.

The earthquake occurred when a 25-mile-long segment of the San Andreas fault ruptured beneath Loma Prieta in the Santa Cruz Mountains, moving 6 feet (2 m) horizontally and 4 feet (1.3 m) vertically (◆ Figure 4.29). Prophetically, this segment had been designated by the U.S. Geological Survey in 1988 as the segment of the northern San Andreas fault most likely to rupture within the next 30 years. The basis of the prediction was the existence of a **seismic gap** along the San Andreas fault in the southern Santa Cruz Mountains (◆ Figure 4.30). Seismic gaps are segments on active faults where there has been little activity in the past and the fault is "stuck," thereby accumulating strain energy. After the Loma Prieta earthquake "filled the gap" in the fault's history, seismologists predicted that the next event would occur farther north along the fault. There was no primary surface rupture from faulting; rather, a

(a)

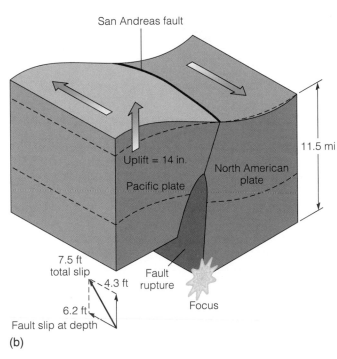

(b)

◆ **FIGURE 4.29** (a) Mainshock and aftershocks of the Loma Prieta earthquake in relation to the surface trace of the San Andreas fault. Size of the symbols denotes the magnitudes of the aftershocks. (b) Inferred slip along the San Andreas fault during the Loma Prieta earthquake. The Pacific plate moved upward and northward relative to the North American plate.

wide zone of large ground cracks occurred parallel to the fault, causing damage to houses and roads.

Most of the damage and loss of life were caused by structural failures due to local amplification of ground shaking or liquefaction in soft sediments or artificial fill surrounding San Francisco Bay. The 30-year-old pillars supporting the upper deck of Oakland's Cypress Street

viaduct, part of Interstate 880, splayed outward, collapsing 2 kilometers (1¼ mi) of the top deck onto the vehicles below (◆ Figure 4.31). Forty-two rush-hour commuters were killed, and 108 were injured in this most deadly disaster of the quake. Although the double-deck structure contained 9,000 tons of steel reinforcing bars, it could not resist the highly amplified 2-meter horizontal and 1-meter vertical motion, and the columns simply collapsed.

The area of most serious damage was probably the Marina district of San Francisco, which is underlain by natural and artificial deposits of sand and mud that liquefied during strong shaking (◆ Figure 4.32). This liquefaction resulted in lateral spreading and opening of ground cracks, which in turn caused tilting and collapse of structures. Many of the failures were in the same general locations as those of the great quake of 1906. These failures in deposits around the edge of San Francisco Bay illustrate that bay muds and poorly engineered landfill, such as those in the Marina district and the Kobe waterfront, can fail at much lower levels of shaking than firmer geologic foundation material (◆ Figure 4.33).

NORTHRIDGE, CALIFORNIA, 1994

The largest earthquake in Los Angeles's short history occurred at 4:30 A.M. Monday, 17 January 1994, on a hidden fault below the San Fernando Valley. The $M_w = 6.7$ earthquake started at a depth of 18 kilometers (11 mi) and propagated upward in a matter of seconds to a depth of 5–8 kilometers (3–5 mi, ◆ Figure 4.34). There were thousands of aftershocks, and their clustered pattern indicated the causative fault to be a reverse fault with a shallow dip of 35° to the south (Case Study 4.3 and Appendix 3). Such a low-angle reverse fault is called a *thrust fault*, and, because the thrust did not rupture the ground surface, it is described as "blind" (◆ Figure 4.35). Blind thrusts were first recognized after the 1987 Whittier earthquake 40 kilometers (25 mi) southwest of Northridge, but they were not fully appreciated until after the Northridge event.

Blind thrust faults were a newly identified seismic hazard, and there is evidence that a belt of them underlies the northern Los Angeles basin. They are especially dangerous because they cannot be detected by traditional techniques such as trenching and field mapping (thus they do not fall within the Earthquake Fault Zones Act as it is written) and yet, as we know now, they can generate significant earthquakes. Just as the residents of earthquake country were beginning to feel confident that geologists knew the location and behavior of most active faults, blind thrusts made their presence known.

Case Study | 4.3

Predictable "Future Shocks"

A large earthquake is normally followed by thousands of smaller-magnitude earthquakes known as aftershocks (Figure 1). If the main shock is small, say in the magnitude-4.0–5.0 range, aftershocks are small and nonintimidating. Following Northridge-size and bigger earthquakes, however, the strongest aftershocks ($M = 5.0–6.0$ at Northridge) can cause buildings damaged by the main earthquake to collapse and, even worse, can greatly increase anxiety in the already damaged psyche of the local citizenry. Aftershocks are caused by small adjustments (slips) on the causative fault or on other faults close to the causative one. For example, try this: push the eraser on the end of a lead pencil across a desktop. You'll find that it does not slide smoothly; it moves in jerky jumps and starts. This is called "stick–slip" and it is what happens along faults that are adjusting after a big earthquake.

Los Angeles experienced 2,500 aftershocks in the week following the Northridge earthquake. Three strong ones, magnitude 5 or greater, occurred the first day. The largest ($M = 5.6$) occurred eleven hours after the main shock, causing concern among rescuers digging for victims beneath the rubble and the already traumatized citizens. Aftershocks follow statistically predictable patterns, as exemplified by the Northridge sequence. On the first day there were 188 aftershocks of magnitude 3 or greater, and on the second day

◁▶ **FIGURE 1** Seismogram of early aftershocks following the Northridge earthquake, one of them a magnitude-5.6.

only 56 were recorded (◁▶Figure 2). By fitting an equation to the distribution of aftershocks during the first few weeks, seismologists were able to estimate the number of shocks to be expected in the future. Statistically, there was a 25 percent chance of another magnitude-5 or greater aftershock occurring within the following year, but it did not happen.

◁▶ **FIGURE 2** Daily record of aftershocks of magnitude 3.0 to 5.9 during the three weeks following the main shock at Northridge. Note the sharp drop in aftershock frequency in the first four days.

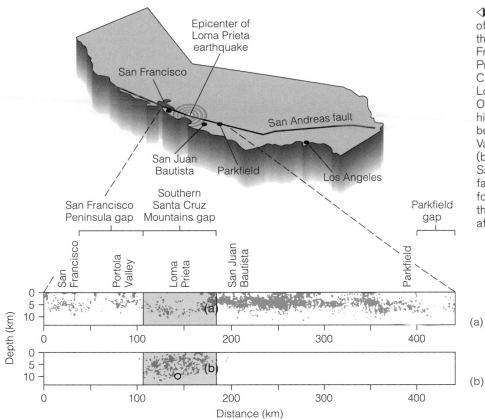

◆▶ **FIGURE 4.30** (a) Historic seismic activity of about 450 kilometers (275 mi) of the San Andreas fault south of San Francisco before the 1989 Loma Prieta earthquake. The southern Santa Cruz Mountains seismic gap, where Loma Prieta is located, is highlighted. Other seismic gaps apparent in this historical record at the time were one between San Francisco and the Portola Valley and one southeast of Parkfield. (b) Seismicity record of the Southern Santa Cruz Mountains portion of the fault after the 1989 earthquake. The former seismic gap had been "filled" by the earth quake (*open circle*) and its aftershocks.

Field inspection of the epicentral region was a depressing experience. Thirteen thousand buildings were found to be severely damaged; 21,000 dwelling units had to be ordered evacuated; 240 mobile homes had been destroyed by fire; 11 major freeway overpasses were damaged at 8 locations (Case Study 4.4). The reason for the extensive damage was the high horizontal and vertical accelerations generated by the earthquake. Accelerations of more than 0.30 *g* are considered dangerous, and the vertical acceleration of 1.8 *g* measured by an instrument bolted to bedrock in nearby Tarzana probably set a world record. The high ground acceleration explains why many people were literally thrown out of their beds

◆▶ **FIGURE 4.32** Life goes on amid damage due to ground failure in the Marina district of San Francisco, 17 October 1989.

◆▶ **FIGURE 4.31** Failure of columns supporting the double-decked Cypress Street viaduct (I-880) caused the upper deck to fall onto the lower one.

◆ FIGURE 4.33 Seismograms for a magnitude-4.1 aftershock as recorded on (a) firm bedrock, (b) alluvium (stream-deposited sediment), and (c) areas of fill and bay muds, where the most damage occurred in the San Francisco earthquakes of 1906 and 1989.

and objects as heavy as television sets were projected several meters from their stands. Although the damage was not as visually spectacular as that at Mexico City in 1985, it was heartbreaking to many residents—brick chimneys, both reinforced with steel bars and unreinforced, came down; houses moved off their foundations; concrete-block walls crumbled; gaps broke open in plaster walls; the number of shattered storefront windows was beyond counting (◆ Figure 4.36). The horizontal ground motion was directional; that is, it was strongest in the north–south direction. With few exceptions, block walls oriented east–west tipped over or fell apart, whereas those oriented north–south remained standing. Many two- and three-story apartment buildings built over open first-floor garages collapsed onto the residents' cars. These open parking areas' lack of shear resistance led to failure of the vertical supports, and everything above came down.

California State University at Northridge sustained almost "textbook" earthquake damage. In its library is "Leviathan II," recognized as one of the most advanced automated book-withdrawal systems in the country. On this day it recorded a record withdrawal of about 500,000 books—all of them onto the floor of the library. Bottled chemicals fell off their storage shelves (a major concern at any university) and caused a large fire in the Chemistry Building. In addition, one wall of a large open parking structure collapsed inward to produce the stunning art deco architecture seen in (◆ Figure 4.36a).

KOBE, JAPAN, 1995

Disaster Prevention Day is observed yearly in Japan on September 1, the anniversary of the disastrous 1923 earthquake that took 143,000 lives. Schoolchildren perform safety drills in earthquake simulators, room-size cubicles furnished with traditional furniture, gas cooking stoves, and kerosene heaters. When the rooms are shaken, the children practice earthquake survival by ducking under tables, turning off gas and kerosene appliances, and remaining calm. Ironically, 90 percent of the deaths attributed to the 1995 earthquake in Kobe and its suburbs occurred in structures almost identical to the simulators.

Kobe, a city of 1.5 million on the island of Honshu, was struck at 5:46 A.M. a year to the day after the Northridge earthquake. The $M_w = 6.9$ earthquake killed 5,378 people, damaged or destroyed 152,000 buildings, and incinerated 6,913 buildings in an area equivalent of 70 U.S. city blocks. Officially known as the Hanshin–Awaji earthquake, its epicenter was off the northeast tip of Awaji Island in Osaka Bay south-

◆ FIGURE 4.34 In the 1994 Northridge earthquake, the fault rupture progressed up the fault plane from the focus at the lower right of the figure to the upper left in 8 seconds. Rather than rupturing smoothly, like a zipper opening, it moved in jerks along the fault plane, as shown by the pink patches. Total displacement was about 4 meters (12 ft).

Seconds after initial shock

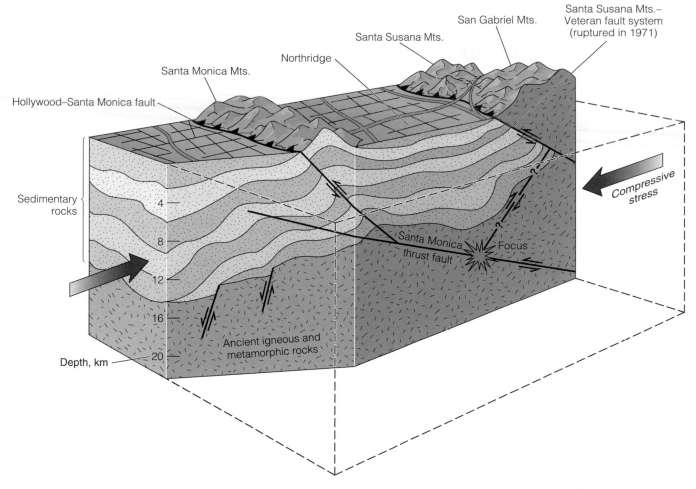

◆ **FIGURE 4.35** Interpretation of the fault system in the San Fernando Valley and surrounding area of Southern California. Compressive stresses built up over a long period, causing the subsurface blind thrust fault to rupture. The ensuing Northridge earthquake impacted the entire valley, and extended from the Santa Susana Mountains on the north to the city of Santa Monica on the south.

west of Kobe. The islands comprising Japan are at a convergent plate boundary, and they have many active faults paralleling the boundary (◆ Figure 4.37). The January 17 earthquake occurred on the northeast-trending Nojima fault. A 50-kilometer (30-mi) rupture on it passed directly under the city of Kobe. Ground rupture with a right-lateral displacement of 1.7 meters (5½ ft) was found on Awaji Island.

The city of Kobe rests on soft river sediments and 27 square kilometers (10½ mi²) of reclaimed land in Osaka Bay. When the sand and gravel fill was dumped into the bay no real attempt was made to compact or drain it. Measured ground accelerations during the earthquake were as high as 0.5 *g* at ten sites, and the strong shaking lasted 10–15 seconds. Not surprisingly, the artificial fill liquified—about 17 square kilometers (6½ mi²) of it—and in so doing it expelled huge amounts of water and sand and settled as much as a meter in places. Splatter marks on walls show that water spouts

from the ground reached heights up to 2 meters and typically reached 0.5 meter.

Nearly 90 percent of the fatalities occurred in crowded older neighborhoods when wood-frame houses of traditional *shinkabe* and *okabe* construction collapsed (◆ Figure 4.38). These are post-and-beam construction methods; unbraced or lightly braced vertical posts are spanned at the top with a thick beam and crowned with a heavy tile roof. "It's like putting a heavy book up on top of a frame of pencils," according to one engineer. There are no interior shear walls; lateral and diagonal bracing are minimal; wood pieces are not nailed, but fitted together by tongue and groove. The collapse of these structures was so extensive that witnesses said nothing remained to suggest that they had ever been anything but a pile of rubble (◆ Figure 4.39). In spite of the survivors' misery, their watchword was the stoic *"Shoganai"* ("It can't be helped").

Case Study | 4.4

Rx for Failed Freeways

Extensive damage to freeway bridges and overpasses typically accompanies earthquakes in large urban areas. Such damage has occurred in Alaska, California, and Japan. Overpass damage commonly is due to failure of the shorter columns, which lack the flexibility of longer ones. During an earthquake the tall columns supporting a bridge or overpass system bend and sway with the horizontal forces of the quake. Because the parts of the overpass system are tied together, the stresses are transferred through the structure to the short columns, which are designed to bend only a few centimeters (◆ Figure 1). Failure causes the short columns to bulge just above ground level, which breaks and pops off the exterior concrete, exposing the warped, "birdcaged" steel in the interior. In addition, high vertical accelerations can cause some columns to actually punch holes through the platform deck. With excessive horizontal motion some deck spans may slip off their column caps at one end and fall to the ground like tipped dominoes.

The California Department of Transportation (CalTrans) performed an extensive engineering study after the 1971 San Fernando earthquake and decided to retrofit 122 overpasses to alleviate these problems. One aspect of the retrofitting was jacketing the short columns with steel or a composite substance and filling the space between the jacket and the original column with concrete (◆ Figure 2a). This allows the columns to bend 12.5 centimeters (5 in), instead of 2.5 centimeters (1 in), without shattering. Another solution to elevated roadway failures is to increase columns' horizontal strength by wrapping heavy steel rods around their clusters of vertical support bars, particularly on short columns

(◆ Figure 2b). This allows the columns to bend but prevents "birdcaging" or permanent bending. To prevent the decks from slipping off their supports and dropping to the ground, steel straps or cables are installed at the joints (◆ Figure 2c). This limits the motion between the deck and supporting columns. Of the 122 overpasses that CalTrans retrofitted, not one collapsed in the 1994 Northridge earthquake. Ten of the eleven that collapsed were slated for future retrofitting.

The experience at Kobe was much the same, but column failures there were mostly due to inadequate steel reinforcement, particularly the wrapping of steel rods around the vertical supports (Figure 2b). In 1994 Japan's engineers had told the public that expressway failures such as those in California could never occur in Japan, and that their roadways would withstand even stronger earthquakes than the one at Northridge. That belief was shattered in Kobe the following January. Some of the problem there was unstable local geology (soft river deposits and bay fill), as was also the case in the I-880 failure in Oakland during the 1989 earthquake (water-saturated bay muds). Most of Japan's elevated expressways are supported by short columns, and nearly every column along the elevated Hanshin Expressway through Kobe was damaged. The most spectacular failure was a 600-meter (2,000-ft) elevated section that simply rolled over (see ◆ Figure 4.39b).

Postmortem studies of expressway failures demonstrate that, like geologists, engineers also learn things from disasters that no theory, mathematical formula, or computer model could have predicted.

◆ FIGURE 1 The difference in long and short columns' flexibility results in failure of the short ones.

In a quake, long columns survive because they sway.

Short columns, unable to bend, absorb horizontal energy produced by longer columns and blow out.

(a)

Roadway

Jacket

Column

Concrete backfill

Soil

(b)

Pre-1971 construction   New method

(c)

Straps keep joints from coming apart.

◀▶ FIGURE 2 Earthquake-resistant design for bridges and overpasses. (a) Short columns are retrofitted with steel or composite jackets that allow them to bend five times as much. (b) Comparison of old and new methods of constructing highway support columns. Wrapping the vertical steel supports with steel rods allows the column to bend without "birdcaging." (c) Steel straps are used to hold deck sections together, preventing them from falling off their support columns.

(a)

(b)

(c)

 FIGURE 4.36 (a) The partial collapse of this $11.5 million, 2,500-car parking structure at California State University–Northridge, was among the most dramatic failures in the Northridge area. (b) The left side of this apartment building collapsed into a lower-level garage. The stairway then led nowhere, because the second floor had dropped several meters. (c) This Reseda apartment building "jumped" onto cars that had been parked outside.

Some of the most spectacular damage in Kobe was to the transportation system. Two elevated structures serving Kobe, the Hanshin and Harbor expressways, were both badly damaged. A 600-meter (⅓-mi) section of the Hanshin expressway was overturned. Supported by single short, rigid columns without sufficient reinforcing bars to contain the concrete under the dynamic earthquake forces, the elevated highways ruptured (see ◆ Figure 1 in Case Study 4.4).

In Kobe's downtown area 1,558 reinforced-concrete buildings were damaged, and 80 collapsed. Some high-rise structures collapsed at mid-height due to design problems that led to a lessening in building rigidity in the higher stories. Not surprisingly, most of the severe damage and collapse occurred in pre-1980 buildings, as there was a major revision of the building code in 1981. ◆ Figure 4.40 makes this very clear. Even though engineering design and construction are differ-

◀▶ **FIGURE 4.37** Mapping of seismicity in part of Japan from 1961 to 1994 and the epicenter of the 1995 Hanshin–Awaji earthquake. Also shown are surface projections of the subduction-zone earthquakes that shook Kobe in 1944 and 1946.

earthquake was caused by a sudden rupture on the North Anatolian fault, one of the longest (about 1,200 km or 750 mi) and best-studied lateral-slip faults in the world (◀▶ Figure 4.41). Displacements up to 4.9 meters (16 ft) were measured, with most fault slips in the 2.5-to 3.0-meter range (8–10 ft). A small tsunami followed the event. Strong motion lasted 45 seconds, and remarkably, the center of Istanbul—with its historic Hagia Sophia (A.D. 537), walled Topkapi Palace Museum (1454), and Blue Mosque (1616)—escaped serious damage. But in the city's down-scale Avcilar district, 15 apartment buildings collapsed, killing hundreds. The final death toll of this deadly earthquake may reach 30,000, and it caused economic impacts in excess of $10 billion, and possibly $20 billion. The crowning blow was a magnitude-7.2 second earthquake on 12 November, about three months later, that extended the Izmit rupture 30–40 kilometers to the east and killed hundreds more.

The high death toll was directly attributed to shoddy construction practices. Due to them, 120,000 households were damaged beyond repair, and 2,000 apartment and office buildings collapsed like stacks of pancakes, leaving little if any space between floors.

ent in Japan than in the United States and other parts of the world, the Kobe experience served to demonstrate the effectiveness of state-of-the-art seismic codes. This lesson should not be lost on other urban areas close to tectonic plate boundaries.

IZMIT (KOCAELI), TURKEY, 1999

At 3:30 A.M. local time on 17 August 1999 a magnitude-7.4 earthquake struck northern Turkey. Its epicenter was 90 kilometers (55 mi) southeast of Istanbul, across the Bosporus Strait, near the town of Izmit (renamed *Kocaeli,* but the Turks still use its earlier name). The

(a)

(b)

◀▶ **FIGURE 4.38** Plans of (a) a typical Japanese residential area (Osaka) and (b) a typical U.S. urban residential area (Oakland, California) drawn to the same scale.

(a)

(b)

(c)

◁▶ **FIGURE 4.39** (a) This traditional Japanese wood-frame house collapsed so completely that it is hardly recognizable as a former home. About 90 percent of the fatalities occurred in houses such as this. (b) A 0.6-kilometer-long (⅓ mi) section of the Hanshin expressway in eastern Kobe was overturned, and railway transportation was totally disrupted by the earthquake. (c) Comparison of the geologic settings of Oakland, California, and Kobe, Japan. Both are on bays, both have areas of extensive fill, and both are underlain by active faults. An earthquake of magnitude 6.7 or more has a 32 percent probability of occurring on the Hayward fault underlying Oakland between 2000 and 2030, according to the U.S. Geological Survey Earthquake Working Group (see Figure 4.48).

◁▶ **FIGURE 4.40** Correlation between year of construction and damage to reinforced concrete buildings in downtown Kobe. The percentage of buildings of varying ages in each damage-level category is shown. Kobe's building code for such buildings was revised in 1981, 14 years before the earthquake.

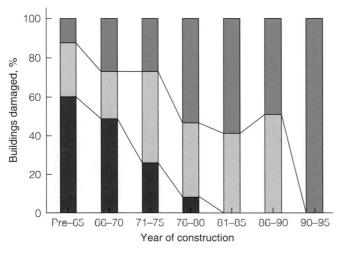

■ Severe damage or collapse
▨ Minor to moderate damage
▨ Slight or no damage

More than 600,000 people were left homeless. One government official stated, "Everyone knew builders cut corners, but nobody paid any attention until now." Corroborating evidence of construction deficiencies was found in many communities where very old buildings outperformed modern buildings surrounding them (◁▶ Figure 4.42). Most of the collapsed buildings were in the 4- to 8-story category. Contributing factors were substandard concrete and reinforcing steel, open ("soft") first floors, added floors, and poor joint connections between walls and floors. By any standard, this event was one of the major natural disasters of the twentieth century. The global humanitarian effort on behalf of the victims was heart warming. The governments of Switzerland, France, Germany, Japan, Israel, the United States, and neighboring Greece, Turkey's traditional antagonist, sent rescue teams.

The movement of the Arabian plate against the Eurasian plate explains the high seismicity in Turkey. The Anatolian "microplate" that contains Turkey is being shoved westward between the two plates, movement that is complicated by the northward-impinging African plate to the south. Thus, the Anatolian plate and Turkey are caught in a "vise"—between the proverbial rock and a hard place—and exist in a very active and complicated plate-tectonic setting.

FIGURE 4.41 The North Anatolian fault. (a) Historic earthquake epicenters (red dots) and their magnitudes. Lengths of the surface ruptures and years of the events produced by the rupture arc also shown. The Izmit earthquake epicenter is represented by a black star. (b) Rupture lengths on the fault causing earthquakes between 1939 and 1999. Note that earthquakes have migrated sequentially westward along the fault toward Istanbul like a line of falling dominoes. The 1999 event filled a seismic gap between the 1967 event and two earthquakes of 1963–64 in the area off the map to the west.

Turkey has a long written history of earthquakes going back to one in A.D. 358. The fourth-century writer Ammianus Marcellinus left a vivid account of this earthquake, which must have had many similarities to the 1999 event, including severe shaking, landslides, and dust clouds forming above fallen debris. Since 1939 the North Anatolian fault has produced a sequence of earthquakes moving eastward and westward along the fault line (Figure 4.41). The westward migration has been more aggressive, with 600 kilometers (370 mi) of contiguous fault rupture occurring between 1939 and 1944. The 1939 earthquake was a magnitude-7.9 that killed 30,000 people. Between 1942 and 1967 six large shocks occurred on five segments of the fault with epicenters and ground rupture moving progressively westward toward Istanbul like a line of falling dominoes. This sequence left a 150-kilometer-long (90-mi) seismic gap between the 1967

event to the east and several earthquakes that occurred to the west during 1963 and 1964. The 17 August 1999 earthquake "filled" this gap (just as Loma Prieta did the Southern Santa Cruz Mountains gap in 1989; see Figure 4.30). Scott Stein and his U.S. Geological Survey group at Menlo Park, California, had earlier calculated that nine out of ten large recorded earthquakes on the North Anatolian fault had occurred where fault movement on one segment of the fault increased the stress on an adjacent segment. They identified the 1999 Izmit rupture segment as one of the two worrisome parts of the fault, but they did not predict the event.

As described in the Chapter 1 opening material (see Figure 1 there), there are many similarities between the North Anatolian fault and the San Andreas fault in California, including their long-term rates of motion, sense of slip, and length. Another similarity is

(a)

(b)

(c)

◆ **FIGURE 4.42** The Izmit earthquake was the most deadly event to strike Turkey in sixty years. (a) The Anatolia fault is directly beneath this four-story apartment building that came apart and "pancaked." This kind of structural failure is not survivable. Note the 2-meter (6 ½-ft) right-lateral offset of the block wall to the left of the building. (b) Damaged buildings in Akyazy, Turkey. Open first floors, typically for shops or parking, is a design feature that creates structural weakness. It contributed to many building failures during the earthquake. (c) Many centuries-old religious structures, such as this mosque, survived the earthquake. Tall, slender minarets, towers attached to mosques from which Muslims are called to prayer, were not so sound.

the segmented nature of the two fault systems; that is, both of these very long faults rupture in short sections. The similarities have led to scientific collaborations between U.S. and Turkish scientists aimed at better understanding the earthquake hazards the two nations and their people face. In late 1999 Turkish officials were demonstrating their improved understanding of earthquake hazards. They vowed that they would institute and enforce stricter building codes in the rebuilding process.

DOES EARTHQUAKE COUNTRY INCLUDE IDAHO, MISSOURI, AND NEW YORK?

Although the vast majority of the world's earthquakes occur at plate boundaries, areas hundreds and even thousands of kilometers away are not free of seismic activity. In the United States the five most seismically active states between 1980 and 1991 were

STATE	EARTHQUAKES 1980–1991 Number Recorded	Largest, M_w
Alaska	10,253	9.2
California	6,732	7.2
Washington	615	5.5
Idaho	536	7.3
Nevada	398	5.6

Although Alaska and California continue to lead the United States in the number of shakers, the earthquakes that have been felt over the largest area occurred in Missouri in 1811 and 1812, and significant seismic hazards are recognized in 39 states. No state is earthquake proof, as the seismic-risk map of the United States shows (◆ Figure 4.43). It appears that U.S. residents who want to be seismically on the "safe" side should move to Texas, Florida, or Alabama.

Why do such strong earthquakes occur *intraplate*— that is, far from a plate boundary (◉ Table 4.2)? Intraplate earthquakes have several characteristics in common:

1. The faults causing them are deeply buried and have not broken the ground surface.
2. Because the rocks of the continental interior are stronger than those at plate boundaries, which are laced with faults, they transmit seismic waves better, causing ground motion over a huge area. In the United States this is usually many states (◆ Figure 4.44).
3. They do not appear to be random events.

TABLE 4.2 **Selected North American Intraplate Earthquakes**

LOCATION	YEAR	MOMENT MAGNITUDE*	IMPACT
New Madrid, Missouri	1811	8.2	Reelfoot Lake formed in NW Tennessee.
New Madrid	1812	8.3	Elevation changes caused Mississippi River to reverse its course locally.
New Madrid	1812	8.1	
Charleston, South Carolina	1886	7.6	Felt from New York to Chicago; 60 killed.
Charleston, Missouri	1895	6.8	Damage in 6 states (see Figure 4.43).
Grand Banks, Newfoundland	1929	7.4	Submarine landslides broke trans-Atlantic cable, disrupting communications.

* Moment magnitude is used in reconstructing the strength of "preinstrument" earthquakes, because we know something of fault length, rupture length, and area felt.

The Charleston, South Carolina, earthquake of 1886 was larger than the 1989 San Francisco shaker. It killed scores of people, ruined the city, and slowed the South's recovery from the Civil War. The New Madrid (pronounced "mad´rid") earthquakes of 1811 and 1812 caused extensive topographic changes, locally reversed the course of the Mississippi River, and may have been felt over a larger area than any other earthquake in recorded history. Ground motion was felt as far away as Washington, D.C., where it caused church bells to ring and scaffoldings on the Capitol to collapse.

Recent studies suggest that intraplate earthquakes are concentrated in areas where normally stable crust has been stretched and faulted. Such zones form where continents have been split apart (*rifted*), forming two continents, as occurs at divergent plate boundaries (see Chapter 3). When North America was separated from Africa at the Mid-Atlantic Ridge about 180–200 million years ago, the continental crust being rifted at the ridge was thinned, stretched, and faulted. The North American continent was transported westward with the plate, and the weakened part of the crust along its eastern edge was hidden by a thick cover of younger sedimentary rocks. Buried rifted crust under the Eastern seaboard states forms the intraplate earthquake zone from the Carolinas to Canada.

◆ **FIGURE 4.43** Seismic-risk map of the 48 contiguous states based upon historical records and intensities collected by the U.S. Coast and Geodetic Survey. The Coast and Geodetic Survey gathers intensity data from questionnaires after earthquakes.

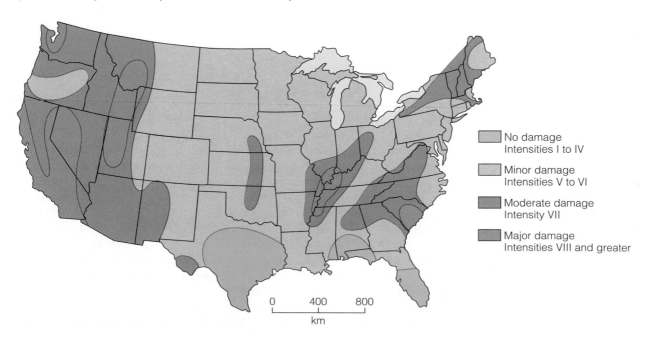

No damage
Intensities I to IV

Minor damage
Intensities V to VI

Moderate damage
Intensity VII

Major damage
Intensities VIII and greater

0   400   800
km

(a)

◇▶ **FIGURE 4.44** Mid-continent earthquakes. (a) Comparison of the effects of two earthquakes of similar magnitude. The 1895 Charleston, Missouri, earthquake impacted a relatively huge area in the central United States, whereas the 1994 Northridge earthquake's effects were limited to Southern California. Red indicates areas of minor-to-major damage to buildings and their contents. Yellow indicates areas where shaking was felt but there was little or no damage to objects, such as dishes. (b) Damage from the 1886 earthquake along East Bay Street in Charleston, South Carolina.

(b)

The Midwest earthquake zone through Arkansas, Missouri, and Illinois is believed to be where an ancient (pre-Pangaea) divergent boundary started to form but "failed" for some reason, leaving behind a significant buried fault zone. Known as the *Reelfoot Rift,* the buried block of crust is down-dropped between faults (the hachured lines on the map of ◇▶ Figure 4.45). The rift is 60 kilometers wide and 300 kilometers long (roughly 40 mi × 190 mi) and formed at least 500 million years ago. The linear trend of earthquakes from Marked Tree, Arkansas, northeastward to Caruthersville, Missouri, reveals upwarped sedimen-

tary rocks along the rift axis. Detailed analysis of the geology across the Reelfoot fault scarp by trenching revealed evidence of three large earthquakes within the past 2,000 years, which yields a recurrence-interval estimate of 600–900 years for the fault, but the recurring earthquakes are not necessarily of the same magnitude as the 1811–1812 events. In fact, scientists estimate that a magnitude-6–7 earthquake will occur in the New Madrid seismic zone between 2000 and 2050 with a 90 percent probability.

Nobody knows how many people died in the 1811–1812 earthquakes. An amateur observer in

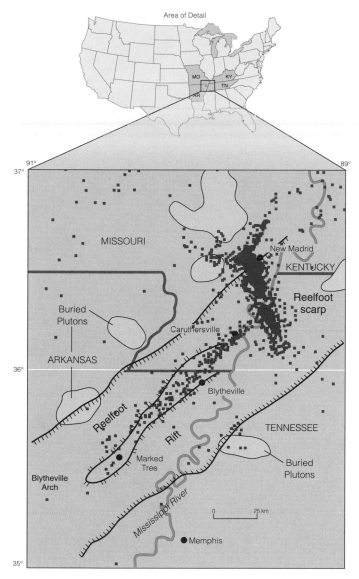

◆ **FIGURE 4.45** The New Madrid seismic zone covers parts of four states and consists of an ancient, fault-bounded depression (rift) buried below 1.5 kilometers (almost a mile) of sedimentary rocks. An area of major seismic hazard, it is studied intensively.

Louisville, Kentucky, recorded almost 2,000 shocks with a homemade pendulum. Liquefaction was widespread along the Mississippi River, and the town of New Madrid, Missouri, was completely destroyed when it settled from 8 meters (25 ft) above sea level to only 4 meters (12 ft). Subsidence caused some swamps to drain and others to become lakes; an example is Reelfoot Lake in northwest Tennessee, which today is more than 50 feet deep. The New Madrid seismic zone is a major geologic hazard in the United States, and efforts are being made to reduce the impact of a future large earthquake there.

Iben Browning, a scientist with a Ph.D. in physiology but who is best-known for his work on climate, predicted there would be a repeat of the 1811 New Madrid earthquake on 3 December 1990. His prediction was based upon alignment of the planets and consequent gravitational pull, which he believed would be sufficient to trigger an earthquake on that date. Although the prediction was discounted by scientists, it created considerable anxiety in Arkansas, Tennessee, Alabama, and Missouri. Months before "the day," earthquake insurance sales boomed, moving companies were booked up, and bottled-water sales rose dramatically—as did sales of quake-related souvenirs. One church sold "Eternity Preparedness Kits," and "Survival Revivals" were held. On the day of the scientifically discredited prediction nothing earthshaking occurred.

Many earthquakes with epicenters in weakened intraplate continental crust have shaken parts of the Midwest, eastern Canada, New England, New York, and the East Coast of the United States. New York does not have a major fault such as California's, but it has many faults. In 1984 the state experienced two earthquakes, one in Westchester ($M = 4.0$) and the other off Coney Island ($M = 5.0$). The Westchester quake toppled chimneys, shattered windows, and caused some panic. As a result, the state implemented new seismic codes that add 2–5 percent to the cost of building, and the response has been favorable. By the mid 1990s New York and Massachusetts were the only Northeastern states that had earthquake building codes.

THE PACIFIC NORTHWEST

In years past, earthquake hazards were considered minor in Oregon and Washington. During the 1980s research changed the perception, however; it revealed geologic evidence that "major" (≥ 7.0 but <8.0) or "great" (≥ 8.0 but <9.0) subduction-zone earthquakes have occurred in the past and that they can occur in the future. The Cascadia subduction zone (see Figure 5.3), extending 1,200 kilometers (740 mi) from northern California to Vancouver Island in Canada, has destructive earthquake potential. Geological evidence, consisting of carbon-dated tsunami deposits and drowned red-cedar forests, suggests that great earthquakes strike the Pacific Northwest roughly every 500 years, the last one in A.D. 1700, approximately 300 years ago. Minor earthquakes occur daily, however, and the cities of Portland, Seattle, and Vancouver, British Columbia, are in earthquake country (◆ Figure 4.46). The states now are working to minimize damage should the "big one" occur on the subduction zone or on the Seattle fault, which runs through downtown Seattle.

Earthquake Prediction

*A guy oughta be careful about making predictions.
Particularly about the future.*

YOGI BERRA, BASEBALL PLAYER

Earthquake prediction has great potential for saving lives and reducing property damage. A good prediction would give the location, the approximate time, and the approximate magnitude of the future earthquake with a high probability of correctness. Prediction was the hottest area of geophysical and geological research from the 1970s to the early 1990s, based upon the belief that measurable phenomena occurring before large and small earthquakes—**precursors**—could be identified. Researchers compiled earthquake case histories where presumed precursors were seen. Such things as changes in the ratio of P-wave to S-wave velocities, ground tilt, electrical resistivity, water-well levels, and emission of noble gases were measured, and piles of data were amassed. Unfortunately, the hope of finding precursors that would lead to reliable predictions seems to have evaporated. Investment of research dollars and time in such inquiries has not been heavy since the early 1990s.

When we think of "earthquake prediction"—of the time, place, and magnitude—we usually think of the short term, a warning a few days to a few hundred days before the event. Scientists doubt that short-term prediction will become a reality any time soon, if ever. This is because current knowledge of the short-term mechanical behavior of faults is inadequate. In addition, a useful short-term prediction, by necessity, is *local* and thus would require siting a huge number of measuring instruments, and luck as well as talent in positioning them.

Intermediate-term prediction (warning time of 1–10 years) is useful, because it provides lead time for making building and construction improvements and for educating the public about seismic safety. Such predictions should really be called *forecasts,* as they do not specify precise time and place. This research has centered on identifying patterns of earthquake occurrences prior to large earthquakes and it, also, relies on well-documented case histories.

Long-term forecasts (warnings of more than a decade) are based on identifying particular faults that rupture with some kind of periodicity, or regularity. Knowledge of this comes from studying earthquake catalogs and investigating ancient fault movements by digging trenches across faults. From these efforts, a new area of study called *paleoseismology* has arisen. It is discussed later in this section. Whether the goal be short-term or long-term prediction, studying earthquake case histories, such as those of Izmit, Loma Prieta, and Northridge, has proved fruitful.

◆▶ **FIGURE 4.46** Seismic-hazard map of downtown Portland, Oregon. The hazard zones are based upon liquefaction potential, landslide potential, and the probable degree of ground shaking. Red denotes greatest hazard; pale yellow, the lowest hazard.

STATISTICAL METHODS

By compiling statistical evidence pertaining to past earthquakes in a region, we acquire basic data for calculating the statistical probability for future events of given magnitudes. These calculations may be done on a worldwide scale or on a local scale, such as the example in ◆▶ Figure 4.47. Analysis of the graph indicates that for the particular area in Southern California, the statistical **recurrence interval**—that is, the length of time that can be expected between events of a given magnitude—is 1,000 years for a magnitude-8 earthquake (0.1/100 years), about 100 years for a magnitude-7 earthquake (1/100 years), and about 10 years for a magnitude-6 earthquake (almost 10/100 years). The probability of a magnitude-7 occurring in any one year is thus 1 percent, and of a magnitude-6, 10 percent. On an annual basis worldwide, we can expect at least two magnitude-8 earthquakes, 20 magnitude-7 earthquakes, and no less than 100 earthquakes of magnitude 6. Thus for seismically active regions, historical seismicity data can be used to calculate the *probability* of damaging earthquakes. Although these numbers are not really of predictive value as we defined it, they can be used by planners for making zoning recommendations, by architects and engineers for designing earthquake-resistant structures, and by others for formulating other life- and property-saving measures.

A new approach is to evaluate the probability of a large earthquake occurring on an active fault during a given time period. This falls under the heading of long-term forecasting. An example of this is the U.S. Geological Survey and other scientists' conclusion that there is a 70 percent probability of at least one magnitude-6.7 or greater earthquake striking the San Francisco Bay region between 2000 and 2030. Such an event would be capable of causing widespread damage (◆ Figure 4.48).

GEOPHYSICAL METHODS

Geophysical and seismological precursors, as noted previously, have not proved reliable predictors of earthquakes. Seismic gaps, however, can indicate stretches of future fault activity. Seismic gaps are stretches along known active fault zones within which no significant earthquakes have been recorded. It is not always clear whether these fault sections are "locked" and thus building up strain energy, or if motion (creep) is taking place there that is relieving strain. Such gaps existed and were "filled" so to speak during the Mexico City (1985), Loma Prieta (1989; see Figure 4.30), and Izmit (1999) earthquakes. Seismic gaps serve as alerts or warnings of possible future events and can be used as forecasting tools.

GEOLOGICAL METHODS

Research suggests that active faults and segments of long, active faults tend to have recurring earthquakes of characteristic magnitude, rupture length, and displacement. For example, six earthquakes have occurred along the San Andreas fault at Parkfield since 1857 (see Figure 4.30). They had Richter magnitudes of around 5.6, rupture lengths of 13–19 kilometers (8–12 mi), and displacements averaging 0.5 meter. These earthquakes have recurred about every 22 years since the earliest recorded earthquake. A 95 percent probability of an earthquake was predicted for this segment between 1988 and 1993, the first prediction in the United States to be endorsed by scientists and subsequently issued by the federal government. The prediction was a failure.

Geologists now believe that stable rough spots along fault planes, called **asperities,** determine earthquake characteristics along fault segments. Where asperities are large in size and far from other asperities, earthquakes tend to be large and infrequent; that is, to have long recurrence intervals. Where asperities are close together and relatively small, shorter recurrence intervals are the rule. Where there are no significant rough spots, creep dominates, and no earthquakes occur. Somewhat like knots in a rope, these rough spots "lock" the fault and serve as stopping points of fault rupture. The larger the

◆ **FIGURE 4.47** A graphing of 463 earthquakes of magnitude 4 or greater in a small area near a nuclear reactor in Southern California over a period of 44 years. Statistically, the graph shows that 600 magnitude-4 earthquakes can be expected in 100 years, or 6 per year on average. The probability of a magnitude-8 earthquake in 100 years is 0.1, which would be 1 in 1,000 years. Such plots can be constructed for a region or for the world. The graphs are all similar in shape; only the numbers differ.

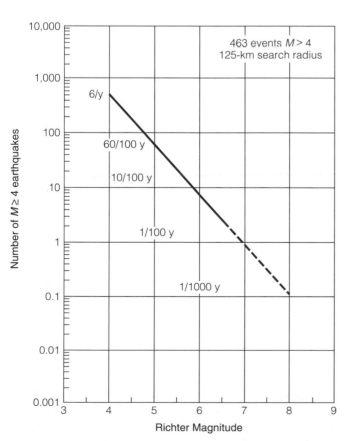

◁▷ **FIGURE 4.48** Probabilities that one or more magnitude-6.7 or greater earthquakes will strike on specific faults in the San Francisco Bay region between 2000 and 2030. The San Andreas, Rogers Creek, and Hayward faults have the highest probabilities. Total probability for the region is computed as 70% (+ or − 10%). These probabilities were developed by scientists with the U.S. Geological Survey, part of the U.S. Department of the Interior, and thus constitute official long-term *forecasts*. The message is that all communities in the Bay region should keep preparing for earthquakes.

"knot" (asperity), the longer the time interval during which strain accumulates and the larger the earthquake that is produced. Smaller, unstable asperities are associated with small earthquakes and shorter time intervals between earthquakes. Using asperities as working models for characteristic earthquakes gives us a basis for studying fault behavior. Two large asperities that appear to govern the behavior of the San Andreas fault in Southern California are shown in Figure 4.49.

Since recorded history in North America is short, geologists need other means of collecting frequency data of large prehistoric earthquakes. One method is to dig trenches into marsh or river sediments that have been disrupted by faulting in an effort to decipher a region's **paleoseismicity,** its rock record of past earthquake events. Kerry Sieh of the California Institute of Technology has done this across the San Andreas in Southern California (◁▷ Figure 4.50). Sieh found an intriguing history of seismicity recorded in disrupted marsh deposits at Pallett Creek (see Figure 4.49 map) and liquefaction effects extending from the seventh century to a great earthquake in 1857. Ten large events, extending from A.D. 650 to 1857, were dated using ^{14}C. The average recurrence interval for these ancient earthquakes is 132 years, but they are clustered in four groups. Within each cluster, the recurrence interval is less than 100 years, and the intervals between the clusters are two to three centuries in length. The last big one, also the last one of a cluster, was in 1857. Thus it appears that this section of the San Andreas may remain dormant until late in this century or beyond.

An old saying among geologists is that "the longer it's been since the last one, the closer we are to the next one." Unfortunately, this is about the status of our present predictive ability. We remain uncertain that an earthquake will indeed follow well-identified percursory phenomena, such as anomalous animal behavior (see Case Study 4.5). It seems that no one precursor is a predictor, and that many physical changes will have to be measured over a long period of time before our knowledge of fault behavior is refined enough for reliable prediction.

Mitigation

Reducing earthquake risks for the general public is an admirable goal of lawmakers and scientists. Law-

Consider This... **?** Humans have caused earthquakes by injecting fluids, such as oil-field waters and chemical warfare wastes, into deep wells (read Case Study 15.4). How does this open the possibility of controlling earthquake activity, and what might be a major deterrent to the effort? ∎

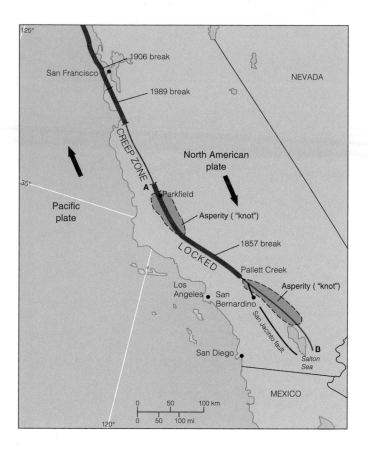

◀▶ **FIGURE 4.49** The San Andreas fault in Southern California. Segment A–B is the part of the fault where the greatest amount of plate motion has occurred. The purple lines represent the rupture zones of the 1906, 1857, and 1989 great earthquakes.

SURVIVAL TIPS

Knowing what to do before, during, and after an earthquake is of utmost importance to you and your family.

Before an Earthquake:

◀▶ Figure 4.51 provides the Federal Emergency Management Agency's suggestions for minimizing the possibility of damage, fire, and injuries in the home.

During an Earthquake:

- Remain calm and consider the consequences of your actions.
- If you are indoors, stay indoors and get under a desk, bed, or a strong doorway.
- If you are outside, stay away from buildings, walls, power poles, and other objects that could fall. If driving, stop your car in an open area.

makers set policies, and scientists and engineers develop the technical expertise to recommend these policies and to carry them out. It has been shown that strict building codes in seismic regions, such as the Pacific Rim, reduce damage and loss of life. Laws passed after the 1933 and 1971 Southern California earthquakes have proven the effectiveness of strict earthquake-resistant design.

The primary consideration in earthquake design is to incorporate resistance to horizontal ground acceleration, or "base shear." Strong horizontal motion tends to topple poorly built structures and to deform more flexible ones. In California, high-rise structures are built to withstand about 40 percent of the acceleration of gravity in the horizontal direction (0.4 *g*), and single-family dwellings are built to withstand about 15 percent (0.15 *g*). *Base isolation* is now a popular design option. The structure, low- or high-rise, is placed upon Teflon plates, seismic-energy dissipators that are similar to auto shock absorbers, or even springs, which allows the ground to move but minimizes building vibration and sway.

◀▶ **FIGURE 4.50** Disrupted marsh and lake deposits along the San Andreas fault at Pallett Creek near Palmdale, California. Sediments range in age from about A.D. 200 at the lower left to A.D. 1910 at the ground surface. Several large earthquakes are represented here by broken layers and buried fault scarps.

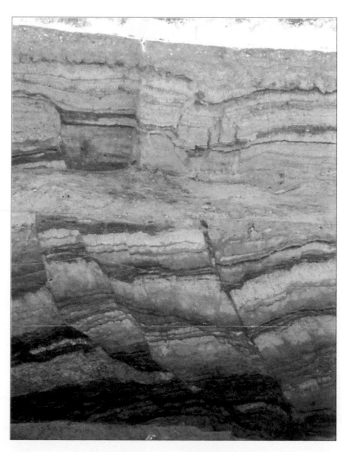

Case Study | 4.5

Depressed Tigers, Restless Turtles, and Earthquakes!

Anomalous animal behavior preceding earthquakes is well documented. Domesticated animals such as barnyard fowl, horses, cats, and dogs have been known to behave so peculiarly before big events that they attracted the notice of people not knowledgeable in what is normal *or* abnormal animal behavior (◆ Figure 1). It should be remembered that although anomalous animal behavior precedes many earthquakes, it does not appear to preccde every earthquake, and abnormal animal behavior is not always followed by an earthquake. Other natural phenomena, such as atmospheric disturbances, also can cause animals to behave strangely. Here are some reported incidents of unusual animal behavior noted before earthquakes:

■ Tientsin Zoo, China, 1969: 2 hours before the magnitude-7.4 earthquake the tiger appeared depressed, pandas screamed, turtles were restless, and the yak would not eat.

■ Haicheng, China, 1975: 1½ months before the magnitude-7.3 earthquake snakes came out of hibernation; 1–2 days before, pigs would not eat and they climbed walls; 20 minutes before, turtles jumped out of the water and cried.

■ Tokyo, Japan, 1855: 1 day before the magnitude-6.9 earthquake wild cats cried, and rats disappeared.

■ Concepción, Chile, 1835: 1 hour and 40 minutes before the earthquake flocks of sea birds flew inland, and dogs left the city.

■ San Francisco, 1906: dogs barked all night before the magnitude-8.3 earthquake.

■ Friuli, Italy, 1976: 2–3 hours before the magnitude-6.7 earthquake cats left their houses and the village, mice and rats left their hiding places, and fowl refused to roost.

◆ FIGURE 1 Anomalous animal behavior. An earthquake may be in the offing if your dog dons a hard hat.

- Do not use elevators, and if you are in a crowded area, do not rush for a door.

After an Earthquake:

- Turn off the gas at the meter.
- Use portable radios for information.

- Check water supplies, remembering that there is water in water heaters, melted ice, and toilet tanks. Do not drink waterbed or pool water.
- Check your home for damage.
- Do not drive.

1 Nail plywood to ceiling joists to protect occupants from falling chimney bricks.

2 Anchor hanging lamps with closed hooks or relocate.

3 Secure top-heavy furniture to wall studs with metal braces.

4 Keep all breakables in low or secure cabinets.

5 Know the locations of main electrical and gas switches for emergency shutoff.

6 Stabilize water heater with metal straps to wall studs.

7 Use flexible connectors where gas lines meet appliances, if local codes permit.

8 Keep fire extinguisher in accessible place.

9 Keep emergency supplies on hand, including drinking water, canned or dried foods, first-aid kit, flashlight, and portable radio with extra batteries.

10 Place secure latches on cupboards to prevent doors from swinging open.

11 Keep heavy, unstable objects away from exit routes, and anchor wheels.

12 Replace heavy hangings over beds with lightweight alternatives.

(a)

Shut-off valve positions

On   Off

(b)

◁▶ **FIGURE 4.51** (a) How to minimize earthquake damage in the home in advance. (b) Keep a small crescent wrench at the gas meter. Turn off the gas by turning the valve end 90°.

Gallery

Gallery

Seismic Photo-Ops

Sometimes an unusual event or damage lightens the tragedy of an earthquake. The 9 October 1995 magnitude-8 earthquake near Manzanillo, Mexico, caused extensive damage and was felt strongly in Mexico City. It set off a moderate tsunami that affected about 200 kilometers (125 mi) of coastline. Although lives were lost in the earthquake and tsunami damage was severe in places, some photos taken at Tenacatita Bay tell a story that has a happy ending. Water advanced 200 meters (650 ft) through the small village of La Manzanilla, stopping only when it reached a hill. ◆ Figure 4.52a is one of the few photos ever taken of the arrival of a tsunami's initial, leading trough. As the photo shows, the trough sucked Tenacatita Bay dry. Faced with onrushing waters, Jose "Pepe" Martinez stood his ground and captured the sensational advance of water onto the land on film (◆ Figure 4.52b, c). The other photos illustrate some humorous and puzzling results of seismic activity. Please view them in this light (◆ Figures 4.53 to 4.55).

(a)

(b)

(c)

◆ **FIGURE 4.52** Tsunami in Mexico, 9 October 1995. There were no injuries at this site. (a) Tenacatita Bay minutes before the tsunami struck. This area is normally covered by several meters of water, even at low tide. (b) Water just beginning its advance onto the land. Note the small white structure in right center of photo. (c) I guess we better get going! The photographer holds his ground, but people are now jogging to stay ahead of the water. Witnesses said the water advanced at a slow run.

◇▶ **FIGURE 4.53** A waterfall now exists where there was once only a pristine trout stream. The waterfall is the result of a 6-meter (19-ft) scarp formed by slip on the Chelungpu fault during the Chi-Chi, Taiwan, earthquake of 21 September 1999.

◇▶ **FIGURE 4.54** Statue of Louis Agassiz, the great geologist, naturalist, and educator, at Stanford University after the 1906 earthquake. Strong horizontal motion knocked the statue off its pedestal, prompting a noted scientist of the day to remark, "I have read Agassiz in the abstract. Now I see him in the concrete."

◇▶ **FIGURE 4.55** The Hector Mine earthquake of 16 October 1999 was a real bomb! The magnitude-7.0 event occurred in California's remote Mojave Desert. The fault scarp, shown here, forms a gash across the desert for 40 kilometers (25 mi), but the earthquake did little damage. This area happens to be a U.S. Marine Corps bombing range, as you can see, which made geologists' field work interesting.

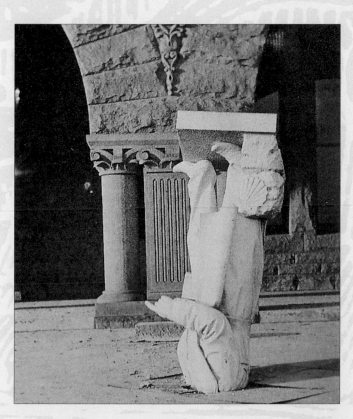

Summary

EARTHQUAKES

CAUSE

Movements on fractures in the crust known as *faults* that result in three types of wave motion: P- and S-waves, which are generated at the focus of the earthquake and travel through the earth, and surface waves.

DISTRIBUTION

Most (but not all) large earthquakes occur near plate boundaries, such as the San Andreas fault, and represent the release of stored elastic strain energy as plates slip past, over, or under each other. Intraplate earthquakes can occur at locations far from plate boundaries where deep crust has been faulted, probably at "failed" continental margins.

MEASUREMENT SCALES

Some of the scales used to measure earthquakes are the modified Mercalli intensity scale (based on damage), the Richter magnitude scale (based on energy released as measured by maximum wave amplitude on a seismograph), and the moment magnitude (based on the total seismic energy released as measured by the rigidity of the faulted rock, the area of rupture on the fault plane, and displacement).

EARTHQUAKE-RELATED HAZARDS AND MITIGATION

GROUND SHAKING

Damaging motion caused by shear and surface waves.

Ways to reduce effects seismic zoning; building codes; construction techniques such as shear walls, seismic joints, and bolting frames to foundations.

LANDSLIDES

Hundreds of landslides may be triggered by an earthquake in a slide-prone area.

Ways to reduce effects—proper zoning in high-risk areas.

GROUND FAILURE (SPONTANEOUS LIQUEFACTION)

Horizontal (lateral) movements caused by loss of strength of water-saturated sandy soils during shaking and by liquefaction of quick clays.

Ways to reduce effects—building codes that require deep-drilling to locate liquefiable soils or layers.

GROUND RUPTURE/CHANGES IN GROUND LEVEL

Fault rupture and uplift or subsidence of land as a result of fault displacement.

Ways to reduce effects—geologic mapping to locate fault zones, trenching across fault zones, implementation of effective seismic zonation like the Earthquake Fault Zones Act in California.

FIRE

In some large earthquakes fire has been the biggest source of damage.

Ways to reduce effects—public education on what to do after a quake, such as shutting off gas and other utilities.

TSUNAMIS

Multidirectional sea waves generated by disruption of the underlying sea floor.

Ways to reduce effects—early warning systems, building design that lessens impact of wave runup in prone areas.

EARTHQUAKE PREDICTION

STATISTICAL METHODS

Historical information on earthquakes in a region is used to calculate recurrence intervals and probabilities of damaging earthquakes.

Upside/downside—good for planning purposes, but not for short-term warnings.

GEOPHYSICAL METHODS

Measurements of physical changes in the earth are used to identify seismic precursors.

Upside/downside—Research continues but results are disappointing.

GEOLOGICAL METHODS

Active faults are studied to determine the characteristic earthquake magnitudes and recurrence intervals of particular fault segments; sediments exposed in trenches may disclose historic large fault displacements (earthquakes) and, if they contain datable C-14 material, their recurrence intervals.

Upside/downside—useful for long-range forecasting along fault segments and for identifying seismic gaps; not useful for short-term warnings.

◆ Key Terms

aftershock	epicenter	focus (pl. *foci*)	liquefaction
asperity	fault	intensity scale	longitudinal wave
body wave	fault creep	isoseismal	magnitude
elastic rebound theory			

modified Mercalli scale
  (*MM*)
moment magnitude (M_w
  or *M*)
paleoseismicity
precursor

P-wave
recurrence interval
resonance
Richter magnitude scale
seismic gap

seismograph
seismogram
spontaneous liquefaction
stress
strain

S-wave
transverse (shear) wave
tsunami (pl. *tsunamis*)
wave period (*T*)

◀ Study Questions

1. What is "elastic rebound," and how does it relate to earthquake motion?

2. Distinguish among earthquake intensity, Richter magnitude, and moment magnitude. Which magnitude scale is most favored by seismologists today? Why?

3. Why should one be more concerned about the likelihood of an earthquake in Alaska than of one in Texas?

4. In light of plate tectonic theory, explain why devastating shallow-focus earthquakes occur in some areas and only moderate shallow-focus activity takes place in other areas.

5. What are some earthquake precursors, and how useful is each as a predictor?

6. What should people who live in earthquake country do before, during, and after an earthquake (the minimum)?

7. Describe the motion of the three types of earthquake waves discussed in the chapter and their effects on structures.

8. Why do wood-frame structures suffer less damage than unreinforced brick buildings in an earthquake?

9. Explain why structures close to the epicenter of the 1989 Loma Prieta earthquake suffered less damage than buildings tens of miles away in the San Francisco and Oakland Bay areas.

◀ For Further Information

BOOKS AND PERIODICALS

Atwater, Bryan, and others. 1999. *Surviving a tsunami: Lessons from Chile, Hawaii, and Japan.* U.S. Geological Survey circular 1187, 19 pp.

Bolt, Bruce. 1998. *Earthquakes, newly revised and expanded.* New York: W. H. Freeman.

Dolan, James F.; Kerry Sieh; Thomas Rockwell; Robert Yeats; John Shaw; John Suppe; Gary Juftile; and Eldon Gath. 1995. Prospects for larger or more frequent earthquakes in the Los Angeles metropolitan region. *Science* 267:199–205.

Earthquake Engineering Research Institute (EERI). 1995. The Hyogo–Ken Nambu earthquake, January 17, 1995; Preliminary reconnaissance report. Oakland, Calif.: EERI, 116 pp.

González, Frank I. 1999. Tsunami. *Scientific American,* May: 56–65.

Gore, Rick. 1995. Living with California's faults. *National geographic,* April:2–34.

Holzer, Thomas L. 1995. The 1995 Hanshin–Awaji (Kobe), Japan, earthquake. *GSA today* (Geological Society of America), August.

Iacopi, Robert. 1971. *Earthquake country.* Menlo Park, Calif.: Lane Books.

Kockelman, William J. 1984. *Reducing losses from earthquakes through personal preparedness.* U.S. Geological Survey open file report 84-765.

Kovachs, Robert. 1995. Earth's fury: An introduction to natural hazards and disasters. Englewood Cliffs, N.J.: Prentice-Hall, Inc.

Knopoff, L. 1996. Earthquake prediction: The scientific challenge. In *Earthquake prediction: Proceedings of the National Academy of Sciences,* pp. 3719–3720.

Ma, Kuo-Fong, and others. 1999. The Chi-Chi, Taiwan, earthquake: Large surface displacements on an inland thrust fault, *EOS* 80, no. 50 (December 14).

Mori, James J. 1994. Overview: The Northridge earthquake: Damage to an urban environment. *Earthquakes and volcanoes* 25, no. 1 (special issue).

Reid, T. R. 1995. Kobe wakes to a nightmare. *National geographic,* July:112–136.

Richter, C. 1958. *Elementary seismology.* New York: W. H. Freeman.

U.S. Department of the Interior, U.S. Geological Survey. 1996. *U.S.G.S. Presponse to an urban earthquake, Northridge '94.* Open file report 96-263, 78 pp.

Wuethrich, Bernice. 1995. Cascadia countdown. *Earth: The science of our planet,* October:24–31.

Yeats, Robert S. 1998. *Living with earthquakes in the Pacific Northwest.* Corvalis, Oreg.: Oregon State University Press.

INTERNET SITES

Cal Tech. Earthquake site
  http://www.gps.caltech.edu/seismo/seismo.page.htm
Earthquakes, Canada
  http://www.seismo.emr.ca/welcome/html
Earthquake, Izmit, Turkey
  http://www.koeri.boun.edu.tr/earthqk/earthquake.htm

Earthquakes, Mammoth Lakes, California
    http://wwwneic.usgs.gov/neis/bulletin/980609052
National Earthquake Information Center
    http://wwwneic.cr.usgs.gov/
New Madrid Seismic Zone
    http://www.nytimes.com/yr/mo/day/news/national/
    science/sci-midwestquake.html
NOAA. Tsunami site

http://www.pmel.noaa.gov/tsunami/
U.S. Geological Survey. Hazards site
    http://www.usgs.gov/themes/hazard.html
Latest earthquake information
    http://quake.wr.usgs.gov/QUAKES/CURRENT/
    current.html

◀▶ World Wide Web Activities

Information on the Taiwan Earthquake of 1999
http://www.eeri.org/Reconn/Taiwan1299/TaiwanFinal.html

This multi-page Web site has both text and pictures detailing the geological setting and damage from the Chi-Chi earthquake of September 1999. Contributors are seismologists, engineering geologists, and structural engineers.

1. What kind stress from plate tectonics is acting on the island of Taiwan?
2. What kinds of damage did manufactured structures sustain from the earthquake? What effects did it have on natural ground?

Factors Affecting Tsunami Generation in the Pacific Northwestern United States
http://walrus.wr.usgs.gov/tsunami/research1.html

The Cascadia subduction zone offshore of the Pacific Northwest is being intensely studied for earthquake and tsunami risk. This Web site examines the fault movement parameters that affect the generation of tsunami.

1. What three factors do the authors conclude affect the energy and direction of earthquake-generated tsunami?
2. Read the text and run the animation of an oblique slip seafloor fault and the tsunami it generates. Which movement component of oblique slip generates the secondary tsunami wave that threatens the Oregon coast in the animation?

Fault Rupture of the Landers Earthquake of 1992, Mojave Desert, California
http://www.scecdc.scec.org/landersq.html

The Landers, California earthquake of 1992 in the Mojave Desert was the second greatest quake to hit that state in the twentieth century, exceeded only by the 1906 San Francisco event. Read the earthquake "specifications" on this master page, and then navigate to linked sites to answer the following questions.

1. Link to *Rupture Movie*. How many faults on the cross section were involved in this complex seismic event? In which direction did earthquake energy propagate during the quake?
2. Link to *Aftershock Animations*. Describe the relationship of the faults in the vicinity of the main shock to the San Andreas Plate Boundary fault.

Details on How and Why Structures Fail during Earthquakes
http://www.eerc.berkeley.edu/bertero/

Designing structures to withstand earthquakes is a major engineering discipline. This Web site has an extensive collection of slides showing structural failure of various kinds, with explanations.

Earth Systems Today CD Correlations

CHAPTER CONCEPT: The Pacific Rim Is Vulnerable to Tsunamis

Because 85 percent of earthquake energy release is along the margins of the Pacific Ocean, coastal areas around the basin face continual tsunami threat from quakes both nearby and at a distance.

SECTION MODULE: Earthquakes and Tsunami, Main Activity: Seismic Case Study of Alaska, 1964

Read the **Overview** on this powerful quake, and then navi-

gate to **Tsunami,** and read text and maps related to the tsunami generated by this quake.

QUESTION: *What was the effective distribution of this tsunami? Why did it reach northern California hours before hitting the western Alaska coast in the Bering Sea?*

CHAPTER CONCEPT: The Epicenter of a Quake Can Be Found Using Seismogram Data

The difference in arrival time of P and S waves will fix the distance of the seismograph to the earthquake epicenter, and three reporting stations will triangulate the point.

SECTION MODULE: Earthquakes and Tsunami, Main Activity: Seismic Case Study of Alaska, 1964

Read the **Overview** to this procedure in the CD-ROM; here you will begin the process of finding the epicenter of the Alaska, 1964 event.

QUESTION: *Which station used was the farthest from the epicenter? How do you know?*

Volcanoes

FIGURE 1

"Ashes, Ashes, All Fall Down"

One night in 1982 a British Airways Boeing 747 on a routine flight from Kuala Lumpur, Malaysia, to Australia cruised at an altitude of 12,000 meters (37,000 ft). Just before midnight, sleeping passengers were awakened by a pungent odor. Through the windows, they saw the huge plane's wings lit by an eerie blue glow. Suddenly the number-4 engine flamed out, followed almost immediately by the other three. The plane glided silently for an agonizing 13 minutes. At 4,500 meters (14,500 ft), the number-4 engine was restarted, then numbers 2, 1, and 3. Nonetheless, an emergency was declared, and the plane landed in Jakarta, Indonesia, with only three engines operating.

This near-death experience gained the attention of the world's airline passengers and pilots. Before this, such a failure seemed virtually impossible in modern aircraft, which have backup systems for almost every contingency. Flying air-gulping jet engines through clouds of volcanic ash, however, is not one of them.

In December 1989, a KLM Boeing 747 encountered airborne ash from Redoubt Volcano at about 8,500 meters (28,000 ft) during a descent to land at Anchorage, Alaska (Figure 1). All four engines flamed out, and the large aircraft suddenly became a glider. After a 4,100-meter fall (13,300 ft, more than 2 mi), the engines were restarted, and the plane went on to make what was described as an "uneventful" landing. All four engines required replacement, as did the windshield and the leading edges of the wings, flaps, and vertical stabilizer, which had all been "sandblasted." One may wonder what it takes to make a landing "eventful." The plane's interior was filled with ash, and the seats and avionic equipment had to be removed and cleaned. It cost $80 million to return the aircraft to service.

Ash clouds are difficult to distinguish from rain clouds, both visually and with radar. The ash cloud from Mount Pinatubo's 1991 eruption traveled westward more than 5,000 miles in three days from the Philippines to the east coast of Africa. Twenty aircraft were damaged by the cloud, most of them while flying more than 1,000 kilometers (600 mi) from the eruption.

North Pacific air routes are some of the busiest in the world. Since 1980, flying through ash clouds on those routes has caused damage to at least 15 aircraft (including the KLM flight), and worldwide there have been more than 100 inadvertent ash-cloud entries. The Alaskan Peninsula and the Aleutian Islands have 40 historically active volcanoes, and the Kamchatka Peninsula has even

(a)

(b)

◆ FIGURE 2 More than 10,000 passengers and millions of dollars in cargo are flown daily on these North Pacific and Far East air routes. (a) The routings (red lines) pass over or near more than 100 active volcanoes (red triangles) on this subduction-zone complex. (b) Air traffic to the Far East and Asia was disrupted in September 1994 when Mount Kliuchevskoi (4,750 m, 15,580 ft) on the Kamchatka Peninsula erupted, spewing ash into high-altitude airlanes. This photo was taken by astronauts on board the *Endeavor*.

(a)

(b)

◆ FIGURE 3 Mount Spurr Volcano, a potential hazard to aviation. (a) Scientists from the Alaskan Volcano Observatory and the University of Alaska install seismometers at Mount Spurr to detect magma activity below the volcano. (b) The ash cloud from Mount Spurr's 1992 eruption traveled across Canada and the United States on prevailing westerly winds.

more (◆ Figure 2). In 1992 Mount Spurr Volcano on the Alaskan Peninsula disrupted air traffic several times in the United States and Canada (◆ Figure 3).

Historically, there are an average of 5 eruptions per year along the 2,400-nautical-mile great circle from Alaska to northern Japan. On average, volcanic ash is present 4 days each year above 9,300 meters (30,000 ft), where passenger jets fly. A massive effort in place to accurately forecast eruptions—the best way to address this potential hazard—involves the Alaskan Volcano Observatory (AVO), the U.S. Geological Survey (U.S.G.S.), University of Alaska, Federal Aviation Administration (FAA), National

Oceanic and Atmospheric Administration (NOAA) Weather Service, Michigan Technological University, and the Russian Institute of Volcanic Geology and Geochemistry in Petropavlovsk-Kamchatka. Forecasting data are collected by monitoring seismic activity in the volcanic arc, reviewing satellite imagery to detect ash plumes, and making overflights of the 40 active volcanoes to measure gas emissions, which precede many eruptions—all to help make "flying the friendly skies" of the North Pacific and Far East safe for the traveler. (For more information see http://www.avo.alaska.edu/) ■

Although people commonly regard volcanoes and volcanic activity as environmentally harmful and life-threatening, volcanoes are not all bad. Many of them, such as Japan's Mount Fujiyama, are esthetically beautiful, some offer outstanding skiing and hiking opportunities, as does Oregon's Mount Hood and California's Mammoth Mountain; and volcanic soils, such as those on the Island of Hawaii, are highly productive. The eruption of Italy's Mount Vesuvius in A.D. 79 illustrated why volcanoes present a problem to humans (Case Study 5.1). An active volcano may be dormant for hundreds or even thousands of years, and human settlement may occur nearer and nearer to the volcanic cone during that time. Like a hibernating bear, the sleeping giant may suddenly awaken and vent an explosive eruption that wreaks havoc upon the surroundings, as did Mount St. Helens in 1980. Today, however, we can forecast an eruption with a good deal of accuracy, as explained later in the chapter. Knowing that a potential volcanic hazard exists allows policy makers and emergency managers to plan for various eruption and evacuation scenarios, thus reducing casualties when an eruption does occur.

Who Should Worry

Most geologic activity that is dangerous to humans occurs along plate boundaries. This is where we find earthquakes and active volcanoes. Explosive volcanic activity, which presents the greatest challenge to life and property, occurs mostly at convergent plate boundaries. Inspection of the Pacific Ocean basin shows that it is surrounded by trenches, which form at convergent plate boundaries. Associated with these subduction zones is the so-called Ring of Fire, the location of two thirds of the world's active violent volcanoes. A recent catalog of the 1,350 volcanoes that have been active within the last 10,000 years (*Holocene time* to the geologist) shows 900 of them around the Pacific Rim. The largest number are found in New Zealand, Japan, Alaska, Mexico, Central America, and Chile (◈ Figure 5.1). Mount Erebus in Antarctica is the southernmost active volcano in this belt. About 250 active volcanoes are found in the Mediterranean, including the famous volcanoes Vesuvius, Etna, and Stromboli. Stromboli has been called the "lighthouse of the Mediterranean" because of the nearly continuous activity in its crater for more than a century.

The most productive volcanic centers, which exhibit less spectacular, comparatively mild activity, occur along divergent plate boundaries at mid-ocean ridges. Hot-spot volcanism may be explosive or mild and may occur on continents or in ocean basins (see also Chapter 3).

There are twice as many volcanoes north of the equator as there are south of it. This is particularly interesting to meteorologists, because large amounts of volcanic ash in the stratosphere result in a net heat loss for the earth, which affects climate. Any concentration of active volcanoes at a particular latitude must be considered as a potential agent of change in global climate. Although most climatic variability is related to causes other than volcanism, historical examples of volcanic eruptions affecting weather are those of Tambora (Indonesia, 1815), El Chichón (Mexico, 1982), and most recently, Mount Pinatubo (the Philippines, 1991). Because two thirds of the world's active volcanoes and land area are north of the equator, the Northern Hemisphere is more vulnerable to climatic impact by volcanic eruptions than is the southern half of the world.

Mapping of the potential volcanic risk in the United States, such as ◈ Figure 5.2, reveals that the risk is seen as nil east of New Mexico. This is understandable in light of plate tectonics. The odd-shaped projections from the high-risk area on the map are due to the estimated transport of ash on the prevailing winds. The northeast–southwest trending high-risk zone through Idaho and into the corner of Wyoming is the trace of a southwestward transport of North America over a stationary hot-spot. A line of calderas and lava flows mark the path, with Yellowstone being the currently active center over a fixed hot-spot plume (◈ Figure 5.3). This sequence is similar geologically to that of the Hawaiian chain of volcanoes.

The Cascadia subduction zone is responsible for the many active volcanoes and volcanic centers in Washington (6), Oregon (21), and California (16; see Figure 5.3). The zone of active volcanoes extends 1,100 kilometers (700 mi) from Mount Lassen in California northward into British Columbia. Alaska ranks as the second-most active volcanic region in the world, and Hawaii is not far behind. Thus the study of volcanoes is important to citizens of the United States.

◆ FIGURE 5.1 Distribution of earth's active volcanoes at plate boundaries and hot spots. Note the prominent "Ring of Fire" around the Pacific Ocean, which contains 900 (66 percent) of the world's active volcanoes. The remaining 450 are in the Mediterranean belt (subduction zones) and at mid-ocean-ridge spreading centers (divergent boundaries). A few important volcanic centers are related to hot spots, such as those in the Hawaiian and Galápagos Islands.

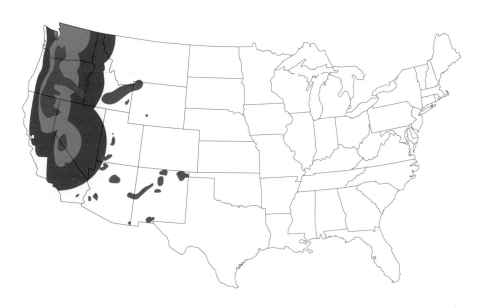

◆ FIGURE 5.2 Map of potential volcanic hazards in the 48 contiguous states. The red areas are subject to lava flows, lateral blasts, ash falls, and mudflows. The severity of risk decreases from red (greatest), to green, to blue (least).

Case Study | 5.1

New Zealand's Blast in the Past, a Skifield, and Fluidization

The two islands of New Zealand are clearly distinguishable geologically. Whereas North Island exhibits explosive volcanism and some of the most spectacular volcanic landforms in the world, South Island lacks volcanism and is more alpine in nature. North Island's volcanic zone includes Lake Taupo, the site of a cataclysmic volcanic eruption in A.D. 186 that was arguably the largest eruption on earth in 5,000 years (◆ Figure 1). A volcanic source, now beneath Lake Taupo, vented a pyroclastic column more than 50 kilometers (30 mi) high, which spread downwind, forming light-yellow ash-fall deposits more than 5 meters (8 ft) thick in some places. At 100 kilometers (60 mi) distant the deposits are 25 centimeters (10 in) thick. Toward the end of the eruption sequence the hot column became so heavy that it collapsed and flowed outward in all directions as a pyroclastic flow. Flow velocity is estimated to have been 600 kilometers per hour (375 mph), and the material traveled more than 80 kilometers (50 mi) from its source, overwhelming all the topography and vegetation in its path. Fortunately, the native New Zealanders, the Maori, had not yet settled the island when the eruption occurred. When they did come to the island about A.D. 1000, they regarded the Tongariro–Taupo area as sacred. They believed that their gods and demons had brought volcanic fire with them from Hawaiki, the Maori's Pacific homeland.

The aftermath of this eruption was a huge explosion caldera, which is now filled with water and known as Lake Taupo. The area is still hot, and north of the lake are the Rotorua thermal springs and geysers, which attract many tourists, and the Wairaki geothermal energy complex (see Chapter 14).

How can a pyroclastic material move at such a phenomenal speed and travel such a great distance, seemingly defying friction and gravity? Some pyroclastic flows are **fluidized,** a process by which solids are transported that is well known in chemical engineering. A "fluidized bed" is a body of fine-grained solids through which a stream of high-pressure gas flows so that the grains separate and move as fluid. Many chemical engineering professors demonstrate this by forcing air through a powder on the bottom of a see-through box. As the air pressure is increased, the powder begins to quiver; then particles start dancing around, and eventually the whole mass is in motion. When the professor has it just right, he or she places a plastic duck (professor's choice) into the container, and it literally floats on the air–solid mixture. If "ducky" is pushed down, it immediately pops up, proving that it is floating on a denser medium. Part of the weight of the solids is taken by the air, reducing the interparticle friction to such a degree that the particles will not stand in a heap. The particles are not *entrained* (carried away bodily) like snowflakes in a blizzard; rather, the mass behaves as a fluid substance would. The point of this is that a well-understood process demonstrates how pyroclastic flows can fluidize and travel far from their source, destroying everything in their path.

At the south end of the Taupo Volcanic Zone is a cluster of stratovolcanoes in the Tongariro National Park. Two of the volcanoes are active and are popular playgrounds for New Zealanders, primarily for skiing and tramping (hiking). Mount Ruapehu, with a summit elevation of 2,797 meters (9,100 ft), is the highest point on North Island and has a lake in its summit crater. It is also the location of the Whakapapa Skifield, a very popular, well-equipped ski area. The terrane at Whakapapa (Maori, pronounced "whack-a-papa") is bare aa lava and requires a lot of snow to be skiable (◆ Figure 2). Throughout the skifield (or lava field, depending on the time of year) signs alert skiers and trampers to stay on high ground and out of the canyons should Ruapehu erupt, which it did in 1969, 1971, 1982, and sensationally in 1995. The eruptions are *phreatic;* that is, superheated water at the bottom of the lake flashes to steam and overflows the crater. The lake's surface-water temperature, which ranges from 20°C to 60°C (68°–140° F), is monitored as a potential eruption predictor. When the water heats up, some kind of activity can be expected. A lahar warning system at Whakapapa is based on detection of earthquake swarms, which precede phreatic and magmatic eruptions. Tongariro (Maori *tonga-,* "south wind," and *-riro,* "carried away") National Park is another case of humans putting nature to good use.

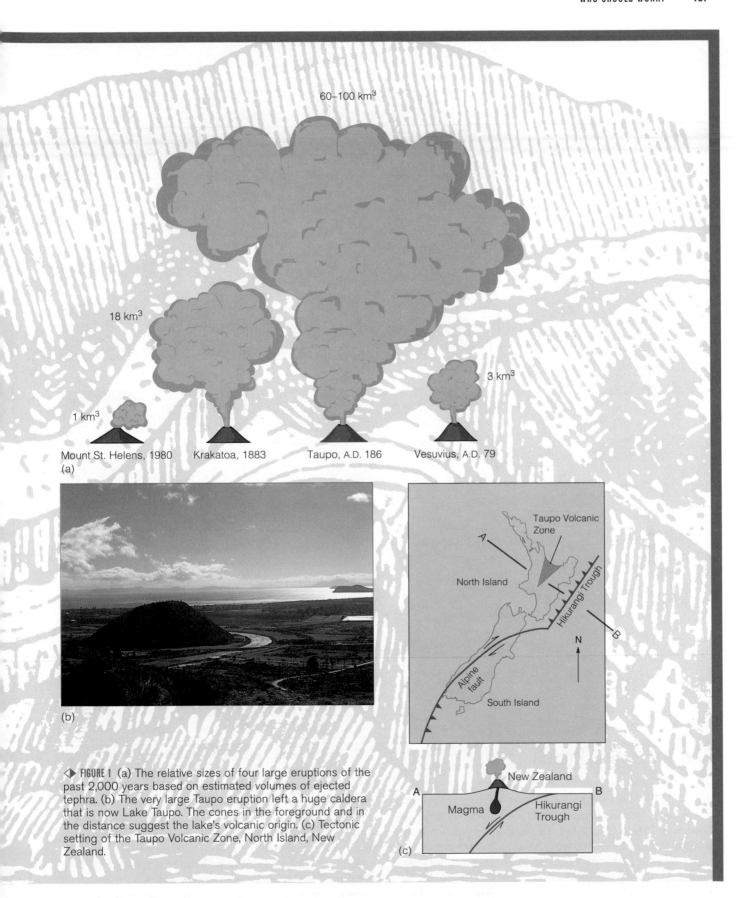

60–100 km³

18 km³

3 km³

1 km³

Mount St. Helens, 1980   Krakatoa, 1883   Taupo, A.D. 186   Vesuvius, A.D. 79
(a)

(b)

Taupo Volcanic Zone

A

North Island

Hikurangi Trough

B

N

Alpine fault

South Island

New Zealand

A   B

Magma   Hikurangi Trough

(c)

◆▶ FIGURE 1 (a) The relative sizes of four large eruptions of the past 2,000 years based on estimated volumes of ejected tephra. (b) The very large Taupo eruption left a huge caldera that is now Lake Taupo. The cones in the foreground and in the distance suggest the lake's volcanic origin. (c) Tectonic setting of the Taupo Volcanic Zone, North Island, New Zealand.

◆ **FIGURE 2** (a) Whakapapa Skifield is on volcanic Mount Ruapehu, the highest point on New Zealand's North Island and the site of a beautiful summit crater lake. Signs on the slopes warn skiers to stay out of canyons and low spots should a phreatic eruption create a dangerous lahar. (b) On 25 September 1995 a combination of hot magma and cold water produced a violent steam eruption that hurled boulders over the upper slopes of the Whakapapa Skifield. The eruption did little damage, but projectiles as big as cars were thrown over the rim. The Maori name for Ruapehu means "exploding pit."

(a)

(b)

The Nature of the Problem

How a volcano erupts determines its impact on humankind. A cataclysmic eruption, such as that of Mount St. Helens in 1980, has serious results. Simple outpourings of lava such as at Kilauea in Hawaii, on the other hand, can be good for tourism and business (except agriculture). Unfortunately, there is no scale for describing the "bigness" of volcanic eruptions as there

is for earthquakes. ◆ Figure 5.4 shows some of the criteria commonly used to describe an eruption's "explosivity." **Volcanic Explosivity Index (VEI)** values range from 0 to 8 according to the volume of material ejected, the height to which the material rises, and the duration of the eruption. Its "Classification" scale associates the particular eruption with a well-known volcano that exhibited the same kind of activity. The "Description" scale employs adjectives such as those used in newspaper headlines to describe the eruption. The

Active volcanoes and volcanic centers in the continental United States. The volcanoes of the Cascades are geologically related to the Cascadia subduction zone.

- Volcanoes that have short-term eruption periodicities (100–200 years or less) or that have erupted in the past 200–300 years, or both.

- Volcanoes that appear to have eruption periodicities of 1,000 years or greater and that last erupted 1,000 years or more ago.

- Volcanic centers that are greater than 10,000 years old, but beneath which exist large, shallow bodies of magma that are capable of producing exceedingly destructive eruptions. The Snake River plain is the track of the Yellowstone hot spot.

1980 eruption of Mount St. Helens, for example, could be indexed as a *4* and appropriately described as an *explosive-to-cataclysmic* eruption.

A volcano's potential explosivity has been found to increase with time since the start of its previous eruption (◀▶ Figure 5.5). This fact allows a region's risk of volcanic eruptions to be assessed when the date of the volcano's last eruption is known. Some cataclysmic eruptions, such as the one that took place at Yellowstone hot spot 600,000 years ago, are considered a minor threat because of their rarity. Thus, both magnitude and frequency must be considered in evaluating volcanic hazards.

A lava's *viscosity*—its resistance to flow—is another determinant of a volcano's explosivity, and viscosity is a function of lava's temperature and composition. Viscosity varies inversely with temperature for most fluids,

including lava: the higher the temperature, the less viscous; the lower the temperature, the more viscous, other factors being equal. A lava's composition, most importantly its silica (SiO_2) content, also influences its viscosity. The silicon-oxygen bond is very strong; considerable heat energy is required to break it. The fluidity of siliceous (containing silica) lavas depends on the continual breaking and remaking of this bond. Thus, the higher the SiO_2 content, the more viscous is the lava—again, other factors being equal.

Viscous lavas with a high gas content can build up such high gas pressures as to be explosive, whereas gases escape readily from fluid lavas, and hence, they are much less dangerous. This fact gives us another method of categorizing large volcanoes and their activity: the nature of the magmas they tap. Magmas form at depths of 50 to 250 kilometers (30–150 mi), where

◀▶ **FIGURE 5.4** Graphic representation of the Volcanic Explosivity Index (VEI). As an example, a VEI 5 would be a very large eruption, described as *cataclysmic, Vulcanian* in its type of eruption, and have an ash plume up to 25 kilometers (15 mi) high. There have been 19 historic VEI-5 eruptions.

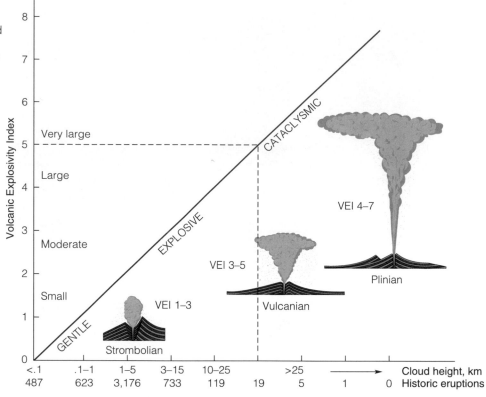

◀▶ **FIGURE 5.5** The longer the time interval since an eruption, the greater is the next eruption's potential explosivity. These data for 4,320 historic eruptions relate known intervals between eruptions to the Volcanic Explosivity Index (Figure 5.4). Also shown is the percentage of eruptions in each VEI that have caused fatalities.

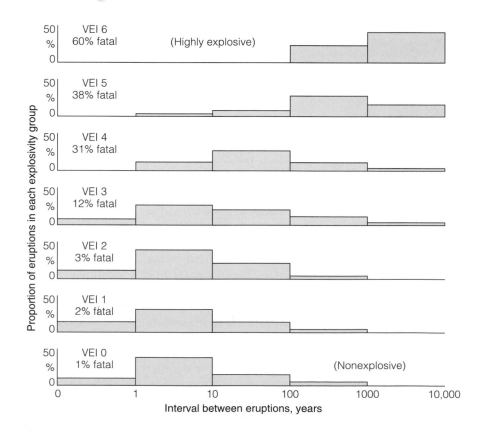

temperatures are sufficiently high to melt rocks completely or partially. At mid-ocean ridges and hot spots, volcanoes draw from magmas in the upper mantle that are high in iron and magnesium and low in silica (50 percent or less). These are **mafic** magmas (*ma-* for magnesium and *-f-* for the Latin word for iron, *ferrum,* plus *-ic*), which yield fluid (low-viscosity) lavas that retain little gas and hence do not erupt violently. Volcanoes that are adjacent to subduction zones, on the other hand, tap magmas that are a mixture of oceanic crust and sediment, upper-mantle material, and melted continental rocks. These are **felsic** magmas (*fel-* for feldspar and *-s-* for silica, plus *-ic*), which yield thick, pasty lavas with SiO_2 contents up to 70 percent. Even though hot lavas have lower viscosities and flow more readily than do cool ones, just as with honey and various other common fluids, the major determinant of lava fluidity is silica content.

High-silica, high-viscosity lavas retain gases, which leads to violent, explosive-type eruptions. The geologically important boundary between mild oceanic eruptions and the more explosive continental ones of the Pacific basin is called the **andesite line** (after the intermediate rock andesite from the Andes Mountains). The line is generally drawn southward from Alaska to east of New Zealand by way of Japan, and along the west coasts of North and South America. The andesite line is also a petrologic boundary between mafic magmas, which yield basalt, and felsic magmas, which yield andesite or some other high-silica rock like rhyolite. Thus, explosive volcanoes are always found on the continental side of the andesite line, and gentle volcanoes are found within the Pacific Ocean basin.

Types of Eruptions and Volcanic Cones

The type of volcanic eruption determines the shape of the structure or cone that is built, and by appearance alone, one can get an idea of a volcano's hazard potential (◈ Figure 5.6).

EFFUSIVE ERUPTIONS

Shield Volcanoes

Shield volcanoes are built by gentle outpourings of low-silica fluid lavas from a central vent or conduit. The lavas cool to form basalt, the most common volcanic rock. A shield volcano's profile is gently convex upward like that of a shield laid on the ground. Mostly oceanic in origin, shield volcanoes are found in Iceland (where the *shield* name was first applied), the Galápagos Islands, and the Hawaiian

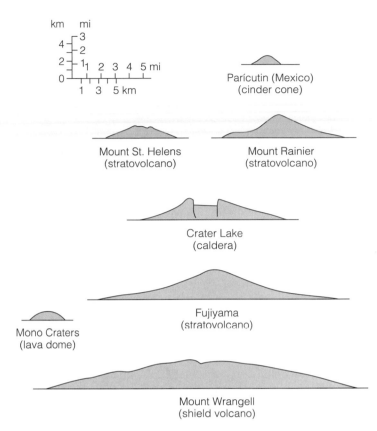

◈ **FIGURE 5.6** Comparative profiles and relative sizes of some Pacific rim volcanoes and the Mono Craters in the western interior of the United States.

Islands (◈ Figure 5.7). They are built up from the sea floor, layer upon layer, to elevations thousands of meters above sea level. Mauna Loa ("long mountain") and Mauna Kea ("white mountain") volcanoes on the Island of Hawaii are shield volcanoes that project 4.5 kilometers (2.8 mi) above sea level, and their bases are in water about 5 kilometers (3 mi) deep. With a total height of 9.5 kilometers (31,000 ft), these are the highest mountains on earth, exceeding Mount Everest by about 650 meters (2,150 ft).

The Island of Hawaii (the "big" island to locals) is composed of five separate volcanoes that have built the island over the past 700,000 years (◈ Figure 5.8). Kilauea (Hawaiian, "much spreading") is the most easterly of the group and is the most active. This is exactly what one would predict as sea floor moves over the fixed Hawaiian hot spot.

A new volcano, Loihi ("the long one"), is growing on the flanks of Kilauea 28 km southeast, providing further evidence of the northwest motion of the Pacific plate over a hot spot. The summit of Loihi is about 1,000 meters below sea level, and it rises 3,500 meters from the Pacific floor (◈ Figure 5.9). In 1996 it gave rise to the largest swarm of earthquakes ever

◁▶ **FIGURE 5.7** Shield volcano Fernandina in the Galápagos Islands. This active volcano is related to the Galápagos "hot spot." Its convex profile is similar to those of the volcanoes of Hawaii.

recorded near any Hawaiian volcano—more than 4300 earthquakes in a month. Loihi is believed to be fed by the same 50-km-wide hot spot that is the source of Mauna Loa and Kilauea volcanic activity. Loihi will probably pop its head above water in about 50,000 years and become a tourist attraction.

It is not uncommon for a shield volcano to erupt from a *fissure,* or crack, on its flanks, rather than from a central vent. This is typical of Kilauea's east fissure zone, where the countryside has been flooded under a sea of lava. Pu'u O'o ("hill of the o'o bird") is a large crater built upon a vent in the east fissure zone. Pu'u O'o's crater has reached impressive proportions as a result of scores of eruptions since 1983. During Pu'u O'o's eruptions, Kilauea's summit deflates slightly, rising again or reinflating between eruptive episodes. This indicates that the fissure zone and Kilauea's vent plumbing are connected with the main lava reservoir. Lava flows associated with Pu'u O'o caused significant damage in the Royal Gardens subdivision near Kalapana about 8 kilometers (5 mi) from the vent. Flows destroyed at least 75 homes and covered 10 kilometers (6 mi) of residential streets (◀▶ Figure 5.10). In addition, lava has built out hundreds of meters into the sea, creating new lands.

Spectacular lava fountains 400 meters (1,300 ft) high have occurred with fissure eruptions at Pu'u O'o (◀▶ Figure 5.11). Some of the more fluid droplets set into glassy, tear-shaped blobs or tiny threads, known, respectively, as "Pele's tears" and "Pele's hair," named after the Hawaiian goddess of volcanoes (◀▶ Figure 5.12). Birds have been known to make nests from some of the more pliable threads of volcanic glass.

(a)

◀▶ **FIGURE 5.8** The Island of Hawaii. (a) Satellite photograph with north at the top. Mauna Kea and Mauna Loa are clearly visible. (b) The volcanoes that form the island have grown over 700,000 years in this order: Kohala (indicated by K), Mauna Kea (MK), Hualalai (H), Mauna Loa (ML), Kilauea (KI), and Loihi (L; submarine). Solid stars denote vigorous growth; open stars, waning growth; open circles, little activity. The island to the northwest is Maui. Its Haleakala Volcano is shown.

(b)

(a)

(b)

◇▶ **FIGURE 5.9** Locations of earthquakes near submarine Loihi Volcano, 1971–1972 and 1975. Similar earthquake swarms occurred in 1984–1985, providing further evidence that Loihi is an active volcano.

Continental Flood Basalts

The landscapes of some areas of the world are dominated by thick, flat-lying basalt flows that form immense plateaus. The Columbia River plateau in Washington, Oregon, and Idaho, and the Deccan plateau in northwestern India to name two (◇▶ Figure 5.13). In scale they cover many times the area of the large volcanoes. When deeply eroded, these "flood basalts" form stairstep topography consisting of flat "treads" (soft rock) and vertical "risers" (hard rock). This geomorphic expression led early geologists to call the rocks collectively *traps,* the Swedish word for "staircase." When you

◇▶ **FIGURE 5.10** (a) A lava flow ignores a stop sign in the Royal Gardens subdivision below the east fissure zone. (b) A flow set this house afire at Kalapana near the Royal Gardens. (c) The historic church at Kalapana was moved just before lava engulfed the area.

(c)

◀▶ **FIGURE 5.11** Lava fountaining at Pu'u O'o, Kilauea Volcano east fissure zone, Island of Hawaii.

(a)

(b)

◀▶ **FIGURE 5.12** (a) Artist's rendition of the revered Hawaiian volcano goddess Pele. She lives in Kilauea Volcano, and her hair resembles pahoehoe lava. (b) "Pele's hair," thin filaments of glassy lava that have been carried by wind.

hear or read terms like "the Deccan Traps" or "the Siberian Traps," this is the reason.

What is the explanation for these great outpouring of lavas—the Deccan Traps are 1–2 kilometers (3,000–6,000 ft) thick and cover a half-million square kilometers? The process is believed to be a superplume (see Chapter 3) originating at the mantle–core boundary that is several hundred kilometers in diameter. When the superplume rises to the base of the lithosphere it is 200°–300° C hotter than the surrounding mantle, and the rock at the mantle–lithosphere boundary melts. Superplumes are now seen as the basalt "fire hydrants" that form the plateaus. Some traps can be matched across continents—the Parana of Brazil and the Etendeka of Namibia, for instance—giving more credence of Wegener's original ideas on drifting continents. The Deccan Traps have been linked by some to the extinction of the dinosaurs 65 million years ago (late Cretaceous time), because the massive outflow of lava would have been accompanied by sulfurous fumes sufficient to kill large animals.

EXPLOSIVE ERUPTIONS

Stratovolcanoes

A combination of effusive and explosive volcanic activity builds **stratovolcanoes,** sometimes referred to as

composite cones, because they are composed of layers of both pyroclastic material and lava. Typically thousands of feet high and 10–20 kilometers (6–12 mi), across at the base, they have a concave upward profile and a central vent. They are stratified, thus the name *stratovolcano,* consisting of alternating layers of ash, cinders, and lava.

The upper, steep slopes are formed mostly of pyroclastic volcanic ejecta, and the less steep, lower slopes are composed of alternating layers of lava and pyroclastics. These cones are some of the most beautiful tourist attractions in the world. Noteworthy examples are Mount Vesuvius in Italy, Mount Fujiyama in Japan, Mount Hood in Oregon, Mount Rainier in Washington, and Volcan San Cristobal in Nicaragua (◀▶ Figure 5.14). They occur on the landward side of

(a)

(b)

◆▶ **FIGURE 5.13** (a) Fissure eruptions forming a lava plateau. (b) Antoja caves, Buddhist shrines excavated into the basalts of the Deccan lava plateau, India.

(a)

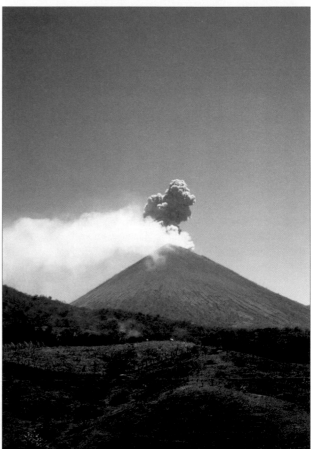

(b)

◆▶ **FIGURE 5.14** Two beautiful examples of stratovolcanoes from widely separated geographic regions. (a) Mount Rainier, Washington, labeled by many volcanologists "the most dangerous volcano in the United States." (b) Volcan San Cristobal in Nicaragua; 1997 eruption.

subduction zones, where melting of descending oceanic crust and mantle forms magma. The magmas rise because, as they melt, they expand, becoming more buoyant than the surrounding lithosphere. As the magma penetrates the continental rocks, partial melting occurs, and the magma becomes enriched in potassium and silica. This mixing and partial melting changes the volcanism from undramatic basalt to explosive andesitic and high-silica rhyolitic volcanism. As the magma rises it becomes gas-charged, and the high-silica lavas, rhyolites, dacites, and the like, are very viscous and thus do not flow easily. They may congeal in the volcano's central conduit (vent), which can cause gas pressures to build to explosive proportions. For this reason, stratovolcanoes present the most immediate threat to humans. Some eruptions have been so great

that large cones have simply disappeared in the explosion. Crater Lake in Oregon represents the stump of the former Mount Mazama, a very large stratovolcano; it was probably the size of Mount Rainier. About 6,900 years ago, 70 cubic kilometers (17 mi³) of Mount Mazama disappeared, due in part to explosive activity, but mostly due to the collapse of the remaining cone into the deflated magma chamber beneath it. The "crater" of Crater Lake is a **caldera,** defined as a volcanic crater that is many times larger than the vent that feeds it (◆ Figure 5.15). Calderas may form by explosive disintegration of the top of a volcano, by collapse into the magma chamber, or by both mechanisms (see Case Study 5.1). Yellowstone National Park occupies several huge calderas, as does Long Valley in the eastern Sierra Nevada, both of them lying above shallow bodies of magma that are capable of producing destructive eruptions.

Lava Domes

Lava domes are formed when bulbous masses of lava pile up around the vent because the lava is too thick and viscous to flow any significant distance from its source. Sometimes called *volcanic domes,* they usually grow by expansion from within. As the outer surface cools, the brittle crust breaks and tumbles down the sides, as at the Mono Craters on the eastern Sierra Nevada of California (◆ Figure 5.16). Relatively small domes may form in the crater of a larger composite cone, such as the one that formed in the crater of Mount St. Helens after the

1980 eruption (◆ Figure 5.17). The source vents of large domes may be substantially plugged, which offers the potential for explosive eruptions, particularly where the lavas have access to underground water or seawater. Mount Pelée on Martinique in the West Indies and Mount Lassen and Mammoth Mountain in California are all dormant lava domes. They extrude lavas such as rhyolite with silica (SiO_2) contents of 65–75 percent, and they also extrude glassy rocks that cool quickly, such as obsidian and pumice.

Cinder Cones

Cinder cones are the smallest and most numerous of volcanic cones (◆ Figure 5.18). They are built of pyroclastic material of all sizes—from blocks and bombs to the finest ash (◉ Table 5.1). Pyroclastic material of all shapes and sizes is collectively known as **tephra.** *Bombs* are blobs of still-molten lava that assume an aerodynamically induced spindle shape and solidify before striking the ground (◆ Figure 5.19). They can be found in great numbers around certain cinder cones and are prized by collectors and geologists. Cinder cone activity is local, within a few kilometers of the source vent, and is usually short-lived. The only new volcano to form in historic time in North America erupted in a farm field in 1943 near the village of Parícutin in the state of Michoacán, Mexico (Figure 5.18). Eruptions caused by frothing gases hurled blobs of lava into the

◆ **FIGURE 5.15** Crater Lake is a caldera, resulting from the explosion–collapse of a huge stratovolcano known as Mount Mazama, about 6,900 years ago.

◆ **FIGURE 5.16** Lava domes; Mono Craters, east-central California. The Sierra Nevada range is in the background.

	TABLE 5.1	**Classification of Pyroclastic Ejecta (Tephra)**

NAME	SIZE	CONDITION WHEN EJECTED
Blocks	>32 mm	Cold, solid
Bombs	>32 mm	Hot, plastic
Lapilli (cinders)	4–32 mm	Molten or solid
Ash	¼–4 mm	Molten or solid
Dust	<¼ mm	Molten or solid

air that became cinders and then fell back around the vent, eventually building a cone nearly 400 meters (1,300 ft) high. An observatory was established on a nearby hill, and the volcano's every burp and belch was recorded for nine years. After the cone-building stage had passed, basaltic lava flowed from the crater, inundating the nearby pueblo of San Juan and removing about 160 square kilometers (100 mi²) from agriculture. Fortunately, no lives were lost. Cinder cones are generally thought to be one-shot events; that is, once the eruption sequence ends, it is rarely reactivated. Nonetheless, exceptions are known, and one usually finds many cinder cones within a volcanic area.

Benefits of Volcanic Action

Most of the atmosphere and hydrosphere that we know on the earth today has come from the interior of the earth via volcanic activity. The earth's primitive atmosphere was probably composed of hydrogen and helium, the two most abundant gases found in the universe. Because the young planet was still very hot,

these gases most likely escaped to outer space. A second atmosphere came into being, sweated out from the earth's interior by volcanic eruptions and fumeroles (steam vents, see Figure 5.37). Working on the principle that "the present is the key to the past," we assume that volcanic action spewed out the same gases then as it does today: water vapor (80%), carbon dioxide (10%), and the remainder mostly nitrogen and some rare gases. The predominantly water and carbon-dioxide atmosphere was continually modified by volcanic **outgassing** until sufficient water was supplied to form clouds, from which rain fell to form the oceans. From this start, plants evolved two to three billion years ago, and they consumed much of the carbon in the atmospheric CO_2. As the proportion of

◆ **FIGURE 5.17** (a) Steaming lava dome in the crater of Mount St. Helens, October 1981. (b) Oblique map of the same lava dome, 8 October 1981.

(a)

(b)

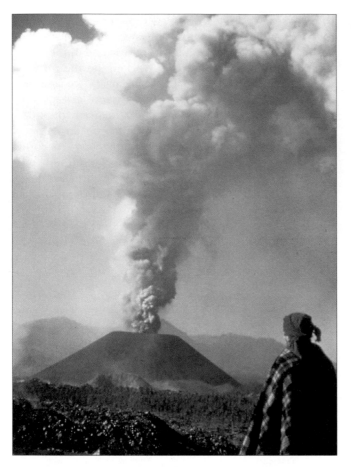

◀▶ **FIGURE 5.18** Parícutin Volcano, Michoacán, Mexico, in 1943. The only totally new volcano in North American history, it began as a fuming crack in a farmer's field. In a short time it was 400 meters (1,312 ft) high, and lava from it covered two nearby villages, leaving only a church steeple protruding through the lava. Volcanic activity decreased rapidly, and the volcano was inactive by 1952.

CO_2 decreased, the atmosphere became richer in oxygen and nitrogen. So we can credit volcanoes for our oceans and atmosphere.

Volcanic activity also provides products that we utilize in many ways. The final polish your teeth receive when you get them cleaned is an example. Dental pumice is refined and flavored volcanic pumice with a hardness just slightly less than tooth enamel. "Lava" brand hand soap, a rough, abrasive bar soap, contains the same powdered rock. Light-weight bricks, cinder blocks, and many road-foundation and decorative stone products originated deep within the earth. Where volcanic cinders are mined for road base, they may also be mixed with oil to form asphalt pavement. The reddish pavement of state highways in Nevada and Arizona contains basalt cinders that are rich in oxidized iron, which gives it the red color. Pieces of rock pumice are sold in drugstores, supermarkets, and hardware stores for use as a mild abrasive for removing skin cal-

luses and unsightly mineral deposits from sinks and toilets. Powdered, it is used in abrasive cleaners and in furniture finishing.

Glassy volcanic rock, such as obsidian, is easy to chip and form. Hence, Native Americans and stone-age people in many parts of the world used it to make tools and arrowheads. Today obsidian and many similar volcanic rocks are the raw material of "rock hounds" and artisans for producing polished pieces and decorative art.

Volcanoes provide opportunities for recreational activities, ranging from mountain climbing and skiing to more passive activities such as photography and birdwatching. Of course, skiing on an active volcano poses some measure of risk, as skiers at Mount Ruapehu in New Zealand found out in 1995 (see Case Study 5.1, p. 136).

GEOTHERMAL ENERGY

By far the most important and beneficial volcanic product is also the most hazardous; it is heat. The same heat energy that causes eruptions also drives geysers and hot springs and, when controlled, it can be converted

---

Consider This… ? Although a few areas on the Island of Hawaii are regarded as relatively free from geologic hazards, you might not want to live in one of those "safe" places. Select an area on the island that appeals to you (for its view, beach access, culture, or some other amenity) and determine what geologic hazards you could encounter there. ■

◀▶ **FIGURE 5.19** A volcanic bomb; Galápagos Islands, Ecuador. Note the spindle shape that the bomb acquired as it was flung through the air.

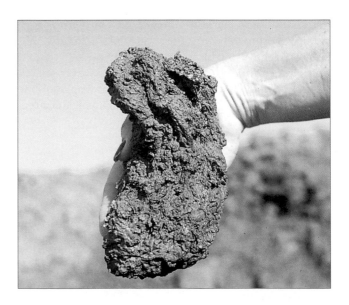

to other uses. Not surprisingly, the prospects for geothermal energy are best at or near plate boundaries, where active volcanoes and high heat flow are found. The Pacific Rim (Ring of Fire), Iceland on the Mid-Atlantic Ridge, and the Mediterranean belt offer the most promise. Energy from earth heat is discussed at length in Chapter 14.

Volcanic Hazards

Sicily's Mount Etna (from the Greek *aitho*, "I burn"), in the Mediterranean volcanic belt, exhibits almost continuous activity in its crater. In 1992 eruptions and lava flows were threatening several villages. Empedocles (circa 490–430 B.C.), Greek philosopher and statesman, is perhaps most notable for throwing himself into the crater of Mount Etna to convince his followers of his divinity. His dramatic suicide inspired Matthew Arnold's epic poem "Empedocles on Etna," which bears no resemblance to the little rhyme attributed to Bertrand Russell:

Empedocles that ardent soul—
Fell into Etna and was roasted whole!

Volcanoes comprise the third-most dangerous natural hazard in terms of loss of life, after coastal flooding (hurricanes and typhoons) and earthquakes. Following the Mount St. Helens eruption of 1980, U.S. civil defense agencies published suggestions for what to do when a nearby volcano erupts. The Federal Emergency Management Agency's suggestions appear later in the chapter. For now, however, let us examine the various types of hazards associated with eruptions, several of which might occur during a single eruptive phase (◆ Figure 5.20).

LAVA FLOWS

Most hazards, natural and human-made, decrease in severity with distance from the point of origin. This is true of earthquakes generally, tornadoes, and falls of volcanic ash, for example. Lava flows may be the

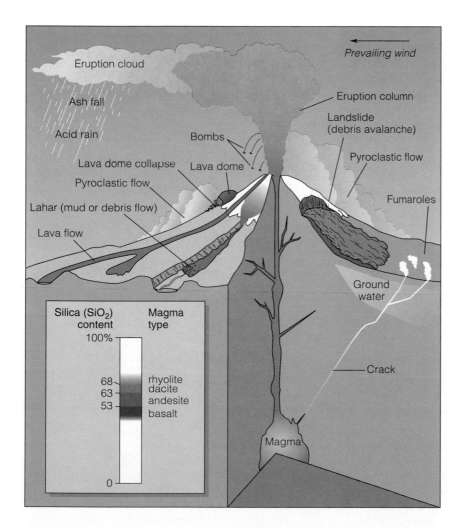

◆ **FIGURE 5.20** Volcanic hazards typical of those found in the western United States and Alaska. Some hazards, such as lahars and landslides, can occur even when a volcano is not erupting. Table 5.1 provides descriptions of pyroclastic ejecta portrayed here.

(a)

(b)

◆▶ **FIGURE 5.21** (a) Smooth, ropy pahoehoe lava; Galápagos Islands, Ecuador. (b) Rough, blocky aa flow in Hawaii. Human approach is possible because the flow moves very slowly.

exception to this, because they generally burn or bury everything in their path, even to their farthest limit. Holocene lava flows in Queensland, Australia, traveled 100 kilometers (62 mi) from their vents down riverbeds with very low gradients. Knowledge of the paths that flows might take from given vents makes it possible to delineate low- and high-risk volcanic-hazard areas. Basaltic lavas, such as those from Mount Etna in 1992, may emerge from a vent or fissure at temperatures in excess of 1,100° C and flow rapidly down-slope like a river because of their low viscosity. As a flow cools, its viscosity increases, and then it may only "chug" along slowly. In either case, the end product is a layer of rock of varying thickness that covers everything in its path.

Two kinds of basalt flows are recognized. **Pahoehoe** ("pa-ho´-e-ho´e") forms when a skin develops on a flow and buckles up as the flow moves, making a smooth or ropy-looking surface (◆▶ Figure 5.21a). Other basalt flows develop upper and lower surfaces that are rough and blocky. This rough, angular lava is called **aa** ("ah´-ah") and it is characterized by a very slow advance in which the top, cold rubble rolls over the front of the flow—moving much like the track of a Caterpillar tractor (◆▶ Figure 5.21b).

A volcanic eruption cannot be prevented, but in some cases, it is possible to divert or chill a flow so as to keep it from encroaching onto croplands or structures. Flows have been diverted in Hawaii and Italy by constructing earthen dikes (◆▶ Figure 5.22). Icelandic fire fighters have successfully chilled flow fronts by spraying them with large volumes of seawater from fire hoses. During World War II, the U.S. Army Air Corps tried, unsuccessfully, to divert lava flows by bombing them.

ASH FALLS

Roman naturalist and historian Pliny the Elder (born A.D. 23) died during the eruption of Vesuvius in A.D. 79 from complications brought on by inhaling noxious fumes and overexertion. He is revered as geology's first martyr, having been quoted by his nephew, Pliny the Younger, as saying, "Fortune favors the brave." Bravery was to be his undoing, as corpulent Pliny died in an area of heavy fallout from the eruption. Fortunately, however, Pliny had enjoyed life: he is credited with coining the phrase *In vino veritas* ("In wine there is truth"). Before Gaius Plinius (Pliny the Elder) died, he

◆▶ **FIGURE 5.22** A dike was constructed to divert a lava flow and save the village of Kapoho, Island of Hawaii, January 1960. Lava fountains are visible in the background.

◀▶ **FIGURE 5.23** Plinian ash cloud over Mount Pinatubo.

◀▶ **FIGURE 5.24** A pyroclastic flow (nuée ardente) hurling down the slope of Augustine Volcano in Alaska, April 1986.

may have seen a great vertical plume of ash with a mushroom- or anvil-shaped head rising from Vesuvius's Crater. This high-velocity column of steam, gas, and fragmented lava is called, for Pliny the Elder, a **Plinian eruption.** Plinian eruptions were seen at Mount St. Helens in 1980, Mount Pinatubo in 1992, and during the Soufrière Hills Volcano eruption sequence on the island of Montserrat in the Antilles from 1995 to 1999 (◀▶ Figure 5.23 and Case Study 5.2). Such an ash cloud, consisting of tephra (jagged pumice, rock fragments, and bits of glass), may be carried away from the crater by prevailing winds and deposited as an **ash fall.** Ash falls, like the ones at Pompeii and Mount St. Helens, bury objects close to their source. At Mount St. Helens, westerly winds carried ash eastward, causing difficulties in Yakima and Spokane, where several inches of ash were deposited. At Butte, Montana, about 750 kilometers (465 mi) east, the fine ash and dust in the atmosphere forced the closure of the airport. People in Yakima wore masks to filter out the choking dust.

Bankers politely requested that all patrons remove their masks before entering their banks because of potential identification problems.

PYROCLASTIC FLOWS

Pyroclastic flows are turbulent mixtures of hot gases and pyroclastic material that travel across the landscape with great velocity. They are generated quickly and with such force that obstructions in their path may be blown down or carried away (◀▶ Figure 5.24). The term **nuée ardente,** French for "glowing cloud," is commonly applied to this kind of eruption because of the intense heat in the cloud's interior. A nuée ardente caused most of the fatalities at Mount St. Helens and devastated an area in excess of 160 square kilometers (◀▶ Figure 5.25).

No other eruption is as famous as that at Mount Pelée on the Island of Martinique in the West Indies on

Case Study

Case Study | 5.2

Montserrat, Paradise Lost

The Soufrière Hills Volcano on Montserrat, a tidy little island and British crown colony in the eastern Caribbean Sea, began a series of eruptions in 1995 that continued episodically into 2000. The eruptions began with a dome of andesitic lava in the central crater that periodically collapsed, generating pyroclastic flows (◆ Figure 1). Ash falls followed the flows and smothered most of Plymouth, the resort capital, and St. Patricks, a resort town. The world followed the eruptions on television and on the internet, particularly the evacuation of half of Montserrat's 12,000 inhabitants away from the shadow of the smoldering peak. Volcanic hazard zones were established in September 1997 by the British Geological Survey's Montserrat Volcano Observatory (◆ Figure 2). To add to Montserrat's problems, it was also subjected to the extreme hurricane season of 1999.

◆ **FIGURE 2** Map of Montserrat Island showing Soufrière Hills Volcano and the volcanic hazard zones. *Exclusion Zone:* no admittance except for scientific monitoring and national security matters. *Central Zone:* residents only on a heightened state of alert; all residents to have a rapid means of exit 24 hours per day and to wear hard hats and dust masks when outside. *Northern Zone:* significantly lower risk; suitable for residential and commercial occupation.

◆ **FIGURE 1** Pyroclastic flow moving down the slope of Soufrière Hills Volcano on the island of Montserrat, British West Indies. Seen here is one of many life-endangering flows that occurred during July 1997 and that threatened the city of Plymouth.

(a)

(b)

◇▶ **FIGURE 5.25** Mount St. Helens. (a) A blast erupted from the volcano on Sunday morning, 18 May 1980, and culminated with a powerful pyroclastic flow. (b) The force of the pyroclastic flow was evidenced by the downed trees stripped of their bark in the "blow-down" area. This is the area where most of the fatalities occurred.

(a)

(b)

◇▶ **FIGURE 5.26** Saint-Pierre, Martinique, (a) before and (b) after the 1902 ash-flow eruption of Mount Pelée in which 30,000 to 40,000 people were killed.

8 May 1902. It caused the deaths of at least 30,000 people in Saint-Pierre at the foot of the volcano and initiated the study of this type of eruption and the unique volcanic deposits it forms. Pelée rumbled for some time before its main eruption. Because of this precursor activity, the population of Saint-Pierre swelled from 19,700 to at least 30,000 as refugees fled there from outlying villages. The concentration of sulfurous gases in the air became so great that citizens were forced to cover their faces with wet cloths. On the fateful day there was a tremendous explosion and a lateral blast from Pelée that boiled and rolled down the slopes of the mountain at a velocity estimated at 100 to 150 kilometers per hour (60–90 mph). The searing cloud engulfed the town in seconds and burned or suffocated all its inhabitants except for two survivors, one of them a condemned prisoner who had been in a dungeon (◇▶ Figure 5.26).

Such eruptions are common in the island arcs of the Caribbean Sea (see Case Study 5.2).

The old Roman city of Pompeii was buried under six meters (20 ft) of volcanic ash over a period of several days when Mount Vesuvius erupted in A.D. 79. Uncovering the city and preserving its treasures marked the beginning of modern systematic archeology. Less well known, but really more dramatic, was the eruption's impact on Pompeii's neighboring city Herculaneum, settled by the Greeks and named after Hercules. Herculaneum was overcome in a matter of minutes by a pyroclastic flow that buried the city to a depth of 20 meters (65 ft). The pyroclastic deposits are extremely hard, and a modern city, Portici, is now situated on the old city's grave. For this reason, only the equivalent of eight city blocks there have been excavated (◇▶ Figure 5.27). Until the 1980s it was thought that Herculaneum's residents had escaped, because only 30 bodies had

(a)

(c)

◀▶ **FIGURE 5.27** Herculaneum. (a) Exhumed part of the city overlain by the modern city of Portici. Note the vertical cuts in the background indicating the hardness and stability of the pyroclastic flow deposits. (b) A timber of a nearly 2,000-year-old house burned by the eruption. (c) A pyroclastic-flow deposit in the shape of a column, probably from an earlier eruption. The pebbles and cobbles suspended in a matrix of finer ash is characteristic of pyroclastic-flow deposits.

been recovered there, compared to 3,000 at Pompeii. Excavations since then along the former waterfront have uncovered hundreds of skeletons huddled together in cavelike openings in buildings used to store fishing boats. Apparently, the victims were overcome by the hot flows after running to the beach seeking refuge. More of this sad attempt of humans to survive the overwhelming forces of a volcanic eruption will be revealed in the continuing archeological excavations at Herculaneum.

Suffocating pyroclastic flows are a major volcanic hazard and they have occurred in historic times in Alaska, Washington, and California. Prehistoric but geologically young pyroclastic-flow deposits are found in all the Western states. They are known as **welded tuffs,**

because the minerals and glass composing them were fused together by intense heat.

LAHARS

Water from heavy precipitation, melting snow, a lake, or a river may mobilize debris on the flanks of a volcano and cause it to move a great distance downslope as a thick mush of rock, ash, and cinders. **Lahar** is the Indonesian word for such a fast-moving volcanic debris flow. A lahar accompanied the eruption of Mount St. Helens and filled the north fork of the Toutle River, a pristine mountain valley and stream before the eruption (◀▶ Figure 5.28). Lahars are basically identical to nonvolcanic debris flows except for their volcanic source (see Chapter 7). Lahar-generating volcanoes may show little activity when such flows occur, and there may be little warning.

Another example of a destructive lahar is the one caused by the mild eruption of Colombia's Nevado del Ruiz, 5,432 meters (17,800 ft) high, in November 1985.

◆ **FIGURE 5.28** Lahar-filled north fork of the Toutle River at Mount St. Helens, formerly a pristine V-shaped stream valley. This photograph was taken in July 1980, well after the main eruption on May 18, and the lahar was still steaming.

Although, as one observer put it, the volcano simply "ran a fever and cleared its throat," a lahar generated by melting snow and ice cover devastated the town of Armero 50 kilometers (30 mi) to the east. A wall of mud 40 meters high (130 ft) careened down the narrow canyon of the Lagunilla River and overwhelmed

Armero at its mouth, leaving more than twenty thousand victims entombed in mud. Because the disaster struck at 11 o'clock at night, many of the victims were sleeping and unable to run to higher ground.

The potential for destructive debris flows is predictable. The map of the Mount Rainier area in ◆ Figure 5.29 specifies the various identified potential volcanic hazards from this stratovolcano. Because Mount Rainier is covered with ice and has some areas of escaping steam and hot water, it is considered capable of generating lahars. Even a small eruption would subject the area at its base to inundation. Similar maps have been developed for most of the active volcanoes near urban areas along the west coast of the United States.

TSUNAMIS

Destructive tsunamis associated with volcanic activity are rare and they occur mainly in the western Pacific Ocean and around Indonesia. Of the 405 tsunamis recorded since 1900, 12 were caused by submarine volcanic eruptions, and only two of those resulted in significant damage. The most famous tsunami of all time, however, was caused by Krakatoa's eruption in 1883.

◆ **FIGURE 5.29** Volcanic hazard map of Mount Rainier area in Washington.

Krakatoa Volcano lies in the Sunda Strait between the islands of Java and Sumatra. In 1883 an eruption destroyed the volcano and formed a 7-kilometer-wide (4 mi) caldera. The islands of Krakatoa, Verlaten, and Lang are remnants of this volcano. The eruption was so enormous (VEI = 6) that its explosion was heard 3,000 miles away, and it spewed a dust cloud 80 kilometers (50 mi) into the atmosphere. About 30 minutes after the eruption, a huge tsunami swept the west coasts of Java and Sumatra, rising to a height of 35 meters (115 ft) above sea level and killing an estimated 36,000 people. The wave traveled across the Pacific, still a meter high when it reached South America, and was noticed as far away as the English Channel. Since 1927 small eruptions in the caldera have been frequent, and a new volcanic island has appeared, Anak Krakatoa ("Son of Krakatoa"; ◆ Figure 5.30). It is characterized by explosive eruptions, with turbulent ash clouds rising 1,200 meters (4,000 ft) above the crater.

WEATHER AND CLIMATE

Benjamin Franklin is regarded as the first person to recognize the connection between cold weather and volcanic eruptions while he was serving as U.S. ambassador to France in 1783–84. Laki Fissure in Iceland erupted in June 1783, putting sufficient ash into the atmosphere that the following winter was extremely cold in Europe. The summer of 1783 was bleak in Iceland, resulting in famine and the loss of a fifth of its population and more than half its sheep, cattle, and horses. The eruption of Tambora Volcano in Indonesia in April 1815 hurled so much fine ash into the upper atmosphere that the following year became known as the "year without a summer." This led to a famine that killed no less than 80,000 humans, produced the coldest three years in British history, and brought snow to New England in July. Crop failures caused the price of flour to double in Great Britain during the two crop years of cold weather.

⊙ TABLE 5.2   **Dust Veil Indexes of Selected Volcanoes**

VOLCANO	DVI
Tambora, Indonesia, 1815	1,500
Krakatoa, Indonesia, 1883	1,000
Mount Pelée, Martinique, 1902	500
Mount St. Helens, Washington, 1980	1

SOURCE: Lamb, H. H. Volcanic Dust in the Atmosphere (Phil. Trans. Royal Society of London, Ser. A, 266, 425–533).

The reason for these drastic climate changes is that the tiny particles blown into the upper atmosphere by a volcano are very effective in reducing incoming solar radiation. The average length of time volcanic dust particles (0.0001–0.005 mm) remain in the upper atmosphere, the *residence time,* has been found to be 1 to 2 years. Sulfur dioxide gas emitted by volcanic action produces white coatings on the particles, which makes them superreflectors of solar energy. This loading of the upper atmosphere with volcanic dust reduces the amount of incoming radiation relative to outgoing radiation, resulting in a net heat loss for the earth. An attempt has been made to quantify these so-called dust veils into a Dust Veil Index (DVI) that can be used to relate a specific eruption to its impact on climate. Indexes for selected volcanoes are shown in ⊙ Table 5.2. As apparent in the table, Tambora and Krakatoa have had potentially the most significant effects on climate in recent history, whereas Mount St. Helens, with a DVI of 1, had little effect. Dust veils are obviously sporadic and transient, and many volcanoes would have to erupt in concert to produce significant long-term effects on climate.

GASES

Denver, Mexico City, and Los Angeles have their smog; the Island of Hawaii, the "big island," has its *vog*–short for "*vo*lcanic *g*as." Since 1986 Kilaeua Volcano near the island's southeast coast has been working almost nonstop producing lava and, as a byproduct, 1,000 tons of sulfur dioxide (SO_2) per day. This sulfurous gas quickly combines with water to form sulfuric acid, which is hazardous to health and corrosive to metallic components of structures and machinery. There are reports of the nails and hinges of houses being eaten away to such an extent that the structures have almost collapsed and of damage to plants and crops by the acidic gases. During most of the year trade winds blow the vog toward the west side of the island, where it swirls behind massive Mauna Loa and gets trapped over the island's main tourist area, the city

◀▶ FIGURE 5.30 Anak Krakatoa, "Son of Krakatoa," in Indonesia. Krakatoa's 1883 eruption caused one of the largest tsunami in history and left a huge caldera. This "son" volcano formed in the caldera. It had nine episodes of explosive activity between 1963 and 2000. This view (1993) shows a rising ash cloud and dark-colored ash falling into the sea.

◀▶ FIGURE 5.31 During trade-wind conditions the almost continuous stream of vog produced by Kilauea (white plume in figure) is blown to the west side of the Island of Hawaii. Traces have been found on Johnston Island 1,600 kilometers (1,000 mi) to the southwest. Onshore (daytime) and offshore (nighttime) winds, indicated by double-headed arrows, trap vog on the island's popular Kona Coast. When the light Kona winds blow (red arrows) the vog is concentrated on the east side of the island.

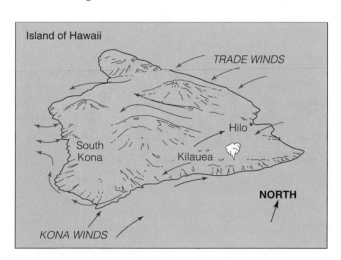

of Kailua Kona and the Kona coast (◀▶ Figure 5.31). Health officials and citizens are concerned, and there is much debate about vog as a health hazard. However, the volcano itself will close the debate when this activity cycle ends.

Tree-killing emissions of CO_2 were discovered in 1990 at Mammoth Mountain, California. These emissions continue to this day and pose a serious hazard to campers and hikers (Case Study 5.3). The intimate association of lethal carbon dioxide gas and volcanism yielded tragic results in 1986. A cloud of carbon dioxide gas emitted from a crater lake in Cameroon, West Africa, killed more than 1,700 people (see Case Study 5.4).

Selected historic volcanic eruptions and prehistoric eruptions indicated by thick, widespread tephra deposits of Holocene age are listed in ● Table 5.3. Although other significant eruptions are known, those in the table were chosen to illustrate the volcanic hazards and impacts discussed in this chapter.

Mitigation and Prediction

DIVERSION

As mentioned earlier in the chapter, various tactics have been used in attempts to prevent lava flows from overwhelming the land. At various times, people have dammed them, diverted them, and even bombed them. Diversion barriers, high earth and rock embankments piled up by bulldozers, have been constructed on Mauna Loa to protect the city of Hilo, and also to

Case Study | 5.3

Carbon Dioxide, Earthquakes, and the Los Angeles Water Supply

The Eastern Sierra Nevada is a mecca for tourists who want beautiful mountains and outstanding recreation. Much of the area's appeal results from volcanic action. About 760,000 years ago a huge eruption blew out 617 cubic kilometers (150 mi³) of magma that spread ash over nine states and as far to the east as present-day Nebraska (◆ Figure 1). The earth surface above the magma chamber sank 2 kilometers, forming the huge, oval caldera (16 km × 33 km) that is now called *Long Valley*. The eruption was 2,000 times the size of the 1980 Mount St. Helens eruption and it left hard welded tuff all over today's east-central California. Mammoth Mountain and Mono Craters, a linear chain of glassy domes (see Figure 5.16), formed within the last 400,000 years, and the people of Long Valley are living with the caldera's continuing restlessness.

The restlessness is manifest in hot springs, carbon dioxide emissions, and swarms of earthquakes, the most intense of which began in 1980 after two decades of quiescence. There were four strong, magnitude-6 earthquakes, three of them occurring on the same day. These quakes prompted studies by the U.S. Geological Survey that found uplift of 60 centimeters (about 2 ft) in the caldera center. Uplift, earthquake swarms, and gas emissions leave little doubt that magma is rising and getting closer to the ground surface. Late in 1997, just after the volcano movie *Dante's Peak* was released, the real volcano here acted up with more than 8,000 earthquakes larger than magnitude-1.2, and a couple of rattlers approached magnitude-5.0. More than 1,000 earthquakes occurred one day, and many scientists believe the volcano was waving a red flag signaling "danger."

Much of the water for the City of Los Angeles comes from Sierra Nevada snowmelt and travels from Mammoth Lakes to the city via open aqueduct. Should the volcano erupt, it would jeopardize or totally cut off this invaluable resource. Furthermore, the area's main highway access is aligned in the direction of the prevailing wind from the mountains. For this reason, an "escape road" was built in the opposite direction, toward the north, to provide an alternate route for the area's residents and visitors.

Mammoth Mountain, a large rhyolitic volcano, is still hot. Tree-killing emissions of CO_2 were discovered at the base of

the mountain in 1990. This indicated magma activity at shallow depth (◆ Figure 2). By 1995 about 30 hectares (75 acres) of pine and fir trees had suffocated, and CO_2 concentrations of up to 90 percent of total gas were found in the trees' soil and root systems. Where CO_2 concentrations exceeded 30 percent, most of the trees were dead. One large campground had to be closed because park rangers found high levels of CO_2 in restrooms and cabins after campers exhibited symptoms of asphyxia (◆ Figure 3). It appears that carbon dioxide degasses from the magma and migrates into a deep gas reservoir. When earthquake swarms cause fractures to develop in the reservoir rock, the gas finds its way to the surface. It is estimated that between 200 and 1,000 tons of carbon dioxide flows into the soil and atmosphere per day. The good news is that gas emissions have been declining, which may indicate that the magma has exhausted its gas supply.

◆ FIGURE 1 Mammoth Mountain and the Mono–Inyo Craters volcanic chain (red). The cataclysmic eruption 760,000 years ago that formed Long Valley ejected hot, glowing pyroclastic flows that cooled to form the Bishop Tuff (yellow). The inset map shows the distribution of Bishop Tuff and ash.

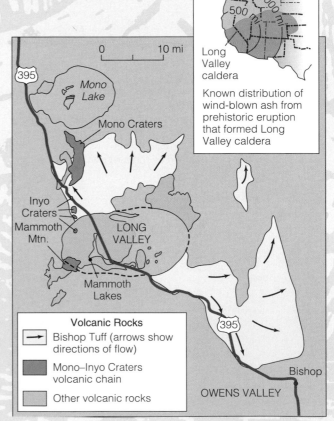

Long Valley caldera

Known distribution of wind-blown ash from prehistoric eruption that formed Long Valley caldera

Volcanic Rocks

→ Bishop Tuff (arrows show directions of flow)

Mono–Inyo Craters volcanic chain

Other volcanic rocks

(a)

(b)

◆▶ FIGURE 2 The Horseshoe Lake tree-kill area, Mammoth Lakes, California. (a) Aerial view looking north toward Mammoth Mountain. Inyo–Mono Craters and Mono Lake are seen in the right background. The light colored area (foreground) at the base of the Mammoth Mountain and on the shore of Horseshoe Lake is the tree-kill area. (b) Close-up of trees killed by leaking volcanogenic carbon dioxide gas.

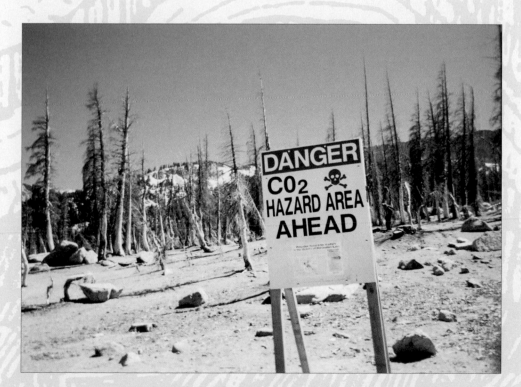

◆▶ FIGURE 3 Looking south through the Horseshoe Lake treekill area on the south side of Mammoth Mountain. The trees have died from high levels of CO_2 in the root zone.

Case Study | 5.4

Volcanic CO₂ and Fountaining Lakes

Lake Nyos is contained within a volcanic crater, one of 40 such craters that are roughly aligned southwest to northeast in Cameroon, West Africa. Specifically, the lake is in a **maar,** a low-relief crater produced by explosive interaction of magma with underground water. Such a high-pressure steam explosion, called a **phreatic eruption,** does not produce much new magmatic material, which is why the resulting crater is relatively shallow. Lake Nyos and several nearby crater lakes and springs are known to be highly charged with carbon dioxide, CO_2, just as is a bottle of soda water.

On the evening of 21 August 1986, almost noiselessly and without warning, the lake erupted, discharging 80 million cubic meters (100 million yd³) of CO_2 dissolved in its waters into the atmosphere. So much gas escaped from the lake that its level dropped 1 meter. Because carbon dioxide weighs about one and a half times as much as air, the suffocating gas descended downslope, following stream valleys and drainages for 16 kilometers (10 mi). The CO_2 cloud was 50 meters (160 ft) thick and it killed all animal life in its path, including 1,200 residents of Nyos village and 500 persons in nearby communities. The entire event took less than an hour (◁▷ Figure 1).

What happened the evening of the tragedy that caused the gas in the lake to explode catastrophically? Was it a small volcanic disturbance, or did some other physical phenomenon upset the lake? It is known that lakes periodically *overturn;* that is, their surface water sinks to the bottom and is replaced by bottom water rising to the surface. This usually occurs in temperate climates in the fall, because as the surface water cools, it becomes denser than the bottom water and displaces it. Overturn usually requires a trigger, such as strong winds, a landslide, or a drastic change in atmospheric pressure. The best hypothesis appears to be that winds associated with the rainy season blew the lake's surface water to one side of the crater. This reduced the hydrostatic (confining) pressure on the bottom waters, which allowed the highly charged gas to come out of solution and form bubbles. The formation of bubbles further reduced the density of the water, and the gas in solution frothed and rose with dramatic speed. That this sequence of events indeed happened is supported by stripped vegetation along the lake shore. The leaves and other parts of plants were most likely removed by a surge of water associated with the explosion—just as a fountain of foam is associated with popping the top of a warm can of soda.

Bantu folklore is filled with stories of exploding lakes and "rains" of dead fish. According to legend, there have been at least three earlier episodes of catastrophic lake explosions. Dead fish found floating on the water surface during minor degassing events are believed by the local people to be gifts from their ancestors who live in the lake.

Since 1986 the gas content of the lake has increased by additions from subterranean springs and magmatic sources. Engineering studies indicate that Lake Nyos and other dangerous crater lakes in the region could be defused by installing pipes in them to bring charged water from the lake bottom to the surface as fountains. Once started, the fountains would be driven entirely by the lifting force of the released gas.

◁▷ **FIGURE 1** (a) Deadly carbon dioxide belched from the bottom of volcanic Lake Nyos, which stood muddy and full of debris for several days. The gas settled into the valleys below, tragically suffocating 1,700 people. (b) Cattle suffocated by carbon dioxide at Lake Nyos.

(a)

(b)

○ TABLE 5.3	Selected Notable Worldwide Volcanic Eruptions*			
YEAR	**VOLCANO NAME AND LOCATION**	**VEI**	**COMMENTS**
4895 B.C.±	Crater Lake, Oregon	7	Posteruption collapse formed caldera
1390 B.C.±	Santorini (Thera), Greece	6	Late Minoan civilization devastated
79	Vesuvius, Italy	5	Pompeii and Herculaneum buried; at least 3,000 killed
186	Taupo, New Zealand	7	16,000 km² devastated; largest eruption in last 5,000 years
1631	Vesuvius, Italy	4	Modern Vesuvius eruptive cycle begins
1783	Laki, Iceland	4	Largest historic lava flows; 9,350 killed
1792	Unzen, Japan	2	Debris avalanche and tsunamis killed 14,500
1815	Tambora, Indonesia	7	Most explosive eruption in history; 92,000 killed; weather changed
1883	Krakatoa, Indonesia	6	Caldera collapse; 36,000 killed, mostly by tsunamis
1902	Mount Pelée, Martinique, West Indies	4	Saint-Pierre destroyed; 30,000–40,000 killed; spine extruded from lava dome
1912	Katmai, Alaska	6	Perhaps largest 20th-century eruption; 33 km³ of tephra ejected
1914–17	Lassen Peak, California	3	California's last historic eruption
1943	Parícutin, Mexico	3	New cone formed; event observed and documented from first eruption
1959	Kilauea, Hawaii	2	Lava lake formed that is still cooling
1963	Agung, Bali	4	1,100 killed; climatic effects
1968	Fernandina, Galápagos	4	Caldera floor drops 350 meters
1980	Mount St. Helens, Washington	4–5	Ash flow, 600 km² devastated
1982	El Chichón, Mexico	4	Ash flows kill 1,877; climatic effects
1991	Mount Pinatubo, Philippines	5–6**	Probably the second largest eruption of the 20th century; huge volume of SO_2 emitted
1991	Unzen, Japan	4**	Pyroclastic flows killed 41 people including 3 volcanologists; lava dome
1991–1993	Etna, Italy	1–2**	Longest continuous activity (473 days) in 300 years ended; 300 million m³ lava extruded
1983–	Kilauea, Hawaii	1–2**	Longest continuing eruption, with more than 50 eruption events
1995–	Montserrat, West Indies	1–2**	Pyroclastic flows and ash falls that have devastated this idyllic island

*Historic lava or tephra volume ≥2 km³, Holocene >100 km³, fatalities ≥1,500.
**Author's estimate.
SOURCE: Data adapted from Smithsonian Institution, Global Volcanism Program.

divert potential flows away from the National Oceanographic and Atmospheric Administration (NOAA) observatory at Mauna Loa (Figure 5.22). Practically, diversion can be considered only if the topography allows the diverted lava to flow onto unimproved lands whose owners do not object to it. Flows have been slowed and even stopped by spraying them with seawater at Heimaey Island, Iceland (◆ Figure 5.32). This was done in 1973 at the fishing village of Vestmannaeyjar. Both damming and diversion were used in

◆ **FIGURE 5.32** Lava flows on Heimaey Island off the south coast of Iceland were successfully chilled by spraying them with seawater in 1973.

◆ **FIGURE 5.33** Lava flows on Kilauea Volcano's east rift zone. By late 1992 more than 83 square kilometers (33 mi²) had been covered by the 1983 to 1992 eruptive sequence shown here, and activity continued into the twenty-first century.

Sicily in 1992 when eruptions of Mount Etna generated extensive lava flows.

VOLCANO HAZARD ZONES AND RISK

The purpose of a hazard-zone map of any type is to give accurate information on the frequency and severity of the particular hazard. In the 1950s, the area below Kilauea's east rift zone on the Island of Hawaii was developed. The volcano had been dormant for 17 years and it had not erupted on the lower east rift zone since 1840. Kilauea became active in 1952, however, and by 1996 there had been 13 separate rift-zone eruptions, the last of which has been almost continuous since 1983 (◆ Figure 5.33). In 1990 lava-flow hazard maps were prepared for the island, delineating zones ranging from low risk (9) for Kohala at the north end of the island to high risk (1 and 2) within and below the east rift zone (◆ Figure 5.34). The Royal Gardens residential subdivision, a development project of the 1950s, has been almost totally

buried under lava (see Figure 5.10), and soon after the hazard-zone maps were published, the town of Kalapana was covered by aa flows. Had the developers known that 90 percent of the land surface of Kilauea had been covered by lava since the arrival of the first Hawaiians 1,500 years earlier, they may have reconsidered building there. May have, that is!

◆ **FIGURE 5.34** Lava-flow hazard zones for the Island of Hawaii. The zones range from low risk (9) to high risk (1). Volcano boundaries are denoted by dashed lines.

PREDICTION

The ability to predict when and where a volcanic eruption will occur has attained a high degree of refinement at the U.S. Geological Survey's Hawaiian Volcano Observatory. It is located at Kilauea crater in Hawaii Volcanoes National Park on the east slope of Mauna Loa, Island of Hawaii. Kilauea and Mauna Loa are two of the most active volcanoes in the world. These volcanoes and the magma reservoir that feeds them can be viewed as balloons buried below thin layers of sand and clay. As magma enters the reservoir prior to an eruption, the volcano's surface layers are pushed upward and outward—just as if gas were being blown into the balloon. This expansion is measurable with tiltmeters. As magma works its way to the volcano surface, earthquakes are so numerous that they are referred to as *swarms*. Earthquake swarms have been found to be quite reliable precursors of an impending eruption. ◆ Figure 5.35 illustrates the three stages of an eruption and the accompanying changes in tilt and seismicity. Increasing tilt indicates that something is

(a)

(b)

(c)

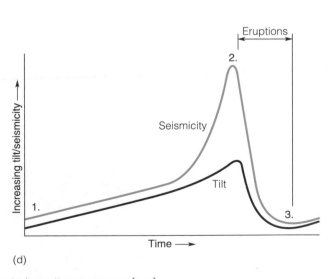

(d)

◆ **FIGURE 5.35** Three stages are apparent in Hawaiian volcanoes' typical eruption sequence. (a–c) Inflation and deflation of the cone are accompanied by (d) measurable changes in slope and seismicity. Eruptions occur after peak inflation and continue until magma is exhausted.

Gallery

Volcanic Wonders

Volcanism creates landscapes that are attractive to humans. One of the most eye-catching volcanic features is *columnar jointing,* which forms as cooling volcanic rock contracts around many equally spaced centers, causing the lava to split into polygonal columns. Six-sided columns are most common, but four-, five-, seven-, and eight-sided columns also occur. An excellent example is seen at the Devil's Tower, Wyoming, a shallow intrusion (◇ Figure 5.36).

◇ **FIGURE 5.36** Devil's Tower National Monument in northeastern Wyoming is an exposed volcanic neck with spectacular columnar jointing. Also known as "Grizzly Bear Lodge" and "Mato Tepee," it rises 865 feet above the surrounding landscape.

◇ **FIGURE 5.37** Nature's foot-warmer; Sarangson fumerole field, Minhasa, Celebes (Sulawesi). Fumerole temperatures range from 30° to 90° C.

FIGURE 5.38 Call the automobile club!

FIGURE 5.39 A DC-10 pointed skyward due to Mount Pinatubo ash on its tail; Cubi Point Naval Air Station, Philippines.

The columns form perpendicular to the main cooling surfaces of the flows, which is why they are sometimes curved. The 264-meter-tall (865 ft) Devil's Tower, called "Grizzly Bear Lodge" and "Mato Tepee" by early Native Americans, is surrounded by colorful Cheyenne and Lakota Sioux legends. According to one legend, a number of children were being chased by a grizzly bear. The Great Spirit saw their plight and raised the ground they were on. The bear scratched deep grooves into the rocks in his attempts to reach the frightened children. Imagine the tribal elders passing on this legend to wide-eyed children. It is a much more memorable explanation for the origin of the columns than the scientific one.

Fumeroles represent the dying stages of volcanism. They emit heat, steam, and sulfurous gases and build brightly colored surface mounds. They also have other uses (◁▷ Figure 5.37). The other gallery photos speak for themselves.

FIGURE 5.40 A mud volcano formed by volcanic gases escaping through fine-grained clay deposits; Imperial Valley, California.

FIGURE 5.41 Surrealistic lava tree molds form when fluid lava surrounds a tree trunk and cools, leaving a cylinder of solidified rock; east rift zone, Kilauea Volcano, Hawaii.

about to happen, and the locations of earthquake epicenters and foci indicate where the outbreak is likely to occur.

Observations of similar phenomena at Mount St. Helens led geologists to expect an eruption—though not such a cataclysmic one. The swelling there produced a noticeable change in the shape of the cone, and the harmonic tremors (caused by magma moving continuously within the reservoir) caused such surface motion on the cone that many persons were uncomfortable. Unlike earthquakes, the eruptions of active volcanoes can be predicted with fair accuracy as to time and place; the magnitude of the pending eruption, however, is more difficult to assess. Predicting the behavior of dormant volcanoes is much more difficult; much remains to be learned about predicting their future activity.

WHAT TO DO WHEN AN ERUPTION OCCURS

Based upon the experience of people at Mount St. Helens, the Federal Emergency Management Agency (FEMA) made the following suggestions to ease the trauma of a volcanic eruption.

At Home

- Stay calm and get children and pets indoors.
- If outside, keep eyes closed or use a mask and seek shelter.
- If indoors, stay there until heavy ash has settled.
- Close doors and windows, and place damp towels at door thresholds and other draft openings.

Consider This... ? Volcanic activity is perhaps the most spectacular of all geologic hazards, and the destructive potential of many volcanoes is immense. Fortunately, hazards associated with volcanoes are not spread equally around the earth's surface. What areas in the United States are most likely to be affected by future volcanic eruptions, and how have they been identified? ■

- Do not run exhaust fans or use clothes dryers.
- Remove ash accumulations from low-pitched roofs and gutters.

In Your Car

- Drive slowly because of limited visibility.
- Change oil and air filters every 50–100 miles (80–160 km) in heavy dust, or every 500–1,000 miles (800–1,600 km) in light dust (visibility up to 200 feet, or 60 meters).
- Use both windshield washer and wipers.
- Volcanic ash is abrasive and can ruin engines and paint.

Pets

- Keep pets indoors.
- Brush or vacuum them if they are covered with ash.
- Do not let them get wet, and do not try to bathe them.

Summary

VOLCANOES

DEFINED
A vent or series of vents that issue lava and pyroclastic material.

DISTRIBUTION
Adjacent to convergent plate boundaries around the Pacific Ocean (the Ring of Fire); in a west–east belt from the Mediterranean region to Asia; along the mid-ocean ridges; and in the interior of tectonic plates above hot spots in the mantle. In the U.S., a 1,100-kilometer (680-mi) belt extends northward from northern California through Oregon and Washington.

MEASUREMENT
The Volcanic Explosivity Index assigns values of 0 to 8 to eruptions of varying size.

EXPLOSIVITY
Depends upon the lava's viscosity and gas content. Viscosity, a function of temperature and composition, determines the type of eruption: felsic (high-SiO_2) lavas are viscous and potentially explosive; mafic (low in SiO_2, high in magnesium and iron) lavas are more fluid and less explosive.

TYPES OF VOLCANOES AND VOLCANIC LANDFORMS

SHIELD (HAWAIIAN OR "EFFUSIVE" TYPE)
Gentle outpourings of lava produce a convex-upward edifice resembling a shield.

CONTINENTAL FLOOD BASALTS
Lava erupts from long cracks, building up broad lava plateaus such as those found in the Columbia River Plateau (Oregon–Washington–Idaho), Iceland, and India. They are fed by mantle plumes.

STRATOVOLCANOES (EXPLOSIVE TYPE)
Characterized by concave-upward cone thousands of meters high. Vesuvius (Italy), Fujiyama (Japan), and Mount Hood (Oregon) are examples. Crater Lake (Oregon) is the stump of a stratovolcano that exploded and collapsed inward, leaving a wide crater known as a *caldera*.

LAVA DOMES
Bulbous masses of high-silica, glassy lavas, such as found at Mono Craters (California). Domes may form in the crater of a composite cone after an eruption, such as at Mount St. Helens (Washington).

CINDER CONES
The smallest and most numerous of volcanic cones, composed almost entirely of tephra.

BENEFITS OF VOLCANIC ACTIVITY

ATMOSPHERE AND HYDROSPHERE
Outgassed from the earth's interior.

BUILDING MATERIALS
Cinder blocks, road-base materials, pumice, light-weight concrete, decorative stone.

GEOTHERMAL ENERGY
Cooling magma at depth heats water that can be converted to steam to drive electrical generators.

HAZARDS

LAVA FLOWS
Destroy or burn everything in their path, can travel a distance of 100 kilometers or more.

ASH FALLS
Heavy falls of volcanic ash from Mount Vesuvius buried Pompeii in A.D. 79. There were heavy ash falls from Mount Pinatubo, the Philippines, 1991. Vertical plumes of ash may rise many kilometers from a vent, which are then carried by the wind and blanket the terrain. Such plumes rising from a vent are known as *Plinian eruptions*.

PYROCLASTIC FLOWS
Pyroclastic flows are hot, fluid masses of steam and pyroclastic material that travel down the flanks of volcanoes at high speeds, blowing down or suffocating everything in their path. Such flows have killed thousands of people.

LAHARS
Catastrophic mud flows down the flanks of a volcano. They cause the most volcanic fatalities, more than pyroclastic flows.

TSUNAMIS
Submarine volcanic eruptions (such as that of Krakatoa in 1883) create enormous sea waves that may travel thousands of miles and still do damage along a shoreline.

WEATHER
Sulfur dioxide coatings on ash and dust particles increase their reflectivity and cause cooling of weather, such as occurred after the eruption of Mount Pinatubo, 1991–1993. The effects of ash and dust in the stratosphere are short-term—a few years.

GASES
Corrosive gases emitted from a volcano can be injurious to health, structures, and crops. Such gases are called *vog* (*vo*lcanic *g*as) on the island of Hawaii. Carbon dioxide emitted from the bottom of volcanic Lake Nyos in Cameroon formed a cloud that sank to the ground and suffocated 1,700 people in 1986. Treekills due to CO_2 have occurred at Mammoth Mountain, California.

MITIGATION AND PREDICTION

DIVERSION AND CHILLING
Flows have been diverted in Hawaii and elsewhere and have been chilled with seawater in Iceland.

PREDICTION
Prediction in Hawaii, based upon seismic activity (earthquake swarms and harmonic tremors) and the tilt of the cone as magma works its way upward, has become quite accurate. Similar phenomena were observed at Mount St. Helens and Mount Pinatubo.

◆ Key Terms

aa
andesite line
ash fall
caldera
cinder cone
felsic
fluidization

lahar
lava dome (volcanic dome)
lava plateau
maar
mafic
nuée ardente
outgassing

pahoehoe
phreatic eruption
Plinian eruption
pyroclastic flow
shield volcano
stratovolcano (composite
   cone)

tephra
Volcanic Explosivity Index
   (VEI)
welded tuff

◆ Study Questions

1. What are the four kinds of volcanic cones, and what type of eruption can be expected from each kind? What are flood basalts, and what have they produced to shape the earth's landscape?

2. How would you expect the eruption of a volcano that taps mafic magma to differ from the eruption of one that taps felsic magma? What kind of rocks will result from each eruption?

3. Explain the distribution of the world's 1,350 active volcanoes in terms of plate tectonic theory.

4. List the six geologic hazards connected with volcanic activity in decreasing order of their threat to human life and limb.

5. Where in the United States are the most dangerous volcanic hazards encountered? Which geologic hazard related to volcanic activity presents the greatest danger to humans, and how can this hazard be mitigated, or at least minimized?

6. Why can eruptions on the Island of Hawaii be predicted more accurately than volcanic eruptions at subduction zones?

7. How may calderas be formed? What geologic evidence is found at Crater Lake, Oregon, that indicates it was formed by a special set of circumstances?

8. In what ways are volcanic activity and its products useful to humankind?

9. Name five volcanic areas of great scenic wonder, and indicate the geologic feature found there; for example, Crater Lake, caldera.

10. How do volcanic eruptions affect climate and weather? What impact does this have on populations, particularly Third World peoples?

11. In what ways do gases, including water vapor, emitted by a volcano impact the nearby area and people living there?

12. Define or sketch these volcanic features: aa, pahoehoe, lahar, Dust Veil Index, lapilli, tephra, welded tuff.

◆ For Further Information

BOOKS AND PERIODICALS

Decker, Robert, and Barbara Decker. 1981 and 1997. *Volcanoes.* New York: W. H. Freeman.

Duennebier, F. K. (contact person) and the Loihi Science Team. 1997. Researchers rapidly respond to submarine activity at Loihi Volcano, Hawaii. EOS 78, no. 22 (3 June).

Evans, W. C. ; L. D. White; M. L. Tuttle; G. W. Kling; G. Tanyileke; and R. L. Michel. 1994. Six years of change at Lake Nyos, Cameroon, yield clues to the past and cautions for the future. *Geochemistry* 28: 139–162.

Fierstein, Judy, and others. 1997. *Can another great volcanic eruption happen in Alaska?* U.S. Geological Survey fact sheet 039-97.

Francis, Peter. 1993. *Volcanoes.* New York: Oxford University Press.

Hazlett, Richard W. 1987. *Kilauea volcano: Geological field guide.* Hilo, Hawaii: Hawaiian Natural History Association.

Hill, David, and others. 1996. *Living with a restless caldera, Long Valley, California.* U.S. Geological Survey fact sheet 108-96.

——————. 1997. *Future eruptions in California's Long Valley area: What's likely.* U.S. Geological Survey fact sheet 073-97.

Kling, George W., and others. 1987. The 1986 Lake Nyos gas disaster in Cameroon, West Africa. *Science* 236: 169–174.

McDonald, G. A.; A. T. Abbott; and F. L. Peterson. 1983. *Volcanoes in the sea: The geology of Hawaii,* 2d ed. Honolulu: University of Hawaii Press.

McGee, Kenneth A., and others. 1997. *Impacts of volcanic gases on climate, the environment, and people.* U.S. Geological Survey open file report 97-262.

McPhee, John. 1989. *The control of nature.* New York: Farrar, Straus, and Giroux.

Myers, Bobbie, and others. 1997. *What are volcano hazards?* U.S. Geological Survey fact sheet 002-97.

Neal, Christina A., and others. 1997. *Volcanic ash: Danger to aircraft in the North Pacific.* U.S. Geological Survey fact sheet 039-97.

Schuster, Robert L., and J. P. Lockwood. 1991. Geologic hazards at Lake Nyos, Cameroon, West Africa. *Association of Engineering Geologists News,* April.

Sigvaldason, G. E. 1989. International conference on Lake Nyos disaster, Yaounde, Cameroon, 16–20 March 1987: Conclusions and recommendations. *Journal of volcanology and geothermal research* 39: 97–109.

Simkin, Tom, and others. 1981. *Volcanoes of the world,* Stroudsberg, Penn.: Hutchinson Ross Publishing Co.

Smith, A. L., and M. J. Roobol. 1991. *Mt. Peleé, Martinique: An active island-arc volcano.* Geological Society of America memoir 175.

Sorey, Mike; Bill Evans; Mack Kennedy; John Rogie; and Andrea Cook. 1999. Magmatic gas emissions from Mammoth Mountain. *California geology,* September/October.

Stager, Curt. 1987. Killer lake. *National geographic,* September: 404–420.

Sutton, Jeff, and others. 1997. *Volcanic air pollution: A hazard in Hawaii.* U.S. Geological Survey fact sheet 169-97.

Tilling, Robert I. 1984. *Monitoring active volcanoes.* Denver, Colo.: U.S. Geological Survey.

Williams, A. R. 1997. Monserrat: Under the volcano. *National geographic,* July.

Wright, Thomas L., and T. C. Pierson. 1992. *Living with volcanoes.* U.S. Geological Survey circular 1073. U.S. Geological Survey, Box 25425, Federal Center, Denver, CO 80225.

INTERNET SITES

Alaska Volcano Observatory
http://www.avo.alaska.edu/
Cameroon, Nyos
    http://www.volcano.und.edu/vwdocs/current_volcs/cameroon/cameroon.html
Hawaii
    http://www.hawaii.volcanoes.national-park.com/
Loihi
    http://www.soest.hawaii.edu/GG/HCV/loihi.html
Long Valley, California
    http://www.usgs.gov/VOLCANOES/Long Valley/
Montserrat
    http://www.geo.mtu.edu/volcanoes/west.indies/soufriere/govt/
U.S.G.S. volcano site
    http://www.usgs.go
Volcano imagery
    http://www.ssec.wisc.edu/data/volcano.html
Volcanoes, world
    http://volcano.und.nodak.edu/
Yellowstone
    http://volcano.und.edu/vwdocs/current_volcs/yellow/yellostone.html

◆ World Wide Web Activities

Volcanic Hazards in the United States
http://volcanoes.usgs.gov/

This master page on volcanic hazards sponsored by the U.S. Geological Survey has links to current volcanic activity in the United States, information on how the USGS and other agencies are working together to reduce hazards from volcanoes, and some basic information about volcanoes and how they work.

1. Navigate to **Current Activity.** Where are the hot spots for volcanic activity in the U.S.? Relate these to plate boundaries or hot spot locations.
2. Navigate to **Warning Schemes.** Why are alert level schemes different for the various volcanically active sites?

Carbon Dioxide Leakage at Mammoth Mountain
http://wrgis.wr.usgs.gov/fact-sheet/fs172-96/

Since 1989, carbon dioxide gas has been leaking from a magma chamber beneath Mammoth Mountain, California, killing trees and endangering people. Geologists are also concerned that the gas may be a precursor to an eruptive event.

1. How are the magma chamber, area faults, and tree kills related? Why is carbon dioxide especially dangerous in low-lying, closed spaces?

The Lahar at Nevado del Ruiz, Colombia
http://volcanoes.usgs.gov/Hazards/What/Lahars/RuizLahars.html

This Web site has photos and text of the 1985 lahar series from the composite Andean volcano Nevado del Ruiz in Colombia that killed 23,000 people. USGS scientists have studied the event in detail because Mt. Rainier and other Cascade volcanoes threaten many areas with potential lahars.

1. Are lahars restricted to major volcanic eruptions? Could lahars happen even without an eruption?
2. Why would a winter eruption of a composite volcano be more likely to produce lahars than a summer event?

Earth Systems Today CD Correlations

CHAPTER CONCEPT: How a Volcano Erupts Determines its Impact on Humankind

This segment of the CD-ROM shows how high-silica, gas-rich lavas tend to be explosive and to clog outlet vents.

SECTION MODULE: Volcanism Introductory Animation: Magma Chemistry and Explosivity

View the pictures and video clips of various eruptive events.

QUESTION: *Which of these eruptions or volcanic settings poses the least risk to humans? What is the general chemistry of the parent magma?*

CHAPTER CONCEPT: Volcano Hazard Zones and Risk

Long Valley, California caldera has been a focus for USGS geologists to produce a model that anticipates likely risks to humans. The magma type is high in silica, many people live in the immediate region, and some area lava flows are only 200 years old. Read the **Overview,** and select the **Long Valley Response** Plan button.

SECTION MODULE: Volcanism Main Activity: Long Valley Caldera, California Overview

The Long Valley Response Plan has been devised as a uniform code for USGS to use as an action guideline, given particular geologic indicators.

QUESTION: *What action would USGS scientists take, given a code yellow alert?*

chapter six

Soils, Weathering, and Erosion

◆ **FIGURE 1** A dust storm hits a southwestern Great Plains village in October 1937. Storms like this one destroyed crops and buried pasturelands. Minutes after this photograph was taken the village was completely engulfed in choking dust.

The Wind Blew and the Soil Flew

The stock market crash of 1929 ushered in the Great Depression of the 1930s, which brought widespread unemployment without unemployment benefits or entitlements such as welfare, social security, and free health care. It was a bad time, and to make things worse, there was protracted drought in the Great Plains, an area that was totally dependent upon rainfall for crop production at the time. Thousands of acres of marginally productive land, mostly grasslands, had been tilled because the government had begun guaranteeing wheat prices. Farmers thought they couldn't lose—but they were banking on rain. The entire United States except New England dried up, but between 1932 and 1940 the southern Great Plains was hit the worst (◆ Figure 1).

A longstanding agricultural practice is to plow up cropland after the fall harvest and allow the land to stand fallow through the winter. Plowing turns crop stubble into the soil, gets rid of weeds, and enhances soil water and air uptake. This is a fine practice as long as it doesn't rain too much—or too little. If too much rain falls, the soil is subject to severe sheet and rill erosion. If the pulverized soil dries out and the wind blows, the result is the same: extreme soil erosion.

The wind did blow in the southern Great Plains in the mid 1930s, and huge quantities of topsoil were simply blown away. There were many dust storms during this period, but the one of April 1935 was the worst. Soil silt, clay, and organic matter were lifted 5 kilometers (16,000 ft) into the air and carried eastward to Washington, D.C., and even 500 kilometers (300 mi) beyond the east coast into the Atlantic Ocean. The source area of the dust storms became known as the "Dust Bowl." It is a zone roughly 900 kilometers long by 450 kilometers wide (540 mi × 270 mi) in Colorado, Kansas, New Mexico, Texas, and Oklahoma (◆ Figure 2).

◆ FIGURE 3 Dust Bowl America. In 1938 Dorothea Lange, who worked for the Farm Security Administration, took this historic photograph of a homeless migrant family of seven on a highway in Pittsburgh County, Oklahoma. By 1939, half of the families in Oklahoma were on relief.

◆ FIGURE 2 The area most seriously affected by blowing dust during the Dust Bowl years, 1935–1936, and 1938. Kansas was the state most devastated in terms of percentage of cultivated area degraded by strong winds.

The wind and dust made farming impossible, and farm families, burdened by debt for equipment, seed, and supplies, left their farms to become migrant workers. Many traveled to California, where their lot was hardly improved, and the entire country was soon referring to this huge unfortunate group, derisively, as "Okies" (◆ Figure 3). This subculture of the 1930s, innocent victims of severe soil erosion, was the subject of John Steinbeck's touching novel *The Grapes of Wrath*.

Even today wind erosion exceeds water erosion in Texas, Colorado, Montana, New Mexico, Oklahoma, and North Dakota. Mismanagement of plowed lands and overgrazing are making parts of these six states potential dust bowls. In fact, the worst wind erosion since the Dust Bowl years took place in the southwest Great Plains in 1996. ■

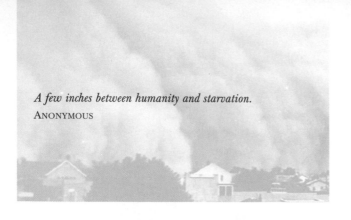

Soil is the solid earth's most fundamental resource. Through it we produce much of our food, and it supports forest growth, which gives us essential products. Soil material, organisms within the soil, and vegetation constitute an ecological system in which the four ingredients needed for plant growth are recycled. These ingredients are water, air, humus (organic matter), and mineral matter (◆ Figure 6.1).

The multiple uses of **soil** are reflected in the many definitions and classification schemes that have been developed for soils. To the engineer, a soil is the loose material at the earth's surface; that is, material that can be moved about without first being dynamited and upon which structures can be built. The geologist and the soil scientist see soil as weathered rock and mineral grains that are capable of supporting plant life. A farmer, on the other hand, is mostly interested in which crops a soil can grow and in whether the soil is rich or depleted with respect to humus and minerals. Soils form by the weathering of **regolith,** the fragmental rock material at the earth's surface. By studying soils we can make inferences about their parent materials and the climate under which they formed and can determine their approximate age.

The carrying capacity of our planet—that is, the number of people the earth can sustain—depends on the availability and productivity of soil. This is why re-

ducing soil erosion is so important. For all practical purposes, once productive topsoil is removed, it is lost to human use forever.

Five environmental factors determine the development of soils: (1) climate, (2) organic activity, (3) relief of the land, (4) parent material, and (5) the length of time that soil-forming processes have been acting. These factors can be entered into the "CLORPT equation," a memory aid for these five factors:

$$Soil = f(Cl, O, R, P, T)$$

where f is read "a function of" and Cl, O, R, P, and T represent climate, organic activity, relief, parent material, and time, respectively. The rate at which a soil weathers is a function of climate. In order to understand how soils form, change with time, and are distributed on the surface of the earth, we must first understand how rocks, minerals, and even the structures we place on the earth are broken down by weathering—their interaction with water and air.

Weathering

Weathering is a destructive process by which rocks and minerals are broken down by exposure to atmospheric agents. Specifically, **weathering** is the physical disintegration and chemical decomposition of earth materials at or near the earth's surface. **Erosion,** on the other hand, is the removal *and transportation* of weathered or unweathered materials by wind, running water, waves, glaciers, underground water, and even gravity.

PHYSICAL WEATHERING

Physical weathering makes little rocks out of big rocks by processes such as wedging. Rocks are wedged apart along planes of weakness by thermal expansion and contraction and by the activities of organisms. **Frost wedging** occurs when water freezes in joints or other rock openings and expands. This expansion, which amounts to almost ten percent in volume, can exert tremendous pressures in irregular openings and joints. As one would suspect, this process operates only in temperate or cold regions that have seasonal or daily freeze and thaw cycles. Roots of even the smallest plants can act as wedges in rock in any climate (◆ Figure 6.2). The most dramatic example of this is probably the damage to paving and structures caused by tree roots. Although this process is independent of climate, it operates in tandem with the effects of climate.

Thermal expansion and contraction may loosen mineral grains and contribute to rock disintegration in

◆ **FIGURE 6.1** Composition of a typical soil. The organic component includes humus (decomposed plant and animal material), partially decomposed plant and animal matter, and bacteria.

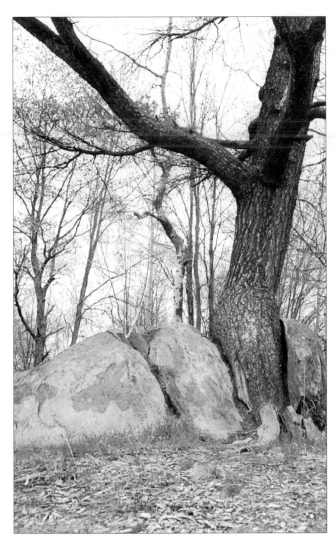

◇▶ FIGURE 6.2 A tree rooted into a crack in granite, and wedging it apart, is an example of mechanical weathering by an organism.

extreme climates. The quantitative importance of this process is difficult to evaluate, and experiments suggest it is probably not great. Salt-crystal growth is known to pry loose individual minerals in a rock, even granites. It is particularly effective in porous, granular rock such as sandstone. It occurs mainly in arid regions.

Animals, on the other hand, contribute significantly to weathering and soil formation by aerating the ground and mixing loose material. Insects, worms, and burrowing mammals, such as ground squirrels and gophers, move and mix soil material and carry organic litter below. Some tropical ants' nests are tunnels hundreds of meters long, for example, and common earthworms can thoroughly mix soils to depths of 1.3 meters (4¼ ft). Earthworms are found in all soils where there is enough moisture and organic matter to sustain them, and it is estimated that they eat and pass their own

weight in food and soil minerals each day, which amounts to about ten tons per acre per year on average. An Australian earthworm is known that grows to a length of 3.3 meters (11 ft) and is a real earth mover!

CHEMICAL WEATHERING

Chemical reactions also work on rock and mineral debris at the earth's surface. These reactions are complex and they involve many steps, but the fundamental processes are reactions between earth materials and atmospheric constituents such as water, oxygen, and carbon dioxide. The products of chemical weathering are new minerals and/or dissolved minerals. The specific reactions are *solution,* the dissolving of minerals; *oxidation,* the "rusting" of minerals; *hydration,* the combining of minerals with water; and *hydrolysis,* the complex reaction that forms clay minerals and the economically important aluminum oxides. ◉ Table 6.1 shows the formulas for the typical chemical-weathering reactions that are explained in the following discussion.

Solution

Carbon dioxide released from decaying organic matter and from the atmosphere combines with water to form weak carbonic acid. This natural weak acid attacks solid limestone, dissolving it and yielding an aqueous solution of calcium and bicarbonate ions. Solution in limestone terrane creates caverns, such as the Mammoth Caves in Kentucky and Carlsbad Caverns in New Mexico. (Chapter 7 explains this.) Acid rain, which forms when sulfur dioxide or nitrogen oxides combine with water droplets in the atmosphere to form sulfuric and nitric acids, takes its toll on rock monuments (◇▶ Figure 6.3).

Oxidation and Hydration

Oxidation produces iron oxide minerals (hematite and limonite) in well-aerated soils, usually in the presence of water. It is the "rusting" commonly seen on metal objects that are left outdoors. Pyroxene, amphibole, magnetite, pyrite, and olivine are most susceptible to oxidation because they have high iron contents. It is the oxidation of these and other iron-rich minerals that provides the bright red and yellow colors seen in many of the rocks of the Grand Canyon and the Colorado Plateau.

Hydrolysis

The most complex weathering reaction, hydrolysis, is responsible for the formation of clays, the most important

◯ TABLE 6.1 **Examples of Chemical-Weathering Reactions**

REACTION	EXAMPLE
Solution	
Step 1: A natural acid forms.	$CO_2 + H_2O \rightarrow H_2CO_3$ carbon dioxide water carbonic acid
Step 2. Acid dissolves limestone, forming calcium ions and bicarbonate ions in solution.	$CaCO_3 + H_2CO_3 \rightarrow Ca^{++} + 2(HCO_3)^-$ limestone carbonic acid calcium ions bicarbonate ions
Oxidation and Hydration	
Oxygen and water combine with iron, producing hydrated iron oxide ("rust").	$4Fe^{++} + 3O_2 + 6H_2O \rightarrow 2(Fe_2O_3 \cdot 3H_2O)$ iron minerals oxygen water limonite (rust)
Hydrolysis	
Orthoclase feldspar combines with acid and water, forming clay minerals and potassium ions.	$2KAlSi_3O_8 + 2H^+ + 9H_2O \rightarrow$ orthoclase feldspar acid ions water $Al_2Si_2O_5(OH)_4 + 2K^+ + 4H_4SiO_4$ clay minerals potassium ions soluble silica

minerals in soil. A typical hydrolytic reaction occurs when orthoclase feldspar, a common mineral in granites and some sedimentary rocks, reacts with slightly acidic carbonated water to form clay minerals, potassium ions, and silica in solution.

The ions released from silicate minerals in the weathering process are salts of sodium, potassium, calcium, iron, and magnesium, which become important soil nutrients. Clay minerals are important in soils, because their extremely small grain size—less than 4 microns (0.004 mm)—gives them a large surface area per unit of weight. Soil clays can adsorb significant amounts of water on their surfaces, where it stays within reach of plant roots. Also, because the clay surfaces have a slight negative electrical charge, they attract positively charged ions, such as those of potassium and magnesium, making these important nutrients available to plants. Other soil components being equal, the amounts of clay and humus in a soil are what determine its suitability for sustained agriculture. Sandy soils, for example, have much less water-holding capacity, less humus, and fewer available nutrients than

◁▷ FIGURE 6.3 Comparative weathering of granite obelisks, both believed to have been shaped 3,500 years ago (1500 B.C.). (a) Obelisk amid the ruins of Karnak near Luxor, Egypt. The hieroglyphs are sharp, and the polished surface is still visible. (b) The red-granite obelisk known as "Cleopatra's Needle" that stands in New York's Central Park was a gift from the government of Egypt in the late 19th century. New York City's humidity and other environmental conditions have caused chemical weathering of the incised hieroglyphs that had endured several thousand years in Egypt's arid climate.

(a)

(b)

do clayey soils. But clay-rich soils drain poorly and are hard to work.

THE RATE OF WEATHERING

The rate of chemical weathering is controlled by the surface environment, grain size (that is, the surface area), and climate. The stability of the individual minerals in the parent material is determined by the pressure and temperature conditions under which they formed. Quartz is the last mineral to crystallize from a silicate magma, and the temperature at which it does so is lower than the temperatures at which olivine, pyroxene, and the feldspars crystallize. For this reason, quartz has greater chemical stability under the physical conditions at the earth's surface. The relative resistances to chemical weathering of minerals in igneous rocks thus reflect the opposite order of their crystallization sequence in a magma. Most stable to least stable, the sequence is

quartz (most stable; crystallizes at lowest temp.)

  orthoclase feldspar

   amphibole

    pyroxene

     olivine (least stable; crystallizes at highest temp.)

Micas and plagioclase feldspars form over a range of temperatures and thus vary in weathering stability. Note that ferromagnesian (iron and magnesium) minerals—olivine, pyroxene, and amphibole—are least stable and most subject to chemical weathering. Igneous rocks made up of iron-rich minerals weather readily in humid climates.

A monolithic mass of rock is less susceptible to weathering than an equal mass of jointed and fractured rock. This is because the monolith has less surface area on which atmospheric processes can act. A hypothetical cubic meter of solid rock would have a surface area of 6 square meters. If we were to cut each face in half, however, we would have 8 cubes with a total surface area of 12 square meters (◆ Figure 6.4). Thus smaller fragments have more area upon which chemical and physical weathering can act. For example, whereas soils have developed on volcanic ash in the West Indies after only 30 years, some nearly 200-year-old solid basalt flows in Hawaii have yet to grow a radish. This reflects differences in particle size and surface area.

The effect of climate has already been discussed. It is important to remember that chemical weathering dominates in humid areas, and physical weathering is most active in arid or dry alpine regions. (Figure 6.3

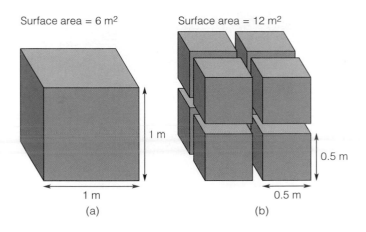

◆ **FIGURE 6.4** Surface area increases with decreasing particle size. A cube that is 1 meter on each side has a surface area of 6 m². If the cube is cut in half through each face, 8 cubes result that have a total surface area of 12 m².

may help you remember this.) However, humans have accelerated chemical weathering by polluting the atmosphere with acid-forming compounds. Limestone monuments in English cemeteries are reported to require 250–500 years to weather to a depth of 2.5 centimeters and thus to obliterate inscriptions. Buildings in northern England are known to weather more on the side that receives sulfur-laden winds from a nearby industrial area. A sixteenth-century monastery in Mexico that was built of volcanic tuff has been ravaged by chemical weathering to the extent that some of the carved statues on its exterior are barely recognizable (◆ Figure 6.5). Salt used to remove ice from Michigan Avenue in Chicago has deeply corroded (9 mm) the granite facing of the Intercontinental Hotel. A rock wall protecting Chicago's Lakeshore Drive was built of Indiana limestone. Chemical weathering by water overtopping the wall has dissolved clearly visible pits in a pattern known as "honeycomb" weathering. You can see examples of weathering in your own community by visiting a cemetery. Compare the legibility of the inscriptions on monuments of various ages and materials. Slate and granite monuments last much longer than those of limestone or marble, as you might suspect.

GEOLOGIC FEATURES OF WEATHERING

Spheroidal weathering is chemical weathering in which concentric shells of decayed rock are separated from a block bounded by joints or other fractures (◆ Figure 6.6). It occurs when a rectangular block is weathered from three sides at the corners and from two sides along its edges, while the block's plane faces are weathered uniformly. It is also called "onion-skin"

(a)

(b)

(c)

◈ **FIGURE 6.5** (a) A 16th-century monastery in Mexico shows the ravages of weathering mostly from wind and wind-driven rain. The rock is volcanic tuff. (b) Deep pitting (9 mm) in Mount Airy granite due to corrosion of feldspars by street salt and water; Intercontinental Hotel, Chicago. (c) A Lake Michigan seawall built of Indiana limestone is chemically weathered by waves overtopping the structure. Solution of the limestone has caused "honeycomb" weathering.

Consider This... ? Tennis is a sport that is played throughout the world on a variety of court surfaces. For example, the five most prestigious professional tennis tournaments are played on grass (in England), clay (in France and Italy), and concrete (in the United States and Australia). Why do you suppose clay-surface courts are more common in the eastern United States, and in tropical and humid countries, than in the southwestern United States, where concrete is the preferred surface? ■

weathering, since the layers of an onion provide a good analogy. **Exfoliation** domes (Latin *folium,* "leaf,"

and *ex-,* "off") are the most spectacular examples of physical weathering. These domes result from the unloading of rocks that had been deeply buried. As erosion removes the overburden, it reduces pressure on the underlying rocks, which in turn begin to expand. As the rocks expand, they crack and fracture along **sheet joints** parallel to the erosion surface. Slabs of rock then begin to slip, slide, or spall (break) off the host rock, revealing a large, rounded domelike feature. Sheeting along joints occurs most commonly in plutonic rocks like granite and in massive sandstone. Half Dome in Yosemite Valley, California, and Stone

FIGURE 6.6 Spheroidal weathering in jointed granite; Alabama Hills, California. Mount Whitney, just right of the twin spires, and the Sierra Nevada are in the background.

Mountain in Georgia are granite exfoliation domes (◆ Figure 6.7).

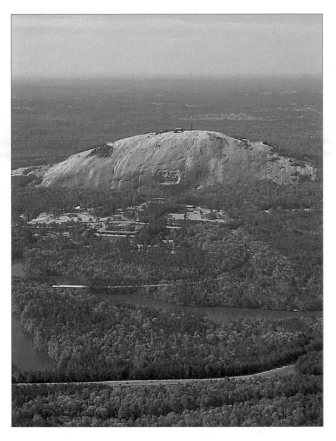

FIGURE 6.7 Stone Mountain, Georgia, is an excellent example of an exfoliation dome.

Soils

SOIL PROFILE

As we have noted, loose rock and mineral fragments at the surface of the earth are referred to as *regolith*. Regolith may consist of sediment that has been transported and deposited by rivers or wind, for example, or it may be rock that has decomposed in place. Where weathering has been sufficient, a surface layer that can support plant life—soil—has developed. Climate is the most important factor in soil development; given enough time, soils formed on different rocks under the same climate will look much the same. The influence of the parent material is most apparent in young soils such as decomposed granite, which contains quartz and feldspar grains, some of them having been altered to clay minerals.

A mature **soil profile** has several recognizable **soil horizons,** layers roughly parallel with the ground surface that are products of soil-forming processes (◆ Figure 6.8). The uppermost soil horizon, the "A horizon," is the *zone of leaching*. Here, mineral matter is most strongly dissolved by downward-percolating water. It may be capped by a zone of organic matter (*humus*) of variable thickness called the "O horizon," which provides CO_2 and organic compounds that make the percolating water slightly acidic. As the dissolved chemi-

cals move downward, some of them are redeposited as new minerals and compounds in the "B horizon," called the *zone of accumulation*. This horizon is usually lighter in color and harder than the A horizon. Below this is the "C horizon," the weathered transition zone that grades downward into fresh parent material.

The thickness and development of the soil profile are profoundly influenced by climate and time. In desert regions of the Southwest, the A horizon may be thin or nonexistent, and the B horizon may be firmly cemented by white, crusty calcium carbonate known as **caliche.** Caliche forms when downward-percolating water evaporates and deposits calcium carbonate that has been leached from above. Also, if the water table is close to the ground surface, upward-moving waters may evaporate into the layers, leaving deposits of caliche. Some of these deposits are so thick and hard that unusual measures are required to till or excavate through them.

Residual Soils

Residual soils are soils that have developed in place on the underlying bedrock. For example, in temperate

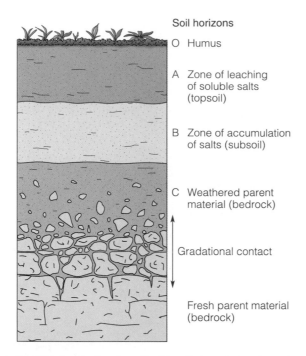

Soil horizons

O Humus

A Zone of leaching
   of soluble salts
   (topsoil)

B Zone of accumulation
   of salts (subsoil)

C Weathered parent
   material (bedrock)

Gradational contact

Fresh parent material
(bedrock)

◆ **FIGURE 6.8** An ideal soil profile. Heavily irrigated soils and very young ones exhibit variations of this profile.

climates granite decomposes into soils that contain quartz, weathered feldspar, and clay minerals formed from the weathering of feldspar. This soil is usually light brown or gray in color and can be mistaken easily for its parent rock. Engineers and geologists call slightly weathered granite *decomposed granite,* or simply *d.g.* It is an excellent foundation material for structures and roads. Heavy clay soils may develop on shale bedrock. The expansive nature of some of these clays makes them troublesome. (This is discussed in the next section of this chapter.) Residual soils may take hundreds or even thousands of years to develop and they are always subject to erosion and removal by one of many geologic processes.

Transported Soils

Some soils have formed on transported regolith. These transported soils are designated by the geologic agent that is responsible for their transportation and deposition: *alluvial* soils by rivers, *eolian* soils by the wind, *glacial* soils, *volcanic* soils, and so on. ◆ Figure 6.9 illustrates a soil that developed on volcanic ash overlying an older basalt on which little or no soil had developed. Known as the *Pahala ash,* this soil supports such exotic crops as macadamia nuts, sugar cane, and papaya on the Island of Hawaii.

As Pleistocene glaciers expanded and moved over northern North America, they scraped off Canada's

soils and redeposited them in a band across the United States from Montana to New Jersey. Soils formed on these widespread deposits—collectively known as *glacial drift*—are extremely variable; they may be thin or thick, bouldery or fine-grained, fertile or relatively barren. Some areas of drift are so bouldery that they are impossible to farm. This is one of the reasons that pastures for grazing dairy cattle, rather than cultivated grain crops, are the agricultural mainstay of such states as Wisconsin and New Hampshire.

Loess (approximately "luss," to rhyme with *cuss*) are wind-blown silt deposits composed of angular grains

◆ **FIGURE 6.9** Volcanic ash (horizontal layers) overlying weathered basalt; Pahala, Hawaii. This volcanic soil supports rich crops of sugar cane, pineapple, and macadamia nuts.

of feldspar, quartz, calcite, and mica. Covering about 20 percent of the United States and about 10 percent of the earth's land area (◆ Figure 6.10), loess is arguably the earth's most fertile soil. The sources of loess are glacial deposits and deserts. Strong winds blowing off Ice Age glaciers that covered most of Canada and the northern United States swept up nearby meltwater deposits and redeposited them far away as loess in valleys and on ridge tops. In the United States loess has produced especially rich agricultural lands in the Palouse region of eastern Oregon and Washington, and in the central Great Plains, sometimes called the "breadbasket" of the United States. The fertility of the Ukraine, formerly part of the Soviet Union but now a republic, is due to a favorable climate and its loessal soils. So highly regarded is this soil that Adolph Hitler had large quantities of it moved to Germany during World War II.

Without doubt, the largest loess deposits are in China, where they cover 800,000 square kilometers (312,500 mi²) and are as thick as several hundred meters. These deposits were derived from the Gobi and Takla Makan deserts (see Figure 12.8) to the north and west. On windy days airborne silt from these deserts is noticeable in Beijing, hundreds of miles away. Simi-

larly, the loess in Africa's eastern Sudan was carried there by winds from the Sahara Desert.

Loess has several noteworthy physical properties. It is tan to almost yellow in color and rather loosely packed (low density), which makes it easily excavated and eroded by running water. Its interlocking angular grains give it cohesive strength, however, and it is capable of standing in almost-vertical slopes without falling. These unusual properties served the Union and the Confederacy well in Mississippi during the Civil War. Both sides excavated tunnels, trenches, revetments, and other earthworks in the loess soil prior to the siege and battle at Vicksburg, which ultimately was won by U. S. Grant's Union forces. In northern China's Shansi Province in the 1500s, elaborate caves scraped out of loess provided refuge for more than a million people (◆ Figure 6.11). Unfortunately for the cave residents, loess is subject to collapse under dynamic stresses such as those of an earthquake, and an estimated 800,000 people were buried in their homes during the earthquake of 1556. The load of loess carried by the Huang Ho River as it flows through this area gave the river its other name, the *Yellow River*—which empties into the Yellow *Sea*.

◆ **FIGURE 6.10** Loess-covered regions of the world. Note that the largest areas are in China, central and western Asia, and North America.

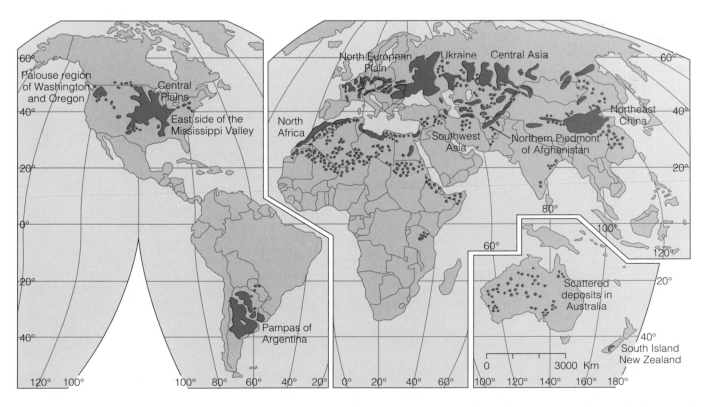

◆▶ **FIGURE 6.11** Almost-vertical slopes supported by loess in People's Republic of China. The small holes are probably insect nests and are testimony to loess's easily eroded nature.

SOIL COLOR AND TEXTURE

Color is a good indicator of a soil's *humus* (organic material) and iron content. As the proportion of humus increases, soil color darkens from brown to nearly black. Oxides of iron impart a reddish or yellowish color found mostly in semitropical and tropical areas (see Figure 6.14a). Bright red soils occur where there is aeration and good drainage, whereas yellow hydrated-iron compounds form where there is poor drainage. In arid climates white B horizons and white-coated soil fragments indicate calcium carbonate or other salts. Because these compounds are water-soluble, they do not collect where rainfall is high.

A soil's texture—size of the individual grains (particles) composing it—is a major determinant of its utility. A sample of any particular soil, excluding any gravel or boulders in it, consists of three basic grain sizes—sand, silt,

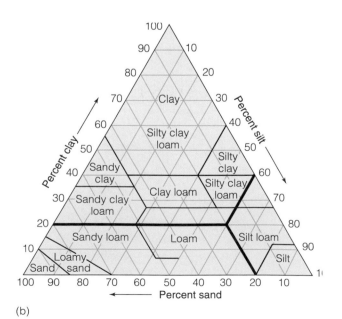

(a)

(b)

◆▶ **FIGURE 6.12** (a) Grain-size scale adopted by the U.S. Department of Agriculture and used by most soil scientists. (b) Soil texture classification triangle. Each side of the triangle is a scale for the percentage content of one of the three textural grades—sand, silt, or clay. A hypothetical soil sample has been found to be 60% silt, 20% sand, and 20% clay. Using this triangle, lines are followed inward from the respective percentages to learn the textural class of the soil, "silt loam."

and clay. The grain-size range used by the U. S. Department of Agriculture for each texture category is shown in ◆ Figure 6.12a. Based upon the percentages of sand, silt, and clay in a soil sample, the soil can be classified as loam, sandy loam, and so forth (◆ Figure 6.12b).

The texture of a soil has practical significance, as it determines the soil's properties—its looseness, workability, drainage, and so on. For example, coarse-textured soils are easily worked, readily penetrated by plant roots, and easily eroded. Fine-textured (clayey) soils tend to be heavy, harder to work, and sticky when wet. Water retention and infiltration are also functions of soil texture. Loose, sandy soils tend to allow water to pass through them readily, but they have low water-retention capacity. The opposite is true of fine-textured soils; they have large water-retention capacity but low filtration rates. This condition leads to poor drainage and excessive water runoff. Loamy soils are generally considered most favorable for agriculture, because they provide a balance between the coarse- and fine-textured soils, as shown in Figure 6.12b.

SOIL CLASSIFICATION
Zonal Classification

Pedologists (Greek *pedon,* "soil," and *logos,* "knowledge of") have long known that soils that form in tropical regions are different from those that form in arid or cold climates. Where rainfall is heavy, soils are deep, acidic, and dominated by chemical weathering. Soluble salts and minerals are leached (removed) from the soil, iron and aluminum compounds accumulate in the B horizon, and the soil supports abundant plant life. In arid regions, on the other hand, soils are usually alkaline and thinner, coarse-textured or rocky, and dominated by physical weathering. Because of low precipitation there, salts are not leached from soils, and when soil moisture evaporates, it leaves additional salts behind (Case Study 6.1). These salts may form crusts and lenses of caliche ($CaCO_3$ and other salts) in the B horizon.

These understandings led to the development of a "zonal" classification of soils based upon climate, which then evolved into a system of Great Soil Groups. These groups may be simplified into four major types (◆ Figure 6.13):

- **pedalfers,** soils that are high in aluminum (Al) and iron (Fe) and that are characteristic of humid regions (rainfall >50 cm/year, or >20 in/y);
- **pedocals,** soils that are high in calcium and found in desert or semiarid regions (rainfall <50 cm/y);
- **laterites,** the brick-red soils of the tropics that are enriched in hydrated iron oxides (rust); and
- **tundra soils,** the soils of severe polar climates, found mostly in the Northern Hemisphere.

Pedalfers are the rich soils of the prairies and steppes that support forest and grassland growth. Pedocals form where there is little rainfall, resulting in desert or scrub vegetation. Pedocals and pedalfers are midlatitude, temperate-climate soils. In contrast, laterites (Latin *later,* "brick, tile") are the residual soils of the tropics, where heavy rainfall leaches out all soluble materials, leaving behind the clay minerals and hydrated aluminum and iron oxides that impart a red color (◆ Figure 6.14a). Laterite is not uniquely identified with any particular rock type, but it does require an iron-containing parent, abundant rainfall, and well-drained terrain. Laterites are currently forming in humid tropical and subtropical regions. Leaching of these soils may be so extensive that with the right parent rock, economically valuable aluminum-rich *bauxite* deposits are the end products of the soil-forming process (◆ Figure 6.14b). Finally, tundra soils form in arctic and subarctic regions and support only such vegetation as mosses, sedges, lichens, and dwarf shrubs, making for bleak, forestless landscapes (◆ Figure 6.15).

U.S. Comprehensive Soil Classification System

Because of dissatisfaction with existing classification schemes, the U.S. Soil Conservation Service instituted the first of seven trial classification systems in 1952.

◆ FIGURE 6.13 The Great Soil Groups and their climatic environments. Note that the temperature scale is reversed on the left and right sides of the diagram.

Case Study

Case Study | 6.1

Salinization and Waterlogging

Saline soils contain sufficient soluble salts to interfere with plant growth. Salinization is the oldest soil problem known to humans, dating back at least to the fourth century B.C. At that time a highly civilized culture was dependent upon irrigation agriculture in the southern Tigris–Euphrates Valley of Iraq, the "Cradle of Civilization." By the second century B.C., the soils there were so saline that the area had to be abandoned. Soils in parts of California, Pakistan, the Ukraine, Australia, and Egypt are now suffering the same fate.

Salinization can be human-induced by intensive irrigation, which raises underground water levels very close to the soil. Capillary action then causes the underground water, containing dissolved salts, to rise in the soil, just as water is drawn into a paper towel or sugar cube. The soil moisture is then subject to surface evaporation, and as the water in the soil evaporates, it leaves behind salts that render the soil less productive, and eventually barren (see ◁▷ Figure 1). The productivity of 15 percent of all U.S. cropland is totally dependent upon irrigation, and 30 percent of U.S. crops are produced on this land.

For millennia, annual flooding of the Nile River added new layers of silt to the already rich soil and removed the salts. After completion of the Aswan Dam in 1970, however, the yearly flooding and flushing action stopped, and salts have now accumulated in the soils of the floodplain. Salinization is particularly evident in the arid Imperial Valley of California. Soils of the Colorado Desert, or any desert for that matter, can be made productive if enough water is applied to them. Because deserts have low rainfall, irrigation water must be imported, usually from a nearby river. The All-American Canal brings relatively high salinity water from the Colorado River to the Imperial Valley, where it is used to irrigate all manner of crops from cotton to lettuce. In places the water table has risen to such a degree that the cropland lies barren, its surface covered by white crusts of salt. The Imperial Valley's Salton Sea is one of the largest saltwater lakes in the United States, and it grows saltier and larger every year by the addition of saline irrigation water. The return of degraded irrigation water in canals to the Colorado River in southern Arizona has required that a desalinization facility be established near Yuma to make the water acceptable for agricultural use in Mexico.

The good news is that soil salinization can be reversed. It can be done by lowering the water table by pumping and then applying heavy irrigation to flush the salts out of the soil. Salinization can also be mitigated, or at least delayed, by installing perforated plastic drain pipes in the soil to intercept excess irrigation water and carry it offsite. Neither remedy is an easy or inexpensive solution to this age-old problem.

◁▷ FIGURE 1 Accumulation of salts at the soil surface due to evaporation of irrigation and ground water; Imperial Valley, California.

(a)

(b)

(a) Laterite, shown here in Madagascar, is a deep, red soil that forms in response to intense chemical weathering in the tropics. (b) Bauxite, the ore of aluminum, forms in horizon B of laterites derived from aluminum-rich parent materials.

Each trial system was sent to a select group of international pedologists for their comments. After incorporating the suggestions and findings of these experts, the "Seventh Approximation," as it is called, was officially adopted by the Soil Conservation Service in 1965. The distinguishing features of this system are that:

1. Soils are classified by their physical characteristics, rather than their origin; that is, the system is nongenetic.
2. Soils that have been modified by human activities are classified along with natural soils.
3. The soils' names convey information about their physical characteristics.

The system subdivides soils into a hierarchy of six levels—orders, families, series, and so forth, with ten orders at the highest level in the system. With a practiced eye, each soil order can be easily identified. Because the number of soil entries increases astronomically at the lower levels (illustrated by the fact that 14,000 soil

series are recognized in the U.S.), this classification is for the professional, not the layperson.

Soil Problems

SOIL EROSION

Although soil is continually being formed, it is for practical purposes a nonrenewable resource, because hundreds or even thousands of years may be necessary for soil to develop. Ample soil is essential for providing food for the earth's burgeoning population, but yet, throughout the world it is rapidly disappearing. Although the United States has one of the world's most advanced soil conservation programs, soil erosion remains a problem after 60 years of conservation efforts and expenditures of billions of dollars. Erosion removed an estimated two billion tons of topsoil from

◆ FIGURE 6.15 Treeless tundra landscape near Kotzebue, Alaska. Most of the soil in tundra regions is frozen much of the year.

◄▶ FIGURE 6.16 World soil degradation, 1945–1991. The areas degraded are in millions of hectares. (A hectare, 10,000 square meters, equals 2.47 acres.) Table 6.3 summarizes the relative contributions of overgrazing, deforestation, and bad agricultural practices to world soil degradation.

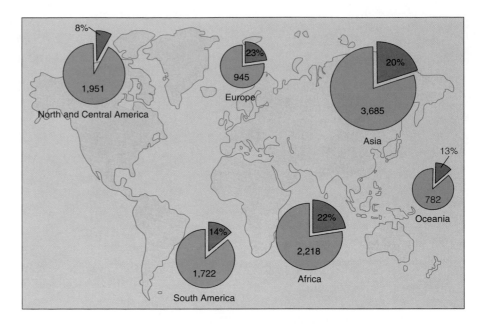

U.S. farmlands in 1992—down from slightly more than three billion tons a decade earlier. With the average annual soil loss estimated as 10–12 tons per hectare (4–4.8 tons/acre) and an assumed annual soil-formation rate of 2–4 tons per hectare (0.8–1.6 tons/acre), three to five times as much soil is being lost as is being formed in the United States. Not only does soil erosion destroy fertility, but tons of soil settle in lakes and clog waterways and drainages with sediment, pesticides, and nutrients each year. Losses of only 2–3 centimeters (1 in) of topsoil represent about 200 tons per hectare. Such erosion cuts crop yields and may render the land unproductive.

On a global basis, an area the size of China and India combined has suffered *irreparable degradation* from agricultural activities and overgrazing—mostly in Asia,

Africa, and Central and South America (◄▶ Figure 6.16). Lightly damaged areas can still be farmed using the modern conservation techniques explained later in this chapter, but "fixing" serious erosion problems is beyond the resources of most Third World farmers and their governments.

Erosion Processes

The agents of soil erosion are wind and running water. Heavy rainfall and melting snow create running water, which removes soil by sheet, rill, and gully erosion. **Sheet erosion** is the removal of soil particles in thin layers more or less evenly from an area of gently slop-

◄▶ FIGURE 6.17 Rill erosion in cultivated soil.

◄▶ FIGURE 6.18 Extreme gullying in soft sediments encroaching on tilled land near Lumpkin, Georgia. The width of the area in the photo is approximately 1¾ kilometers (1 mi).

ng land. It goes almost unnoticed. **Rill erosion,** on the other hand, is quite visible in discrete streamlets carved into the soil (◆ Figure 6.17). If these rills become deeper than about 25–35 centimeters (10–14 in), they cannot be removed by plowing, and gullies form. Gullies are created in unconsolidated material by widening, deepening, and headward erosion of rills (◆ Figure 6.18). **Gullying** is a big problem in the easily eroded loess soils of the Palouse region of eastern Washington. Snow melting there saturates the loess soil, causing it to become fluid and then to flow like wet concrete across the underlying still-frozen ground. Estimated annual U.S. cropland soil loss from sheet and rill erosion is shown by state in ◆ Figure 6.19a. Tennessee headed the list in 1980, followed by Hawaii, Missouri, Mississippi, and Iowa, as shown in ◉ Table 6.2.

Wind erodes soils when tilled land lies fallow and dries out without vegetation and root systems to hold it together. The dry soil breaks apart, and wind removes

the lighter particles. If the wind is strong, a dust storm may occur, as it did in the 1930s, creating the "Dust Bowl" (described at the beginning of the chapter). Much of the reason for the Dust Bowl was the widespread conversion of marginal grasslands, which receive only 25–30 centimeters (10–12 in) of rainfall per year, to cropland. Unfortunately, this is still occurring some places in the United States, and the rest of the world. ◆ Figure 6.19b presents estimates of the amount of soil lost each year to wind erosion in the Great Plains states. When the 50 states' estimated wind and water erosion losses are combined (Table 6.2), the state with the greatest annual loss is Texas, followed by Tennessee, Hawaii, and New Mexico, respectively. ◉ Table 6.3 summarizes the recognized causes of accelerated soil erosion worldwide and assesses their relative impact.

In addition to the soil erosion caused by overgrazing, deforestation, and bad agricultural practices, some recreational pursuits also cause soil erosion, particularly those that involve offroad vehicles (ORVs). A 1974 study indicated that recreational pressure on semiarid lands of Southern California by ORVs was "almost completely uncontrolled" and that those desert areas had experienced greater degradation than had any other arid region in the United States (◆ Figure 6.20). In the two decades since that study, much has been accomplished in controlling offroad traffic in California and other states, but ORVs continue to exert

◉ TABLE 6.2 Significant U.S. Soil Erosion by State, 1980

RANK	STATE	TONS/ACRE/YEAR
Sheet and Rill Erosion		
1	Tennessee	14.12
2	Hawaii	13.71
3	Missouri	11.30
4	Mississippi	10.92
5	Iowa	9.91
Wind Erosion		
1	Texas	14.9
2	New Mexico	11.5
3	Colorado	8.9
4	Montana	3.8
5	Oklahoma	3.0
*Total Soil Erosion, Water and Wind**		
1	Texas	18.40
2	Tennessee	14.12
3	Hawaii	13.71
4	New Mexico	13.50
5	Colorado	11.40

*The U.S. average annual rate of total soil erosion is estimated at 4.0–4.8 tons/acre (10–12 tons/hectare).
SOURCE: *Environmental Trends* (Washington, D.C.: Council on Environmental Quality, 1981), cited in Sandra Batie, *Soil Erosion: A Crisis in America's Cropland?* (Washington, D.C.: The Conservation Foundation, 1983).

◉ TABLE 6.3 Causes of World Soil Erosion and Deterioration

CAUSE	ESTIMATED % OF TOTAL
Overgrazing	35
■ reduced vegetation cover	
■ trampling that leads to decreased waterholding capacity	
Deforestation	30
Bad agricultural practices	28
■ salinization due to improper draining of irrigated land	
■ fertilization that results in acidification	
■ lack of terraces on tilled sloping land	
■ wind erosion of fallow lands	
Other causes, natural and human	7

SOURCE: World Resources Institute, United Nations Environment and Development Program, *World Resources: A Guide to the Global Environment.* (New York: Oxford University Press, 1992).

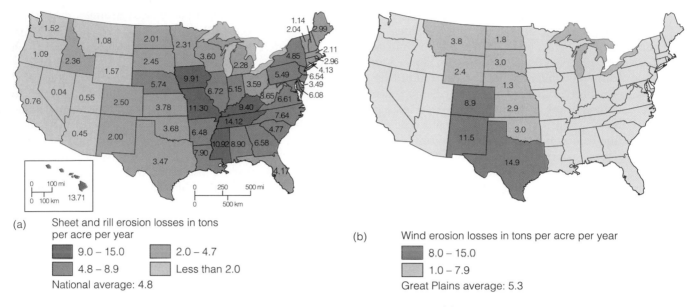

(a)   Sheet and rill erosion losses in tons
per acre per year

■ 9.0 – 15.0	■ 2.0 – 4.7
■ 4.8 – 8.9	☐ Less than 2.0

National average: 4.8

(b)   Wind erosion losses in tons per acre per year

■ 8.0 – 15.0	
☐ 1.0 – 7.9	

Great Plains average: 5.3

◆ **FIGURE 6.19** Estimated annual rates of cropland soil loss in the United States, 1980. (a) Sheet and rill erosion losses. (b) Wind erosion losses in the Great Plains. (Light green represents <1 ton/acre per year.)

severe impacts upon areas established (translate *sacrificed*) for their use.

THE IMPACT OF CROPLAND LOSS ON HUMANS

The main concern when addressing soil loss—whether the loss is due to erosion or to conversion of cropland to nonagricultural uses—is that soils are biological and chemical factories that produce food. Increasingly, the U.S. population is moving into urban areas, and to accommodate them, cities are expanding outward onto agricultural land (◆ Figure 6.21). About 20 percent of the United States was under cultivation in 1992 (◆ Figure 6.22), and each year some more of that land is urbanized or converted to forestland, particularly in

◆ **FIGURE 6.20** Hill-climbing motorcyclists have devastated this slope in Jawbone Canyon in the Mojave Desert, California. The bikers have compacted the soil surface, causing reductions in vegetation and soil permeability. The reduced soil moisture content has seriously harmed the local plant and animal ecology. Desert rains have eroded the tire tracts, transforming them into gullies, which increases water runoff and sediment yield (erosion).

◆ **FIGURE 6.21** Total U.S. cropland, 1982 to 1997. Reductions were due to urbanization, conversion to other uses, and soil erosion.

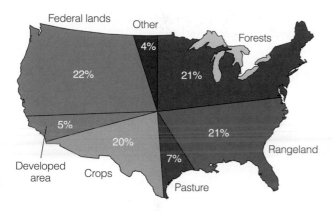

◆ **FIGURE 6.22** U.S. land use, 1992. Note that 20 percent was cropland and 28 percent was rangeland and pasture.

the northeastern and southeastern states. This trend is also seen in India and other nations with high growth rates. Between 1950 and 1998 the amount of grain-growing land per person worldwide declined from .23 hectare to .012 hectare (.57A and .03A). World crop-land decreases had not been regarded as a major problem till the late 1900s, because increases in productivity had been great. Increases in productivity leveled off in 1990, however, and it is doubtful that future increases will keep pace with increasing population and decreasing cropland soils.

Some of the many ways soils are important to humans are shown in ◆ Figure 6.23. When soil is lost, our beneficial use of the resource is reduced.

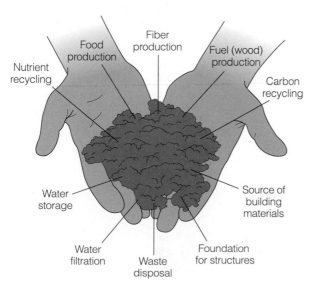

◆ **FIGURE 6.23** Some of soil's beneficial uses.

◆ **FIGURE 6.24** Terracing steep hillsides is an ancient soil conservation practice; China.

Mitigation of Soil Erosion

Thomas Jefferson was among the first people in the United States to comment publicly on the seriousness of soil erosion. He called for such soil conservation measures as crop rotation, contour plowing, and planting grasses to provide soil cover during the fallow season. These practices are now common in industrialized nations, but unfortunately, only minimally employed in Third World countries. Proper choices of where to plant, what to plant, and how to plant are the components of soil conservation. The major erosion-mitigation practices are described here.

- *Terracing*—creating flat areas, terraces, on sloping ground—is one of the oldest and most efficient means of saving soil and water (◆ Figure 6.24).
- *Strip-cropping,* by which close-growing plants are alternated with widely spaced ones, efficiently traps soil that is washed from bare areas and it also supplies some wind protection. An example of this is alternating strips of corn and alfalfa.
- *Crop rotation* is the yearly alternation of soil-depleting crops with soil-enriching crops. Rotating corn and wheat with clover in Missouri, for example, has been found to reduce soil runoff from 19.7 tons/acre for corn to 2.7 tons/acre, a huge loss reduction. Groundcovers such as grasses, clover, and alfalfa tend to be soil-conserving, whereas row-crops like corn and soybeans can leave soil

vulnerable to runoff losses. Retaining crop residues, such as the stubble of corn or wheat, on the soil surface after harvest has been found to reduce erosion and to increase water retention by more than half.

- *Conservation-tillage* practices minimize plowing in the fall and encourage the contour plowing of furrows perpendicular to the field's slope so that they will catch runoff, increasing water infiltration.

- *No-till* and *minimum-till* practices also conserve soil. With no-till farming, seeds are planted into the soil through the previous crop's residue, and weeds are controlled solely with chemicals. Specialized no-till equipment is required. "No-till" has been especially successful in soybean farming. In 1993 U.S. farmers left 30 percent of their crop debris on 37 million cultivated hectares (92 million acres), nearly a third of all cultivated land in the United States. Ten percent of the U.S. planted acreage was not plowed at all. No-till agriculture is most beneficial during times of drought, when plowed fields dry out and topsoil is more vulnerable to wind erosion. No-till practices saved much of the agricultural soil that would have otherwise been eroded by floodwaters of the Great Flood of 1993 in the Mississippi River valley.

Wind-caused soil losses can also be reduced by planting windbreaks of trees and shrubs near fields and by planting row crops perpendicular to the prevailing wind direction. Wind losses are almost zero with cover crops such as grasses and clover.

EXPANSIVE SOILS

Certain clay minerals have a layered structure that allows water molecules to be absorbed between the layers, which causes the soil to expand. (This process differs from the *adsorption* of water to the surface of nonexpansive soil clays.) Soils that are rich in these minerals are said to be **expansive soils.** Although this reaction is somewhat reversible, as the clays contract when they dry, soil expansion can exert extraordinary uplift pressures on foundations and concrete slabs, with resulting structural and cosmetic damage (◆ Figure 6.25). Damage caused by expansive soils costs about $6 billion per year in the United States, mostly in the Rocky Mountain states, the Southwest, and Texas and the Gulf Coast states (◆ Figure 6.26). The swelling potential of a soil can be identified by several standard tests, usually performed in a soils engineering laboratory. An obvious test is to compact a soil in a cylindri-

> **Consider This… ?** Expansive soil is the soil problem most likely to be encountered in maintaining a home. Assuming that you, as a homeowner, find that your newly purchased dream house is built on expansive soil, what could you do to minimize its impact on your home? ■

◆ **FIGURE 6.25** This well-built concrete-block wall in Colorado was uplifted and tilted by stresses exerted by the underlying expansive soil.

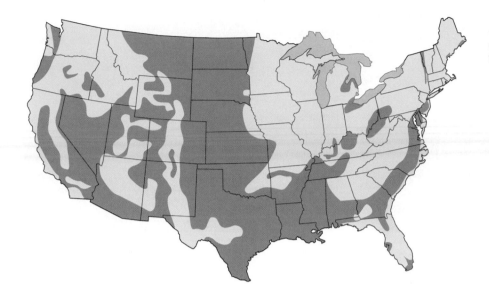

◀▶ **FIGURE 6.26** Distribution of expansive soils (shown in brown) in the United States.

cal container, soak it with water, and measure how much it swells against a certain load. Clays that expand more than 6 percent are considered highly expansive; those that expand 10 percent are considered critical. Treatments for expansive soils include (1) removing them, (2) mixing them with nonexpansive material or with chemicals that change the way the clay reacts with water, (3) keeping the soil moisture constant, and (4) using reinforced foundations that are designed to withstand soil volume changes. Identification and mitigation of damage from expansive soils is now standard practice for U.S. soil engineers.

PERMAFROST

The term **permafrost,** a contraction of *permanent* and *frozen,* was coined by Siemon Muller of the U.S. Geological Survey in 1943 to denote soil or other surficial deposits in which temperatures below freezing are maintained for several years. More than 20 percent of the earth's land surface is underlain by permafrost, and it is not surprising that most of our knowledge about frozen ground has been derived from construction problems encountered in Siberia, Alaska, and Canada. Permafrost becomes a problem for us humans when we change the surface environment by our activities in ways that thaw the near-surface soil ice, since this results in soil flows, landslides, subsidence, and related phenomena. Permafrost forms where the depth of freezing in the winter exceeds the depth of thawing in the summer. If cold ground temperatures continue for many years, the frozen layer thickens until the penetration of

surface cold is balanced by the flow of heat from the earth's interior. This equilibrium between cold and heat determines the thickness of the permafrost layer. Thicknesses of 1,500 meters and 600 meters (almost 5,000 and 2,000 ft) have been reported in Siberia and Alaska, respectively. Such thick layers are very old; they must have formed during the Pleistocene Epoch of geologic time, the time known as the Great Ice Age.

The top of the permanently frozen layer is called the **permafrost table** (◀▶ Figure 6.27). Above this is

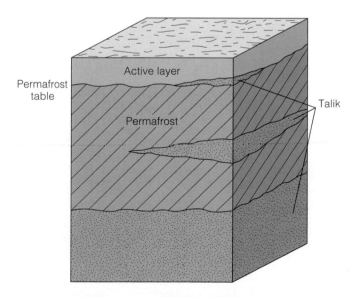

◀▶ **FIGURE 6.27** Permafrost zones. Talik, unfrozen layers and "lenses," may occur in the active layer and in the permanently frozen layer.

◀▶ **FIGURE 6.28** Utility poles extruded from permafrost ground; Alaska. One solution is to place the poles in tripods on the ground surface.

◀▶ **FIGURE 6.29** Scars made by vehicles on tundra soil remain for many years; Alaska.

the active layer, which is subject to seasonal freezing and thawing and is always unstable during summer. When winter freezing does not penetrate entirely to the permafrost table, water may be trapped between the frozen active layer and the permafrost table. These unfrozen layers and lenses, called *talik* (a Russian word), are of environmental concern, because the unfrozen ground water in the talik may be under pressure. Occasionally, pressurized talik water bursts through to the surface, forming a domical mound 30–50 meters (100–160 ft) high called a *pingo*. The presence of ice-rich permafrost becomes evident when insulating veg-

etation, usually tundra, is stripped away for a road, runway, or building. Thawing of the underlying soil ice results in subsidence, soil flows, and other gravity-induced mass movements (◀▶ Figure 6.28). Even the casual crossing of tundra on foot or in a jeep can upset the thermal balance and cause thawing. Once melting starts, it is impossible to control, and a permanent scar is left on the landscape (◀▶ Figure 6.29).

Structures that are built directly on or into permafrost settle as the frozen layers thaw (◀▶ Figure 6.30). Both active and passive construction practices are used to control the problem. Whereas active methods in-

(a)

(b)

◀▶ **FIGURE 6.30** Two casualties of improper construction for permafrost conditions in Alaska, (a) a modern home in Fairbanks and (b) Bert and Mary's Roadhouse on the Richardson Highway. The front center portion of the log-construction roadhouse was the area of greatest subsidence, because that was the location of the furnace. The 14-year-old structure had to be removed in 1965 because of distress due to continued sinking.

(a)

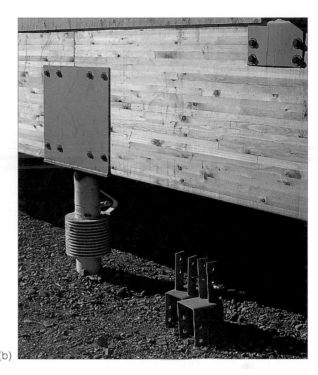

(b)

◆ **FIGURE 6.31** Passive construction methods for mitigating damage due to permafrost. (a) Raised WW II–era Quonset hut; Nome, Alaska. (b) Heat-radiating foundation piers on a modern structure; Kotzebue, Alaska.

volve complete removal or thawing of the frozen ground before construction, passive methods build in such manners that the existing thermal regime is not disturbed. For example,

- a house can be built with an open space beneath the floor so that the ground can remain frozen (◆ Figure 6.31a),
- a building can be constructed on foundation piers whose temperature is held constant by heat dissipators or coolant (◆ Figure 6.31b), and
- roads are constructed on top of packed coarse-gravel fill to allow cold air to penetrate beneath the roadway surface and keep the ground frozen.

The most ambitious construction project on permafrost terrain to date is the 800-mile oil pipeline from Prudhoe Bay on Alaska's North Slope to Valdez on the Gulf of Alaska. About half the route is across frozen ground, and because oil is hot when it is pumped from depth and also during transport, it was decided that the pipe should be above ground so as to avoid the thawing of permafrost soil (◆ Figure 6.32). The four-foot-diameter pipeline also crosses several active faults, three major mountain ranges, and a large river. The pipeline's above-ground placement allows for some fault displacement and for easier maintenance if it should be damaged by faulting. This placement has the added advantage of accommodating the long-standing migratory paths of herding animals such as caribou and elk.

SETTLEMENT

Settlement occurs when a structure is placed upon a soil, rock, or other material that lacks sufficient strength to support it. Settlement is an engineering problem, not a geological one. It is treated here, however, because settlement and poor soil seem inextricably associated. All structures settle, but if settlement is uniform, say two or three centimeters over the entire foundation, there is usually no problem. When differential settlement occurs, however, cracks appear or the structure tilts.

The Leaning Tower of Pisa, Italy, is probably the best-known, most-loved example of poor foundation design. It may have scientific importance as well, as Galileo is said to have conducted gravity experiments at the 54.5-meter (163 ft) structure. Begun in 1174, the tower started to tilt when the third of its eight stories was completed, sinking into a two-meter-thick layer of soft clay just below the ground surface. Initially the tower leaned north, but after the initial settlement and throughout the rest of its history it has leaned south. To compensate for the lean, the engineer in charge had the fourth through eighth stories made taller on the leaning side. The added weight caused the structure to sink even further. It now leans about 5.2 meters (17 ft) from

(a)

(b)

◆ **FIGURE 6.32** The 800-mile Alaskan oil pipeline was constructed to accommodate permafrost and wildlife. (a) The meandering pattern allows for expansion and contraction of the pipe and for potential ground movement along the Denali fault, which it crosses. The pipeline supports feature heat-dissipating radiators. (b) Moose and caribou are able to pass beneath the elevated pipeline.

> **Consider This... ?** It is jokingly said that the twelfth-century architect of the Leaning Tower of Pisa saved money by not having a foundation investigation. Had engineers and geologists with today's knowledge and tools existed at that time, what routine tests would they have performed on bore-hole soil samples that would show settlement was inevitable? What recommendations might they have made to the client? ∎

the vertical, about 5.5°, which gives one an ominous feeling when standing on the south side of the tower (◆ Figure 6.33).

In 1989, a similarly constructed leaning bell tower at the cathedral in Pavia, Italy, collapsed, causing officials to close the tower at Pisa to visitors. In 1990 the Italian government established a special commission of structural engineers, soils engineers, and restoration experts to determine new ways to save the monument. The commission's goal is to reverse the tilt 10–20 centimeters (4–8 in), which they believe will add 100 years to the lifetime of the 800-year-old structure.

It should be noted that tilting towers are not exceptional in Italy. Few of the 170 campaniles (bell towers) still standing in Venice are vertical, for example, and many other towers throughout Italy and Europe tilt. These graceful towers are certainly testimony to past architects' ability and talent, but also to their failure to understand that a structure can be truly beautiful only if local geology permits.

OTHER SOIL PROBLEMS

Laterite soils, the red soils of the tropics, are soft when they are moist but become brick-hard and durable when they dry out (◆ Figure 6.34). The Buddhist temple complex at Angkor Wat in Cambodia is partly constructed of laterite bricks that have endured since the thirteenth century. Deforestation of areas underlain by laterites is an invitation to this hardening, as U.S. forces discovered in Vietnam when they attempted to grade roads and runways on lands that had been defoliated for military reasons. Deforested laterites make for swampy conditions during the rainy season, because they lack permeability; there is little infiltration of the water. This can raise havoc with wheeled vehicles and restrict mobility in such regions.

Although laterite soils support some of the densest rain forests in the world, they are relatively infertile because of extensive leaching of minerals. The nutrients that support the rain forest are in the vegetation, and they are recycled when the plants die and decompose

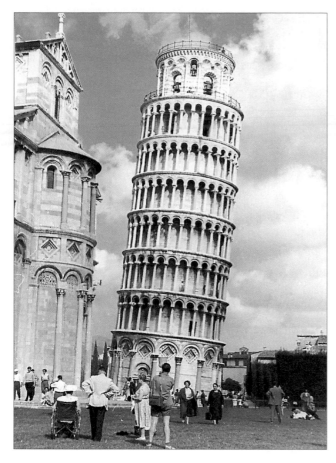

◁▷ **FIGURE 6.33** Inadequate foundation investigation and design; Pisa, Italy.

◁▷ **FIGURE 6.34** Dried and hardened laterite; Brazil. This soil is so hard that soil scientists call it "ironstone."

into the soil. Deforested laterites used for agricultural purposes must be fertilized intensively after a few harvests in order to support productive crops.

In some areas of extensive irrigation, agricultural soils form impervious clayey layers called **hardpans** in the B horizon. Hardpans inhibit waters from percolating and thus reduce crop yields. They must be broken up by deep plowing using large, heavy tractors pulling long, single-tooth rippers that penetrate the subsoil. Hardpans and claypans are the scourge of farmers, as they can render their farmland unproductive.

The presence or absence of beneficial elements in soils can result in various dietary benefits or deficiencies in humans. This is because the soil is the actual beginning of the food chain. Soil elements are taken up by plants (vegetables, fruits, and grains), which are then eaten by animals. Some examples of soil elements and health are discussed in Case Study 6.2. The presence of pollutants in the soil now requires expensive cleanup (Case Study 6.3).

Case Study | 6.2

Soils, Health, and Locoweed

Humans need trace elements in their systems in order to function properly. **Trace elements** occur in small amounts in the body, usually a few parts per million, as opposed to the bulk elements, which are found in large quantities. The body obtains these elements naturally by this sequence: rocks to soils to plants and finally to animals. Humans get trace elements by eating plants or by eating animals that ate plants that grew on soils containing them or by using dietary supplements of them. Some water-soluble trace elements, such as fluoride, are sometimes added to municipal drinking-water supplies.

The best-known relationship between soils and human health is the one between soil iodine deficiency and the enlargement of the thyroid gland known as *goiter*. Besides being cosmetically demoralizing, goiter is debilitating in other ways. Iodine deficiency during pregnancy can lead to "cretinism" in newborns, a severe mental and physical disorder. Endemic goiter has been firmly established in the northern United States in the so-called goiter belt (see ◆ Figure 1). Because of the wide availability of iodized table salt, this disease has practically disappeared except among the poorly educated. Additionally, little of the U.S. diet is "home grown" anymore. This is another reason goiter is no longer an endemic problem. Soil-iodine thyroid diseases are also known in England, Thailand, Mexico, the Netherlands, and Switzerland.

Several trace elements—zinc, copper, iron, cobalt, manganese, and molybdenum—are associated with and stimulate enzymes, the catalysts of biochemical reactions. These enter the body mostly in plant and animal foods. Their importance might be better appreciated when we realize that silver, mer

- • Cobalt deficiency
- • Copper deficiency
- • Molybdenum deficiency
- • Phosphorus deficiency
- • Selenium deficiency
- • Iron deficiency
- • Unknown causes
- "Goiter belt"
- General areas where soil problems exist

◆ **FIGURE 1** Distribution of mineral-nutritional diseases of animals in the United States. Dots show specific locations where problems have been reported.

cury, and lead are toxic because they are enzyme *inhibitors.* Some of the beneficial effects of trace elements are that zinc aids in healing wounds and burns, iron transports oxygen in the blood, and cobalt is a vital component of vitamin B_{12}. Deficiencies of these may result in anemia, slowed growth, vitamin-B_{12} deficiency, and other ailments.

A most interesting trace element is selenium. It is poisonous at high dosages, but deficiencies cause abnormalities in plants and animals. It is abundant in soils of the high plains and western United States, and certain plants there concentrate it in their tissue. Eating these plants, known as *locoweeds,* causes cattle to get the "blind staggers" and to appear drugged. ◆ Figure 1 shows areas where mineral-nutritional diseases are known to occur in animals. These are all localities where soils have deficiencies or excesses of vital or toxic trace elements.

Although molybdenum is probably best known as a metal that strengthens steel, it is also essential to plant and possibly animal health. Many plant enzymes contain molybdenum, two of which are important in metabolizing nitrogen, a major plant nutrient. In large concentrations, molybdenum interferes with the absorption of copper, an essential element, and may be detrimental to certain livestock. *Molybdenosis* can cause anemia, depressed growth, poor reproduction, bone malformation, and loss of normal hair color

(◆ Figure 2). Black Angus cattle with molybdenosis turn a mouse-gray color, for example. If treated early enough, the symptoms of the disease can be reversed by corrective nutrition.

Lignite coal (see Chapter 14) is known to be rich in molybdenum. Ranchers in South Dakota reported problems resembling molybdenosis in cattle pastured near an abandoned open-pit lignite mine. Chemical analysis revealed that grass growing near the exposed lignite contained high levels of molybdenum. This led to the discovery that molybdenum was being leached from the coal and transported by surface-water runoff to nearby pastures, where it was taken up by range grass and thereby entering the food chain.

Molybdenum-related disease has not been found in humans in such countries as the United States where food-production and distribution are widespread. In a region in southern India, however, some individuals developed *genu valgun* (knock-knee), a crippling bone deformity, with a devastating impact on the local population. The outbreak was traced to a high level of molybdenum in sorghum, a staple in the residents' diet. High levels of the metal were then found in the ground water. Ultimately it was discovered that construction of a large dam and reservoir nearby had raised the water table, making the high-molybdenum ground water accessible to the sorghum crop (see Chapter 10).

(a)

(b)

◆ **FIGURE 2** Molybdenosis results from an imbalance of copper to molybdenum in the diet of cattle. The visible symptoms are hair discoloration and loss and weight loss. (a) A healthy Hereford steer. (b) A steer that is suffering from molybdenosis.

Case Study

Case Study | 6.3

Rx for Contaminated Soils

Although spills of hazardous materials frequently contaminate soils, the Environmental Protection Agency (EPA) did not include a soil component in their system for ranking Superfund cleanup sites until 1990 (see Case Study 1.2 table). Soils become contaminated when they are forced to absorb a wide variety of contaminants, the most common being oil-field brines (salty water), heavy metals, pesticides, cleaning agents, and petroleum products. Fortunately, this contamination by absorption is reversible, and soils can be "cleaned," *remediated,* by a variety of methods. A large soil remediation industry has sprung up in response to EPA requirements.

Various soil remediation techniques are used, depending upon the thickness of the soil to be cleaned and the chemical nature of the contaminant. *Dilution,* mixing contaminated soil with clean soil, is one technique. It works, but it is expensive and it contaminates an even larger volume of soil. The old saying "the solution to pollution is dilution" is relevant to many pollution problems, but it is not practical when large volumes of soil are contaminated.

Bioremediation, a natural process, uses microorganisms to degrade and transform organic contaminants. Bacteria and fungi flourish naturally in soils where oil, sewage sludge, or food-production waste has been discarded. Biodegradation can be stimulated by increasing the amount of air in the soil through cultivation, vapor-extraction pumping (see below), or stockpiling the soil and forcing air into it. Adding easily decomposed organic matter such as hay, compost, or cattle manure also speeds up degradation of organic contaminants. Bioremediation is relatively inexpensive, and some contaminated soils simply require turning (disking or plowing) to speed up oxidation. A successful bioremediation project at Mobil Oil Company's natural-gas refinery at Liberal, Kansas, is shown in ◇ Figure 1. The soil contained up to 26,000 parts per million (2.6%) of hydrocarbons. It was decided that the contaminated soil should be spread in a layer less than 20 centimeters (8 in) thick and that a nutrient and microbial amendment (cow manure) should be applied to it at a rate of about 60 tons per hectare (15 tons/acre) (◇ Figure 1b). The soil was periodically disked to promote biological activity and was subsequently reclaimed and planted (◇ Figure 1c). Under careful management even heavy metals such as selenium and arsenic can be converted to gaseous forms and released to the atmosphere. The three-year project was closely coordinated with the Kansas Department of Health and Environment.

Vapor extraction is used for removing volatile (easily evaporated) organic compounds such as cleaning solvents from soils. Air is drawn through the soil by means of dry wells connected to vacuum pumps. The air passing through the soil evaporates the compounds and exhausts them to the atmosphere. This technique is especially desirable for cleaning soils beneath buildings or roads, because no excavation is required except for the drilling of shallow wells. The down side is that it may take 6 to 12 months to sanitize the soil adequately.

Phytoremediation, the use of plants to remove contaminants, is used mainly for extracting salts. This has been used in the Netherlands, where periodic seawater flooding contaminates soils. Plant scientists are developing *hyperaccumulator* plants that clear soils of heavy metals such as cobalt, nickel, and zinc by concentrating them in their tissues. The cost of phytoremediation materials, equipment, and labor is relatively low, but up to a decade can be required to remediate soil to the point that it can be used again.

The justification for soil remediation is obvious. It allows vegetation to be established; it decreases soil erosion, ground-water contamination, and air pollution; and it can restore soils' agricultural productivity. The soil remediation industry is in its infancy, and a considerable amount of research is under way for developing faster and more economical means of "cleaning dirt."

(a)

(b)

(c)

◁▷ FIGURE 1 A successful bioremediation project. (a) Oil-contaminated soil (2.6% hydrocarbons) at a Kansas refinery is removed for bioremediation by a bulldozer. Note the patchy distribution of the oil contamination. (b) The soil is spread in a layer about 20 centimeters (8 in) thick. Nutrient amendment (cow manure) is added to stimulate biological degradation of the hydrocarbons. (c) Remediated soil (<.01% hydrocarbons) is spread and planted with grass. Note the clods of manure and the white chunks of caliche in the redistributed soil. The structure in the background is a grain silo in which wheat is stored.

Gallery

A Few Inches between Humanity and Starvation

Our planet's carrying capacity depends upon soil, and it is being lost faster than weathering processes can produce it. Poor agricultural practices and attempting to make marginal soils productive are major problems (◇ Figure 6.35). Some soils defy conversion to agricultural use because of their origin (◇ Figure 6.36), and some productive soils are damaged by soil inhabitants (◇ Figure 6.37). Other soils are "mined" for clays to make pottery or bricks (◇ Figure 6.38).

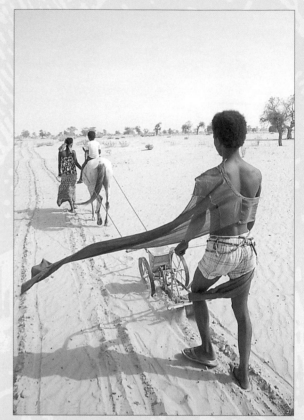

◇ **FIGURE 6.35** The "soil" of the Sahel of Africa is little more than dry, wind-blown sand. A region of severe drought and famine in recent years, the Sahel encompasses an east–west belt that separates the arid Sahara Desert from tropical Africa. It includes parts of Chad, Niger, Mali, Mauritania, Senegal, and Gambia.

◇ **FIGURE 6.36** Remnants of a boulder fence in an area that was once cleared for agriculture; Attleboro, Massachusetts. Rocky glacial soils are difficult to farm, because freezing and thawing cycles continually lift boulders up into the soil root zone. This produces a new "crop" of rocks each spring in some areas. Boulder fences and walls are seen in fields and forests throughout the northeastern United States.

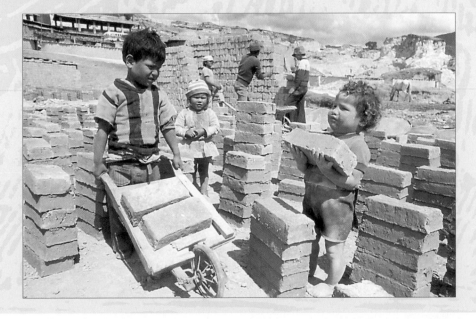

◁▷ FIGURE 6.37 A soil-covered slope that was literally honeycombed by burrowing animals; England. Such burrows lead to heavy soil erosion and sediment loading of nearby streams.

◁▷ FIGURE 6.38 Children in some developing nations do not have time to make mud pies or sand castles, because they are employed in brickyards.

Summary

SOIL
DEFINED
Many definitions depending upon use (engineering, agriculture, geology, etc.). For our purposes, however, soil is defined as weathered rock and mineral grains that can support plant life.

ENVIRONMENTAL FACTORS OF SOIL FORMATION
Climate, organic activity, relief (topography), parent material, and time. Soils form from the weathering of regolith.

WEATHERING
Physical and chemical breakdown of bedrock and regolith.
1. *Physical weathering* is caused by temperature differences and disruption by plants and animals.
2. *Chemical weathering* processes include solution, oxidation, hydration, and hydrolysis. It is important in the formation of clays, the most important mineral group in soils. Clays are important because they hold moisture and exchange nutrients with plants.

RATE OF WEATHERING
Depends upon the rock or regolith's mineral composition and surface area. Smaller particles weather faster than large masses, and ferromagnesian minerals weather faster than quartz or feldspar.

FEATURES AND LANDFORMS
Spheroidal weathering and exfoliation domes are common features.

SOIL PROFILE
DEFINED
Distinguishable layers, or horizons, in mature soils composed of the A horizon, or zone of leaching (where minerals are dissolved), and a lower B horizon known as the zone of accumulation, where salts and clays accumulate. Below this is the C horizon of weathered parent material. The thickness of the horizons and the development rate of soil are profoundly influenced by climate.

RESIDUAL SOIL
Developed in place on underlying bedrock. Laterite, for example, is the brick-red soil of humid (subtropical and tropical) regions that represents intensive weathering of parent rock.

TRANSPORTED SOIL
Developed on regolith transported and deposited by wind (eolian), glaciers, rivers, or volcanic action. Loess, for example, is an eolian soil associated with ice-age continental glaciation. It is highly productive and is found throughout the central U.S. "breadbasket."

SOIL CLASSIFICATION
ZONAL
A climatic classification based upon rainfall and temperature. The four main zonal soils are pedalfers (rich brown soils of temperate regions), pedocals (alkaline soils of arid regions), laterites (tropical and subtropical soils), and tundra soils of polar regions.

U.S. COMPREHENSIVE SOIL CLASSIFICATION SYSTEM
Known as the *Seventh Approximation,* a classification of soils by their physical characteristics.

SOIL PROBLEMS
EROSION
A major problem, because soil determines the earth's carrying capacity. Globally, irreparable soil degradation has impacted an area the size of China and India combined. U.S. soil losses vary greatly with area.

PROCESSES
Sheet erosion by running water (thin layer is removed) can become rill erosion (defined streamlets are carved in the soil), which leads to gullying (the most destructive). Sheet and rill erosion can be halted, but once gullying starts, it is extremely difficult and costly to mitigate. Wind erodes soil on marginal lands that were once cultivated and then dried out. Without plant root systems to hold the soil, wind erosion can lead to "dust bowl" conditions. Off-road vehicles also seriously degrade soils.

MITIGATION
Good agricultural practices such as terracing, no-till and minimum tillage, strip cropping, and crop rotation. Wind losses can be minimized by planting windbreaks, strip crops, and cover crops such as clover.

OTHER SOIL PROBLEMS
EXPANSIVE SOILS
Certain clay minerals absorb water and expand, exerting high uplift pressures that can crack or even destroy structures. Mitigated by removal of soil, mixing, or keeping the moisture content constant.

PERMAFROST
Permanently frozen ground (soil or rock) underly more than 20 percent of the earth's surface. Structures built directly on soil permafrost will settle, so mitigation involves building with an open space below the structure so as not to disturb the thermal regime in the soil and melt the permafrost. Removal of permafrost zones is sometimes possible.

SETTLEMENT
Occurs when the applied load is greater than the bearing strength of the soil; e.g., Leaning Tower of Pisa. An engineer-

ing problem that can be mitigated by proper design or by strengthening of the underlying soil.

MISCELLANEOUS SOIL PROBLEMS
Salinization and waterlogging (high water table), hardening of laterites due to deforestation and desiccation, hardpans (cemented B horizon), and presence or absence of trace elements essential to human health.

CONTAMINATION
Brines, petroleum products, crude oil, and chemicals of all sorts may contaminate soils. Common methods of cleansing the soil are bioremediation (bacteria), phytoremediation (plants), and vapor extraction (oxidation).

◆ Key Terms

active layer	hardpan	phytoremediation	soil profile
bioremediation	laterite	regolith	spheroidal weathering
caliche	loess	residual soil	trace elements
erosion	pedalfer	rill erosion	tundra soils
exfoliation dome	pedocal	sheet erosion	vapor extraction
expansive soils	pedologist	sheet joint	weathering
frost wedging	permafrost	soil	
gullying (erosion)	permafrost table	soil horizon	

◆ Study Questions

1. Sketch a soil profile and describe each soil horizon.
2. Distinguish chemical weathering, physical weathering, and erosion.
3. What kinds of soils would you expect to develop on basalt in Hawaii, New Jersey, and Arizona? Explain the differences in terms of the CLORPT equation.
4. What is permafrost, and where is it found?
5. What soil problem does permafrost present to engineering works, and what techniques are used to prevent damage from this problem?
6. What is expansive soil, and what problems and challenges does it pose to soils engineers? What can be done to prevent distress due to this condition?
7. Soil erosion and degradation are two of humankind's oldest problems. How has human-induced soil loss come about? What agricultural practices have led to salinization and water-logging of soils?
8. An inscribed rock obelisk dating from 1500 B.C. was resurrected from the ruins at Karnak, Egypt. Its inscriptions were crystal clear when it was brought to New York's Central Park as a gift of the Egyptian government in 1880, but today they are totally illegible. Explain the reason for this.
9. Laterite is one of the most interesting of the zonal soil groups. What special problems have laterites presented in deforested areas of Brazil and war-torn Vietnam?
10. Compare the rates of soil erosion and soil formation in the United States.

◆ For Further Information

BOOKS AND PERIODICALS

Batie, Sandra S. 1983. *Soil erosion: A crisis in America's croplands?* Washington, D.C.: The Conservation Foundation.

Birkeland, Peter. 1984. *Soils and geomorphology.* New York: Oxford University Press.

Brady, Nyle C. 1990. *The nature and properties of soils.* New York: Macmillan Publishing Co.

Brown, K. W. 1994. New horizons in soil remediation. *Geotimes* (American Geological Institute), September.

Brown, Lester. 1999. *State of the world.* New York: W. W. Norton & Co.

Ferrians, O. J., and others. 1969. *Permafrost and related engineering problems,* U.S. Geological Survey professional paper 678.

Heiniger, Paolo. 1995. The leaning tower of Pisa. *Scientific American* 273, no. 6: 62–67.

Loynachan, Thomas E.; Kirk W. Brown; Terence H. Cooper; and Murray Milford. 1999. *Sustaining our soils and society.* Alexandria, Virginia: American Geological Institute.

Rahn, Perry H. 1987. *Engineering geology: An environmental approach.* New York: Elsevier Publishing Co.

Winkler, Erhard M. 1998. The complexity of urban stone decay. *Geotimes* (American Geological Institute), September.

World Resources Institute and United Nations Environment and Development Programs. 1992. *World resources, 1992–1993: A guide to the global environment.* New York: Oxford University Press.

World Wide Web Activities

The Risk of Expansive Soils
http://www.dnai.net/~meehan/rlmce/expan/master.html
This site explains the risk of expansive soils to structures and gives engineering approaches to damage mitigation. Several links lead to soil tests, geological origins, engineering solutions, and financial considerations of expansive soils.
1. What is the geologic origin of expansive clay minerals in the San Francisco Bay region?
2. What are some of the consequences to homeowners of expansive soils?

Soil Erosion Problems and Approaches to Soil Stabilization
http://soils.ecn.purdue.edu/~wepphtml/wepp/wepptut/jhtml/intro.html
This Purdue University site uses text and photos to address major soil erosion problems and approaches to their mitigation. Using the blue triangles, navigate through **wind erosion,** and **water erosion,** to **soil erosion pre-**diction technology. You may also link to other concepts that are highlighted within the text of these pages, and view some video clips.
1. What beneficial use of crop residues is described here?
2. What are the main variables affecting water erosion of soil?

Salton Sea Images and Ecological Problems
http://www.sci.sdsu.edu/salton/SaltonSeaHomePage.html
The Salton Sea, a sag pond on the San Andreas transform plate boundary, is California's largest and most polluted lake. View the magnificent satellite image, read the **Brief Description of its Current Condition,** and navigate to other sites of interest from this master page.
1. What is the elevation of the lake's surface? How does the lake's elevation and desert location contribute to toxicity?
2. What types of plans have been proposed to restore health to this lake?

 # Earth Systems Today CD Correlations

CHAPTER CONCEPT: Weathering
This CD section demonstrates that weathering is a set of processes within the rock cycle that break down rocks and create soil and sediment.

SECTION MODULE: The Rock Cycle
Go to the **Introductory Animation** module of **Rocks and the Rock Cycle,** and find **Weathering** on the diagram.

QUESTION: *What processes make rocks vulnerable to weathering; that is, what processes are likely to immediately precede weathering?*

chapter seven

Mass Wasting and Subsidence

A Bad Slice on the Golf Course

Golf's explosion in popularity in the late 1990s led to many new golf courses. Some are in resort communities, but others are in urban and suburban areas, like the beautiful Ocean Trails Golf Course, south of Los Angeles on the Palos Verdes Peninsula. Geologists know the peninsula well for its many landslides, due in part because it is uplifted 370 meters (1,300 ft) above sea level. Its stratified sedimentary rock inclines gently to the sea, in 30- to 50-meter cliffs where the layers are exposed. Just as a slice of bread slides down a breadboard when tilted, some weak layers have slid toward the sea in the past, carrying houses and roads with them. The nearby Point Fermin landslide is an example; see Figures 7.11 and 7.40. In other cases, constant battering by waves at the cliffs' base has created unstable conditions.

Ocean Trails Golf Course was almost ready for its early summer 1999 opening when, without warning, a fissure opened parallel to the cliff, and 300 meters of the 18th fairway slid into the ocean. Local resident Tony Baker and his dog were temporarily stranded on a precarious 215-meter- (700 ft)-long island between fissures. "I heard crumbling earth and started seeing dust rising," Baker said later. "The trail started cracking up. I was doing a little running around, jumping over big cracks . . . I just found a place and hunkered down. I wasn't sure if I was going to make it back."

Damage to the course would be repaired and the $80 million facility's grand opening rescheduled, but the final round would be played out in the courts. ■

Ocean Trails golf course, Palos Verdes Peninsula, California, shortly after the "slice" occurred, disrupting grand opening plans.

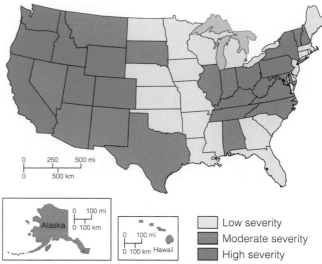

◆ **FIGURE 7.1** Severity of landsliding in the United States. Note that the most serious problems are found in the Appalachian Mountains, the Rocky Mountains, and the coastal mountain ranges of the Pacific Rim.

Low severity
Moderate severity
High severity

Mass wasting is the general term that denotes any downslope movement of soil and rock under the direct influence of gravity. Mass-wasting processes include landslides, rapidly moving debris flows, slow-moving soil creep, and rockfalls of all kinds. Annual damage from landsliding alone in the United States is estimated at between one and two billion dollars. If we include other ground failures, such as subsidence, expansive soils, and construction-induced slides and flows, total losses are many times greater than the annual combined losses from earthquakes, volcanic eruptions, floods, hurricanes, and tornadoes.

Mass wasting is also a global environmental problem. Earthquake-triggered landslides in Kansu Province in China killed an estimated 200,000 people in 1920, and debris flows left 600 dead and destroyed 100,000 homes near Kobe, Japan, in 1938. The largest loss of life from a single landslide in U.S. history, 129 fatalities, occurred at Mameyes, Puerto Rico, in 1985.

The term **landslide** encompasses all moderately rapid falls, slides, and flows that have well-defined boundaries and that move downward and outward from a natural or artificial slope. They occur in all the states and are an economically significant factor in more than 25 of them (◆ Figure 7.1). Wherever unstable slopes exist, some form of mass wasting usually occurs. Areas with the greatest topographic relief are at highest risk, because gravity acts to smooth out topography by reducing the high areas through mass wasting and filling in the low areas with slide debris. For example, more than two million mappable slides exist in the Appalachian Mountains from New England to Alabama. Almost ten percent of the land area of Colorado is landslide terrane, and landslides and debris flows in Utah caused $300 million worth of damage in 1983–84. In Southern California, eight wet years between 1950 and 1993 averaged $500 million in landslide damage. Slope stability problems have such impact that a national landslide-loss-reduction program

was implemented by the U.S. Geological Survey in the mid-1980s. A high priority of the program is to identify and map landslide-prone lands.

The classification of slope movements used in this book is similar to the one widely used by engineering geologists and soil engineers (◆ Figure 7.2). The bases of the classification are

- the type of material involved, such as rock or soil;
- how the material moves, such as by sliding, flowing, or falling;
- the moisture content of the material; and
- how fast the material moves, its velocity.

For instance, soil creep occurs at imperceptibly slow rates and is relatively dry; debris flows are water-saturated and move swiftly; and slides are coherent masses of rock or soil that move along one or more discrete failure surfaces, or **slide planes.** Failure occurs when the force that is pulling the slope downward (gravity) exceeds the strength of the earth materials that compose the slope. Because gravity acting on a given slope is essentially constant, either the earth materials' resistance to sliding must decrease or the gradient of the slope must increase in order for sliding to occur.

Flows

TYPES OF FLOWS

Creep is the slow (a few millimeters per year), essentially continuous downslope movement of soil and rock

MECHANISM		MATERIAL			VELOCITY
		Rock	Fine-grained Soil	Coarse-grained Soil	
SLIDE		Slump	Earth slump	Debris slump	Slow
		Block glide	Earth slide	Debris slide	Rapid
FLOW		Rock avalanche	Mudflow, avalanche	Debris flow, avalanche	Very Rapid
		Creep	Creep	Creep	Extremely slow
FALL		Rockfall	Earthfall	Debrisfall	Extremely rapid

◀ **FIGURE 7.2** Classification of landslides by mechanism, material, and velocity.

on steep slopes. It involves either freezing and thawing or alternate wetting and drying of a hill slope, which causes upward expansion of the ground surface perpendicular to the face of the slope. As the slope dries out or thaws, the soil surface drops vertically, resulting in a net downslope movement of the soil (◀ Figure 7.3). Burrowing animals and other biological processes that produce openings in the soil also contribute to soil creep. Bent trees, leaning fence posts and telephone poles, and bending of tilted rock layers downslope are all evidence of soil creep. Homes built with conventional foundations 18–24 inches deep may develop cracks due to soil creep. This is seldom catastrophic; typically this is a cosmetic and maintenance problem. The influence of soil creep can be overcome by placing foundations through the creeping soil into bedrock (Figure 7.3).

Debris flows are dense, fluid mixtures of rock, sand, mud, and water. They may be generated quickly during heavy rainfall or snowmelt where there is an abundant supply of loose soil and rock. Moving with the consistency of wet concrete, they are very destructive, with velocities up to many meters per second. Because of their high density, commonly 1.5–2.0 times the density of water, debris flows are capable of transporting large boulders, automobiles, and even houses in their mass. The house shown in ◀ Figure 7.4 is in Shields Canyon in the San Gabriel Mountains of Southern California. Thirteen cars were packed around the house, and several others were deposited in the backyard swimming pool. The family was rescued from inside the house, floating on mud that reached nearly to the ceiling.

The areas most subject to debris flows are characterized by sparse vegetation and intense seasonal rainfall, or are in regions that are subject to drenching rains (see Case Study 7.1). Debris flows are relatively common in Canada, the Andes (particularly Peru), and alpine and desert environments worldwide. Debris flows associated with volcanic eruptions (lahars) are potential hazards in all volcanic zones of the world. They pose a particular danger in populated areas near volcanoes in Washington, Oregon, and California, and are the most significant threat to life of all the landslide hazards. An *earthflow* is a sliding mass of soil and weathered bedrock over a discrete basal surface, usually fresh bedrock. Earthflows terminate in lobelike forms and grade into debris flows as water content increases. They may be thought of as embryonic debris flows.

A **debris avalanche** is simply a fast-moving (>4 m/sec) debris flow (◀ Figure 7.5). A structure in the path of one of these flows can be severely damaged or completely flattened with attendant injury or loss of life. During the winter of 1982, debris flows and avalanches in the San Francisco Bay area killed 25 people and resulted in $66 million worth of damage. In Mill Valley, a small town north of San Francisco, a debris flow ripped a house from its foundations and deposited it 45 meters (150 ft) downslope—without serious injury to the residents (passengers). The house effectively dammed the canyon and prevented debris-flow damage farther downslope.

CAUSES OF DEBRIS FLOWS AND DEBRIS AVALANCHES

The combination of heavy rainfall and loose soil on steep slopes promotes debris flows and avalanches. Studies of debris flows indicate that two rainfall

Case Study | 7.1

Dynamic Real Estate in Oregon

By volume, the Columbia River is the largest river flowing into the Pacific Ocean and the second-largest river in the United States. Its narrow gorge is a site of great scenic beauty, with towering basalt cliffs, beautiful river views, and boreal forests beyond description. No wonder people want to live there. There is a problem, however. On a stretch of the gorge between the towns of Warrendale and Dodson, Oregon, massive debris flows have occurred every 50 to 100 years since the gorge formed 13,000 years ago. The record-setting 1996 snowfall of 4 meters (12 ft) and then 32 centimeters (13 in) of rain on the 2,000–4,000-foot cliffs above the towns changed the idyllic environment to an agonizing one. About 3:00 A.M. on February 6, Dodson resident Mark Chandler went outside to call the family dog. He lived on Tumult Creek and could not help but notice it when the creek's rampaging waters suddenly stopped. He stared unbelievingly at the creek and then heard an ominous rumbling in the distance, like "two Union Pacific trains colliding" on the nearby tracks. Realizing that something was amiss, he dashed back to the house with the dog, awakened his family, and got them all out of town immediately.

A giant landslide had dammed Tumult Creek and was mobilized by the water to form a debris flow. With boulders the size of minivans, the flow broke off 20-meter (65-ft) pine trees, filled houses with mud, and blocked I-84 and the railroad. When it was all over, geologists calculated that there was enough debris to cover a football field with mud 215 meters (700 ft) thick (◇ Figures 1 and 2).

In all, seven landslides struck the gorge between the two towns. County officials quickly placed restrictions on building along this stretch of the gorge and designated the towns as "hazardous areas." This means residents will have to hire a geological consultant for any building whatsoever. Property values have plummeted, and some owners think the government should buy them out. Others want to rebuild, but with government help. The last major slide in the area occurred in about 1920, and the lessons learned then apparently have been forgotten. Remarkably, no one was killed, only one person was injured, and a horse and a dog that were caught in the muck were rescued.

◇ FIGURE 1 This home was submerged to its second floor in the Tumult Creek debris flow. The beautiful high cliffs of basalt in the background were the source area of the landslide/rockfall and ensuing debris flow.

◇ FIGURE 2 The thickness and size of boulders in the flow are apparent in this photo. Again, note the precipitous basalt cliffs in the background.

(a)

(c)

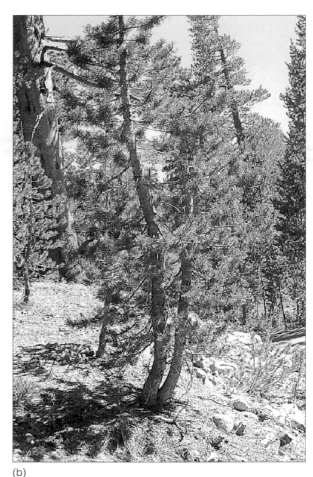

(b)

◆▶ **FIGURE 7.3** (a) Diagram and (b) photograph of soil and rock creep; east slope of the Sierra Nevada, California. (c) Securing a structure on pilings that extend through the creep zone into firm bedrock can prevent damage due to soil motion.

conditions are necessary to cause them: (1) an initial period of rainfall, known as *antecedent rainfall,* that saturates the soil and (2) a subsequent period of intense rainfall that puts even more water into the soil and initiates flowage. Thus the onset of debris flows may be estimated during the rainy season if we have data on the amount of rainfall needed to initiate flowage. Such data have been compiled for flows and slides in the San Francisco Bay area of northern California (◆ Figure 7.6). They indicate that a short period of intense rainfall or less-intense rainfall over a longer period can produce debris flows. Specifically, the curves of Figure 7.6 show that for a rainfall intensity of 0.5 inch per hour, the threshold time for the onset of debris flows is 8 hours in Marin County and 14 hours in Contra Costa County. The differing rainfall thresholds in these relatively close areas are due to the variability of geologic

materials and topography. The lower curve of Figure 7.6 shows that less-intense rainfall will produce debris flows in semiarid and arid areas of California. This is because these dry areas have little vegetation with root systems that retain soils and, therefore, have abundant loose surface debris. Although these curves are only preliminary, they will serve to alert residents in critical areas once the threshold conditions for flows have been attained, and they will be refined as more data become available.

Heavy rains associated with El Niño events have resulted in exceptional landslide and debris flow activity, particularly along the west coasts of North and South America. The major El Niño of 1982–1983 was marked by widespread landsliding (slumps, slides, and flows) in various parts of the Western Hemisphere. The 1997–1998 El Niño, possibly the largest of the twentieth

◇▶ **FIGURE 7.4** Debris flow in Shields Canyon, San Gabriel Mountains, California, 1977. Note that both the car and the man are on the roof of a house.

century, also had exceptional flows and slides. Hard hit was the San Francisco Bay area, where some of the 85,000 landslides shown on the thematic maps of the region were reactivated (◇▶ Figure 7.7). This is an excellent example of interaction between the atmosphere, hydrosphere, and solid earth systems. More on this in Chapter 10.

Mountain slopes burned by range and forest fires are also susceptible to debris flows during the wet season. The loss of active root systems to bind soil particles can result in an extremely dangerous condition. In addition, debris flows from burned slopes have been found to have longer runout distances into foothill areas than those from vegetated slopes.

Not all debris flows and avalanches are triggered by intense rainfall. Slide Mountain, about halfway between Reno and Carson City, Nevada (and between Washoe Lake and Lake Tahoe), is not named for the ski area on its slopes; it is the site of many flows and slides.

◇▶ **FIGURE 7.5** (a) Debris avalanche track and zone of deposition. A debris avalanche may travel thousands of feet beyond the base of the slope.
(b) Debris avalanche scars at Panuelos, Puerto Rico, following heavy rains of October 1985. One of these avalanches flattened the local Lions Club building, fortunately unoccupied at the time.

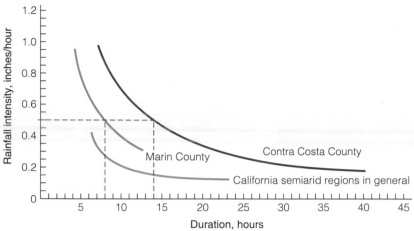

On 30 May 1983, a debris flow was generated on Slide Mountain that killed one person, injured many more, and destroyed a number of homes and vehicles, all in less than 15 minutes (◆ Figure 7.8). About 720,000 cubic meters of weathered granite gave way from the mountain's steep flank and slid into Upper Price Lake. This mass displaced the water in the lake, causing it to overflow into a lower lake, which in turn overflowed into Ophir Creek gorge as a water flood. Picking up sediment as it went, it became a debris flow that emerged from the gorge, spread out, destroyed homes, and covered a major highway. Because this is a popular recreational area and debris flows can be sudden and hazardous to one's health, geologic hazard warning signs are posted on trails throughout the area. The 1983 Slide Mountain debris flow and the one associated with an earthquake at Yungay, Peru, that killed

18,000 people (discussed in Chapter 4) were both initiated by landslides into water masses.

Landslides that Move as a Unit

TYPES

Two kinds of coherent slide masses are recognized, distinguished by the shape of the slide surface. They are **slumps,** or **rotational slides,** which move on curved, concave-upward slide surfaces and are self-stabilizing, and **block glides,** or **translational slides,** which move on inclined slide planes (see Figure 7.2). A block glide moves as a unit until it meets an obstacle or until the slope of the slide plane changes.

◆ **FIGURE 7.7** (a) Rainfall in excess of normal in the San Francisco Bay area during El Niño years 1982–1983 and 1997–1998. (b) Reactivated earth flow near the town of Aromas in southern Santa Cruz County, California, in April 1998. The flow cut natural gas service to 60,000 Santa Cruz County residents.

(a)

(b)

(c)

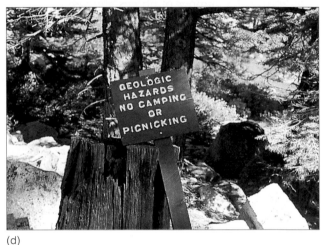
(d)

◆▶ **FIGURE 7.8** Slide Mountain in western Nevada, site of the 1983 disaster. (a) Looking up Ophir Creek and the runout area of the debris flow. The scar of the debris-avalanche source area is the treeless area on the mountain. (b) Debris flow and a damaged home in lower Ophir Creek. (c) Abraded bark on tree records the thickness of the flow. (d) A warning to be taken seriously.

Slumps

Slumps are the most common kind of landslide and range in size from small features a few meters wide to huge failures that can damage structures and transportation systems. They are spoon-shaped, having a slide surface that is curved concave-upward and exhibiting a backward rotation (◆▶ Figure 7.9). They occur in geologic material that is fairly homogeneous, such as soil or badly weathered or fractured bedrock, and the slide surface cuts across geologic boundaries. Slumps move about a center of rotation; that is, as the toe of the slide

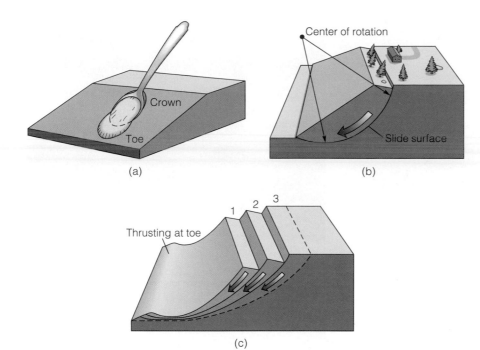

◀▶ **FIGURE 7.9** A rotational landslide.
(a) Stylized oblique view of its spoon
shape, crown, and toe. (b) Cross
section, showing the curved slide
plane, the center of rotation, and
backward-tilting trees at the top of
the slide. (c) Headward growth of
slump blocks results in stair-step
topography at the crown.

rotates upward, its mass eventually counterbalances the downward force, causing the slide to stop (◀▶ Figure 7.9b). Slumping produces repeated uniform depressions and flat areas on otherwise sloping ground surfaces. Slumps grow headward (upslope). This is because as one slump mass forms, it removes support from the slope above it. Thus, a stairstep surface is common in landslide terrain (◀▶ Figure 7.9c). Slump-type landslides are the scourge of highway builders, and repairing or removing them costs millions of dollars every year.

Block Glides

Block glides are coherent masses of rock or soil that move along relatively planar sliding surfaces (failure planes), which may be sedimentary bedding planes, metamorphic foliation planes, faults, or fracture surfaces. For a block glide to occur, it is necessary that the failure plane be inclined *less steeply* than the inclination of the natural or manufactured hill slope. Slopes may be stable with respect to block glides until they are steepened by excavating for building subdivisions or roads, thus leading to landsliding (◀▶ Figure 7.10).

A classic block glide of about ten acres is seen along a sea cliff undercut by wave action at Point Fermin in San Pedro, California. Movement was first detected there in January 1929, and by 1930 the landslide had moved two meters seaward. It has been intermittently active ever since (◀▶ Figure 7.11). The rock of the slide mass is coarse sandstone (not the type of earth material usually involved in block glides), but a thin layer of **bentonite** dipping 15° seaward forms the slide plane. Bentonite is volcanic ash that has chemically weathered to clay minerals, which become plastic and slippery when wet (see Case Study 2.3). Bentonite is very commonly involved in slope failures; addition of water is all that is needed to initiate a landslide where dips are as slight as 5 degrees.

Some of these translational slides are large and move with devastating speed and tragic results. In 1985, a tropical storm dumped a near-record 24-hour rainfall that averaged almost 470 millimeters (18.5 in) on a mountainous region near the city of Ponce on the south coast of Puerto Rico. Rainfall intensities peaked in the early morning hours of October 7, reaching 70 millimeters (2.8 in) per hour in some places. At 3:30 that Monday morning, much of the Mameyes residential district of the city was destroyed by a rock block glide initiated during the most intense period of rainfall. This resulted in the worst loss of life from a landslide in U.S. history—129 deaths. The landslide was in a sandstone whose stratification parallels the natural slope of the slide mass, a condition called a *dip slope*. It moved at least 50 meters (165 ft), probably on a clay layer in the sandstone, before breaking up into the large blocks that destroyed 100 homes (◀▶ Figure 7.12). The scarp at the top of the slide is 10 meters high (32 ft), and the maximum thickness observed at the toe of the landslide is 15 meters (49 ft).

The importance of water in initiating landslides and debris flows cannot be overemphasized. South-central Puerto Rico is the "dry" side of the island, where the average annual rainfall is about 1,000

> ◆▶ FIGURE 7.10 Cutting a natural slope for homesites leads to potential block glides on bedding or foliation planes. (a) Natural slope and proposed cut. (b) After cutting, the slope exhibits unstable conditions. (c) A slide plane in clay shale shortly after the overlying slide mass began moving; Santa Monica Mountains, California. (d) Striations gouged in wet clay by the landslide mass shown in (c).

millimeters (39 in). Ponce received almost half of that amount in the 24 hours preceding the landslide.

In addition to the heavy rainfall and adverse geologic conditions, two other factors probably contributed to the disaster: (1) Mameyes was a densely populated district with no sewer system; sewage was discharged directly into the ground. (2) A water main at the top of the slide was reported to have been leaking for some time. If this is true, the subsurface rocks there were already saturated with water, and the heavy rains of Tropical Storm Isabel were all that was needed to trigger the slide. The slide mass closely followed the boundaries of the Mameyes residential district, testimony to the impact of urban development on the landscape.

A similar landslide occurred in the Alps of northeastern Italy in 1963 at the site of Vaiont Dam, the highest thin-arch dam in the world at the time (275 m;

900 ft). The geology at the reservoir consists of a sedimentary-rock structure that is bowed downward into a concave-upward **syncline** with its axis parallel to the Vaiont River canyon. Limestones containing clay layers dip toward the river and reservoir from both sides of the canyon due to this downfolded structure (◆▶ Figure 7.13). Slope movements and slippage along clayey bedding planes above the reservoir had been observed before the dam was constructed. This condition gave engineers and geologists sufficient concern that they placed survey monuments on the slope above the dam for monitoring such movement. Heavy rains fell for two weeks before the disaster, and slope movements as large as 80 centimeters (31 in) per day were recorded. On the night of October 9, without warning, a huge mass of limestone slid into the reservoir so fast it generated a wave 100 meters (330 ft) high. The wave burst over the top of the dam

(a)

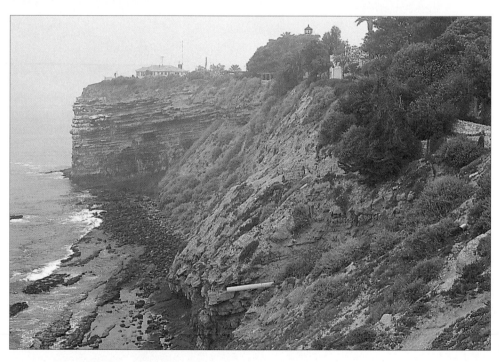

(b)

◆► **FIGURE 7.11** (a) The Point Fermin landslide at San Pedro, California, is a rock block glide that has moved intermittently since 1929. Damaged homes have been removed from the slide area in the foreground. The slide plane emerges just in front of the houses in the background, which are on relatively stable ground. (b) A 50-meter-high sea cliff adjacent to the area in photo (a). The sandstone and shale layers inclined seaward out of the face of the cliff are unstable. A plastic drain pipe was installed into the rocks to remove excess water from the potential slide mass.

and flowed into the Piave River valley, destroying villages in its path and leaving 2,500 people dead. The landslide velocity into the reservoir was so speedy that the slide mass almost emptied the lake. The dam did not fail and it is still standing today—a monument to excellent engineering but poor site selection.

Slump and block glides are the most common types of landslides, but there are also **complex landslides,** combinations of the two types. These landslides have elements of both rotational and translational failure, and there are many examples, most of them large landslides such as the Slumgullion landslide in Colorado (Case Study 7.2).

Consider This... ? You must choose between two hillside homesites. One site is underlain by schist, the other by gneiss, both with foliation dipping out of the hill ("daylighting"). Which site would you choose, and why? (Hint: Examine Figure 7.10, and refer to the discussion of sedimentary and metamorphic rocks in Chapter 2). ■

Case Study | 7.2

Colorado's Slumgullion Landslide—A Moving Story 300 Years Old

The Rocky Mountain states, and particularly Colorado, have some of the nation's highest levels of landslide hazard. Each year landslides in the Rockies take several lives and cause millions of dollars in damage to forests, roads, pipe and electrical-transmission lines, and buildings. What may be the largest active landslide in the United States is between Gunnison and Durango, Colorado, and bears the unusual name *Slumgullion,* a name that was probably bestowed upon it by early prospectors (◁▷ Figure 1).

> **slumgullion** (slum-gul´yən), *n.* 1 *Slang.* a meat stew with vegetables, as potatoes and onion. 2. *Mining.* a muddy red residue in the sluice. *(Webster's New International Dictionary)*

Slumgullion, the stew, was a standard of mining-camp cooks, and slumgullion, the red residue, was what miners usually found instead of gold in the sluice box.

The Slumgullion landslide originated 700 years ago when highly weathered and altered Tertiary volcanic rocks gave way on a ridge above the Lake Fork tributary of the Gunnison River. It is a few kilometers upstream from Lake City, a historic mining town in the San Juan Mountains. The landslide dammed Lake Fork River, forming the largest natural lake in Colorado, Lake San Cristobal. The landslide is huge—6 kilometers (3.6 mi) long, 1 kilometer (0.6 mi) wide, and an average of 40 meters (132 ft) thick. As ◁▷ Figure 2 shows, the Slumgullion is a landslide within a landslide; its 300-year-old active portion is flowing within the larger, 700-year-old mass. The total drop from headscarp to toe is about 762 meters (2,500 ft), and the slide moves at about 6 meters (20 ft) per year. The Slumgullion exhibits three distinct regions of deformation (◁▷ Figure 3):

■ The top of the slide is in tension (extension). Normal faulting dominates this region.
■ The middle region flows as a rigid block, called *plug flow.* Nearly vertical strike-slip faults are found along plug's lateral edges.
■ The toe is a region of compression. It exhibits thrust (reverse) faults.

Within the active landslide the terrane is jumbled, and trees are either highly tilted or dead.

The landslide poses several geologic hazards. If it overruns Colorado Highway 149, it will isolate recreational development upstream. Downstream flooding will result if the lake level rises and overtops the natural dam. For these reasons, the U.S. Geological Survey is conducting multidisciplinary studies of the area. In addition to 30 years' worth of observation data, the USGS is evaluating:

◁▷ **FIGURE 1** The Slumgullion landslide in southwestern Colorado has a tremendous length and a sinewy path. The head scarp is more than 500 meters (1,600 ft) high.

FIGURE 2 Generalized contour map showing the 300-year-old active landslide within an inactive older slide mass. Lake San Cristobal formed when the landslide dammed the Lake Fork River.

■ precise measurements of the landslide's rate of movement using conventional survey techniques as well as Global Positioning System (GPS) technology.

■ old photographs of the area. These enable the researchers to estimate the slide's rate of headward growth. Because a slide's retrogression feeds rock debris and mass to the head of the slide, it serves to stimulate movement and growth of the slide.

■ detailed studies of Lake San Cristobal and its landslide dam. These will provide information that is applic-

able to other landslide dams in Colorado and the Rocky Mountains.

■ three-dimensional models of the active and inactive portions of the slide based upon the accumulated data. These facilitate prediction of the Slumgullion's future behavior.

It is hoped that the results of this study will be applicable to large landslides in other states.

FIGURE 3 Flow regions within the landslide. Extension (pull apart) at the top of the landslide leads to slumping there. The coherent plug flow in the middle of the landslide moves with lateral slip at the sides of the plug. Thrusting takes place at the toe of the landslide as the coherent plug attempts to override the shattered but stable toe.

(a)

(b)

◆ FIGURE 7.12 The 1985 Mameyes landslide near Ponce, Puerto Rico. (a) Disrupted surface and destroyed homes. (b) Two phases of movement are visible from the left side of the photograph to the right side.

FALLS

Rockfalls occur very rapidly and are among the most common of mass-wasting processes. Signs alerting motorists to watch for falling rocks are a common sight (◆ Figure 7.14). Chunks of rock of all sizes simply fall from vertical or very steep cliffs, having separated from the main mass along joints, faults, foliation planes, and other rock defects. The rocks are loosened by root growth, frost wedging, and heavy precipitation. A pile of rock debris may build up at the base of a cliff, forming a slope called **talus** (◆ Figure 7.15). Two huge rockfalls in Yosemite Valley in 1996 killed one visitor and injured 14 others. The force of the shock-wave from the rockfall toppled nearby trees and stripped bark from more distant ones. The geology of this catastrophic event has been studied intensively by Radford University and the U.S. Geological Survey in an attempt to pinpoint areas of future rockfall hazard.

THE MECHANICS OF SLIDES

Landslides do not just happen. They are explainable as a change in the balance between **driving forces,** the gravity forces that pull a slope downward, and **resisting forces,** the cohesive and frictional forces that hold a slope in place. A change in the balance can result from water seepage into slope material, an oversteepening of the slope by natural erosion or artificial cutting, or the addition of weight at the top of a slope. If any of these things happen, the driving forces may

eventually exceed the resisting forces, which will cause a landslide to occur.

This can be demonstrated using a simple model of a sliding block on an inclined plane (◆ Figure 7.16). The driving force, d, is the component of gravity acting parallel to the inclined plane at an angle a with the horizontal. The coefficient of friction between the block and the plane is f. Static friction between the block and slide plane increases as the normal force, n, acting across the slide plane increases. Friction between the block and the plane resists sliding. Thus,

$$\text{resisting forces} = \text{friction coefficient} \times \text{normal force}$$

or

$$r = f \times n$$

A delicate balance exists, and sliding is imminent when

$$d\text{(driving force)} = r\text{(resisting force)}$$

that is, when

$$r/d = 1.0$$

The Factor of Safety (F. S.) relates resisting force to driving force. When the two forces are equal, the F. S. equals 1.0. When resisting forces are greater than driving forces, the F. S. is greater than 1.0; and when resisting forces are less than driving forces, the F. S. is less than 1.0. Most modern building codes require that manufactured or natural slopes have a Factor of Safety of 1.5 or greater. In practice, one simply sums up all the driving forces and resisting forces acting along a potential

(a)

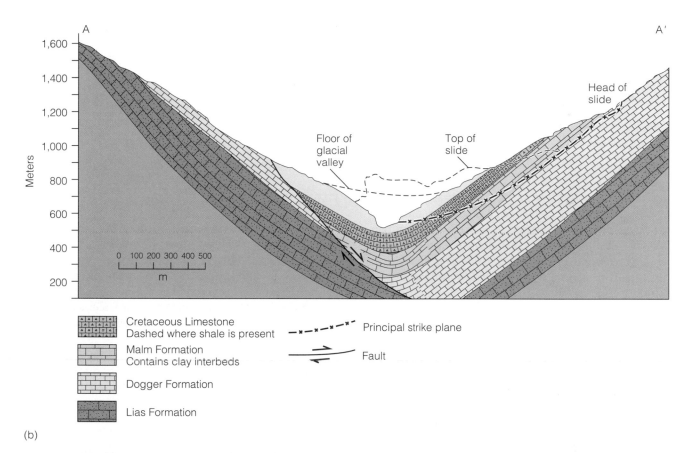

(b)

◇▶ **FIGURE 7.13** The 1963 Vaiont Dam landslide. (a) Map of the slide mass that hurtled into the reservoir and the area that was impacted by the giant wave and flooding. Longarone and several other villages along the Piave River below the dam were devastated. (b) Cross-section through the Vaiont River valley showing a syncline with the sedimentary layers dipping toward the valley axis. The principal slide plane and resultant slide mass are indicated. The cross-section line is shown in (a).

slide plane. From this it can be calculated whether the F. S. of the particular slope is greater than 1.0 (relatively stable) or less than 1.0 (relatively unstable).

If water pressure builds up in the pore spaces between sediment grains, it pushes them apart and reduces the soil's effective grain-to-grain friction—

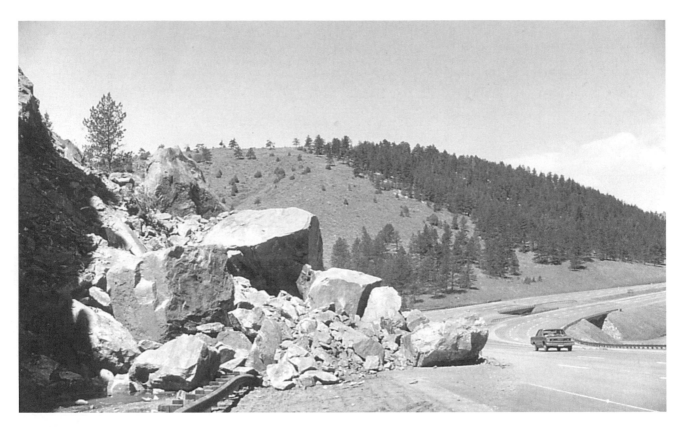

◆▶ **FIGURE 7.14** Rockfall onto I-70 in Jefferson County, Colorado, precipitated by heavy rainfall on weakened rock structure—joints and foliation planes.

referred to as its **effective stress.** This is analogous to injecting water under pressure beneath the block on an inclined plane. The water pressure will support a portion of the block's weight, thereby reducing the normal force n and the sliding resistance of the block. Water in soil or sediment pore space supports part of the grain-to-grain pressure, thereby reducing the frictional resistance across the grain contacts. The effective stress increases as the normal force increases and decreases as the pore-water pressure increases. A "quick" condition exists when the pore pressure is equal to intergranular friction; in other words, when the net effective stress is reduced to zero. At this point, sand becomes a dense sand-water mixture known as *quicksand.* It is often said that water "lubricates" a failure plane, making it easier to slip. It has been demonstrated, however, that a certain amount of moisture actually increases the strength of some materials. Moist sand, for example, builds a better castle and stands on a steeper slope than does dry sand. Water itself has no lubricating qualities. Rather, it is pore-water pressure that reduces grain-to-grain friction and therefore soil strength.

LATERAL SPREADING

Horizontal movement of a mass of soil overlying a liquefied or plastic layer characterizes a form of mass wasting called **lateral spreading.** Such slides are complex, as they involve elements of translation, rotation, and flow. Typically triggered by earthquakes, lateral spreading may result in the spontaneous liquefaction of water-saturated sand layers or in the collapse of **sensitive clays**—also known as **quick clays** (see Chapter 4). Spreading failures in the United States occur mostly in sand layers; however, areas underlain by glacial sediments may spread because of quick clay layers within the deposits.

Much of the damage in the Marina district of San Francisco in the 1906 and 1989 earthquakes was due to

Consider This...? The geologists' report on a building site that you are considering buying states that the sedimentary strata—shale and sandstone—have a "bad attitude." Does this mean they are not nice? Is there some geological interpretation of this term? ■

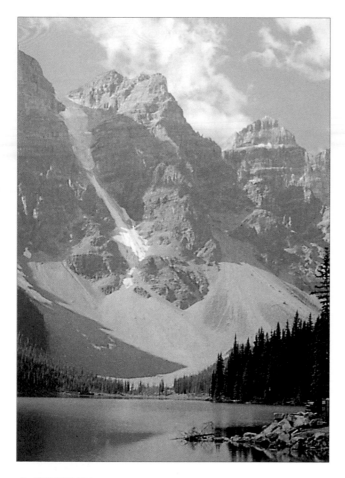

◆ **FIGURE 7.15** Talus slopes caused by frost wedging and other physical weathering processes; Banff National Park in the Canadian Rocky Mountains.

◆ **FIGURE 7.16** Resolution of driving forces (d) and resisting forces (f × n) acting on a slide plane inclined at an angle α.

lateral spreading caused by liquefaction (see Figure 4.22). Most of the damage in Anchorage, Alaska, during the 1964 quake was due to quick-clay-induced lateral spreads in the Turnagain Heights residential neighborhood and the downtown area. The spreading was so extensive that two houses that had been more than 200 meters (more than 2 football fields) apart collided within the slide mass. The Turnagain Heights area is now a tourist attraction known as "Earthquake Park" (◆ Figure 7.17).

Factors that Lead to Landslides

Seldom can a landslide be attributed to a single cause; rather, landslides result from series of events that lead

◆ **FIGURE 7.17** Turnagain Heights after the landslide that accompanied the 1964 Alaska earthquake.

Grains roll down slope
when angle exceeds 34°

Beach sand

34° angle of repose

◆▶ **FIGURE 7.18** The angle of repose for dry sand is 34°. Coarser material, such as that found on a talus slope, will stand slightly steeper, but it also has a specific angle of repose.

to failure. Nonetheless, the weakening of slope materials due to the addition of water is the most important causative factor of all slides and flows. Thus heavy rainfall, rapid snowmelt, leaking water mains, private sewage-disposal (cesspool) inflow, and poor building-pad drainage can all lead to landsliding. Excess water causes a buildup of pore-water pressure, which weakens the materials supporting the slope. This is why mass wasting is closely correlated with a series of heavy-rain years in semiarid Southern California and with torrential flooding in the eastern United States.

Recall that when driving forces exceed resisting forces, a failure is imminent. Factors that increase driving force are

- an increase in the slope angle,
- removal of lateral support at the toe of a slope, and
- added weight at the top of a slope.

The effect of increasing slope angle is best illustrated by the natural **angle of repose** of granular material such as dry sand or gravel. The angle of repose of such material is the maximum slope angle at which it can remain stable. For example, no matter how steeply you try to pile dry sand, it will always form a slope of about 32°–34°, its angle of repose (◆▶ Figure 7.18). Moist sand has a greater angle of repose because of the temporary cohesion (added strength) imparted to it by moisture. This is why a castle built of moist sand collapses when the sand dries and loses its temporary cohesion. The angle of repose represents the point at which the frictional resistance between sand grains (shear strength) and the downslope component of gravity are in balance.

Theoretically, mass movements will not occur as long as the angle of repose of the particular slope material is not exceeded. Because undercutting the toe of a

Consider This... ❓   Suppose you wish to build a structure at the top of a vertical cut that is 3.1 meters (10 ft) high. The site is underlain by poorly cemented river sand. How far back from the top of the slope would you have to build to achieve an F. S. of 1.0? How far back for an F. S. of 1.5? (*Hint:* Consider the material's angle of repose.) ■

slope is equivalent to increasing the slope angle, it causes instability. Adding weight to the slope, as with fill material or an unusually heavy structure, increases the pore pressure and can thus induce failure. Factors that reduce a slope's strength and, therefore, its resistance to sliding are

- infiltration of water underground,
- weathering and breakdown of minerals, especially when clays are formed, and
- burrowing by animals.

In addition, it should be noted that sliding is also facilitated when a slope is cut in such a way that its bedding planes are exposed or inclined (dipping) out of the slope face.

A slope with a Factor of Safety less than 1 is ready to slide, but it might not do so unless it is "triggered" by an earthquake, heavy traffic, sonic boom, detonation of explosives, or another source of energy. Such triggers are split-second stresses that overcome *static friction,* the force resisting sliding between two surfaces at rest. Static friction is always greater than *sliding (kinetic) friction,* the friction existing between two surfaces in relative motion. This explains why landslides are very hard to stop once they start.

Reducing Losses from Mass Wasting

LANDSLIDE HAZARD ZONATION

The first step in reducing losses due to mass-wasting over the long term is to make a careful inventory or map of known landslide areas. Ancient landslides have a distinct topography characterized by hummocky (bumpy) terrain with knobs and closed depressions formed by backward-rotated slump blocks. On topographic maps, slide terrain can be evidenced by curved, closely spaced, amphitheater-shaped contour lines indicating scarps at the heads of landslides. In the field, one may see scarps, trees, or structures that are tilted uphill; unexpected anomalous flat areas; and water-loving vegetation such as cattails where water issues at the toe of a landslide.

Explanation

Stable

Generally stable

Moderately unstable

● Single isolated small landslide deposits

Subject to liquefaction

Moderately stable

Unstable

◆ FIGURE 7.19 Map of the relative slope stability of parts of Contra Costa County and adjacent counties derived by combining a slope map, a landslide-deposit map, and a map of susceptible geologic units. The area shown here is 15 kilometers (9 mi) wide.

With an adequate data base provided by maps and field observations, it is possible to define landslide hazard zones. Maps can be generated that show the *landslide risk* in an area, and then these areas can be avoided or work can be done to stabilize them (◆ Figure 7.19). Mapping programs do not prevent slides, but they offer a means for minimizing slides' impact on humans. The human tragedy of landslides in urban areas is that they cause the loss of land as well as of homes.

BUILDING CODES AND REGULATIONS

Building codes dictate which site investigations must be performed by geologists and engineers and the way structures must be built. Chapter 70 of the *Uniform Building Code,* which deals with slopes and alteration of the landscape, has been widely adopted by cities and counties in the United States. It specifies compaction and surface-drainage requirements and the relationships between such planar elements as bedding, foliation, faults, and slope orientation. Code requirements for manufactured slopes in Los Angeles resemble the specifications found in codes elsewhere (◆ Figure 7.20). The maximum slope angle allowed by code in landslide-prone country is 2:1; that is, 2 feet horizontally for each foot vertically (27°). The rationale for the 2:1 (27°) slope standard is that granular earth materials have a natural angle of repose of about 34°; 2:1 slopes allow for a factor of safety.

Losses have been cut dramatically by enforcement of grading codes and they will continue to decrease as codes become more restrictive in hillside areas. For instance, before grading codes were established in the City of Los Angeles in 1952, 1,040 building sites were damaged or lost by slope failures out of each 10,000 constructed—a loss rate of 10.4 percent. With a new code in effect between 1952 and 1962 that required minimal geologic and soil-engineering investigation, losses were reduced to 1.3 percent for new construction. In 1963, the city enacted a revised code requiring

◆ FIGURE 7.20 Part of a typical grading ordinance (code) that shows the limitation of cut-slope steepness, the required benches on slopes for collecting rainfall and reducing slope-face erosion, and an approved method of constructing a fill slope with bedrock benches for increased stability.

extensive geological and soils investigations. Losses were reduced to 0.15 percent of new construction in the following six years, which included 1969, the largest rain year in recent history. A new international code that is applicable anywhere in the world is in the works.

Control and Stabilization

The basic principles behind all methods of preventing or correcting landslides relate to strengthening the earth materials (increasing the resisting forces) and reducing the stresses within the system (decreasing the driving forces). For a given slide mass, this requires halting or reversing the factors that promote instability, which will probably be accomplished by one or more of the following: draining water from the slide area, excavating and redistributing the slide mass, and installing retaining devices. Debris flows require special devices that divert the path of the flow away from structures.

WATER DRAINAGE AND CONTROL

Water is the major culprit in land instability, and surface water must be prevented from infiltrating the potential slide mass. Water within the slide mass can be removed by drilling horizontal drains, called *hydrauger* (water bore) holes, and lining them with perforated plastic pipe (◀▶ Figure 7.21). This is commonly done on hillside terrain where underground water presents a problem. In some areas the land surface of a building site above a suspect slope is sealed with compacted fill to help keep surface water from percolating underground. Also, plastic sheeting is widely used to cover cracks and prevent water infiltration and erosion of the slide mass (◀▶ Figure 7.22). The efficacy of drainage control and planting is shown in ◀▶ Figure 7.23. The unimproved slope is raveling badly and will undoubtedly be an expensive maintenance headache for the owner. The terraced and planted slope, protected from the elements, is performing well.

EXCAVATION AND REDISTRIBUTION

Recontouring is a textbook method of stabilizing a slide mass. By this method, material is removed from the top of the landslide and placed at the toe. Compacted-earth structures called *buttress fills* are often designed to retain large active landslides and known inactive ones believed to be vulnerable. They are constructed by removing the toe of a slide and replacing it, layer by layer, with compacted soil. Such fills can "buttress" huge slide masses, thus reclaiming otherwise unusable land. The finished product is required to have concrete terraces for intercepting rainwater and preventing erosion and downdrains for conveying the intercepted water off the site (◀▶ Figure 7.24). Buttressing unstable land has aesthetic and economic advantages, in that it can be landscaped and built upon. The cost of the additional work needed to stabilize unstable land is usually offset by a relatively low purchase price, which can make it economically "buildable." Of course, the ultimate mitigating measure is total removal of the slide mass and reshaping of the land to buildable contours.

◀▶ **FIGURE 7.21** Hydrauger drains drilled into serpentine. Perforated plastic drain pipes are connected to a large "manifold" that directs water drained from the cut slope into a natural drainage; San Luis Obispo, California.

◀▶ **FIGURE 7.22** Plastic sheeting placed over the head scarp of a landslide prevents rainwater infiltration. Although this is common practice in landslide-prone areas, plastic sheeting is virtually unavailable in "home" stores during the rainy season in many communities.

◆▶ **FIGURE 7.23** Erosion prevention measures on a graded slope; San Clemente, California. Whereas slope planting and terracing have maintained the middle slope in good condition, the unimproved slope on the left has raveled badly and will need maintenance in the future.

RETAINING DEVICES

Many slopes are oversteepened by cutting back at the toe, usually for the purpose of obtaining more flat building area. The vertical cut can be supported by constructing steel-reinforced concrete-block retaining walls with drain (weep) holes for alleviating water-pressure buildup behind the structure (◆▶ Figure 7.25). Retaining devices are constructed of a variety of materials, including rock, timber, metal, wire-mesh fencing, and a sprayed concrete known as *shotcrete* (◆▶ Figure 7.26).

Steep or vertical rock slopes that are jointed or very seamy are often strengthened by inserting long rock bolts into holes drilled perpendicular to planes of weakness. This binds the planes together much as a beam is bound. Rock bolts are used extensively to support tunnel and mine openings. They add considerably to the safety of these operations by preventing sudden rock "popouts." They are also installed on steep road-cuts to prevent rockfalls onto highways (◆▶ Figure 7.27). Retaining devices also are effective for mitigating rockfalls and debrisfalls.

(a)

(b)

◆▶ **FIGURE 7.24** (a) Buttress fill with horizontal drains for removing water and surface drains for preventing erosion. The fill acts as a retaining wall to hold up unstable slopes such as those that result from excavation. (b) A slope with benches and surface drains built according to accepted standards (see Figure 7.20).

DIVERSION TECHNIQUES

Debris flows present a different challenge, because they can originate some distance away from the site of interest. It is simply not good practice to build at the bottom or the mouth of a steep ravine or gully, especially if loose soils are present on the higher slopes. This is particularly true in areas of high mean annual rainfall, areas that are subject to sudden cloudbursts, and areas that have been burned, such as in the tragic October 1993 firestorms in Southern California. As long as people continue to build houses in the "barrels" of canyons, as is done in the foothills of Southern California, debris flows will continue to take their toll. Debris flow insurance is expensive, but the cost can be less if *deflecting walls* are designed into the house plans (◁▷ Figure 7.28). These walls have proven to be effective in protecting dwellings, but they could have adverse consequences if they divert debris onto neighbors' structures, in which case neither an architect nor a geologist would be as helpful as a lawyer. ◉ Table 7.1 summarizes mitigating measures that can slow or stop landsliding and other mass-wasting processes.

Consider This... ?   In many parts of the world, large numbers of people live on or close to potentially unstable hillslopes. How do grading codes help to reduce the risk of landslides? Do you think they are effective if they are enforced? ■

Snow Avalanches

The mechanics of snow avalanches are similar to those of landsliding, the differences being in the material and the velocity. Although snow avalanches have been around as long as there have been mountains and snow, the relatively recent emergence of the recreational skiing industry as big business has transformed avalanche control from an art to a science. Unfortunately, it's too late for Hannibal. His crossing of the Alps in 218 B.C. was plagued by avalanches. Purportedly 18,000 of his troops and who knows how many of his elephants were killed by them. It's also too late for others. In World War I 6,000 troops were buried in one day by avalanches in the Dolomite Mountains of northern Italy. Avalanches in high mountain regions have wiped out villages, disturbed railroad track alignments, blocked roads, and otherwise made life difficult if not impossible for millennia.

Avalanches are caused by an addition of heavy new snowfall or a weakening of older snow. A **slab avalanche** is a coherent mass of snow and ice that hurtles down a slope of more than 30° as an entity—much like a magazine sliding off a tilted coffee table. It may be a thick, wet snow accumulation sliding on ice, or a coherent slab of snow that detached along a *depth hoar* layer. Depth hoar begins to form at the surface when snow particles evaporate in very cold, calm weather and form a loose, open array of snow crystals. As the layer of crystals is buried, evaporation continues, resulting in a

---

◉ TABLE 7.1   **Summary of Mass-Wasting Processes and Their Mitigation**

LANDSLIDES		
Cause	*Effect*	*Mitigation*
Excess water	Decreased friction (effective stress)	Horizontal drains, surface sealing
Added weight at top	Increased driving force	Buttress fill, retaining walls, decrease slope angle
Undercut toe of slope	Increased slope angle and driving force	Retaining walls, buttress fill
"Daylighted" bedding	Exposure of unsupported bedding in cut or natural slope	Buttress fill, retaining walls, decrease slope angle

OTHER MASS-WASTING PROCESSES	
Process	*Mitigation*
Rockfall	Rock bolts and wire mesh on the slope, concrete or wooden cribbing at bottom of the slope, cover with "shotcrete"
Debris flow	Diversion walls or fences
Soil creep	Deep foundations
Lateral spreading	Dewater, buttress, retain (difficult to mitigate)

◀▶ **FIGURE 7.25** Typical retaining wall with drain (weep) holes to prevent water retention and the buildup of hydrostatic pressure behind the wall. The wall is constructed of concrete blocks with steel reinforcing rods and concrete in their hollow centers.

◀▶ **FIGURE 7.26** "Shotcrete" stabilization of a roadcut in fractured igneous rock; Kobe, Japan.

very weak, porous layer at depth. This hoar layer may collapse when it is disturbed, forming a compressed-air layer upon which the overlying snow can move with great velocity. These avalanches account for the majority of avalanche deaths because of their sheer weight and their tendency to solidify when they stop. About 25 avalanche fatalities occur each year in North America, mostly in backcountry areas, and the victims tend to be cross-country skiers, snowmobilers, climbers, and hikers.

Powder-snow avalanches are more common in the Sierra Nevada and Coast Ranges, where depth hoar is rare and abundant moisture results in deep snow packs. These "dry" avalanches are dangerous because they entrain a great amount of air and behave like a

(a)

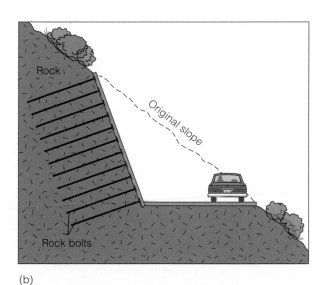

(b)

◀▶ **FIGURE 7.27** (a) Typical rock bolt installation in a tunnel or mine opening. (b) Rock bolts support jointed rocks above a highway.

(a)

(b)

◆ **FIGURE 7.28** (a) The outlet of a debris-flow diversion structure; San Fernando Valley, California. (b) One effect of the May 1968 Heath Canyon mudflow caused by rapid snowmelt; Wrightwood, California. A diversion wall or dike might have saved this mountain cabin in the San Gabriel Mountains. Heath Canyon is now leveed to contain debris flows within the river channel.

fluid. Furthermore, the strong shock wave that precedes them can uproot trees and flatten structures (◆ Figure 7.29). Victims of dry avalanches tend to be buried in light snow close to the surface, which makes their chances of being rescued alive better than those of wet-avalanche victims.

Avalanches can be triggered by skiers, rockfalls, loud noises, or explosives. Avalanche control is thus achieved by triggering an avalanche with explosives or

cannon fire when dangerous snow conditions exist in an area frequented by humans. The explosions collapse the depth hoar or detach powder snow, causing a "controlled" avalanche.

Various means have been developed for mitigating the destructive might of avalanches. Where structures already exist, fences can be built to divert avalanches away from them. In some areas in the Alps, the uphill sides of chalets are shaped similar to

◆ **FIGURE 7.29** The toe of an avalanche reveals a tangled mass of aspen and pine trees ripped out by their roots. The power and danger of this winter phenomenon is obvious. The photo was taken in June 1995 in the San Juan Mountains near Telluride, Colorado.

◆ **FIGURE 7.30** The historic White Pass and Yukon Railroad, near Skagway, Alaska, is protected from avalanches by snowsheds.

a ship's prow so that they will divert avalanching snow. Railroads and highways can be protected by building snowsheds along stretches adjacent to steep slopes (◆ Figure 7.30).

Subsidence

Two million people live below the high-tide level in Tokyo. The canals of Venice overflow periodically and flood its beloved tourist attractions (Case Study 7.3). The Houston suburb of Baytown subsided almost three meters in the 1900s, and 80 square kilometers (31 mi²) of this coastal region is permanently under water. Earthen dikes prevent the sea from flooding the land in some places in California, and Mexico City's famous Palace of Fine Arts rests in a large depression in the middle of the city (◉ Table 7.2). These areas suffer from natural and human-induced **subsidence,** a sinking or downward settling of the earth's surface. Subsidence is not usually catastrophic, and loss of life due to it is rare, but land subsidence is currently observed in 45 states and is estimated to cost $125 million annually. In the United States the area of human-induced subsidence is estimated as at least 44,000 square kilometers (17,000 mi²), about the size of Maryland and New Jersey combined, but the actual area is probably even greater. Planners and decision makers need to be aware of the causes and impacts of subsidence in order to assess the risks and reduce material losses.

HUMAN-INDUCED SUBSIDENCE

Human-induced subsidence occurs when humans extract underground water, engage in mining or oil and gas production, and when they cause loose sediments at the ground surface to consolidate or compress. The effects may be local or regional in scale. In the United States and Mexico alone, an area the size of Vermont has slowly subsided 30 centimeters (1 ft) due to withdrawal of underground water. Sinking of the land changes drainage paths, and it is particularly damaging to coastal areas and lands adjacent to rivers because it increases flood potential.

Along the Texas Gulf Coast, for example, annual losses from subsidence due to fluid withdrawal were estimated at $109 million between 1943 and 1973. Lowering of the water table in areas of cavernous limestone and gypsum has caused surface subsidence and collapse costing many millions of dollars each year (see Chapter 9 discussion of hydrologic hazards). Each year subsidence of the land surface into abandoned coal mines costs the United States $30 million. About a quarter of the seven million acres that have been mined for coal has subsided, and some of this area underlies cities.

NATURAL SUBSIDENCE

Natural subsidence is caused by earthquakes, volcanic activity, and the solution of limestone, dolomite, and gypsum. Earthquake-related subsidence occurs rapidly (◆ Figure 7.31), and although it is best known in

◯ **TABLE 7.2** **Subsiding Cities, 1986**			
WORLD		**UNITED STATES**	
City	Maximum Subsidence, cm*	City	Maximum Subsidence, cm
Mexico City	850	Long Beach	900
Tokyo	450	(San Joaquin Valley)	880
Osaka	300	San Jose	390
Shanghai	263	Houston	270
Niigata, Japan	250	Las Vegas	66
Bangkok	100	Denver	30
Taipei, Taiwan	190	New Orleans	22
London	30	Savannah	20

* Since measurement began
SOURCE: R. Dolan and H. Grant Goodell, "Sinking Cities," *American Scientist* 74 (1986), no. 1.

◆ **FIGURE 7.31** Area of subsidence and uplift (tilt) that resulted from the Alaska earthquake, Good Friday 1964. As much as a meter of subsidence occurred.

Case Study | 7.3

A Rescue Plan for the Merchants of Venice

alled "the city where streets are paved with water," Venice has special interest because of its historical importance, priceless antiquities, and beauty, as well as its charming canals. The city has a major flooding problem, however. St. Mark's Square and other squares are flooded during periods of extremely high tides the locals call the *aqua alta* (Italian, "high water"). It is said that the first Venetians settled here in A.D. 421 "in order to escape the barbarian Goths" that were sweeping down from the north. Sitting on a low mud bank in a lagoon sheltered by barrier islands (see Chapter 10), the city has subsided about three meters (10 ft) relative to sea level since it was founded. The sinking is due in part to the existence of 20,000 water wells in the region (now water is brought in by aqueduct), settlement into the soft sediments upon which it was built, and rising sea level. The traditional building practice was to drive wooden piles into the soft sediment to form level foundations on which to build structures, and of course, timbers rot and settle. According to records, one cathedral is founded on more than a million timbers.

During periods of *aqua alta* the avenues and squares of Venice become flooded and gondola traffic comes to a screeching halt (◆ Figure 1). The *aqua alta* occurs when an onshore flow of water (a storm surge) from the Adriatic Sea coincides with exceptionally high tides. There were an average of 30 high waters per year between 1957 and 1967, and the number had increased to 40–60 floods annually by 1990 (◆ Figure 2). Dan Stanley of the Smithsonian Institution in Washington, D. C. likens Venice to a man standing in a swimming pool: His feet are embedded in lead, and the water level is rising. It's already to his nose. He'd better not make any waves.

The solution? Install movable floodgates at the entrances to the 90-mile-perimeter lagoon that can temporarily cut the lagoon off from the sea. The cost? Two billion dollars (yes, that's $2,000 millions). The idea is to raise the floodgates like dams in anticipation of high water, and then to reopen them when the water level drops. There are envi-

◆ FIGURE 1 When the *aqua alta* floods St. Mark's Square in Venice, raised boardwalks are provided for pedestrians.

Alaska and California, it also occurs in a number of other states. Displacement along large faults can raise or lower the land surface over a large area. For instance, subsidence of one meter over 180,000 square kilometers (70,000 mi²) took place during the Alaskan earthquake of 1964. Much of the subsidence was along the coast, and the subsided area is now flooded at high tide (Figure

7.31). In fact, the land actually tilted, and a large area of the Gulf of Alaska was uplifted several meters.

Severe ground shaking that leads to liquefaction may also lower the ground surface. This happened along the valley of the Mississippi River during the New Madrid earthquakes of 1811 and 1812 and in the Marina district of San Francisco in 1989. A very small-scale ronmental reasons as well as financial ones for not embark-ing on the project right away. Assuming a sea level rise of 5 centimeters over the next century, which is a conservative estimate, the gates would have to be raised 70 times a year, and that would interfere with boat traffic and tourism. According to British expert Edmund Penning-Rowsell, "The problem is that as sea level rises, they will have to be closed off so often that they will be quite ridiculous." However, the Italians love their antiquities and do a superior job of pre-serving them, but if the money is to be raised by taxing the city's merchants and citizens you may hear *"basta!"*

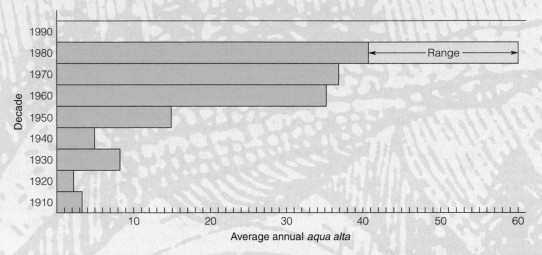

◆ FIGURE 2 Frequency of *aqua alta* flooding in Venice since 1910. The increase in frequency is due to subsidence, settlement, and rising sea level.

subsidence problem is the collapse of roofs of shallow lava tunnels and tubes in volcanic areas. Volcanic activity that empties magma chambers can cause collapse and subsidence over much larger areas when it forms calderas (see Chapter 5). Subsidence caused by tectonic (mountain-building) processes occurs so very slowly that it is not considered an environmental hazard.

To put this geologic hazard into perspective, most subsidence that adversely affects humans is caused by human activity and is irreversible. Once the land surface sinks by any significant amount, it cannot be "jacked" back up to its original level. As noted, most subsidence occurs so slowly that it poses little danger to people. There are some exceptions, however, such as sudden collapses into underground openings and subsidence-related dam failures. Subsidence that has well-identified connections with human activities is emphasized in this chapter.

A Classification of Subsidence

One method of classifying land subsidence is according to the depth at which the subsidence is initiated. *Deep subsidence* is initiated at considerable depth below the surface when water, oil, or gas is removed from there. *Shallow subsidence,* on the other hand, takes place nearer the ground surface when underground water or solid material is removed by natural processes or by humans. In addition, many poorly consolidated shallow deposits such as peat are subject to compaction by overburden pressures, ground-water withdrawal, and in some soils, simply by being saturated with water.

Settlement differs from subsidence. It occurs when an applied load—such as that of a structure—is greater than the bearing capacity of the soil onto which it is placed. Settlement is totally human-induced and it is a soils- and foundation-engineering problem. The Leaning Tower of Pisa is a classic example of foundation settlement; its interesting construction history is discussed in Chapter 6.

DEEP SUBSIDENCE

Removal of fluids—water, oil, or gas—confined in the pore spaces in rock or sediment causes deep subsi-

dence. Pore-water pressure—that is, the hydrostatic pressure of water in the pores between sediment grains—helps to support the overlying material. As pressure is reduced by extraction, the weight of the overburden gradually transfers to mineral-and-rock-grain boundaries. If the sediment was originally deposited with an open structure, the grains will reorient into a closer-packed arrangement, thus occupying less space, and subsidence will ensue (◆ Figure 7.32). Because clays are more compressible than sands, most compaction takes place in clay strata.

At least 22 oil fields in California have subsidence problems, as do many fields in Texas, Louisiana, and other oil-producing states. A near world record for subsidence is held by the Wilmington Oil Field in Long Beach, California, where the ground has dropped 9 meters (about 30 ft). This field is the largest producer in the state, with more than 2,000 wells tapping oil in an upward-arched geologic structure called an *anticline* (see Chapter 14). The arch of the Wilmington field's anticline gradually sagged as oil and water were removed, causing the land surface to sink below sea level. Dikes were constructed to prevent the ocean from flooding the adjacent Port of Los Angeles facilities and the naval shipyard at Terminal Island. People who have their boats moored on Terminal Island must actually walk *up* to board them at sea level (◆ Figure 7.33)! When it rains it is necessary to pump water out of low spots upward to the sea, and oil-well casings and underground pipes have "risen" out of the ground as the land has sunk. To remedy the situation, in 1958 the City of Long Beach initiated a program of injecting water into the ground to replace the oil being withdrawn. Subsidence is no longer a problem, and the surface has even rebounded slightly in places.

Visible effects related to ground-water withdrawals are water-well damage, cracking of long structures such as canals, and development of tension cracks and fissures. In the Antelope Valley, California, home of Edwards Air Force Base, water levels beneath the dry lake have been lowered by more than 60 meters (200 ft). The dry lakebed at Edwards is the landing strip for the space shuttle and other high-performance aircraft. A ground fissure appeared in the lakebed that, initially only a few centimeters wide, became greatly enlarged by surface-water runoff and erosion. This halted operations on the lakebed until repairs could be made. Similar fissuring has occurred in Arizona, New Mexico, Texas, and Nevada. In Arizona ground-water withdrawal in some places has caused subsidence of 5 meters (more than 15 ft) and the development of cracks and earth fissures, particularly along the edges of some basins (◆ Figure 7.34). The subsidence has changed

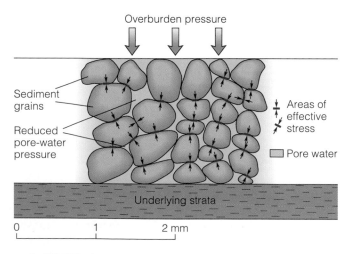

◆ **FIGURE 7.32** A reduction of pore pressure causes increased effective stress and rearrangement and compaction of sediment particles.

(a)

(b)

◆► **FIGURE 7.33** The effects of subsidence on the coastal communities of Wilmington and Long Beach, California. (a) The power plant (left) is at the point of maximum subsidence in the Wilmington Oil Field, about 9 meters (30 ft). Well below sea level, the building is protected from the sea by dikes. The yachts are moored at sea level. (b) This fire hydrant extruded as the ground around it sank; Long Beach.

◆► **FIGURE 7.34** Earth fissure due to subsidence caused by withdrawal of underground water near Picacho, Arizona.

◆ **FIGURE 7.35** Subsidence due to excessive ground-water withdrawal for agriculture in the San Joaquin Valley, California. The numbers on the pole indicate the ground level for the years indicated. The area is still subsiding, but at a much slower rate.

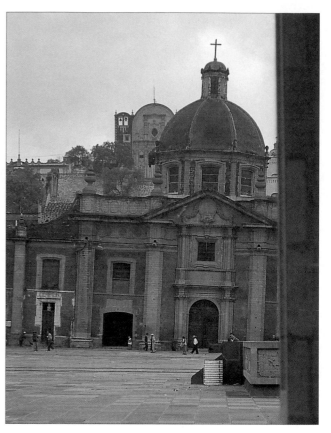

◆ **FIGURE 7.36** Differential settlement at Our Lady of Guadalupe church, Mexico City. The foundation rests across both solid basalt and a subsiding lakebed.

natural drainage patterns and caused highways to settle and crack. It is influencing the siting of waste-treatment and solid-waste disposal sites and generally impacting all development activity in the state. The problem is so serious that the Arizona Department of Water Resources and the Arizona Geological Survey have established the Center for Land-Subsidence and Earth-Fissure Information, with the catchy acronym *CLASEFI,* to provide information and assistance to the public.

Coincidentally, the magnitude of the subsidence due to ground-water withdrawal is similar to oil-field subsidence. In the San Joaquin Valley of California, for example, 8.9 meters (29 ft) of subsidence resulted from 50 years' water extraction for agricultural irrigation (◆ Figure 7.35).

Mexico City sits in a fault valley that has accumulated nearly 2,000 meters (6,600 ft) of sediments, mostly of pyroclastic origin. In 1925 it was demonstrated that subsidence was occurring here due to the withdrawal of water from these sediments. Total subsi-

Consider This...? Subsidence of coastal cities such as Venice and Houston is particularly dangerous because it threatens them with reclamation by the sea. The subsidence problems in land-locked areas such as Mexico City and the San Joaquin Valley are a major geological concern. However, these areas do not face catastrophic inundation by the sea as do the major coastal cities of Tokyo, Venice, Houston, and New Orleans, to name but a few. What can be done to prevent or slow such potential catastrophes? ■

dence varies throughout the city, but almost 6 meters (20 ft) of subsidence had occurred by the 1970s in the northeast part of the city, mostly in the water-bearing top 50 meters (160 ft) of sediment. In 1951 an aqueduct was completed that brought water to the city, and as a result, a number of wells were closed. The effect on subsidence rates was dramatic, and by 1970 subsidence had decreased to less than 5 centimeters per year. The famous Our Lady of Guadalupe (the "brown Madonna") Cathedral tilts, because its foundation is partly on lava flows and partly on the subsiding lake sediments (◈ Figure 7.36).

SHALLOW SUBSIDENCE
Collapsing Soils

Subsidence that results from heavy application of irrigation water on certain loose, dry soils is called **hydrocompaction.** The in-place densities of such "collapsible" soils are low, usually less than 1.3 grams per cubic centimeter, about half the density of solid rock. Upon initial saturation with water, such as for crop irrigation or the construction of a water canal, the open fabric between the grains collapses and the soil compacts.

Subsidence of monumental proportions has occurred because of hydrocompaction. In the United States, it is most common in arid or semiarid regions of the West and Midwest where soils are dry and moisture seldom penetrates below the root zone. Any sediment that was deposited rapidly and that has an open, granular structure is subject to compaction. These are most commonly debris flows or wind-blown (eolian) deposits. Known areas of hydrocompaction are the Heart Mountain and Riverton areas of Wyoming; Denver, Colorado; the Columbia Basin in Washington; southwest and central Utah; and the San Joaquin Valley of California.

Collapsing soils were a major problem for designers of the California Aqueduct in the San Joaquin Valley. In order to maintain a constant ground slope for the flow of water, it was necessary to identify those areas of loose soil that could not be avoided. Sediments were field-tested by saturating large plots and measuring subsidence at various levels beneath the surface. Subsidence was found to average 4.2 meters (close to 14 ft) over a period of 484 days, during which thousands of gallons of water had been applied to the test plots (◈ Figure 7.37). On the basis of the field tests, 180 kilometers (112 mi) of the aqueduct alignment in the

(a)

◈ **FIGURE 7.37** (a) Location of the San Joaquin Valley. Collapsible soils occur almost entirely along the western border of the valley, whereas subsidence due to underground-water withdrawal is found throughout the valley. (b) Water-soaked test plot in the valley's collapsible soils. Total subsidence at this plot is on the order of 3 meters (10 ft).

(b)

Consider This... ? When a delta complex is drained for urban development, as in New Orleans, subsidence accelerates in the region. Which of these foundation types—(a) salty and sandy levee sediments, (b) backswamp clays, and (c) peat deposits—possesses the greatest and the least, subsidence potential? (Hint: see also Case Study 8.2) ■

valley was "presubsided" by soaking the ground three to six months before construction. This assured that the gradient would be maintained after construction. Twenty-two million dollars was added to the cost of the project. Water transport systems, unlike highways or utility lines, are totally determined by the slope of the land beneath them. Sediments deposited on floodplains or on the deltas of large rivers, such as the Mississippi River delta, also are subject to compaction and subsidence. (See Chapter 9.)

Mitigation

The best strategy for minimizing losses is to restrict human activities in areas that are most susceptible to subsidence and collapse. Unfortunately, most instances of land subsidence due to fluid withdrawals have not been anticipated up to now. Knowledge of the causes, mechanics, and treatment of deep subsidence has increased with each damaging occurrence, so that we can now implement measures for reducing the incidence and impact of these losses.

Some of the measures now taken to minimize subsidence are controlling regional water-table levels by monitoring and limiting water extraction, replacing fluids by water-injection, and importing surface water for domestic use in order to avoid drawing on underground supplies.

Underground-mining operators are now required to leave pillars to support mined-out coal beds, and mine openings are being backfilled with compacted mine waste to support the overburden. This latter procedure, however, may introduce chemicals that could contaminate underground water unless the backfill is first stripped of pollutants. Pennsylvania's Mine Subsidence Insurance Act of 1962 allows some property owners to buy subsidence insurance after inspection of their property. Another Pennsylvania bill allows homeowners to buy the bituminous coal beneath their property at a fair price and therefore exercise control over mining operations.

Collapsible soils along canals are now routinely precollapsed by soaking, thereby anticipating in advance any adjustments of the canal gradient. Relatively inexpensive geophysical techniques are available for locating shallow limestone caverns beneath potential building sites. Subsidence problems will continue, but we should employ every means possible to foresee the problems and minimize their impact.

Gallery

Slip, Slide, and Fall

Undeniably, the results of mass-wasting processes can be tragic and devastating, as explained throughout this chapter. In some cases, however, the results are whimsical. For example, a block glide occurred along Highway 89 in Montana when the toe of a dormant ancient slide mass was removed to make room for the road (◆ Figure 7.38). This caused the landslide to reactivate, endangering a telephone line across the moving mass. Over time the solution to this problem became increasingly obvious: the telephone pole supporting the line should be cut off near its base. The pole now hangs from the telephone wires, and the base of the pole has moved 70 feet downslope. Clay-rich shale inclined toward the river in the background of the photograph formed the slide plane. A beautifully developed landslide on the planet Mars shows that gravity and water, even though the water is mostly ice, operate on this rocky planet (◆ Figure 7.39). Explanations for a few other "light" landslide examples (◆ Figures 7.40–7.42) are given in their captions.

◆ FIGURE 7.38 An active block glide on U.S. 89 in Montana. The telephone pole, hanging from the lines, was cut from its base, which had moved 23 meters (70 ft) downslope in the three years before this photograph was taken. The person standing downslope is near the "stump" of the pole.

◆ FIGURE 7.39 My favorite Martian landslide. Oblique view from the south toward the Olympus Mons landslide complex, Mars. The arcuate landslide crown scarp in the background is about 25 kilometers (15 mi) wide and 4 kilometers (2.4 mi) high. The landslide debris flowed as far as 65 kilometers (39 mi) from the scarp toward the observer. The volume of the slide is roughly a thousand times that of the 18 May 1980 debris avalanche from Mount St. Helens. Mass wasting is widespread on this rocky planet.

◇▶ **FIGURE 7.40** "Pedestal pavement" formed by landsliding and erosion; Point Fermin, San Pedro, California. This former street was lined with houses, all destroyed by the landslide.

◇▶ **FIGURE 7.41** Honey, please back the car out! Bluebird Canyon in Laguna Beach, California, was the site of a large and rapid translational landslide that destroyed many homes. The movement was so rapid that even some emergency vehicles became trapped.

◇▶ **FIGURE 7.42** Anything for an ocean view! A 1998 rock fall/slide removed about 20 meters (65 ft) of the sea cliff in front of these beach homes; Leucadia, California.

Summary

MASS WASTING

DEFINED
Downslope movement of rock and soil under the direct influence of gravity. The term *landslide* includes falls, flows, and slides that have well defined boundaries.

CLASSIFICATION BY
1. Type of material—rock or earth.
2. Type of movement—flow, slide, or fall.
3. Moisture content.
4. Velocity.

FALLS

DEFINED
Relatively free fall of earth materials from a steep cliff or slope.

TYPES
1. Rockfall—involves hard bedrock of any size.
2. Debris fall—the collapse of weathered rock material and/or soil from a steep slope or cliff.

FLOWS

DEFINED
Mass movement of unconsolidated material in the plastic or semifluid state.

TYPES
1. Creep—slow, imperceptible downslope movement of rock and soil particles.
2. Debris flow—dense fluid mixture of rock, sand, mud, and water that is fast-moving and destructive.
3. Debris avalanche—fast-moving (75 km/h) debris flow.

CAUSES
Heavy rainfall on steep slopes with loose soil and sparse vegetation. Most common in arid, semiarid, and alpine climates and where dry-season forest fires expose bare soil.

SLIDES

DEFINED
Movement of rock or soil (landslide) as a unit on a discrete failure surface or slide plane.

MECHANICS
Driving force (gravity) exceeds resisting forces (friction and force normal to the slide surface).

TYPES
1. Rotational slide (slump) on a curved, concave-upward slide surface.
2. Translational slide (block glide) on an inclined plane surface.
3. Complex landslide, a combination of (1) and (2).
4. Lateral spreads of water-saturated ground (liquefaction of sand and quick clays).

CAUSES
1. Weakening of the slope material by saturation with water, weathering of rock or soil minerals, and burrowing animals.
2. Steepening of the slope by artificial or natural undercutting at toe.
3. Added weight at the top of the slope such as by a massive earth fill.
4. A trigger such as an earthquake, vibrations from explosives, heavy traffic, or sonic booms.

REDUCTION OF LOSSES

GEOLOGIC MAPPING
To identify active and ancient landslide areas.

BUILDING CODES
To limit the steepness of manufactured slopes and specify minimum soil and fill conditions and surface-water drainage from the site.

CONTROL AND STABILIZATION

DRAIN INSERT PIPES
To drain subsurface water.

EXCAVATION
To redistribute soils and rock from the head of the potential slide to the toe. This decreases driving forces and increases resisting force.

BUTTRESS FILLS AND RETAINING DEVICES
To retain active slides and potentially unstable slopes.

ROCK BOLTS
To stabilize slopes or jointed rocks.

DEFLECTION DEVICES
To divert debris flows around existing structures.

SNOW AVALANCHES

TYPES
Slab avalanche is old snow and ice that moves as a coherent mass. Powder snow avalanche is composed of new snow that moves like a fluid at high velocities.

CAUSES
Weak boundary layer between unstable layer above and stable mass below.

MITIGATION
Controlled avalanching by shooting with light cannons, diversion structures, and snow sheds.

SUBSIDENCE

DEFINED
A sinking of the earth's surface caused by natural geologic processes or by human activity such as mining or the pumping of oil or water from underground.

TYPES

1. Deep subsidence caused by the removal of water, oil, or gas from depth.
2. Shallow subsidence of ground surface caused by underground mining or by hydrocompaction, the consolidation of rapidly deposited sediment by soaking with water.
3. Collapse of ground surface into shallow underground mines or into natural limestone caverns, forming sinkholes.
4. Settlement occurs when the load applied by a structure exceeds the bearing capacity of its geologic foundation.

CAUSES OF DEEP SUBSIDENCE

Removal of fluids confined in pore spaces at depth transfers the weight of the overburden to the grain boundaries, causing the grains to reorient into a closer-packed arrangement.

CAUSES OF SHALLOW SUBSIDENCE AND COLLAPSE

The ground surface may collapse into shallow mines (usually coal). Sinkhole collapse is due to a lowering of the water table, which reduces the buoyant forces of underground water. Most sinkholes result from the collapse of soil and sediment cover into an underground limestone or dolomite cavern (see Chapter 9). Heavy irrigation causes some loose, low-density sediments and soils to consolidate (hydrocompaction). Compaction due to dewatering of peat and cypress-swamp (delta) clays and oxidation of the highly organic sediments causes subsidence.

SUBSIDENCE PROBLEMS

Ground cracking, deranged drainage, flooding of coastal areas, and damage to structures, pipelines, sewer systems, and canals. Whole buildings may disappear into sinkholes.

MITIGATION

Underground water can be monitored and managed. Extracted crude oil can be replaced with other fluids. Mine operators are now required to leave pillars to support the mine roof, and rock spoil can be returned to the mine to provide support. Collapsible soils along canals can be precollapsed by soaking. Foundations on organic deltaic sediments should be on piles seated on firm geologic material. This protects structures from damage if the surrounding land subsides.

 Key Terms

angle of repose	driving force	resisting force	slump
bentonite	effective stress	rotational landslide	subsidence
block glide	hydrocompaction	quick clay	syncline
complex landslide	landslide	sensitive clay	talus
creep	lateral spreading	settlement	translational landslide
debris avalanche	mass wasting	slab avalanche	
debris flow	powder-snow avalanche	slide plane	

Study Questions

1. What grading and land-use practices can reduce the likelihood of landslides?
2. How can vegetation, topography, and sediment characteristics aid in identifying old or inactive slumps, slides, and debris flows?
3. Name the processes that move earth materials downslope (gravity is the *cause,* not the *process*). On what factors does each depend?
4. Modern building codes require a Factor of Safety of 1.5 or greater in order to build on or at the top of a cut slope. What is the Factor of Safety, and why should it be 1.5?
5. What are some common, cost-effective methods of landslide control and prevention?
6. How did geologic structure factor into the disaster at Vaiont Dam?
7. What is the difference between settlement and subsidence? What can be done in advance to prevent settlement?
8. Distinguish between subsidence and collapse that are due to natural causes and due to human causes.
9. What can be done to slow or stop subsidence that is caused by withdrawal of water or petroleum?
10. What engineering problems make constructing canals and aqueducts in areas of hydrocompaction such an insidious geologic hazard?
11. What surface manifestations of subsidence due to ground-water withdrawal make it a serious problem, particularly in California and Arizona?
12. What hazard does subsidence present in coastal areas?

◆ For Further Information

BOOKS AND PERIODICALS

Cannon, S. H., and others. 1998. Slope failure and shoreline retreat during northern California's latest El Niño. *GSA today*. Boulder, Colorado: Geological Society of America.

Dolan, R., and H. Grant Goodell, 1986. Sinking cities. *American scientist* 74, no. 1:38–47.

Dolan, R., and H. Grant Goodell, 1976. Sinking cities. *American scientist*, no. 1.

Fellows, Larry D. 1995. Land-subsidence and earth fissure information. *Arizona geology* 25, no. 2.

Fleming, R. W., and T. A. Taylor. 1980. *Estimating the costs of landslides in the United States*. U.S. Geological Survey circular 832.

Halliday, William R., M.D. 1976. *Depths of the earth*. New York: Harper and Row.

Highland, Lynn M. 1993. Slumgullion: Colorado's natural landslide laboratory. *Earthquakes and volcanoes* 24, no. 5: 208.

Highland, Lynn; J. Godt; David Howell; and W. Z. Savage. 1998. *El Niño 1997–1998: Damaging landslides in the San Francisco Bay area*. U.S. Geological Survey fact sheet 089-98 (free).

Holzer, T. L., ed. 1984. Man-induced land subsidence. *Geological Society of America reviews in engineering geology* VI.

Kolb, C. R., and R. T. Saucier. 1982. Engineering geology of New Orleans. *Geological Society of America reviews in engineering geology* V.

Krohn, J. P., and James Slosson. 1976. Landslide potential in the United States. *California geology*, October.

McPhee, John. 1989. Los Angeles against the mountains. *The control of nature*, ch. 3. New York: Farrar, Straus & Giroux.

Menard, H. W. 1974. *Geology, resources, and society*. New York: W. H. Freeman and Co.

Monastersky, Richard. 1999. Against the tide: Venice's long war with rising water. *Science news* 156, no. 4 (24 July): 63.

National Academy of Sciences. 1978. *Landslides: Analysis and control*. Transportation Research Board special report 176. Washington, D.C.: National Academy Press.

Nilsen, Tor H., and others. 1979. *Relative slope stability and land-use planning in the San Francisco Bay region, California*. U.S. Geological Survey professional paper 944. Washington, D.C.: U.S. Government Printing Office.

Pearson, Eugene. 1995. Environmental and engineering geology of central California: Salinian block to Sierra foothills. National Association of Geology Teachers, Far Western Section, Fall Conference, University of the Pacific, Stockton, California.

Saucier, Robert T., and Jesse O. Snowdon. 1995. Engineering geology of the New Orleans area. *Geological Society of America annual meeting field trip guidebook*, nos. 6a and 6b.

Schultz, Arthur, and Randall W. Jibson (eds.). 1989. *Landslide processes of the eastern United States and Puerto Rico*. Geological Society of America special paper 236, 102 pp.

Slosson, J. E.; A. G. Keene; and J. A. Johnson. 1992. Landslides/landslide mitigation. *Reviews in engineering geology* IX. Boulder, Colorado: Geological Society of America.

Smelser, M. G. 1987. Geology of the Mussel Rock landslide, San Mateo County. *California geology* 40, no. 3: 56.

State of California, Resources Agency. 1965. *Landslides and subsidence*. Geologic Hazards Conference, Los Angeles.

Tremper, Bruce. 1993. Life and death in snow country. *Earth: The science of our planet*, no. 2.

U.S. Geological Survey. 1982. *Goals and tasks of the landslide part of a ground-failure hazards reduction program*. U.S. Geological Survey circular 880. Washington, D.C.: U.S. Government Printing Office.

Varnes, D. J., and W. Z. Savage (eds.). *The Slumgullion earth flow: A large-scale natural laboratory*. U.S. Geological Survey bulletin 2130.

———————. 1995. *Debris-flow hazards in the San Francisco Bay region*. U.S. Geological Survey fact sheet 112-95 (free).

Varnes, David. 1984. *Landslide hazard zonation: A review of principles and practice*. Paris: UNESCO.

White, W. B.; D. C. Culver; J. S. Herman; T. C. Kane; and J. E. Mylroie. 1995. Karst lands. *American scientist* 83, no. 5 (September–October): 450.

Wilshire, Howard; Keith A. Howard; C. M. Wentworth; and Helen Gibbons. 1996. *Geologic Processes at the land surface*. U.S. Geological Survey bulletin 2149.

◆ World Wide Web Activities

Recent Landslides in the United States

http://landslides.usgs.gov/html_files/landslides/newsinfo.shtml

This USGS master site features links to reports on very recent landslide events in the United States and elsewhere.

1. Landslides often are triggered by other natural disasters. What other geologic events triggered some of the recent events you see reported here?
2. Why does the USGS, a federal agency for the United States, become involved in geological disasters in other countries?

Real-time Monitoring of a Landslide near Lake Tahoe, Nevada/California

http://vulcan.wr.usgs.gov/Projects/CalifLandslide/framework.html

The USGS has installed monitors in an active landslide that threatens Highway 50 between Placerville and Lake Tahoe.

Go to the **Real Time Data Interactive Imagemap** to see the instrumentation set-up, and click on each instrument for data output. You may also go to **Current Landslide Status** for a text description of current activity.

1. What are the current water pore pressure indicators for the slide? How recently has it moved?

Shoreline Landslide at Lake Michigan
http://wrgis.wr.usgs.gov/fact-sheet/fs020-98/

A large, sudden landslide removed 1,600 feet of shoreline at Sleeping Bear Dunes, Michigan, a popular bathing beach, in 1995. Other historic landslides in the area tell of long-term instability. Look at the Web site text, photos, and diagrams.

1. What weather condition(s) seem to trigger the landslides at this site?
2. How did scientists establish that wave erosion was not a factor in these slides?

 # Earth Systems Today CD Correlations

CHAPTER CONCEPT: **Landslide Risk in the United States Varies Regionally**

The text (Figure 7.1) shows that the first and second largest river valleys in the contiguous 48 United States, the Mississippi and the Columbia, have very different levels of landslide risks.

SECTION MODULE: **Plate Tectonics**

Included in this CD module Introductory Animation is an interactive map with plate boundaries, earthquake epicenters, and volcanic activity sites. Compare the situations regarding these geologic phenomena for the Columbia River Gorge region and the Mississippi River Valley, and then answer the following questions.

QUESTIONS: *How are the Columbia River Gorge and Mississippi River Valley situated with respect to plate boundaries? How might location on a convergent plate boundary affect slopes? What geologic events near the Columbia River Gorge might trigger wet, steep slopes to landslide?*

chapter eight

Fresh-Water Resources

Whiskey is for drinking.
Water is for fighting.
MARK TWAIN (1835–1910)

Keeping the Peace on the High Plains Aquifer

The next war in the Middle East may well be over water, not oil! Most water in that arid region comes from three shared rivers, the Jordan, the Tigris–Euphrates, and the Nile. Water there is in short supply, and the region's population is expected to double in only 25 years, further stressing water supplies for agriculture and domestic use. You can be sure that water sharing will be a major item on the agenda at any peace talks involving Iraq, Syria, Jordan, Israel, and Egypt.

The same might be true, without the threat of military action, in the water wars over the High Plains aquifer. The High Plains aquifer contains as much water as Lake Huron and underlies 480,000 square kilometers (147,000 mi²) of parts of Kansas, Colorado, New Mexico, Wyoming, South Dakota, Nebraska, Oklahoma, and Texas. Water distributed by the center-pivot irrigation system in the photo is from a well drilled into the aquifer's water-bearing strata that can produce 4,000 liters per minute (about 1,000 gals/min) 24 hours a day. When the amount of water withdrawn from an aquifer exceeds the amount replenished to it by rainfall and rivers, an overdraft condition exists, a shortfall, and as with shortfalls of all kinds, it will have to be reckoned with later if not sooner. This overdraft condition is called ground-water *mining,* as the resource is being exploited without replenishment, and water levels in the strata have been dropping dramatically (see Figures 8.17 and 8.18). Rather than fight, however, farmers in these states are supporting federal government plans for managing the resource and are withdrawing water from the aquifer conservatively. ■

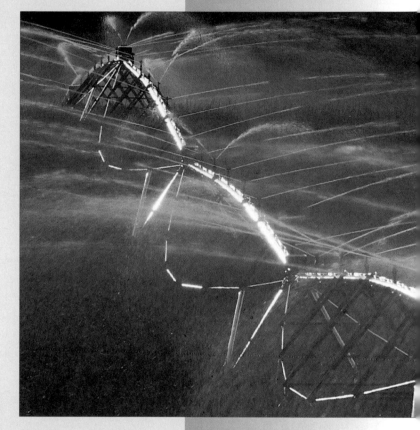

For decades, ground water from the High Plains aquifer has irrigated our nation's "bread basket." This center-pivot irrigation system draws water from a well that taps the aquifer. The well's potential yield was estimated at 4,000 liters per minute (about 1,000 gallons per minute), 24 hours a day.

If extraterrestrials exist and are able to view earth from afar, they cannot help but be struck by our planet's heavy cloud cover and its expanse of ocean, clear indications that earth is truly "the water planet." Water is indeed what distinguishes our planet from other bodies in our solar system. Three quarters of the earth's surface is covered by water, and even the human body is 65 percent water. Water provides humans with means of transportation, and much human recreation is in water—and *on* it where it exists in solid form as ice and snow. Since it is not equally distributed on the earth's surface, we have droughts, famine, and catastrophic floods. The absence or abundance of water results in such marvelous and diverse features as deserts, rain forests, picturesque canyons, and glaciers. In this chap-

ter we investigate the reasons for this uneven distribution and its consequences. Water is involved in every process of human life; it is truly our most valuable resource.

Water as a Resource

Water is the only common substance that occurs as a solid, a liquid, and a gas over the temperature range found at the earth's surface. About 97 percent of the earth's water is in the oceans, and 2 percent is in ice caps and glaciers; about 0.6 percent is readily available to humans as underground or surface fresh water (◀▶ Figure 8.1). Fortunately, water is a renewable resource, and this is illustrated by the **hydrologic cycle,** the earth's most important natural cycle (◀▶ Figure 8.2). Water that is evaporated from the ocean or land surface goes back into the atmosphere and forms clouds. Precipitation from these clouds that falls over land areas eventually flows in streams back to the sea, and the cycle is repeated. Remarkably, very little water is in the atmosphere at any one time, about 0.001 percent of the total. Because of this, we know that water must recycle continuously in order to supply the precipitation we measure. The average annual rainfall over the surface area of the 48 contiguous states is about 75 centimeters

◀▶ **FIGURE 8.1** The planet's water budget. Only a tiny fraction by volume of the world's water supply is fresh water available for human use.

◀▶ **FIGURE 8.2** The hydrologic cycle.

(30 in or 2.5 ft). **Evapotranspiration** is the term for both the direct return of surface water to the atmosphere by evaporation and its indirect return through transpiration from the leaves of plants. It accounts for the return of about *55 centimeters* (22 in) of the average annual precipitation. River runoff to the sea and infiltration into the ground account for the remaining 20 centimeters, but it varies from 2.5 centimeters per year in the dry West to more than 50 centimeters per year in the East. Even though runoff varies greatly with location, the net water transport from land to sea must over time balance the net transport from sea to land.

Planners and forecasters recognize several categories of water use:

- *Instream use*–all uses that occur in the channel itself, such as hydroelectric power generation and navigation.
- *Offstream use*–the diversion of water from a stream to a place of use outside it, such as to your home. The amount of water available for offstream use along any stretch of a river is the total flow *minus* the amount required for instream purposes and for maintaining water quality. ◀▶ Figure 8.3 is a flow diagram of public water supplies and self-supplied water. Note that the amount of water that is consumed is the difference between the amount deliv-

ered (*B*) and the amount that is returned to the stream system (*D*).

- *Consumptive use*–water that evaporates, transpires, or infiltrates and cannot be used again immediately. Forty-four percent of all water that is withdrawn is used consumptively, and agriculture accounts for about 90 percent of that.
- *Nonconsumptive use*–water that is returned to streams with or without treatment so that it can be used again downstream. Interestingly, domestic (household) water is used nonconsumptively, as it is returned to the cycle through sewers and storm drains.

The sources and users of public water supplies are shown in ◀▶ Figure 8.4. The data do not include water used in agriculture or water that is self-supplied.

In many areas, particularly the western United States, consumption exceeds local water supplies. This necessitates importing water from other areas or mining ground water. For example, about 75 percent of Southern California's water is imported from sources more than 200 miles away (◀▶ Figure 8.5). Two aqueducts import water from west and east of the Sierra Nevada, and a third aqueduct brings water from the Colorado River. Irrigation utilizes about 40 percent of the total water withdrawn in the United States, and a startling 80

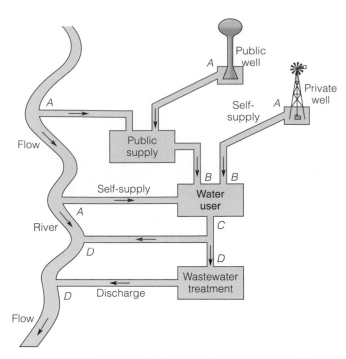

◆ **FIGURE 8.3** Water is supplied to users from streams and from public and private wells (*A*). Consumptive use (*C*) is the difference between the amount delivered to users (*B*) and the amount returned to the system (*D*).

percent in California. It is not surprising then that California, with the largest population and a huge irrigation system, is the largest U.S. consumer of water, using almost as much as the two next-largest consumers, Texas and Illinois, combined (◆ Figure 8.6).

Total world withdrawals of water for offstream purposes amount to about 2,400 cubic kilometers per year (1.7 trillion gal/day, *gpd*), of which 82 percent is for agricultural irrigation. This is expected to increase manyfold to 20,000 cubic kilometers (14.5 trillion gpd) by the year 2050, which is beyond the firm flow of the world's rivers. Availability of water will limit population growth in the twenty-first century, either by statute

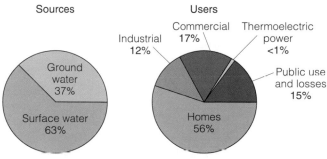

153,000 million liters (40,400 million gallons) per day

◆ **FIGURE 8.4** The sources and users of the U.S. public water supply in 1995. This does not include fresh water supplied for agricultural use or water that is self-supplied.

in developed countries or by natural means, most likely drought and famine, in developing nations. Repetition of the terrible droughts experienced in Africa in the 1970s and 1980s, compounded by overtaxed water supplies, would be disastrous for those countries and their people.

Fresh Water Underground

Some of the water in the hydrologic cycle infiltrates underground and becomes **ground water,** one of our most valuable natural resources. In fact, most of the earth's liquid fresh water exists beneath the land surface, and its geologic occurrence is the subject of many misconceptions. For example, it is commonly believed that ground water occurs in large lakes or pools beneath the land. The truth is that almost all ground water is in pore spaces and fractures in rocks. Another misconception is that deposits of ground water can be found by people with special skills or powers, people referred to as *water witches* or *dowsers*. They locate underground water using two wires bent into L shapes, forked branches, or other forked devices called *divining rods*. A dowser locates an underground "tube" or lake of water by walking over the ground, holding the di-

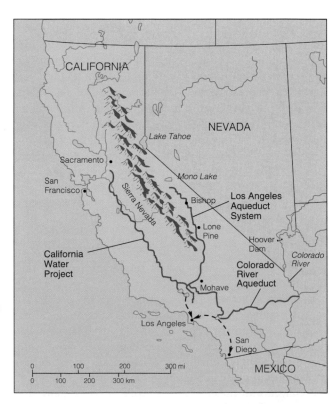

◆ **FIGURE 8.5** Aqueducts bring water to Southern California from three source areas: the Colorado River, the east slope of the Sierra Nevada, and the west slope of the Sierra Nevada.

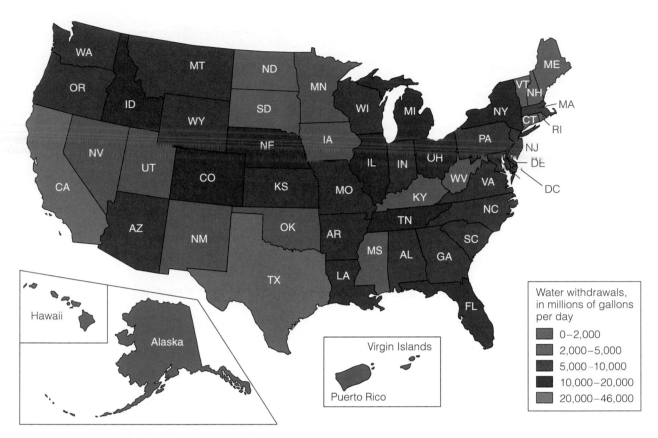

◀▶ **FIGURE 8.6** Total water withdrawals by state, 1995.

vining rod in a prescribed manner until it jerks downward, pointing to the location of ground water, a mineral deposit, or a lost set of keys. Some dowsers claim they can locate good places to drill water wells using only a map of the land. The divining rod is passed over the map until it jerks downward where presumably a well should be located (X marks the spot). Water dowsers are found throughout the world, particularly in rural areas, and many of them will guarantee finding water. This is one difference between dowsers and ground-water geologists, or **hydrogeologists.** Hydrogeologists will not make this guarantee.

Water Supply

Ground water provides 40 percent of public water supplies and 30 to 40 percent of all water used exclusive of power generation in the United States. It is by far the cheapest and most efficient source of municipal water, because obtaining it does not require the construction of expensive aqueducts and reservoirs. Thirty-four of the largest 100 cities in the United States depend entirely upon local ground-water supplies; Miami Beach, San Antonio, Memphis, Honolulu, and Tucson are just a few. Ground water provides 80 percent of the water

for rural domestic and livestock use and it is the only source of water in many agricultural areas. It is not surprising that some of the largest agricultural states—such as Texas, Nebraska, Idaho, and California—are the largest consumers, accounting for almost half of all the ground water produced. Ground water is important even to communities that import water from distant areas, because in almost every case, local ground water provides a significant low-cost percentage of their water supply.

Although it is not generally known, ground water sustains and maintains streamflows during periods when there is no rain. In arid climates where there are permeable water-saturated rocks beneath the surface, ground water seeping into stream channels may be the only source of water for rivers. The fabled oases of the Sahara and Arabian deserts occur where underground water is close to the surface or where it intersects the ground surface, forming a spring or watering hole.

Location and Distribution

Between the land surface and the depth at which we find ground water is the *zone of aeration,* a zone where voids in soil and rock contain only air and water films.

◁▶ FIGURE 8.7 Schematic cross section of ground-water zones showing a static water table with a capillary fringe. A perched water table exists where a body of clay or shale acts as a barrier to further infiltration.

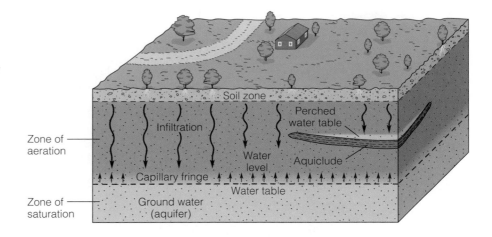

The contact between this zone and the *zone of saturation* below it is the **water table** (◁▶ Figure 8.7). Above the water table is a narrow moist zone, the *capillary fringe*. In the saturated zone, free (unattached) water fills the openings between grains of clastic sedimentary rock, the fractures or cracks in hard rocks, and the solution channels in limestones and dolomites (◁▶ Figure 8.8).

Water-saturated geologic formations whose porosity and permeability are sufficient to yield significant quantities of water to springs and wells are known as **aquifers.** A rock's **porosity** (root word *pore*) is a func-

tion of the volume of openings, or void spaces, in it and is expressed as a percentage of the total volume of the rock being considered. Without such voids, a rock cannot contain water or any other fluid. **Permeability** is the ease with which fluids can flow through an aquifer; it is thus a measure of the connections between pore spaces. **Hydraulic conductivity** is the term hydrologists prefer, but *permeability* will suffice for our purposes, as it has a long history of use and is easy to remember. **Aquicludes** are rocks or sediments such as shale and clay that lack permeability and hence will not transmit water. In some places aquicludes occur as *lenses* that trap water and prevent it from percolating down to the water table. Such a condition creates a **perched water table** (see Figure 8.7).

Aquifers are important because they transmit water from *recharge areas*—that is, areas where water is added to the zone of saturation—to springs or wells. In addition, they store vast quantities of drinkable water. In terms of the hydrologic cycle, we can view aquifers both as transmission pipes and as storage tanks for ground water. The Nubian Sandstone in the Sahara Desert, for instance, is estimated to contain 600,000 cubic kilometers (144,000 mi^3) of water, mostly untapped. It is the world's largest known aquifer and is a resource of great potential for North Africa.

STATIC WATER TABLE

Unconfined aquifers are formations that are exposed to atmospheric pressure changes and that can provide water to wells by draining adjacent saturated rock or soil (◁▶ Figure 8.9). As the water is pumped, a **cone of depression** forms around the well, creating a gradient that causes water to flow toward the well (◁▶ Figure 8.10). A low-permeability aquifer will produce a steep cone of depression and substantial lowering of the water table in the well. The opposite is true for an aquifer

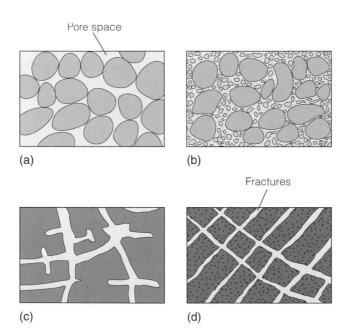

◁▶ FIGURE 8.8 A rock's porosity is dependent on the size, shape, and arrangement of the material composing the rock. Whereas (a) a well-sorted sedimentary rock has high porosity, (b) a poorly sorted one has low porosity. (c) In soluble rocks such as limestones, porosity can be increased by solution, whereas (d) crystalline metamorphic and igneous rocks are porous only if they are fractured.

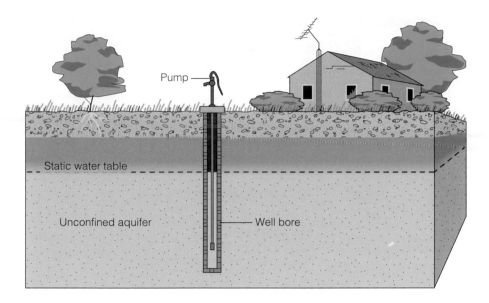

in highly permeable rock or soil. This lowering, or the difference between the water-table level and the water level in the pumping well, is known as **drawdown.** Stated differently, low permeability produces a large drawdown for a given yield and increases the possibility of a well "running dry." A good domestic (single-family) well should yield at least 11 liters (2.6 gal) per minute, which is equivalent to about 16,000 liters (3,700 gal) per day, although some families use as little as 4 liters per minute (1,350 gal/day). As pointed out, a pumping well in permeable materials creates a small drawdown and a gentle gradient in the cone of depression. Some closely spaced wells, however, have overlapping "cones," and excessive pumping in one well lowers the water level in an adjacent one (◀▶ Figure 8.11). Many lawsuits have arisen for just this reason. Water law is discussed briefly in Case Study 8.1.

The water table rises and falls with consumption and the season of the year. Generally, the water table is lower in late summer and higher during the winter or the wet season. Thus water wells must be drilled deep enough to accommodate seasonal and longer-term, climatic fluctuations of the water table (◀▶ Figure 8.12). If a well "goes dry" during the annual dry season, it definitely needs to be deepened.

Streams that are located above the local water table are called **influent,** or "losing," streams, as they contribute to the underground water supply (◀▶ Figure 8.13a). Beneath influent streams there may be a mound of water above the local ground-water table known as a *recharge mound.* A stream that intersects the water table and which is fed by both surface water and ground water is known as an **effluent,** or "gaining," stream (◀▶ Figure 8.13b). A stream may be both a

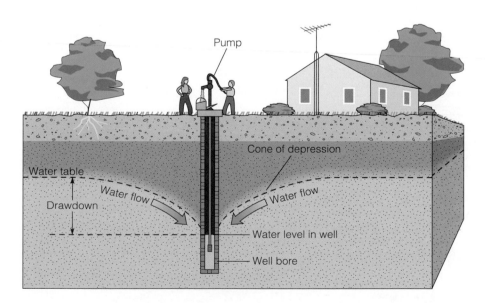

◀▶ **FIGURE 8.10** When pumping begins, the water table is drawn down an amount determined by the pumping rate and the aquifer's permeability. The development of a cone of depression creates a hydraulic gradient, which causes water to flow toward the well and enables continuous water production. The amount of water the well yields per unit of time pumped depends upon the aquifer's hydrologic properties.

Case Study | 8.1

Ground-Water Law

Ownership of ground water is a property right and has traditionally been regulated as such by individual states. Each state applies one or more of four legal doctrines to ground-water rights, none of which considers the geological relationship between surface-water infiltration and ground-water flow. The doctrines are

- *Riparian doctrine* (Latin *ripa,* "river or bank") holds that landowners overlying well-defined underground "streams" have absolute right to that water. They may exercise that right whenever they wish without limitation.
- *Reasonable-use doctrine* restricts a landowner's right to ground water to "reasonable use" on the land above that does not deprive neighboring landowners of their rights

to "reasonable use." This doctrine results in a more equal distribution of water when there is a shortage.

- *Prior-appropriation doctrine* holds that the earliest water users have the firmest rights; in other words, "first-come, first-served."
- *Correlative-rights doctrine* holds that landowners own shares to the water beneath their collective property that are proportional to their shares of the overlying land.

In the San Joaquin Valley of California, for example, surface-water rights are governed by the doctrine of prior appropriation (first-come, first-served), whereas ground-water rights are governed by the doctrine of correlative rights (proportional shares). In the late nineteenth century the Sierra Nevada to the east of the valley and the lower ranges to the west maintained high water tables, and they both contributed water to the flow of the San Joaquin River (◀ Figure 1). By 1966, however, heavy ground-water withdrawals on the west side of the San Joaquin Valley had reversed the regional ground-water flow, so that the river was losing water underground, rather than gaining. The San Joaquin had become an influent stream. Imagine a courtroom scenario in which hundreds of people with first-come, first-served rights to *surface* water sue thousands of people holding proportional shares rights to *underground* water for causing their surface water to go underground.

As explained throughout this chapter, it is impossible to treat ground water and surface water as separate entities. Surface water contributes to underground water supplies, and ground water contributes to streamflow and springs.

◆ **FIGURE 1** Ground-water flow in the central San Joaquin Valley in (a) 1906 and (b) 1966. Note that the San Joaquin River was an effluent (gaining) stream in 1906 and an influent (losing) stream is 1966. The change reflects intensive development and extraction of underground water.

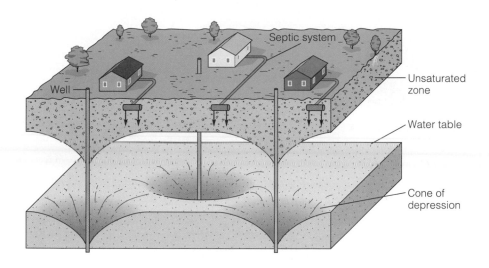

◆ **FIGURE 8.11** Cones of depression overlap in an area of closely spaced wells. Note the contributions to ground water from septic tanks.

gaining stream and a losing stream, depending upon the time of year and variations in the elevation of the water table.

Where water tables are high, they may intersect the ground surface and produce a spring. Springs also are subject to water-table fluctuations. They may "turn on and off" depending upon rainfall and the season of the year. ◆ Figure 8.14 illustrates the hydrology of three kinds of springs.

PRESSURIZED UNDERGROUND WATER

Pressurized ground-water systems cause water to rise above aquifer levels and sometimes even to flow to the ground surface. Pressurized systems occur where water-saturated permeable layers are enclosed between aquicludes, and for this reason, they are called

confined aquifers (◆ Figure 8.15). A confined aquifer acts as a water conduit, very similar to a garden hose that is filled with water and situated such that one end is lower than the other. In both cases a water-pressure difference known as a *hydraulic head* is created. Water gushes from the low end until there is no longer a pressure difference. At that point the water level stabilizes, becomes *static*.

Wells that penetrate confined aquifers are called *artesian wells,* after the province of Artois, France, where they were first described. Artesian systems are distinguished by a water-pressure gradient called the **artesian-pressure surface** (see ◆ Figure 8.15b). This "surface" is the level to which water will rise in a well at a given point along the aquifer. It is important to note that the slope of the water table of an unconfined aquifer also is a pressure surface. Underground water moves from areas of high hydraulic head to areas of

◆ **FIGURE 8.12** The effect of seasonal water-table fluctuations on producing wells. A shallow well may go dry in the summertime.

(a) Influent (losing) stream

(b) Effluent (gaining) stream

◆ **FIGURE 8.13** Hydraulic gradient (*v/h*) in (a) an influent (losing) stream that contributes water to the water table, and (b) an effluent stream that gains water from the water table.

low hydraulic head (pressure). Contour maps of the pressure surface can be very useful, because they indicate the direction of ground-water flow and the slope of the hydraulic gradient (◆ Figure 8.16). If an aquifer's hydraulic conductivity can be estimated or measured, the rate and volume of ground-water flow can be determined using Darcy's Law (see Appendix 5).

Artesian wells differ from water-table wells in that the latter depend upon pumping or topographic differences to create the hydraulic gradient that causes the water to flow. Because the energy required to raise water to the surface is the greatest expense in tapping underground water, artesian systems are much desired and exploited (see Case Study 8.2). Not all artesian wells flow to the ground surface, because the pressure surface lowers as water is extracted; water may rise naturally only part way up the well bore. Nevertheless, a natural rise of water any distance reduces the cost of extracting ground water.

Finally, many bottled water and beverage producers attempt to lead us to believe that deep-well or artesian water is purer than other kinds of water. *Artesian* refers only to an aquifer's hydrogeology and has no water-quality connotation. Do not be misled by advertisements attributing greater purity to artesian or deep-well water. Such waters may or may not have low levels of dissolved contaminants.

(a)

(b)

(c)

◆ **FIGURE 8.14** Geology and hydrology of three kinds of springs. (a) Springs form where a high water table in cavernous limestone intersects the ground surface. (b) Artesian springs. A fault barrier can cause pressurized water in a confined aquifer to rise as springs along irregularities in the fault. (c) Springs form where a perched water table intersects the ground surface. Such springs are intermittent, because the small volume of perched water is quickly depleted.

Consider This...? Some hot springs form geysers that are celebrated for their regularity. Old Faithful in Yellowstone National Park is an example. Cold springs, on the other hand, do not generate attention-grabbing geysers, and their flow is often far from regular. Indeed, many springs operate on a sort of stop-and go-basis. How can we explain this intermittent flow? ■

Consider This...? As explained in Case Study 8.1, the legal issues surrounding the withdrawal and use of ground water are complex, and several legal doctrines apply to them. From a philosophical viewpoint, which of the four doctrines seems fairest for multiple users of an aquifer? Please explain your choice. ■

Ground-Water Storage and Management

In order to manage underground water supplies in a ground-water basin intelligently, the hydrologist or planner must have two measurements:

- the quantity of water stored in the basin, and
- the **sustained yield,** the amount of water the aquifers in the basin can yield on a day-to-day basis over a long period of time.

(a)

(b)

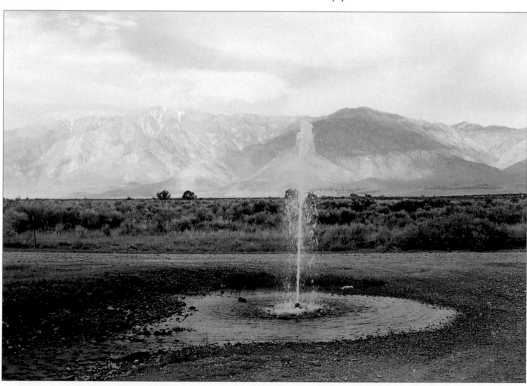

(c)

◇▶ **FIGURE 8.15** Water under pressure. (a) The hydraulic head in a hose lowers as water flows out the low end. Eventually a static condition prevails. (b) A confined aquifer, commonly called an *artesian* aquifer. The artesian-pressure surface is the level to which water will rise in wells at specific points above the aquifer. (c) A flowing artesian well in the Owens Valley near Bishop, California.

Case Study | 8.2

The Edwards Aquifer, A Texas Bonanza

The Edwards aquifer in south-central Texas is one of the most prolific aquifers in North America (◆ Figure 1). It provides water for two million people, including the City of San Antonio. A well drilled for the city in 1991 is reportedly the world's greatest flowing well, yielding 95,000 liters per minute (25,000 gal/min). The aquifer is composed of limestone of Cretaceous age that is a *karstified* flow system; that is, it is honeycombed with solution cavities, caves, and caverns and feeds many springs. About 80 percent of the aquifer's recharge is from losing (influent) streams that flow over its exposed outcrops; the rest is from precipitation. Where it dips down below the ground surface toward the south, it is under pressure, and water rises in well bores and sometimes to the ground surface (see ◆ Figure 8.14). At depth, the water quality and potability deteriorate to a zone marked on maps by the "bad-water line."

The region served by the aquifer is one of the fastest growing areas in the United States, and water supply is critical to this growth. Recharge areas are environmentally sensitive to pollutants, and the resource is threatened by disregard for this potential problem (◆ Figure 2). Threats are from human sewage, highway construction, agricultural chemicals, runoff from paved surfaces, and leaking underground storage tanks, to name a few. Government agencies at all levels are committed to protecting the aquifer. The federal government designated the aquifer a Sole Source Aquifer, which makes it eligible for funding for some protec-

◆ FIGURE 1 Area of south-central Texas's Edwards aquifer.

Recharge zone
Extent of Edwards aquifer potable water
Bad water line

tion projects. The state declared the Edwards aquifer the most susceptible to pollution in the state and established conservation districts for overseeing its protection.

◆ FIGURE 8.16 Hypothetical pressure-surface contour map based upon water-well data near a city. The dots indicate the locations of water wells, and the arrows show the direction of ground-water movement in the saturated zone. Elevations are in feet above sea level, rather than meters, because English units are used almost exclusively on pressure-surface contour maps in the United States.

◆▷ FIGURE 2 Midnight Cave in the Edwards aquifer before cleanup, 1993; southern Travis County near Austin, Texas. The trash is household garbage, oil filters, pesticide bottles, partially filled turpentine cans, and automobile parts. Trash had been floated up to higher ledges of the cave during high-water aquifer conditions. Cleanup efforts involved conservation groups, government agencies, and cavers. An estimated 3,000 cubic feet of trash was removed from the cave.

Consider This... ? Ground-water pollution by surface-water seepage through waste-disposal sites into the water table is a major problem. Where would you locate a facility for solid-waste disposal in the setting of Figure 8.16 to minimize this hazard? Designate the site you choose by relating it to the letters at the corners of the diagram. Also, which three wells would you expect to provide the best-quality water for domestic use in the city? Briefly explain your answers. ■

A ground-water basin consists of an aquifer or a number of aquifers that have well-defined geological boundaries and *recharge areas*—places where water seeps into the aquifer (see Figure 8.15b). To determine the amount of usable water in the basin, we rely on the concept of specific yield of water-bearing materials. **Specific yield** is the ratio of the volume of water an aquifer will give up by gravity flow to the total volume of material, expressed as a percentage. For example, a saturated clay with a porosity of 25 percent may have a specific yield of 3 percent; that is, only 3 percent of

the saturated porosity will drain into a well. The remaining water in the pore spaces (22 percent) is held by the clay-mineral surfaces and between clay layers against the pull of gravity. This "held" water is its *specific retention*. Similarly, sand with a porosity of 35 percent might have a specific yield of 23 percent and a specific retention of 12 percent. Note that the sum of specific yield and specific retention equals the porosity of the aquifer material.

Thus the total water-storage capacity of an aquifer may be calculated by multiplying the specific yield by the volume of the aquifer determined from water-well data. The process is a bit more complicated for artesian aquifers, but it follows the same procedure. The amount of water available to us in a ground-water basin can be determined fairly accurately if the basic data are available from numerous wells.

As noted, sustained yield is the amount of water that can be withdrawn on a long-term basis without depleting the resource. Sustained yield is more difficult to assess than aquifer storage capacity, as it is affected

◈ **FIGURE 8.17** The effect of ground-water mining on the water table. The drop from the initial level to the present level is the *overdraft*, the amount of water that has been mined.

by precipitation, runoff, and recharge. Nevertheless, it must be at least estimated for intelligent management of the resource.

GROUND-WATER MINING

As explained in the chapter opening, when the amount of water withdrawn from an aquifer exceeds the aquifer's sustained yield, an overdraft condition called *ground-water mining* exists (◈ Figure 8.17). The Ogallala Sandstone makes up about 80 percent of what is collectively known as the High Plains aquifer. The Ogallala's average thickness is 60 meters (about 200 ft), but thicknesses as great as 425 meters (1,400 ft) have been measured. In general, water levels in this important aquifer are dropping. In Texas, overdrafts have exceeded 65 percent, and overdrafts of 95 percent are known in a few places. In the latter case, 20 times more water was extracted from the aquifer than was recharged. This magnitude of overdraft is a serious problem because much of the water in the aquifer was acquired during wetter glacial climates thousands of years ago.

The High Plains aquifer was not massively tapped until the 1950s, and by 1980 the water-saturated zone had decreased an average of 3 meters, and as much as 30

meters under some parts of Texas and Nebraska (◈ Figure 8.18). Government subsidies have encouraged growing water-gulping corn, rather than less-water-intensive wheat, sorghum, and cotton.

In the early 1980s water decline rates decreased because of heavy rain and snow, better water management, and new technologies. New *LEPA* (low-energy, precision-application) irrigation nozzles use less water and less energy and decrease evaporation by as much as 98 percent from the amounts lost with spray-type irrigation. Furthermore, treated wastewater is now being recycled on fields, and rainfall has been induced by cloud seeding in attempts to recharge the aquifer in Kansas.

A 1982 Department of Commerce study found that 6 million hectares (15 million acres) were being irrigated with water drawn from 150,000 wells in the Ogallala and other High Plains aquifers. By the year 2020, according to the study report, about a fourth of the aquifer's water will have been mined at the present rate of withdrawal. A 1995 estimate of the total overdraft of the Ogallala is that it equals one year's flow of the Colorado River. The Ogallala aquifer is certainly

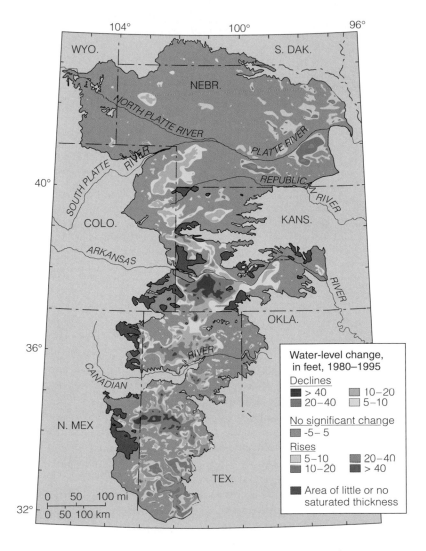

◈ **FIGURE 8.18** Water-level change in the High Plains aquifer, 1980–1995.

the life-blood of High Plains agriculture, and most users of its water are aware that "mining" the resource must be discontinued. Strict management that includes conservative usage, monitoring, and utilization of new technologies is required in attempting to balance its recharge and withdrawals. As a postscript to the Ogallala story, the aquifer's name is derived from that of the Oglala Tribe of the Sioux Nation. Led by Chiefs Red Cloud and Crazy Horse, the tribe fought to retain tribal lands in the late 1800s.

Ground-water mining causes shallow wells to go dry, increases the cost of lifting water to the surface as water levels drop, and may eventually cause ground subsidence (see Chapter 7). Currently ground-water pumping is twice the volume of natural recharge in Arizona's Tucson ground-water basin, and it has been necessary to import water from the Central Arizona Project to end the overdraft. One product of nuclear testing in the 1950s and early 1960s was radioactive tritium. This isotope of hydrogen has been used as a tracer of ground water recharged into the Tucson basin on alluvial fans of the adjacent Catalina and Rincon Mountains. The study indicates that a significant amount of the basin's ground water is very young and has infiltrated since the atom-bomb tests. This does not bode well for bringing the water table back to predevelopment levels in the basin.

The Santa Clara Valley of central California is a classic case of regional overdraft. Water levels there declined 46 meters (150 ft) between 1912 and 1959, and downtown San Jose dropped 2.76 meters (9 ft). As a final example, overpumping in Alabama dropped the pressure surface in one aquifer 62 meters (200 ft); it is now below sea level more than 80 miles inland from the sea. The solution to such a problem is *artificial recharge*, or "water spreading," which is accomplished by importing or diverting surface water and ponding it where it can percolate into the aquifer (◆ Figure 8.19). Such a program was initiated in the Santa Clara Valley in 1959, which, in combination with decreased pumping, has restored the water table to 1912 levels. Raising the water table does not bring the land back to its original level, however, because pore spaces are lost as subsidence occurs.

GROUND WATER–SALTWATER INTERACTION

Aquifers in coastal areas may discharge fresh water into the ocean, creating bodies of diluted seawater. If the

◆ **FIGURE 8.19** Water-spreading grounds for artificial recharge; Santa Ana River, Southern California. The dam is inflatable and may be raised or lowered to impound or release water as needed.

water table is lowered by pumping to or near sea-level elevation, however, saltwater invades the freshwater body, causing the water to become saline *(brackish)* and thus undrinkable. The direction of flow between the two water masses is determined by their density differences. Seawater is 2.5 percent (¹⁄₄₀th) heavier than fresh water. This means that a 102.5-foot-high column of fresh water will exactly balance a 100-foot-high column of seawater (◆ Figure 8.20a). Fresh water floats on seawater in the Hawaiian Islands, the Outer Banks of North Carolina, Long Island, and many other coastal areas (Case Study 8.3).

This relationship is known as the *Ghyben–Herzberg lens*—named for the two scientists who independently discovered it and for the lenslike shape of the freshwater body. In theory, the mass of fresh water extends to a depth below sea level that is 40 times the water-table elevation above sea level. If the water table is 2.5 feet above sea level, the freshwater lens would extend to a depth of 100 feet below sea level. Any reduction of water-table elevation would cause the saltwater to migrate upward into the freshwater lens until a new balance is established. If the water table drops below the level necessary for maintaining a balance with denser seawater, wells will begin to produce brackish water and eventually saltwater. Thus it is important to maintain high water-table levels in coastal zones so that the resource is not damaged by *saltwater encroachment*. This is accomplished by good management combined, when necessary, with artificial recharge using local or imported water (water spreading), or with injection of imported water into the aquifer through existing wells.

GROUND WATER–SURFACE WATER INTERACTION AND USAGE TRENDS

As has been emphasized thus far, surface water and ground water are not separate entities. Nearly all surface-water features—lakes, streams, rivers, and artificial lakes, and wetlands—interact with water underground. Some surface-water bodies gain water and acquire a different chemistry from ground water, and other surface waters contribute to underground water and may even pollute it. Thus, it is incumbent in water planning to have a clear understanding of the linkages between ground water and surface water and of the geology of both.

The greatest volume of fresh water is used for irrigation, and the greatest total volume of water, both fresh and saline, is used for power generation. However, 98 percent of the water used for power generation is returned to streams or other reservoirs; only 2 percent is consumed. How water is used affects the reuse potential of returns flows. For example, irrigation water may be too salty or contaminated by pesticides and insecticides to have any reuse potential unless it is purified naturally or artificially. On the other hand, water used in power generation has great reuse potential, because the principal change is usually just an increase in its temperature.

Since 1950 national water-use statistics have been compiled every five years (see Solley, W. B., For Further Information). There was a steady increase in water usage from 1950 to 1980, the all-time high, as population grew at a steady pace. Between 1980 and 1995, there was a 10 percent decrease in water use, even though the U.S. population grew by 37 million people

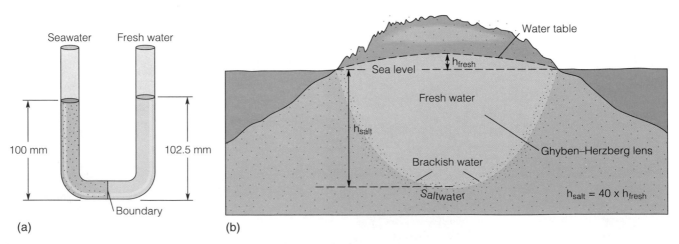

◆ FIGURE 8.20 (a) Illustration of the density balance between fresh water and seawater. Because seawater is 2.5% heavier than fresh water, 102.5 millimeters of fresh water is required to balance 100 millimeters of salt water. (b) Hypothetical island and Ghyben–Herzberg lens of fresh water floating on seawater. The freshwater lens extends to a depth 40 times the height of the water table above sea level. Not to scale.

Case Study | 8.3

Long Island, New York—Saltwater or Fresh Water for the Future?

Long Island is an informative case study of the cause and remedies of saltwater encroachment. By far, Long Island's most important source of fresh water is its aquifers, which are estimated to hold 10–20 trillion gallons of recoverable water (◁▷ Figure 1). This ground water infiltrated underground over centuries before development, and the excess was discharged naturally to the sea. Ground water was exploited as the island was urbanized, causing the water table in Kings County, the county closest to the mainland, to drop below sea level. By 1936 saltwater had invaded the freshwater aquifer (◁▷ Figure 2a). Pumping wells were then abandoned or were converted to recharge wells using imported water, and by 1965 the water tables had recovered to acceptable elevations above sea level (◁▷ Figure 2b).

Meanwhile, in adjoining Queens County, pumping had been increasing and recharge was diminishing due to the construction of sewers and paving of streets. Treated wastewater was thus being carried to the sea, rather than infiltrating underground as it would with cesspools and septic tanks. With lowered water tables, saltwater invaded the aquifers below Queens County (Figure 2), and the State of New York in 1970 asked the U.S. Geological Survey to conduct ground-water studies.

In order to preserve the fresh water below Long Island, it was necessary to determine the *safe yield,* the amount of water that could be withdrawn from Long Island aquifers without causing undesirable results. It was also necessary to achieve a balance between total freshwater outflow and recharge. Several methods of managing the water resource and preventing saltwater encroachment were developed, and they are offered here as examples.

The basic challenge was to offset ground-water withdrawals of 1.7 million cubic meters per day with equal amounts of surface infiltration or injection. A line of injection wells was proposed for replacing the decreased natural recharge. These wells would inject treated sewage effluent or imported water and build a freshwater barrier against saltwater intrusion (◁▷ Figure 3a). Spreading highly treated sewage effluent and natural runoff into recharge basins could be carried out along with injection into the deeper aquifer. In both the injection-well and the spreading-basin alternatives, the treated sewage is purified to drinking-water

◁▷ FIGURE 1 Hydrologic relationship between fresh and salty underground water; Long Island, New York. Arrows indicate direction of water movement in the hydrologic cycle.

(a) 1936

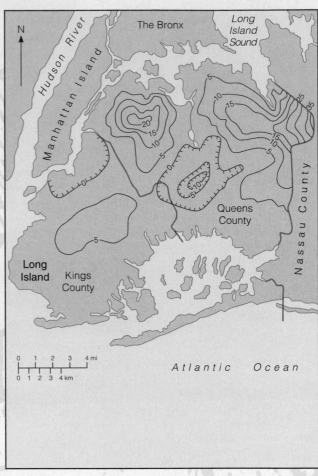

(b) 1965

◆▶ FIGURE 2 Water-table levels in (a) 1936 and (b) 1965 in Kings and Queens Counties, Long Island, New York.

standards so that it does not affect local shallow wells. Recharge, like injection, builds a freshwater ridge that keeps saltwater at bay. An unusual but viable method is to allow controlled saltwater intrusion so that a true Ghyben–Herzberg lens of fresh water floating on saltwater develops (◆▶ Figure 3b). Treated sewage effluent would be discharged into the ocean or injected underground as the need arose. This method has the distinct advantage of salvaging much of the fresh water that otherwise would flow through aquifers into the ocean. In addition, this alternative would increase the yield of the aquifer by several hundred million cubic meters per day, although it would decrease the total volume of fresh water in the reservoir.

Each method of balancing outflow and inflow—injection wells, spreading basins, and controlled saltwater intrusion—

has advantages and disadvantages. As is the case with many geological problems, applying a combination of methods yields the most fruitful and cost-effective solution. Kings and Queens Counties now rely entirely upon importation for their freshwater needs. Public water in Nassau and Suffolk Counties is entirely underground water, and only a very small amount of seawater intrusion occurs in the southwest corner of Nassau County. Injection of tertiary treated wastewater (see Chapter 15) was tested in the 1980s. Unfortunately, the chemical differences of the injected water, natural fresh water, and saltwater caused the injection wells to clog and require excessive maintenance. Wastewater injection is feasible but at present is too expensive. Currently, there is a network of recharge basins that collect surface-water runoff and allow it to percolate underground into the shallow

◆▶ FIGURE 3 Methods of balancing freshwater outflow and seawater inflow; Long Island, New York. (a) The aquifer is recharged with highly treated wastewater (*T*) using injection wells. This method reverses saltwater intrusion and also improves the outlook for long-term water yield. (b) Saltwater is allowed to move inland, and a lens of fresh water forms, floating on the saltwater. This serves to establish a new equilibrium between fresh water and seawater.

(a) Inject treated water as a barrier.

(b) Permit saltwater intrusion.

aquifer (◆▶ Figure 4). The counties require developers to dedicate land for recharge basins within their housing or commercial projects. This method of recharge has proven effective in maintaining the balance between seawater and fresh water.

◆▶ FIGURE 4 A recharge basin for capturing surface-water runoff; Nassau County, New York.

◆▶ **FIGURE 8.21** Trends in fresh ground- and surface-water withdrawals and population, 1950–1995.

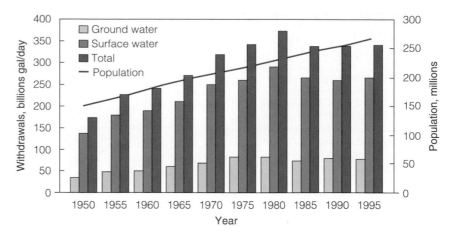

(◆▶ Figure 8.21). It was estimated that the United States was saving 144 billion liters (38 billion gal) per day in 1995, enough to fill Lake Erie in a decade. Some of the reasons cited for this decrease were the development and use of more efficient irrigation methods, conservation programs, increased water recycling, and a downturn in farm economies.

GEOLOGIC WORK BY UNDERGROUND WATER

Solution (a chemical reaction discussed in Chapter 6) of limestone and gypsum by underground water leads to formation of caves, caverns, disappearing streams, and prolific ground-water aquifers such as the Edwards aquifer (Case Study 8.2). Land area exhibiting these features is called **karst terrane** (German *Karst,* for the Kars limestone plateau of northwest Yugoslavia; ◆▶ Figure 8.22). Urbanization on karst terrane typically results in problems such as increased flooding, contamination of underground water, and collapse of cavern roofs to form sinkholes.

The first sign of impending disaster in Winter Park, Florida, occurred one evening when Rosa Mae Owens heard a "swish" in her backyard and saw, to her amazement, that a large sycamore tree had simply disappeared. The hole into which it had fallen gradually enlarged, and by noon the next day her home had also fallen into the breach. Six Porsches in a nearby storage yard suffered the same fate, as did a city swimming pool and two residential streets. Deciding to work *with* nature, the City of Winter Park stabilized and sealed the sinkhole, and converted it to a beautiful urban lake (◆▶ Figure 8.23). Few people who read of this notorious sinkhole in May 1981 realized the extent of the sinkhole problem in the southeastern United States.

Sinkholes are most commonly found in carbonate terranes (limestone, dolomite, and marble), but they are also known to occur in rock salt and gypsum. They form when carbonic acid (H_2CO_3) dissolves carbonate rock, forming a near-surface cavern. Continued solution leads to collapse of the cavern's roof and the for-

mation of a sinkhole. About 20 percent of the United States, and 40 percent of the country east of the Mississippi River, is underlain by limestone or dolomite. The states most impacted by surface collapse are Alabama, Florida, Georgia, Tennessee, Missouri, and Pennsylvania (◆▶ Figure 8.24).

Tens of thousands of sinkholes exist in the United States, and the formation of new ones can be accelerated by human activities. For example, excessive water withdrawal that lowers ground-water levels reduces or removes the buoyant support of shallow caverns' roofs. In fact, the correlation between water-table lowering and sinkhole formation is so strong that "sinkhole seasons" are designated in the Southeast. These "seasons" are declared whenever ground-water levels drop naturally because of decreased rainfall and in the summer, when the demand for water is heavy. Vibrations from construction activity or explosive blasting also can trigger sinkhole collapse. Other mechanisms that have

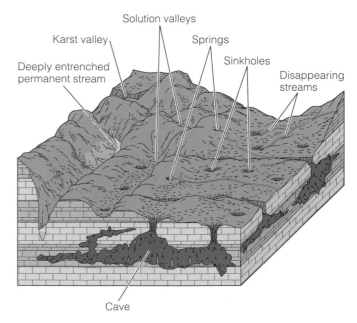

◆▶ **FIGURE 8.22** Surface and subsurface features of karst topography.

(a)

(b)

◁▶ **FIGURE 8.23** Sinkhole in Winter Park, Florida. (a) Shortly after the initial collapse. Note the expensive sports car going down the sink. (b) Air view that shows its almost perfect symmetry. (c) Plugged and landscaped, the sinkhole area is now an urban park.

(c)

been proposed for sinkhole formation are fluctuating water tables (alternate wetting and drying of cover material reduces its strength) and high water tables (which erode the roofs of underground openings).

Although the development of natural sinkholes is not predictable, it is possible to assess a site's potential for collapse resulting from human activities. This requires extensive geologic and hydrogeologic investigation prior to site development. Under the best of geologic conditions, good geophysical and subsurface (drilling) data can tell us where collapse is most likely to occur, but they cannot tell us exactly when a collapse may occur. Armed with this information, planners and civil engineers can implement zoning and building restrictions that will minimize the hazards to the public.

some of these elements and compounds, such as arsenic, are poisonous, others are required for sustaining life and health; common table salt is one and it occurs in almost all natural waters. Years ago the U.S. Public Health Service established **potable** (drinkable) **water** standards for public water supplies based on purity, which is defined as allowable concentrations of particular dissolved substances in natural waters. These concentrations are expressed in weight per volume, which is milligrams per liter (mg/L) in the metric system. The standards can also be expressed in a weight-per-weight system. For example, in the English system, one pound of dissolved solid in 999,999 pounds of water would be

Water Quality

DISSOLVED SUBSTANCES

Dissolved in natural waters are varying concentrations of all the stable elements known on earth. Whereas

Consider This...? You could become a millionaire if you developed a relatively inexpensive method of locating caverns that may collapse and form sinkholes. Got any ideas? If not, what types of more sophisticated (and expensive) methods might be employed to locate caverns? ■

◀▶ **FIGURE 8.24** Satellite image map of central Florida showing its karst terrane with disappearing streams, hundreds of lakes, and thousands of sinkholes. Central Florida is underlain by the Tertiary Ocala Limestone, and the coastal areas by younger, detrital sand deposits. This explains the location of karst terrane along the state's north–south axis.

one part per million (ppm). The Public Health Service standards allow only minuscule amounts of pollutants in public water supplies. For example, 1 ppm or 1 mg/L is equal to about 1 ounce of a dissolved solid in 7,500 gallons of water, or 0.0001 percent. The standard for "sweet" water—the preferable quality classification for domestic use—is less than 500 ppm of *total dissolved solids (tds),* and the standard for "fresh" water is less than 1,000 ppm of total dissolved solids. ◉ Table 8.1 lists the Public Health Service secondary standards, and ◉ Table 8.2 lists some common water problems and suggestions for correcting them.

The quality of ground water can be impaired either by a high total amount of various dissolved salts or by a small amount of a specific toxic element. High salt content may be due to solution of minerals in local geological formations, seawater intrusion into coastal aquifers, or the introduction of industrial wastes into ground water. (Organic contaminants, mainly animal wastes, solvents, and pesticides, also degrade water. They are discussed in the next subsection.) ◉ Table 8.3 lists Public Health Service limits for some of the inorganic substances that occur in ground water and impair its safety.

◉ TABLE 8.1   U.S. Public Health Service Secondary Drinking-Water Standards*

DISSOLVED SUBSTANCE	STANDARD, mg/L	EFFECT
Calcium and magnesium	–	principal cause of water hardness; boiler scale (Ca and Mg salts)
Chloride	250	unpleasant taste; corrodes pipes
Copper	1.0	metallic taste; stains clothes and containers
Fluoride	2.0	may stain tooth enamel
Sulfate	250	bitter taste; laxative effect
Iron	0.3	unpleasant taste
Zinc	5.0	unpleasant taste
Manganese	0.05	unpleasant taste
Total dissolved solids (tds)	500	possible relationship to cardiovascular disease; corrosive to pipes

*Secondary standards are nonenforceable and apply to substances that occur in concentrations below U.S. Public Health Service standards and affect such water qualities as taste and odor. For example, although it is desirable that drinking water contain no iron, sulfate, or zinc, concentrations above these standards are permissible.

Water with high salt content, particularly bicarbonate and sulfate contents, can have a laxative effect on humans. Dissolved iron or manganese can give coffee or tea an odd color, impart a metallic taste, and stain laundry. Concentrations greater than 3 ppm of copper can give the skin a green tinge (it's not easy being green, as Kermit says; the condition is reversible), and lead is both poisonous and persistent. A *persistent pollutant* is one that builds up in the human system because the body does not metabolize it. (The Romans used lead pipes for plumbing, and skeletal materials exhumed from old Roman cemeteries retain high lead concentrations.) Lead concentrations in the water supply greater than about 0.1 ppm can build up over a period of years to debilitating concentrations in the human body. Arsenic is also persistent, and the standards for arsenic and lead are the same—less than 0.05 ppm, preferably absent. Nitrates in amounts greater than 10 ppm are known to inhibit the blood's ability to carry oxygen and to cause "blue baby syndrome" *(infantile methemoglobinemia)* during pregnancy. The element with the greatest impact upon plants is boron, which is virtually fatal to citrus and other plants in concentrations as low as 1.0 ppm.

Dissolved calcium (Ca^{2+}) and magnesium (Mg^{2+}) ions cause water to be "hard." In hard water soap does not lather easily, and dirt and soap combine to form scum. Hard water is most prevalent in areas underlain by limestone ($CaCO_3$) and dolomite ($CaMgCO_3$). A water softener simply exchanges sodium (Na^+) for calcium and magnesium ions in the water using a zeolite (a silicate ion exchanger) or mineral sieve. Water with more than 120 ppm of dissolved calcium and magnesium is considered hard (⊙ Table 8.4). Because sodium is known to be bad for persons with heart problems (hypertension) and water softeners increase the sodium content of water, those people on low-sodium diets should not drink softened water.

The good news regarding dissolved ions is that fluoride ion (F^-) dissolved in water reduces tooth cavities when it is present in amounts of 1.0–1.5 ppm. The explanation for this is that tooth enamel and bone are composed of the mineral apatite, $Ca_5(PO_4)_3(F, Cl, OH)$, whose structure allows the free substitution of fluoride F^-, chloride Cl^-, or hydroxyl $(OH)^-$ ions in the mineral structure. In the presence of fluoride, the crystals of apatite in tooth enamel are larger and more perfect, which makes them resist decay. Too much fluoride in drinking water, however—concentrations greater than 4 ppm—can cause children's teeth to become mottled with dark spots. This cosmetic blemish, among children in particular, led to the discovery of the beneficial effect of fluorine and to the establishment of concentration standards for it. Fluoride in natural waters is rare, but Colorado Springs, Colorado, is a well-known area with very few children's dentists. Dissolved fluoride is also believed to slow osteoporosis, a bone-degeneration process that accompanies aging.

GROUND-WATER POLLUTANTS

The term *water pollution* refers to the introduction of chemical, physical, or biological materials into a body of natural water that affect its future use. It is more widespread than formerly thought, as revealed by studies by

---

⊙ TABLE 8.2   Common Water Quality Problems

PROBLEM	PROBABLE CAUSE	CORRECTION
Mineral scale buildup	water hardness	water softener or hardness inhibitor (such as Calgon)
Rusty or black stains	iron and/or manganese	aeration/filtration; chlorination; remove source of Fe/Mn
Red-brown slime	iron bacteria	chlorination filtration unit; remove source of slime
Rotten-egg smell or taste	hydrogen sulfide and/or sulfate-reducing bacteria; low oxygen content	aeration; chlorination filtration unit; greensand filters
Salty taste	chloride—seawater or other saline water	reduce pumping to raise water table
Gastrointestinal diseases, typhoid fever, dysentery, and diarrhea	coliform bacteria and other pathogens from septic tanks or livestock yards	remove source of pollution; disinfect well; boil water; chlorinate; abandon well and relocate new well away from pollution source
Petroleum smell or film	fuel oil, diesel fuel, or lubricating oil	remove source of pollution

SOURCE: D. Daly, "Groundwater Quality and Pollution" (Geological Survey of Ireland Circular 85–1, 1985), in J. E. Moore and others, *Ground Water: A Primer,* American Geological Institute, 1994.

TABLE 8.3   Health Effects of Significant Concentrations of Common Inorganic Environmental Pollutants

INORGANIC POLLUTANT	STANDARD, mg/L	HEALTH EFFECTS
Arsenic	0.05	dermal and nervous-system toxicity effects, paralysis
Barium	1.0	gastrointestinal effects, laxative
Boron	1.0	no effect on humans; damaging to plants and trees, particularly citrus trees
Cadmium	0.01	kidney effects
Copper	1.0	gastrointestinal irritant, liver damage; toxic to many aquatic organisms
Chromium	0.05	liver/kidney effects
Fluoride	4.0	mottled tooth enamel
Lead	0.05	attacks nervous system and kidneys; highly toxic to infants and pregnant women
Mercury	0.002	central-nervous-system disorders, kidney disfunction
Nitrates	10.0	methemoglobinemia ("blue baby syndrome")
Selenium	0.01	gastrointestinal effects
Silver	0.05	skin discoloration (argyria)
Sodium	20–170	hypertension and cardiac difficulties
	70.0	renders irrigation water unusable

TABLE 8.4   Water Hardness Scale*

CONCENTRATION OF CALCIUM^{2+}, mg/L	CLASSIFICATION
0–60	soft
61–120	moderately hard
121–180	hard
>180	very hard

*Hardness is expressed here as milligrams of dissolved Ca^{2+} per liter. Dissolved calcium and magnesium in water combine with soap to form an insoluble precipitate that hampers water's cleansing action.

knowledge allows us to estimate the approximate time a particular system will take to clean itself naturally and indicates those cases in which a pollutant must be removed from a water body artificially by human means because of a long residence time.

Although most pollution is due to careless disposal of waste at the land surface by humans, some is due to the leaching of toxic materials in shallow excavations or mines. Ground-water pollution occurs in urban environments due to improper disposal of industrial waste and due to leakage from sewer systems, old fuel-storage tanks, wastewater settling ponds, and chemical-waste dumps (Figure 8.25a). In suburban and agricultural areas, septic-tank leakage, fecal matter in runoff and seepage from animal feedlots, and the use of inorganic fertilizers, such as nitrates and phosphates, and of weed and pest-control chemicals all can serve to degrade both surface and subsurface water (Figure 8.25b).

Leaking gasoline-storage tanks are a more recent cause of soil and ground-water pollution (Figure 8.25c). By federal law, all underground tanks now must be inspected for leakage, and if gasoline is found in the adjacent soil, the tanks must be removed and the soil cleaned. (Although the problem is not amusing, the technology created by the need for hydrocarbon cleanup has come to be lightly referred to as *yank-a-tank*.) If gasoline is found floating on ground water, it must be removed (Case Study 8.4). Both the gasoline and the decontaminated ground water are then usable. Soils can be cleansed of hydrocarbons biologically with bacteria, a process known as **bioremediation**, or by aeration and oxidation of the hydrocarbon films (see Case Study 6.3).

Table 8.5 lists some of the detrimental effects of selected organic pollutants. Water standards for common organic pollutants of water bodies are given in Table 8.6. The very low limits allowed for the pesticides lindane and endrin indicate their toxicities. Some of these toxic chemicals are also carcinogenic (cancer-causing), which makes them doubly dangerous.

state and federal agencies and by complaints from users. It is a serious problem because thousands of years may be required to flush contaminated water from an aquifer and replace it with clean water. River water, in contrast, is exchanged in a matter of hours or days. **Residence time,** the length of time a substance remains *(resides)* in a system, is an important concept in water pollution studies. We may view it as the time that passes before the water within an aquifer, lake, river, glacier, or ocean is totally replaced and continues on its way through the hydrologic cycle. The average residence time of water in rivers has been found to be a few days; of water in large lakes, several decades; in shallow gravel aquifers, a few days; and in deeper, low-permeability aquifers, perhaps thousands or even hundreds of thousands of years. This

(a) Urban areas   (b) Rural areas   (c) Density effects

◆ FIGURE 8.25 (a) Urban-area and (b) rural-area pollution sources. (c) Some pollutants float and others sink due to differences in density between ground water and the pollutants.

Density differences between ground water and particular pollutants often govern the remedial measures that can be used; for example, gasoline and oil float on ground water, whereas salt brines from industry or agriculture sink to lower levels. In some cases it is possible to skim a layer of gasoline from the top of the water table. Gasoline or fuel oil that is floating on the capillary fringe may yield vapors that can rise and be trapped beneath or in the walls of buildings. If these fumes contain BTX (*b*enzene, *t*oluene, and *x*ylene), they are carcinogenic. "Air-stripping" is currently the preferred method of removing volatile organic pollutants such as BTX compounds and trichloroethylene (TCE) and tetrachloroethylene (PCE) cleaning solvents from ground water. (The origin of these substances is usually connected to refinery, industrial, or military ac-

tivities.) The polluted water is brought to the earth surface, where the contained volatile pollutants (TCE or BTX) are gasified (air-stripped) and wasted to the atmosphere with no adverse effects.

In Canada 10,000 metric tons of PCE are used per year, primarily in the dry-cleaning and metal-cleaning industries. An exhaustive study mandated and sponsored by the Canadian Environmental Protection Act (CEPA) indicates that PCE is entering the environment in significant quantities. Due to its volatility (it evaporates readily), it is found mostly in the atmosphere, where it has a relatively short residence time and does not contribute to global warming. The study concludes that the present atmospheric concentration of PCE is not expected to have adverse effects on Canada's aquatic and terrestrial animals, including humans, but

⦿ TABLE 8.5	Health Effects of Common Organic Pollutants*	
ORGANIC SUBSTANCE	**HUMAN HEALTH EFFECTS**	**ENVIRONMENTAL EFFECTS**
Dieldrin	convulsions, kidney damage	toxic to aquatic organisms
DDT	convulsions, kidney damage	reproductive failure in animals, eggshell thinning
PCBs	vomiting, abdominal pain, liver damage	eggshell thinning in birds, liver damage in mammals
Benzene	anemia, bone-marrow damage	toxic to some fish and aquatic invertebrates
Phenols	death at high doses	decreased phytoplankton productivity
Dioxin	acute skin rashes, systemic damage, mortality	lethal to aquatic birds and mammals

*Insecticides or compounds used in their manufacture.
SOURCE: U.S. Public Health Service; Environmental Protection Agency.

Case Study | 8.4

DNAPL and LNAPL

A couple of the newest acronyms in the environmental cleanup business are *DNAPL* (for *dense, nonaqueous-phase liquid* hydrocarbons) and *LNAPL* (for *light, nonaqueous-phase liquid* hydrocarbons). With the discovery of crude oil and the explosion (no pun intended) of its use in the reciprocating engine, numerous refineries and above-ground petroleum-storage-tank farms were developed. All the states face the problem of these tanks leaking and polluting the soils beneath them with refined hydrocarbons, including lubricating oils, diesel fuel, jet fuel, and gasoline—all of them LNAPLs (◆ Figure 1).

No place in the United States is the problem more obvious than the Los Angeles coastal plain, where there are 16 major refineries and 33 above-ground petroleum-storage-tank farms. Underlying these refineries are several important aquifers, one of them less than 60 meters (200 ft) below the ground surface. Some of the refineries are on the Environmental Protection Agency (EPA) National Priorities List (Superfund sites), and others are regulated by the Resource Conservation and Recovery Act (RCRA) as hazardous waste sites (see Chapters 1, 13, 14, and 15). The environmental concerns at these sites include:

- Hydrocarbons dissolved in the ground water will limit the water's beneficial use.
- LNAPL hydrocarbon pools occurring as "perched" zones serve as continuous sources of contamination for both soil and ground water.
- Soil contamination might convert the soil to a hazardous material.
- The accumulation of hydrocarbon vapors poses potential fire or explosion hazards.

The combined areal extent of LNAPL pools perched, or floating, on ground water beneath the refineries is estimated as being on the order of 600 hectares (1,500 acres or 2.3 mi^2), and the volume is estimated to be between 1.5 and 7.5 million barrels—the size of a small oil field!

The most sophisticated LNAPL-hydrocarbon recovery is accomplished by a recovery well with a two-pump system (◆ Figure 2). The two pumps are controlled by a series of electrodes set at predetermined depths. The lower, water-recovery pump runs until the drawdown brings the LNAPL–water interface into contact with the electrode just above the lower pump's intake. When the electrode senses the presence of LNAPL hydrocarbons, it shuts down the water pump so that LNAPLs are not ingested and mixed with fresh water. The upper, LNAPL-recovery pump is switched on and off by a probe that measures the electrical resistivity of fluids adjacent to the intake. LNAPLs have high electrical resistivity, and water has a low resistivity. The probe auto-

◆ FIGURE 1 Leaking oil-storage tanks, West Texas. Such leaks lead to soil and ground-water contamination.

◆ FIGURE 2 A two-pump LNAPL-hydrocarbon recovery well. Sensors activate or deactivate pumps so that the upper pump draws only LNAPL hydrocarbons and the lower one draws only water. Recovered oil is reused.

matically shuts down when it senses fresh water. In this manner, water and hydrocarbons are recovered separately, treated, and then recycled to the benefit of all concerned.

TABLE 8.6 Water Standards for Common Organic Pollutants*

SUBSTANCE	MAXIMUM SAFE CONCENTRATION, mg/L
Cyanide	0.05
Lindane	0.004
Endrin	0.0002
2,4-D	0.10
Phenols	0.001

*Insecticides and animal-control compounds.
SOURCE: U.S. Public Health Service; Environmental Protection Agency.

it has the potential of harming trees. Local spills, of course, require cleanup and remediation.

Conservation and Alternative Sources

Drought in the southwestern United States is a persistent and nagging problem. In 1990–91, for example, following several years of below-average rainfall in California, surface-water reservoirs became depleted. This led to excessive ground-water withdrawals and eventual mining. Remedies that have been proposed include desalinizing seawater, recycling wastewater, and even bringing huge Antarctic icebergs to dry coastal areas to melt and provide fresh water. Voluntary water conservation has been quite successful, reducing consumption by as much as 15 percent, and

even greater reductions have been achieved where rationing is mandatory.

Conservation is every citizen's obligation where a resource is in short supply. Table 8.7 provides some examples of how water can be conserved in the home. Using recycled wastewater, so-called gray water, is the simplest conservation measure after rationing. Recycled shower, bath, and laundry water can be used for irrigating lawns and certain plants, especially when biodegradable soaps and detergents are used. Gray water contaminated with chlorine should not be used, however.

Desalinization of seawater is a viable alternative where no other water source is available. The process is expensive, and the water must be pumped uphill to users, which further increases its cost. In most cases, more energy is required to pump the water than to desalinize it. For this reason, desalinization factories are found only on desert coasts (as in some Middle Eastern nations), on small islands that do not receive much precipitation, and on military installations.

Artificial recharge (water-spreading; see Figure 8.19) of aquifers during heavy-rain years—"banking" water, so to speak—may provide longer-term benefits. This is done by diverting water that ordinarily would flow into rivers into manufactured basins surrounded by dikes. This temporarily ponds the running water, giving it time to infiltrate through the streambed and into the aquifer. By this restocking of aquifers when surplus surface water or abundant cheap imported water is available, the people of some drought-prone areas are able to prepare for those episodic droughts that seem to strike all mid-latitude areas sooner or later.

TABLE 8.7 Some Household Water-Conservation Practices

ACTIVITY	NORMAL USE	CONSERVATIVE USE
Shower	5 minutes, 30–40 gallons	Reduce shower length; use water-conserving shower head.
Tub bath	full tub, 36 gallons	Use minimal water level (10–12 gallons).
Teeth brushing	tap running, 2 gallons	Wet brush, rinse briefly.
Shaving	tap running, 5 gallons	Fill sink or basin (1 gallon).
Dishwashing	tap running, 30 gallons	Wash and rinse in sink (5 gallons).
Automatic dishwashing	full cycle, 25 gallons	Wash full loads on short cycle (12 gallons).
Hand washing	tap running, 2 gallons	Fill basin (1 gallon).
Toilet flushing	5–7 gallons per flush	Place bottles or bar dam in tank (4–6 gallons).
Washing machine	full cycle with top water level, 30–35 gallons	Wash full loads on short cycle with minimum water level (15–17 gallons).
Outdoor watering:		
hose	5 minutes, 40–50 gallons	Water in the morning (20 gallons).
lawn sprinklers	20 minutes daily, 1,200 gallons per week	Water as needed.

Gallery

Wonders of Karst—That Sinking Feeling

Karst terrane exists in many parts of the United States, but it is particularly prevalent in parts of Florida and Kentucky. Perhaps the most spectacular karst area in the world is along the Li River in the predominantly limestone region of Kwangsi in southern China near Canton (Guangzhou). Solution of limestone has reduced the landscape there but left enormous pinnacles of resistant limestone, called "tower karst," that present an almost extraterrestrial appearance (◆ Figure 8.26). The towers and other karst features draw tourists from all over the world.

In many places, water containing dissolved calcium carbonate percolates downward through soil and rock into caverns, where evaporation and precipitation form needlelike stalactites—extending down from the ceiling—and the more irregular **stalagmites**—rising up from the floor (◆ Figure 8.27). These fresh-water limestone features in caves are called **dripstone,** because they build up drip-by-drip. Carlsbad Caverns in New Mexico and Mammoth Cave in Kentucky offer spectacular examples, but literally hundreds of other limestone caves also exhibit these features and are open to the public. **Speleology** is the study of caves, and spelunking, or "caving," is an exciting recreational pastime for many people. Cavers explore and map natural underground openings. Much of what we know about caves and caverns is attributable to these hobbyists' efforts and talents.

(a)

(b)

◆ FIGURE 8.26 (a) Residual karst towers; Kwangsi region, southern China. The former limestone plateau has been lowered by solution except for resistant-limestone pinnacles. (b) The Stone Forest, 126 kilometers (80 mi) southeast of Kunming, People's Republic of China, is a high-relief karst landscape formed by the dissolution of carbonate rocks.

Mineralization by underground water is also responsible for transforming organic material into hard mineral water—a process known as *fossilization*. The replacement mineral is typically silica or calcium carbonate, and replacement may take place cell by cell as in the case of petrified wood (◁▷ Figure 8.28).

Geodes (pronounced "gee-odes") are rounded cavities in rocks that are lined with quartz crystals formed from per-colating ground water. They are one of nature's most beautiful geological creations. Geodes from Mexico and Brazil are lined with light-colored agate from which "grow" beautiful purple amethyst or smoky quartz crystals (◁▷ Figure 8.29). They are produced when silica in aqueous solution is deposited directly into cavities in limestone or volcanic rocks.

◁▷ FIGURE 8.28 Petrified wood from Arizona. Woody tissue is replaced cell by cell with mineral matter from ground water.

◁▷ FIGURE 8.27 Dripstone stalactites, stalagmites, and pillars in a limestone cavern.

◁▷ FIGURE 8.29 Amethyst geode from Brazil. Silica dissolved in underground water percolates into an opening in the host rock and is deposited, forming crystals. Most commonly, geode crystals are varieties of quartz, but other minerals dissolved in ground water may also crystallize.

Summary

THE HYDROLOGIC CYCLE

Only .6 percent of the earth's water is readily available to humans. It is accessible because of the hydrologic cycle: continuous evaporation from the oceans, precipitation on land, runoff back to the oceans, and storage of fresh water in lakes and underground.

WATER USE

1. Consumptive use–water that cannot be immediately reused. Where consumptive use exceeds supplies, water must be imported.
2. Irrigation and thermoelectric power plants are the largest consumers. Irrigation water is usually highly degraded, whereas 98 percent of power-plant water is returned to the system, the only change being a higher temperature.
3. Worldwide, about 82 percent of water use is for irrigation. Water availability will limit population growth in the twenty-first century, either by statute in developed countries or by drought and famine in developing nations. The world population projected for 2050 will require water beyond the firm flow of the world's major rivers.

GROUND WATER
DEFINED

Loosely defined as all water underground, as distinct from surface water; the water in the hydrologic cycle that infiltrates underground.

ZONES

1. The *zone of aeration,* where the voids in rock or sediment are filled with air and water films.
2. The lower *zone of saturation,* where voids are filled with water.

WATER TABLE

The contact between the zone of aeration and the zone of saturation.

AQUIFERS
DEFINED

Bodies of porous and permeable rock or sediment that are saturated with water.

1. *Porosity*–The volume of pore spaces in a geologic material, expressed as a percentage of the total volume.
2. *Permeability*–Ease with which fluids flow through a geologic material.
3. *Aquiclude*–A geologic material that lacks permeability and inhibits water flow.

TYPES

1. Unconfined–An aquifer in which the water table is static and that is exposed to atmospheric pressure changes. A slope, or gradient, is required to induce subsurface flowage in a given direction. A well that taps the static water table creates a cone of depression, which provides the necessary gradient for water to flow to the well. Drawdown is the amount of lowering of the water table in a pumping well.
2. Artesian–An aquifer that is confined between aquicludes. The hydraulic head causes water to rise in a well when the aquifer is penetrated. The height to which water will rise at any point along the aquifer is known as the artesian pressure surface.

GROUND-WATER MANAGEMENT
DEFINED

Managing the amount of water contained in a ground-water basin so that it will produce water in the future.

1. *Sustained yield*–The amount of water an aquifer will yield on a day-to-day basis.
2. *Specific yield*–The amount of water a body of rock or sediment will yield by gravity alone.
3. *Specific retention*–The amount of water held by a body of rock or sediment against the pull of gravity. Specific yield *plus* specific retention *equals* porosity.

SEAWATER ENCROACHMENT

The invasion of coastal aquifers by saltwater. Management requires maintaining high freshwater tables to balance the denser saltwater.

GROUND WATER–SURFACE WATER INTERACTION

Surface water and ground water cannot be treated as separate entities; what affects one impacts the other.

SURFACE WATER SOURCES

Lakes, rivers, streams, and artificial reservoirs.

U.S. USE TRENDS

U.S. water use increased from 1950 until 1980. Between 1980 and 1995 it decreased 10 percent, even though population increased by 37 million people. Reasons include more economical application of irrigation water, public conservation programs, recycling, and a downturn in farm economies.

SOLUTION BY GROUND WATER

1. Solution of limestone and gypsum by underground water forms caves, caverns, disappearing streams, sink holes, and prolific ground-water aquifers. When terranes underlain by carbonate rocks exhibit these features, they are said to have *karst* topography.
2. Water and carbon dioxide from carbonic acid, a weak natural acid, which, over long periods of time, will dissolve carbonate rocks and gypsum.
3. Urbanization on karst terrane results in increased flooding, contamination of underground water, and collapse of the ground surface to form sinkholes.
4. Forty percent of the country east of the Mississippi is underlain by carbonates. States most impacted by karst and sinkhole formation are Alabama, Florida, Georgia, Tennessee, Missouri, and Pennsylvania.

WATER QUALITY

DEFINED

Purity or drinkability (potability) of water as established by the U.S. Public Health Service. Dissolved constituents are measured in milligrams per liter (volume) or parts per million (weight).

STANDARDS

Less than 500 ppm total dissolved solids for "sweet" water, and limits on ions and molecules that impair the quality of water.

POLLUTION

DEFINED

Chemical, physical, or biological materials that impair the future use of water and that may be a health hazard.

RESIDENCE TIME

The average length of time a given substance will stay in a system, such as water in a lake or aquifer. Whereas the residence time of human-generated smog in the atmosphere of large cities is on the order of ten days, pollutants may reside in some aquifers for thousands of years.

COMMON POLLUTANTS

Hydrocarbons, carcinogenic BTX solvents (benzene, toluene, and xylene), fertilizers, pesticides, livestock fecal matter, and gasoline.

CONSERVATION AND ALTERNATIVE SOURCES

Conservation means include the use of treated wastewater for irrigation and recharging aquifers, water-saving irrigation devices, water-spreading during wet years, and voluntary water conservation. Desalinization and importation are the only viable alternative sources at present.

Key Terms

aquiclude	dripstone	influent (stream)	specific yield
aquifer	effluent (stream)	karst terrane	speleology
artesian-pressure surface	evapotranspiration	perched water table	stalactite
bioremediation	ground water	permeability	stalagmite
cone of depression	hydraulic conductivity	porosity	sustained yield
confined aquifer	hydrogeologist	potable water	unconfined aquifer
drawdown	hydrologic cycle	residence time	water table

Study Questions

1. Sketch a geologic cross section that shows the ground-water zones and the static water table.
2. Sketch or define a confined aquifer. Explain the artesian-pressure surface and its importance in production of water from an artesian well.
3. How does topography affect the shape of the static water table?
4. How does one determine the amount of water a well that taps the static water table might produce on a long-term basis?
5. Explain the distinction between effluent and influent streams.
6. Explain the relationship between the amount of drawdown in a water well and the aquifer's permeability.
7. What states are the largest consumers of ground water, and why is their usage so heavy?
8. What is meant by ground-water *mining*? Cite the case history of an aquifer where this has happened.
9. Many beverage manufacturers claim the water in their drinks is purer because it comes from artesian wells or springs. Comment on these claims.
10. What are the Public Health Service standards for total dissolved solids in drinking water? What is the maximum amount of dissolved salts for safe drinking water?
11. Name some organic compounds sometimes found dissolved in ground water that are hazardous to human health.
12. Considering that agriculture is the largest consumer of ground water and that aquifers in many agricultural areas are being mined, how might farmers conserve ground water?

◁▷ For Further Information

BOOKS AND PERIODICALS

American Institute of Professional Geologists. 1985. *Ground water: Issues and answers*. Arvada, Colo.: A.I.P.G.

Baldwin, H. L., and C. L. McGuinness. 1963. *A primer on ground water*. Washington: U.S. Geological Survey.

Bull, W. B., and R. E. Miller. 1975. *Land subsidence due to ground-water withdrawal in the Los Banos–Kettleman City area, California*. U.S. Geological Survey professional paper 437-E.

Cohen, P.; O. L. Franke; and B. L. Foxworthy. 1970. *Water for the future of Long Island, New York*. New York Division of Water Resources, Department of Environmental Conservation, in cooperation with the U.S. Geological Survey.

Hauwerrt, Nico M., and Shawn Vickers. 1994. *Barton Springs/Edwards aquifer: Hydrogeology and ground water quality*. Austin, Texas: Texas Water Development Board, Barton Springs/Edwards Conservation District.

Heath, Ralph C. 1983. *Basic ground-water hydrology*. U.S. Geological Survey water-supply professional paper 2220.

Kidd, Mary A. 1996. *Nutrients in the nation's drinking water: Too much of a good thing?* U.S. Geological Survey circular 1136.

Loveland, D. G., and Beth Reicheld. 1987. *Safety on tap: A citizen's drinking water handbook*. Washington: League of Women Voters Education Fund.

Moore, John E.; A. Zaporozed; and James W. Mercer. 1994. *Ground water: A primer*. Environmental awareness series. Alexandria, Va.: American Geological Institute.

Sharp, John M., Jr., and Jay L. Banner. 1997. The Edwards aquifer: A resource in conflict. *G.S.A. Today* (Geological Society of America), August.

Solley, W. B.; R. R. Pierce; and Howard A. Perlman. 1993. *Estimated use of water in the United States in 1990*. U.S. Geological Survey circular 1081.

Solley, Wayne B.; Robert R. Pierce; and Howard A. Perlman. 1998. *Estimated use of water in the United States in 1995*. U.S. Geological Survey circular 1200.

Testa, Stephen. *Ground-water remediation of petroleum-handling facilities, Los Angeles coastal plain*. In Pipkin, Bernard W., and Richard Proctor (eds.). 1992. *Engineering geology practice in southern California*. Association of Engineering Geologists special publication no. 4. Belmont, Calif.: Star Publishing Co.

Todd, D. K. 1980. *Groundwater hydrology,* 2d ed. New York: John Wiley.

Winter, Thomas C.; Judson, O. Harvey; Lehn Franke; and William M. Alley. 1998. *Ground water and surface water: A single resource*. U.S. Geological Survey circular 1139.

Zwingle, Erla. 1993. Ogallala aquifer: Well spring of the High Plains. *National geographic* 183, no. 3 (March).

◁▷ World Wide Web Activities

Gateway Site to U.S. Watershed Information
http://www.epa.gov/surf2/locate/

This Web page is a gateway to detailed information on U.S. watershed areas. Add your zip code below the U.S. map; this will link you to a second page where you will press on the number code that most closely matches the geographic designation of your area. The activated page will give a description of your watershed, with links to regional data and recent news.

1. What are the rankings for commercial, domestic, and industrial water use in your watershed?
2. What is the principal aquifer (if any) in your region? Where are the areas of impaired water quality? What water quality studies and water clean-up activities are being conducted in your watershed?

Overview of the Sacramento Delta Region, California
http://www.bay-delta.org/itis/index.html

This page provides an overview of the Sacramento Delta region: its location and importance to California, which has limited water resources for the number of residents.

1. How many California residents receive water from this site, and what is the furthest large city supplied?
2. The delta region has multiple uses and functions. Name several.
3. What are the current environmental challenges to this region?

Recent Changes in the High Plains Aquifer of the Central United States
http://www-ne.cr.usgs.gov/highplains/hp97_web_report/ precip96.pdf

This Web page has a report and map explaining the changes in the High Plains Aquifer under eight states east of the Rocky Mountains. The aquifer has generally rebounded in recent years, due to better water management and increased precipitation.

1. In 1996–97, which three states did *not* increase water resources within the aquifer region? (See Table 2.)
2. Analyze the data in Table 3 and suggest a reason for the lack of water resource recovery in these three states.

Emerging Threats to Water Quality in the United States
http://toxics.usgs.gov/regional/emc.html
This page introduces the efforts of the U.S. Geological Survey to monitor some of the newer threats to surface and ground-water quality in the United States.

1. What are the emerging chemical contaminants to stream waters that are under study?

2. Monitoring stations are in various types of land-use regimes (i.e., urban, agricultural, mixed use). What is the station type nearest to your college? What category of emerging chemical contaminants might be the most prevalent in this location?

 # Earth Systems Today CD Correlations

CHAPTER CONCEPT: Ground Water Is Stored in Permeable Rock Bodies Called Aquifers
This segment of the CD addresses the properties of natural materials that most commonly provide for underground water storage.

SECTION MODULE: Ground Water: Definition and Storage Conditions
Read the CD text and review the definitions of *ground water* and *aquifer*.

QUESTION: *What specific Earth materials most commonly store underground water in nature—that is, function as aquifers? What specific property do all aquifers share?*

CHAPTER CONCEPT: Ground Water Is Replenished through Precipitation. When Withdrawal Exceeds Replenishment, the Ground Water Table Is Lowered

SECTION MODULE: Ground Water: Main Animation
Work through the interactive main activity for ground water, making special note of the relationship of the ground water table to precipitation, and withdrawal by pumping.

QUESTION: *What is the name for local depression of the ground water table at an overpumped well? What aspect of the aquifer might affect the steepness of this feature?*

chapter nine

Surface Hydrologic Hazards

Rivers, Gorges, and National Parks

Throughout the world national parks are areas of historical and scientific interest as well as places of great beauty. U.S. National Parks exhibit a variety of terranes and origins; for example, Hawaii Volcanoes and Yellowstone are volcanic, Yosemite is glacial, and the Grand Canyon and Bryce have been shaped by river erosion. A definition of what constitutes a national park was supplied by the International Commission on National Parks in 1971:

> ". . . a large area where ecosystems have not been materially altered, where the highest political authorities have taken steps toward protection, and where visitors are allowed to enter."

The U.S. government's designating areas to be protected and preserved began when Abraham Lincoln mandated that the State of California should protect Yosemite Valley in 1864. In 1872 Congress enacted legislation that provided funds for the federal government to protect, preserve, and maintain Yellowstone in its natural condition. Thus was the U.S. national park system created, and Yellowstone is considered to be our first national park. Other areas needing protection for ensuring their enjoyment by the public were added to the National Park roster early in the system's development; most notably, Crater Lake National

Park in 1902, Glacier National Park in 1910, Hawaii Volcanoes National Park in 1916, and Grand Canyon and Zion in 1919. To manage and maintain these parks the U.S. National Park Service was established in 1916.

It has been said that the park system was the United States' only possible response to Europe's marvelous castles and cathedrals. The beauty and attraction of a dozen parks is totally the result of stream and river erosion—Grand Canyon, Bryce Canyon, Zion, Arches, and Canyonlands to name a few. Fourteen new parks were created by Congress in the 1990s, with our fifty-fifth national park, Black Canyon National Park in Colorado (in the figure), the last new park to be established in the twentieth century. The Gunnison River there has cut deeply into imposing Precambrian schist and gneiss laced with veins of granite, and the near-vertical canyon walls are 830 meters (2,700 ft) high. The spectacular Black Canyon gorge is the heart of this park. With National Park status,

the gorge has changed from a "should we stop and look" natural feature to a "must see."

Not far from the Black Canyon National Park is another river gorge with only "state park status," but which is just as spectacular—the Royal Gorge of the Arkansas River near Canon City, Colorado. Only a few hundred feet wide at the top, it narrows to about 15 meters wide at the bottom and is spanned by the highest suspension bridge in the world, 325 meters (1,053 ft) above the river (see Figure 9.3). The bridge construction took seven years at a cost of only $350,000 in 1929. Compare this to restoration costs of $1.6 million in 1997. It takes a long time to make beautiful gorges like those of the Arkansas and Gunnison Rivers. The Royal Gorge, over 300 meters deep, was formed entirely by running water cutting into solid granite. It is estimated that it was 3 million years in the making, yielding a rate of erosion of one foot every 2,500 years. Yes—gorges like these described here take a long, long time to form. ■

Geologic agents are processes that perform the work of erosion and deposition, and which shape and modify the earth's landscape. Rivers, mass-wasting, glaciers, wind, underground water, and waves are examples. Rivers, however, are the most important geologic agent at work in today's world. Their importance is reflected in the fact that river flooding may well be the most globally pervasive, environmentally diverse, and continually destructive of all natural hazards.

Hydrology is the science that deals with liquid and solid global water, its properties, circulation, and distribution on earth and in the atmosphere. It arose from the very practical problems of human water supply, river flooding of human habitat, and water quality. We mainly address hydrologic hazards in this chapter, after a necessary introduction to the geology of river processes.

River Systems

A stream's ability to erode (Latin *erodere,* "to gnaw away") the land and impact humans is a function of the stream-flow velocity and the amount of water discharged by the stream. With equal discharge—that is, the same volume of water flowing past a point in a river channel in a given time period—a fast-flowing stream erodes its banks more significantly than does a slow-moving one. In the United States velocity (V) is usually expressed in feet per second (ft/sec), and **discharge** (Q), the amount of water, in cubic feet per second (cfs). A cubic foot contains about 7.4 gallons of water. A third factor, the cross-sectional area of the stream, also is important. A stream channel that is wide and shallow will have a much lower velocity than a stream with the same discharge that is narrow and deep. This is because the wide, shallow channel has greater area of friction between the running water and its banks (◈ Figure 9.1). Whatever the mechanics of the stream's flow, with increased velocity and discharge, there are increases in the stream's erosion potential and its sediment-carrying capacity (◈ Figure 9.2).

The longitudinal profile, or side view, of a stream channel has a slope known as its **gradient,** which can be expressed as the ratio of its vertical drop to the horizontal distance it travels (in meters per kilometer or feet per mile). The gradient of the Colorado River in the Grand Canyon is about 3.2 meters per kilometer, a moderately steep gradient, whereas in its lower reaches, where it flows into the Gulf of California, its gradient is only a few centimeters per kilometer. A mountain stream may have a gradient of several tens-of-meters per kilometer and an irregular profile with rapids and waterfalls. As the same stream approaches the lowest level to which it can erode, an elevation known as **base level,** its gradient may be almost flat. A stream's ultimate base level is sea level, but it may have intermediate, temporary base levels such as a lake surface or a rock formation that is highly resistant to erosion. A stream whose gradient has adjusted so that an equilibrium exists between its discharge and its sediment supply is known as a **graded stream.** Longitudinal profiles of graded streams are generally concave-

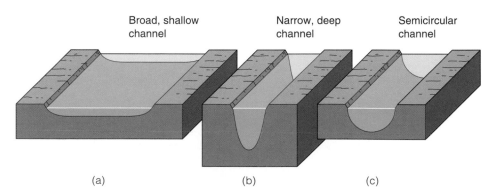

◈ FIGURE 9.1 Three stream channels with the same cross-sectional area but different shapes. The semicircular channel, (c), has the least perimeter in contact with the water and thus the least frictional resistance to flow. If all other variables such as roughness and slope of the channel are equal, flow velocity will be greatest in the semicircular channel.

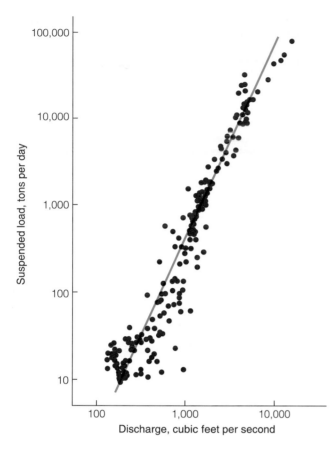

◆ **FIGURE 9.2** The relationship between water discharge (*Q*) and the suspended load for a typical stream. The points represent separate measurements. A tenfold increase in water discharge results in almost a 100-fold increase in sediment load.

find conspicuous depositional features known as **alluvial fans** (◆ Figure 9.4). They are in fact fan-shaped and they build up as the main stream branches, or splays, outward in a series of *distributaries,* each "seeking" the steepest gradient. Alluvial fans can provide beautiful, interesting views of surrounding terrain, but they also can be subject to flash floods and debris flows from intense seasonal storms in nearby mountains.

Deltas form where running water moves into standing water. The Greek historian Herodotus (484–420 B.C.) applied the term to the somewhat triangular shape of the Nile delta in recognition of its resemblance to the Greek letter delta (Δ). Deltas are characterized by low stream gradients, swamps, lakes, and an ever-changing channel system. The "arcuate" (arc-shaped) delta of the Nile and the "bird's foot" delta of the Mississippi also are described by their appearance on maps (◆ Figure 9.5). Deltas also form at the mouths of small streams, as illustrated by the arcuate delta at Malibu Creek—location of the famous Malibu Colony of the rich and famous and the early beach-boy-type surfing movies (◆ Figure 9.6). The tremendous sand output of the Nile has created the huge coastal dunes east of Alexandria that formed the background for the famous World War II battle at El Alamein (see the Chapter 12 opener). The Ganges–Brahmaputra delta of Bangladesh consists of tidal flats and many low-lying sand islands offshore (Figure 9.5c). Because deltas are at or near sea level, they are subject to flooding by tropical storms. Coastal flooding is the most deadly of natural disasters, even more deadly than earthquakes or river floods. This type of flooding was particularly disastrous in Bangladesh in 1970, when more than 500,000 people were drowned by a surge of water onshore from a tropical cyclone. A similar tragedy occurred in 1991, when another onshore water surge left a reported 100,000 dead on the low-lying islands off the delta (see Chapter 10).

Streams and rivers with low gradients and floodplains form a number of features that both aggravate and alleviate flood damage. The courses of such rivers follow a series of S-shaped curves called **meanders** (after the ancient name of the winding Menderes River in

upward from source to end. ◆ Figure 9.3a contrasts the profiles of the Arkansas River (see ◆ Figure 9.3b), which drains the wide east slope of the Rocky Mountains, and the Sacramento River, which drains the much narrower Pacific slope.

Gradients decrease in the lower reaches of a river, and sedimentation may occur as streams overflow their banks onto adjacent **floodplains.** Decreased water velocity on the floodplain results in the deposition of fine-grained silt and clay adjacent to the main stream channel. This productive soil, commonly known as *bottom land,* is much desired in agriculture (Case Study 9.1).

Stream Features

All sediment deposited by running water is called **alluvium** (Latin, a flooded place or flood deposit). In the arid West, where streams lose velocity as they flow from mountainous terrain onto flat valley floors, we

Consider This... ? We generally associate high sediment loads with high river discharge, but there are other variables that determine the sediment transported by a river. With this in mind, would you expect the suspended sediment load per cubic meter of water to be greater in the Columbia River, which has a high-rainfall drainage basin, or in the Colorado River, which has a low-rainfall basin? ■

Case Study | 9.1

The Nile River—Three Cubits between Security and Disaster

Flooding is a natural part of the hydrologic cycle, and as farmers, hydrologists, and geologists know, it can be beneficial. The annual flood on the Nile River, the longest river in the world (depending upon how one measures the Amazon), was a source of life for at least 5,000 years. Eighty percent of the water for these floods came from the faraway Ethiopian Highlands and it replenished the soil moisture and deposited a layer of nutrient-rich silt on the floodplain. The Greek historian and naturalist Herodotus, in 457 B.C., recognized the importance of annual flooding when he spoke of Egypt as "the gift of the Nile." The priests of ancient Egypt used principles of astronomy to calculate the time of the river's annual rise and fall. Records of the height of the flood crests go back to 1750 B.C. (some say to 3600 B.C.), as measured with primitive hydrographs known as Nilometers (◆ Figure 1). Water height was measured in cubits, one cubit being the distance from one's elbow to the tip of the middle finger, approximately 0.5 meter. In Shakespeare's *Antony and Cleopatra,* Mark Antony notes that

> They take the flow o' th' Nile . . . .
> By certain scales . . . they know
> By th' height, the lowness, or the mean, if dearth
> Or foison follow. The higher Nilus swells,
> The more it promises. As it ebbs, the seedsman
> Upon the slime and ooze scatters his grain,
> And shortly comes to harvest.

And, of course, John Lennon of the Beatles supposedly once pointed out that "denial is not a river in Egypt."

The annual flooding cycle has changed since completion of the Aswan High Dam on the Upper Nile in southern Egypt in 1970. The dam's purposes were to increase the extent of arable land, generate electricity, improve navigation, and control the flooding on the Lower Nile, and it is accomplishing all of those things. Now that the flooding is controlled, however, crop maintenance depends upon irrigation. Severe environmental consequences have resulted, including depletion of the soil nutrients that are beneficial to farming (necessitating the use of artificial fertilizers) and the proliferation of parasite-bearing snails in irrigation ditches and canals. *Schistosomiasis* is a human disease found in communities where people bathe or work in irrigation canals containing snails serving as temporary hosts to parasitic flatworms (schistosomes). These revolting creatures, commonly called blood flukes, invade human's skin when they step on the snails while wading in the canals. The flukes live in their human hosts' blood, where the females release large numbers of eggs daily, which cause extensive cell damage. Next

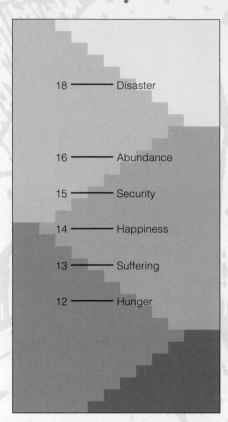

◆ **FIGURE 1** The Nilometers of ancient Egypt were primitive hydrographs used to measure the Nile's flood heights in cubits, a cubit being the distance between one's elbow and the tip of the middle finger (usually about 18 in). The environmental interpretations shown here were used during the time of Pliny the Elder (A.D. 23–79) of Mount Vesuvius fame.

to malaria, *schistosomiasis* is the most serious parasitic infection in humans. Its outbreak in Egypt is a result of controls on the river that prevent the generous floods that formerly flushed the snails from canals and ditches each year. Another negative consequence is the loss of a huge shrimp fishery off the Nile delta; the dam traps the nutrients that encouraged the growth of plankton, upon which the shrimp larvae fed. Every dam has a finite useful lifetime that is determined by how fast the lake it impounds fills with sediment. Lake Nasser, behind the Aswan Dam, is rapidly filling with the (precious) sediment that formerly enriched the floodplain of the lower Nile.

(a)

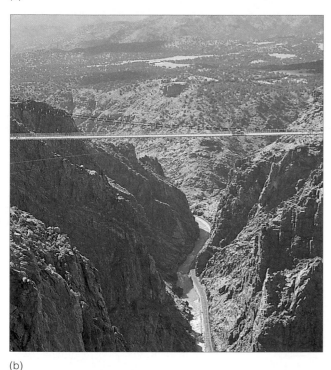

(b)

◁▶ **FIGURE 9.3** (a) Measured longitudinal profiles of the Arkansas and Sacramento Rivers. The vertical scale is exaggerated 275 times. (b) The Arkansas River at Royal Gorge, 45–50 miles upriver from Pueblo, Colorado, near Cañon Springs. The white-water rapids in the photo reflect the steep gradient portrayed in (a). The suspension bridge that spans the gorge is 1,053 feet above the river. Zebulon Pike, for whom Pike's Peak (30 miles northeast) was named, is believed to have been among the first Europeans to see the gorge.

Turkey; ◁▶ Figure 9.7). Straight rivers are the exception, rather than the rule, and it appears that running water of all kinds will meander—ocean currents, rivers on glaciers, and even trickles of water on glass. The key to these regularly spaced curves is the river's ability to erode, transport, and deposit sediments. The meandering process begins when the gradient is reduced to a slope at which a stream stops deepening its valley. As a stream meanders, it erodes the outside curve of each meander and deposits sediment on each inside curve, producing a deposit known as a **point bar.** ◁▶ Figure 9.8 shows the velocity distribution on a stream meander, which explains the erosion–deposition pattern. Meanders also widen stream valleys by lateral cutting of valley side slopes, causing mass wasting and eventually creating an alluvium-filled floodplain adjacent to the river. It is across this surface that the river mean-

Consider This... ? Deltas are deposited where a river flows into a body of standing water, such as a lake or an ocean. Why does one of the great rivers of North America, the St. Lawrence River, have no delta, even though it flows into a landlocked estuary? ■

ders, sometimes becoming so tightly curved that it shortcuts a loop to form a **cutoff** and **oxbow lake** (◁▶ Figure 9.9).

By the time a river has developed a wide floodplain with meanders, cutoffs, and oxbow lakes, a situation of poor drainage exists that increases the impact of overbank flooding. **Natural levees** build up on both

(a)

(b)

◈▶ **FIGURE 9.4** (a) Alluvial fans in Death Valley, California. The Panamint Range is in the background. A fan forms where a mountain stream flows onto an adjacent, flat-floored valley and deposits its sediment load in a system of distributary channels. (b) A fan's symmetrical shape is the result of the active channel "migrating" back and forth across the fan surface.

Distributaries

banks of a river when it overflows and deposits sand and silt adjacent to the channel (◈▶ Figure 9.10). Natural levees, or manufactured ones, **artificial levees,** keep flood waters channelized so that the water level in the stream may be above the level of the adjacent floodplain. This is a tenuous situation, because a breach in the levee would cause flooding of the floodplain, which could force the river into a new course. New Orleans on the Mississippi River is an example of a city on a floodplain that is almost perennially below the level of the river that flows through it (see Case Study 9.2). Although levees may prevent flooding of adjacent floodplains, they can cause flooding downstream in areas where there are no levees.

During the 1993 floods in the Midwest, artificial levees saved some cities in the Mississippi River

(a)

(b)

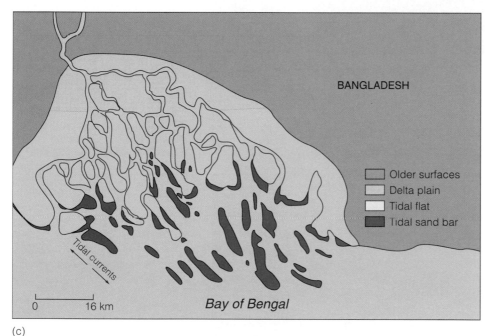

(c)

◁▶ FIGURE 9.5 River deltas. (a) Birdsfoot delta of the Mississippi River. The main channel and subordinate distributaries are separated by swamps and wetlands. (b) Arc-shaped delta of the Nile River is predominantly wetlands and rice-growing areas. The barrier islands are sand and they contribute to large coastal dunes to the west. (c) The Ganges–Brahmaputra delta of Bangladesh. The small islands rise above the tidal flats. Though subject to catastrophic floods, they are densely populated.

drainage basin, notably St. Louis, where the flood crested at 49.4 feet and the levee height is 52 feet. Quincy, Illinois, lacking artificial levees, was inundated, but the levees at Hannibal, Missouri, on the opposite side of the river held and spared the town. A levee on the Raccoon River at Des Moines, Iowa, failed, and floodwaters inundated a city water-purification plant. Thousands of residents were without drinking water for

19 days. Some levees were intentionally breached in order to reduce flood-crest elevation downstream. More than 800 of the 1,400 levees in the nine-state disaster area were breached or overtopped. Most of these were simple dirt berms built by local communities, which explains why so many small towns were almost completely submerged by the floodwaters. However, more than 30 levees built by the U.S. Army Corps of

◆▶ **FIGURE 9.6** The arcuate, wave-dominated delta of Malibu Creek, Malibu, California, after a record rainfall in 1969. Note the suspended sediment in the ocean and the concentration of homes along the delta north (left) of the creek.

◆▶ **FIGURE 9.7** Meanders on the Missouri River at Glasgow, Missouri. Note that the meander belt swings across the full width of the river's floodplain. Lighter color represents a change in vegetation.

Case Study | 9.2

Water, Water, Everywhere

New Orleans, the Crescent City, has the flattest, lowest, and youngest geology of all major U.S. cities. The city's "alps," with a maximum elevation of about 5 meters (16 ft) above sea level, are natural levees built by the Mississippi River, and its average elevation is just 0.4 meters (1.3 ft). No surface deposits in the city are older than about 3,000 years. The city was established in 1717 on the natural levees along the river. The levees' sand and silt provided a dry, firm foun-

◁▶ **FIGURE 1** (a) Major landforms and depositional environments in the New Orleans area before urbanization. The Pleistocene Prairie Formation dips down seaward beneath Lake Pontchartrain and New Orleans, providing a firm foundation for pilings that support the Superdome and other large buildings. The natural levees on both sides of the river also provide a firm, dry foundation. (b) High-altitude false-color image of New Orleans, a city almost completely surrounded by water. The light spot in the lower center is the city. Note that the Mississippi River meanders through it, Lake Pontchartrain is to the north, and smaller Lake Borgne is to the northeast. Low, swampy delta lands appear brownish to the east and south. The land width in the photo is approximately 200 kilometers (125 mi). (c) Cypress swamp on the outskirts of present-day New Orleans.

(a)

(b)

(c)

Case Study

◀▶ FIGURE 2 The French Quarter as seen from atop the natural levee on the Mississippi River. Note that the levee slopes downward toward the French Quarter. The wall and gate are part of a floodwall built to hold back floodwaters should the river overtop the levee.

dation for the original city, which is now called *Vieux Carré* (French "old square"), or "the French Quarter." Farther from the river the land remained undeveloped, because it was mainly water-saturated cypress swamp and marsh formed between distributaries of the Mississippi River's ancient delta. These areas are underlain by as much as 5 meters (16 ft) of peat and organic muck (◀▶ Figure 1). When high-volume water pumps became available about 1900, drainage canals were excavated into the wetlands to the north. Swamp water was pumped upward into Lake Pontchartrain, a cutoff bay of the Gulf of Mexico. About half the present city is drained wetlands lying well below sea and river level. The lowest part of the city, about two meters below sea level, is in the lively, historic French Quarter, noted for its music, restaurants, and other tourist attractions (◀▶ Figure 2).

Subsidence is a natural process on the Mississippi River delta due to the great volume and the sheer weight of the sediment laid down by the river. The sediment compacts, causing the land surface to subside. The region's natural subsidence rate is estimated to have been about 12 centimeters (4 in) per century for the past 4,400 years. This is based upon C-14 dating of buried peat deposits and does not take into account any rise of sea level during the period.

Urban development in the 1950s added to the subsidence. Compaction of peaty soils in reclaimed cypress swamps coincided with the period's construction of drainage canals and planting of trees, both of which lowered the water table. Peat shrinks when it is dewatered, and also oxidation of organic matter and compaction contribute to subsidence. Most homes constructed on reclaimed swamp and marsh soils in the 1950s were built on raised-floor foundations. These homes are still standing, but they require periodic leveling. Unfortunately, homes built on concrete-slab

◀▶ FIGURE 3 Slab-on-pilings construction (a) before and (b) after differential subsidence.

foundations, a technique that had just been introduced, sank into the muck and became unlivable. Other homes' foundations were constructed on cypress-log piles sunk to a depth of 10 meters or more (at least 30 ft). These houses have remained at their original level, but the ground around them has subsided. This *differential subsidence* requires that fill dirt be imported in order to make the sunken ground meet the house level (◀▶ Figure 3). It is very common to see carports and garages that have been converted to extra rooms when subsidence has cut off driveway access to them (◀▶ Figure 4). One also sees many houses with an inordinate number of porch steps; the owners have added steps as the ground has sunk. In 1979, Jefferson and Orleans Parishes (counties are called "parishes" in Louisiana) passed ordinances requiring 10- to 15-meter-deep wooden-pile foundations for all houses built upon former marsh and swamp land. This has been beneficial, but differential subsidence continues to damage New Orlean's sewer, water, and natural-gas lines and streets and sidewalks.

◆ **FIGURE 9.9** (a–d) Meander pattern sequence of erosion leading to a cutoff and an oxbow lake. A cutoff forms in (c) and completely isolates the oxbow lake in (d). (e) Aerial photo of an oxbow lake on the Tallahatchie River in Mississippi. Note the old scars from the migration of meanders.

canyon, destroying structures and endangering lives (◆ Figure 9.12). Lowland streams, in contrast, have low gradients, adjacent wide floodplains, and meandering paths.

A **hydrograph** is a graph of a water body's discharge, velocity, stage (height), or some other characteristic over time, and it is the basic tool of hydrologists (◆ Figure 9.13). The "synthetic" hydrograph in ◆ Figure 9.14 graphically illustrates the basic difference between upland (flash) floods of short duration and the longer-lasting lowland floods. The timing of flood crests on the main stream—that is, whether they arrive from the tributaries simultaneously or in sequence—in large part determines the severity of lowland flooding.

Floods are normal and, to a degree, predictable natural events. Generally, a stream will overflow its banks every two to three years, gently inundating a portion of the adjacent floodplain. Over long periods of time, catastrophic high waters can be expected. Depending upon their severity, these events are referred to as fifty-, one-hundred-, or five-hundred-year floods. The more infrequent the event, the more widespread and catastrophic is the inundation (◆ Figure 9.15). The Midwest flood of 1993 that inundated 421 counties in 9 states, with a peak flow of 720,000 cfs (5.1 million gallons per second) at Boonville, Missouri, is regarded as a 500-year flood. Flood frequency is explained in more detail later in this section.

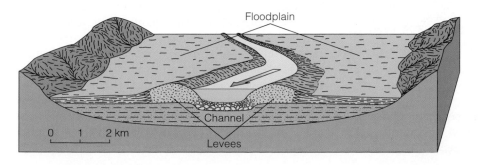

◆ **FIGURE 9.10** Relationship of natural levees and floodplain to a hypothetical stream. The coarsest sediment is deposited in the stream channel, where the stream's velocity is greatest, and finer silt and clay are deposited on levees and the floodplain as velocity diminishes during overflows.

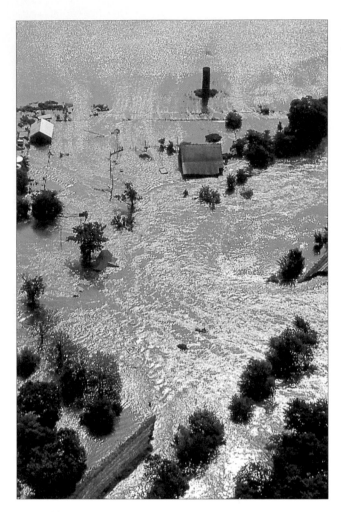

◆ **FIGURE 9.11** One of the 800 levees that failed or were overtopped along the Mississippi, Missouri, and Illinois Rivers during the 1993 floods. This levee is near Davenport, Iowa, on the Mississippi River.

Floods from all causes are the number-one natural disaster in the world in terms of loss of life. Twenty-six disastrous U.S. floods are listed in ⊙ Table 9.1, and their locations are shown on the map of ◆ Figure 9.16. Although by no means a complete listing of severe floods, the table and map show the general distribution of riverine floods, flash floods, and coastal floods in the United States. Coastal flooding due to hurricanes and typhoons is the greatest single killer. The record-holding river floods are due to dam failures. The legendary 1889 Johnstown flood killed 3,000, and the 1928 St. Francis Dam failure in Southern California killed 450 people.

FLOOD MEASUREMENT

As we have seen, the basic unit in river studies is the drainage basin, which can be defined on any scale. A large river's drainage basin includes the drainage

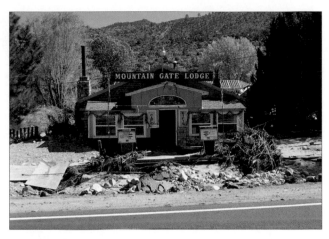

◆ **FIGURE 9.12** The result of a flash flood in the Sierra Nevada Mountains in January 1997, when floodwaters left the channel of the Walker River and inundated U.S. Route 395. This lodge and a portion of the highway were destroyed.

basins of many tributaries, and each tributary's drainage basin includes the drainage basins of all *its* tributaries, and so on. It follows that a large drainage basin has the potential for a larger flood than does a small drainage basin.

Flood studies begin with hydrographs for a particular storm on a river within a given drainage basin. These graphs are generated from data obtained at one or more stream gauging stations, where stage (the water elevation above a previously established datum plane) and velocity are measured and then related to discharge. Rapid runoff in a drainage basin is accompanied by high stream discharge and high water velocity. As discharge increases, the water rises to the *bankfull stage* and then spills onto the adjacent floodplain at *flood stage*. ◆ Figure 9.17 shows such a condition. It is a hydrograph superimposed on the cross section of a hypothetical stream valley. A *stage hydrograph* is most useful for flood planning purposes, because it relates stream discharge to river height, or elevation. One can see at a glance at what point the stream will overflow its banks and at what elevations flood waters will come in contact with specific structures (see Case Study 9.1). Note the contrasting hydrographs in ◆ Figure 9.18. Whereas the flash flood in Tucson, Arizona, lasted only about five hours, the North Dakota flood lasted for almost a month. This reflects the climatic and topographic differences between an area with intense, short-duration, monsoonal rains (Tucson), and an area in the north-central United States where a very slow moving low-pressure system stalled and 43 centimeters (16.91 in) of rain fell in a 24-hour period. The Mississippi River Valley flood of 1993 lasted for three months.

TABLE 9.1 **Selected U.S. Floods since May 1889**

FIGURE 9.16 MAP NUMBER	DATE	LOCATION	LIVES LOST	ESTIMATED DAMAGE, $ MILLIONS
1	May 1889	Johnstown, Pa. (dam failure)	3,000	–
2	8 Sept. 1900	Galveston, Tex. (hurricane)	6,000	30
3	May–June 1903	Kansas and Lower Missouri Rivers	100	40
4	March 1913	Ohio River basin	467	147
5	14 Sept. 1919	South of Corpus Christi, Tex. (hurricane)	600–900	22
6	Spring 1927	Mississippi River valley	313	284
7	12–13 March 1928	California (St. Francis Dam failure)	450	14
8	13 Sept. 1928	Lake Okeechobee, Fla.	1,836	26
9	March 1938	Southern California	79	25
10	21 Sept. 1938	New England	600	306
11	August 1955	Northeastern U.S. (Hurricane Diane)	187	714
12	27–30 June 1957	Texas and Louisiana (Hurricane Audrey)	390	150
13	Jan.–Feb. 1969	California	60	399
14	17–18 August 1969	Mississippi, Louisiana, and Alabama (Hurricane Camille)	256	1,421
15	February 1972	Buffalo Creek, W. Va. (dam failure)	125	10
16	June 1972	Black Hills, S.D.	238	165
17	June 1972	Eastern U.S. (Hurricane Agnes)	105	4,020
18	June–July 1975	Red River of the North basin	<10	273
19	June 1976	Southeast Idaho (Teton Dam failure)	11	1,000
20	July 1976	Big Thompson River, Colo.	139	30
21	Sept.–Oct. 1983	Tucson, Arizona	<10	50–100
22	Summer 1993	Upper Mississippi River drainage basin	50	12,000
23	Jan.–Feb. 1996	Oregon, Washington, Idaho, Montana	1	89*
24	April 1997	Red River, Manitoba (Canada), Minnesota, North Dakota	1	1,000+
25	October 1998	South-central Texas, San Antonio to Houston, Guadalupe River	29	400
26	September 1999	North Carolina, Hurricane Floyd and post-hurricane rains	47	5,000+

SOURCE: Various sources, including U.S. Geological Survey.
*Figure is for Oregon and Washington only.

FLOOD FREQUENCY

One of the most useful relationships that can be derived from long-term flood records is that of **flood frequency,** or **recurrence interval;** that is, how often on average a flood of a given magnitude can be expected in a particular location. With this kind of information we are better equipped to design dams, bridge clearances over rivers, sizes of storm drains, and the like.

The information needed for determining recurrence intervals is a long-term record of annual *peak discharge* (the largest flow) for the particular location on the river. The longer the period of record, the more reliable the statistics will be. The peak discharges are then ranked according to their relative magnitudes—with the highest discharge being ranked as 1, the second-largest as 2, and so forth. To serve as an example, selected annual peak discharges on the Rio Grande

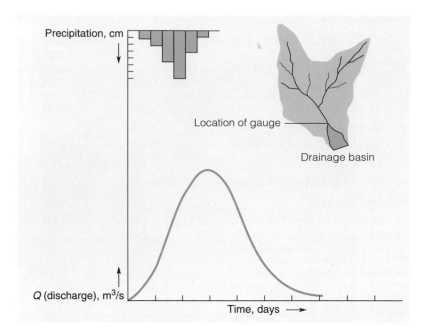

◆ **FIGURE 9.13** A hypothetical hydrograph that relates precipitation and discharge to time. Note that peak discharge occurs after the peak rainfall intensity. The drainage basin includes the main stream and all its tributaries.

near Lobatos, Colorado, are listed in ◉ Table 9.2, with their ranks and recurrence intervals. A recurrence interval, T, is calculated as

$$T = (N + 1) \div M$$

where N is the number of years of record (76 years in our example), and M is the rank of the event. The analysis predicts how often floods of a given size can be expected to occur in a watershed. In this example, a flood the size of the 1906 flood ($M = 10$) would be expected to occur on average once every 7.7 years [$(76 + 1) \div 10$]. If 100 years of records are available, the largest flood to occur during that period ($M = 1$) would have a recurrence interval of about 100 years on average. This is the so-called 100-year flood. The 100-year flood can also be obtained by projecting rainfall or streamflow data beyond the period of record.

Big floods seem to be occurring more often now in the United States, begging the question, How can "the 100-year flood" occur so *often*? The term is misleading, because it implies that a flood of the particular size can happen only once in 100 years. "The 100-year flood" is really a statistical statement of the probability that a flood of a given *rank* will occur in any one year; that is, $1/T$. For the 100-year flood, there is a 1 percent chance ($1/100$) of its occurring in any one year; and for the 50-year flood ($1/50$), a 2 percent chance. The U.S. Geological Survey says a

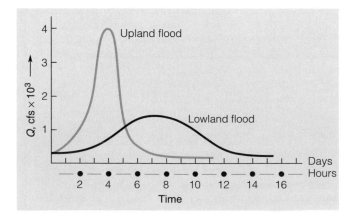

◆ **FIGURE 9.14** Hypothetical hydrographs of an upland flash flood and lowland riverine flood.

◯ TABLE 9.2	Annual Peak Discharges on the Rio Grande River near Lobatos, Colorado, 1900–1975		
YEAR	DISCHARGE, m³/sec	RANK (M)	RECURRENCE INTERVAL (T)
1900	133	23	3.35
1901	103	35	2.20
1902	16	69	1.12
1903	362	2	38.50
1904	22	66	1.17
1905	371	1	77.00
1906	234	10	7.70
1907	249	7	11.00
1908	61	45	1.71
1909	211	13	5.92
1974	22	64	1.20
1975	68	43	1.79

SOURCE: U.S. Geological Survey.

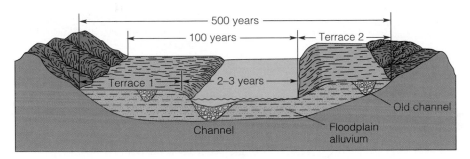

◆ **FIGURE 9.15** The probable extent of flooding over long time intervals on a hypothetical floodplain. The 2–3-year flood inundates the existing floodplain. The 100-year flood inundates both the existing floodplain and a higher one (Terrace 1), which formed when the river stood at a higher level. The 500-year event inundates an even higher terrace (Terrace 2) and all lower terraces and floodplains.

better term would be "the 1-in-100 *chance* flood." Viewed as a probability phenomenon, it is more understandable that we might get big floods in successive years or every few years. Keep in mind that the longer the rainfall or streamflow record, the better the statistic. Look at the streamflow data collected between 1940 and 1996 on the Chehalis River in Washington State in ◆ Figure 9.19. Note that the 1-in-100

chance flood calculated for data from 1940 to 1975 is lower than the same statistical flood based upon data for a period 21 years longer. This illustrates the importance of long-term data collection in assessing the magnitude of the 1-in-100 chance flood. There is, however, no meteorological or statistical prohibition against more than one 100-year flood occurring in a century, or even in one year, for that matter.

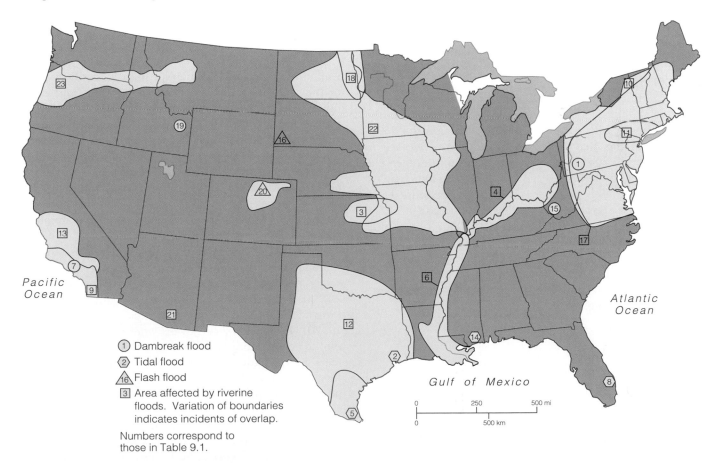

◆ **FIGURE 9.16** Locations of 26 major floods in the conterminous 48 states, 1889–1999. The numbers on the map refer to descriptions of the floods in Table 9.1.

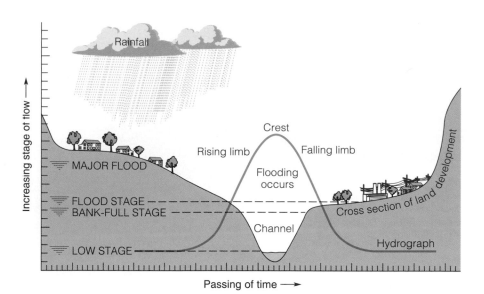

◀▶ **FIGURE 9.17** A stage hydrograph superimposed on a hypothetical stream valley, which illustrates the impact of increasing stage height on the flooded area.

Recurrence intervals are not exact timetables; they merely indicate the statistical probability of a given-size flood. However, knowing the flood discharge that can be expected in a river system every 25, 50, or 100 years enables us to better plan against the flooding that is certain to occur eventually.

Once we have determined recurrence intervals from the historical record, it is possible to construct flood-frequency graphs like the ones in ◀▶ Figure 9.20. Such graphs allow us to estimate at a glance how often a flood-level or discharge of a particular magnitude should occur. For example, if we wish to build a warehouse with a useful life of 50 years on the floodplain of the Red River at Grand Forks, we see in ◀▶ Figure 9.20a that the river reaches a reference elevation of about 49 feet every 50 years on average. We would probably want to build the warehouse at or above that elevation. However, urbanization, artificial levees, and other human-generated changes can alter the amount and timing of runoff and cause higher-discharge flood events to become more common. Consequently, the "100-year flood" may become the 70- or 80-year flood. These concerns are discussed later in the chapter.

THE FLOOD OF 1993

Forty percent of all the precipitation falling on the contiguous 48 states is collected in the Mississippi River drainage basin. Thus it is not surprising that one of the largest riverine floods attributable to rainfall was the great Mississippi Valley flood of 1993. The flood was the result of a wet spring followed by record-breaking rains in June and July. Above-normal precipitation in the six months preceding the flooding filled reservoirs and produced stream flows that were

Consider This... ? Artificial levees proved to benefit some communities at the expense of others during the Great Flood of 1993. What benefits are gained by constructing levees, and what damage may levees cause downstream along a river course? ∎

well above normal. In June and July the cool, dry air of the jet stream dipped abnormally south and collided with moist air from the Gulf of Mexico. This caused heavy rainfall in parts of nine midwestern states. Large streamflows, exceeding those expected only once every 100 years, were recorded at 45 gauging stations in these states (◀▶ Figure 9.21). Flooding along the Mississippi and its tributaries inundated 4 million hectares (10 million acres) in the nine states. At least 50 lives were lost, and damage estimates ranged from 10 to 12 billion dollars.

Most of the rain fell in the upper Mississippi drainage basin, and the flood crests migrated progressively downstream from St. Paul, Minnesota (June 26), to Dubuque, Iowa (July 6), to Quincy, Illinois (July 13), and finally to St. Louis, Missouri (July 19–20; ◀▶ Figure 9.22). Levees were overtopped, and the adjacent floodplains inundated in spite of volunteers' heroic efforts to raise them by sandbagging. The July 19 crest at St. Louis was 46 feet above the datum plane, 3 feet higher than ever before recorded. The next day the river crested at 47.1 feet, setting another record. On August 1, the river surged again to a record 49.4 feet. The Mississippi River is joined by the Missouri and Ohio Rivers at St. Louis and Cairo, Illinois, respectively. Because the Mississippi is deeper and wider below Cairo, it was able to accommodate the flood discharge there.

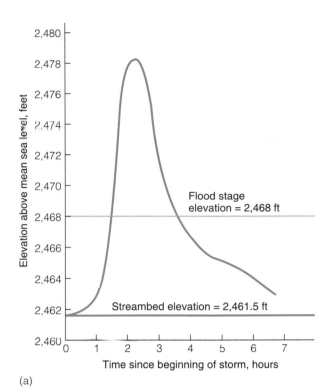

(a)

◆▶ FIGURE 9.18 (a) Stage hydrograph for a 5-hour storm in 1975 on Tanque Verde Creek in Tucson, Arizona. Note that the water rose 10 feet above flood-stage elevation and returned to below flood stage in about 2 hours. All the damage was done in a very short time. (b) Flood discharge of July 1996 on the South Branch Kishwaukee River near Fairdale, Illinois. Of the approximately 120 hours of flooding in 1996, flooding was severe for 3 days.

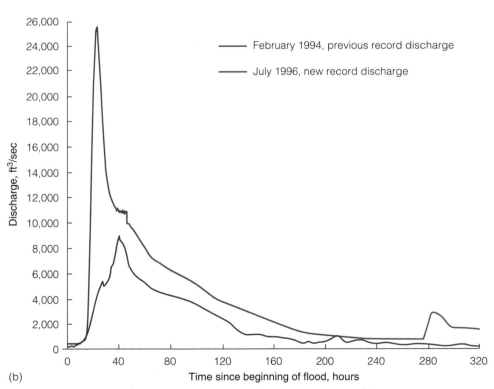

February 1994, previous record discharge

July 1996, new record discharge

(b)

At the peak of flooding the Mississippi River valley was referred to as the sixth Great Lake. Flooding was most severe where artificial levees failed and along stretches of the river where natural levees had not been raised artificially. Rivers flood naturally, and the only way to reduce losses is to limit floodplain development.

Mitigation

OPTIONS

Dams: The Solution or the Problem?

In the contiguous 48 states there are more than 75,000 dams more than two meters high, and nearly every

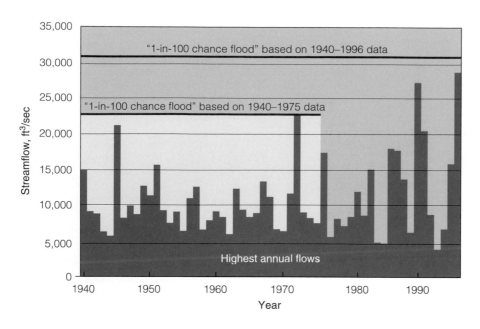

◁▶ **FIGURE 9.19** Streamflow data for the Chehalis River near Doty, Washington. Note the difference in streamflow for the 100-year flood based upon a 35-year data-collection period and one of 56 years in length.

major river is controlled by a dam. No one argues that dams provide such societal benefits as cheap electricity, flood control, recreation, and reduction of fear of drought. However, the negative side is loss of habitats, loss of the natural cycles of flooding and sediment deposition (see Case Study 9.1), changes in aquatic and riverine biology, and destruction of important spawning grounds for migratory fish such as salmon. The halcyon days of dam construction in the United States are over. Dams are too expensive, they displace huge numbers of people, and the reservoir and downstream environmental impacts are too great (see Case Study 9.3). Although more dams are being built in developing countries, only a few sites in the United States, Canada, and Europe are seriously being considered in the early twenty-first century.

The water emerging from a dam is not the same as the natural river water before damming. Sediment-free outflow scours the riverbed below the dam; because the temperature is different, native fish either are stressed by it, or they die; and new **riparian** communities arise in the downstream area. In addition, some dams fail, and dam-failure floods have accounted for the greatest loss of life due to flooding (see Table 9.1 and the Gallery). For all these reasons, dams are being dismantled or "fiddled with" to make them more ecologically acceptable. The Kennebec River in Maine now flows freely, because the 162 year-old Edwards Dam in Augusta was breached in 1999. Other dams scheduled for breaching are the Elwha Dam in Washington (to make way for spawning Pacific salmon) and ten hydroelectric dams on the Deerfield River in Vermont and Massachusetts (to allow passage by Atlantic salmon and other migratory species). This dismantling is occurring because of a little-known 1996 federal law requiring that environmental concerns be addressed before dams can be relicensed. Although the Deerfield River had been cleaned up as mandated by clean water laws, the natural ecosystem needed to be reestablished. It is interesting to note that more than 500 dams are due for relicensing early in the twenty-first century. We can expect to see more "fiddling" with dams to make their aquatic environments closer to pre-dam conditions.

Artificial Levees: The Solution or the Problem?

Dams, retaining basins, floodwalls, and artificial levees are time-honored means of flood control. All of these are attempts to keep floodwaters within the river channel or to store water for slow release at a later time. Unfortunately, population and property values adjacent to and below flood-control structures increase faster than the land can be protected. That is, floodplain development occurs in response to speculation that flood control has been achieved or is not far off. For example, many artificial levees were built along the Mississippi River to protect farmland. The floodplain was later urbanized, and the existing levees, though adequate to protect farmland, are not adequate for protecting homesites (◀▶ Figure 9.23).

An artificial levee is built upward upon a natural levee or riverbank so as to increase the amount of water the channel can accommodate and thus prevent the natural periodic flooding on adjacent land. When streamflow is high, the area's would-be floodwater flows downstream to areas that have lower natural levees or no levees, which then experience greater flooding than they would if the artificial levee had not been built upstream. Some of these downstream communi-

(a)

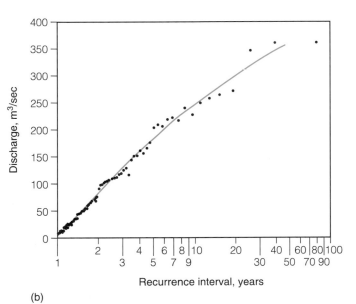

(b)

◆ **FIGURE 9.20** Flood frequency curves for two rivers, (a) the Red River at Grand Forks, North Dakota, using water elevation as the standard, and (b) the Rio Grande River near Lobatos, Colorado, using water discharge as the measure.

ties will build their own artificial levees to prevent this in the future. Then areas without artificial levees downstream from those levees are subject to flooding, so more levees are built there, and so it goes. Artificial levees are thus self-perpetuating, and the Mississippi River today is largely contained within them. Even at nonflood river levels, the water surface is above the

Consider This... ? Artificial levees proved to benefit some communities at the expense of others during the Great Flood of 1993. What benefits are gained by constructing levees, and what damage may levees cause downstream along a river course? ■

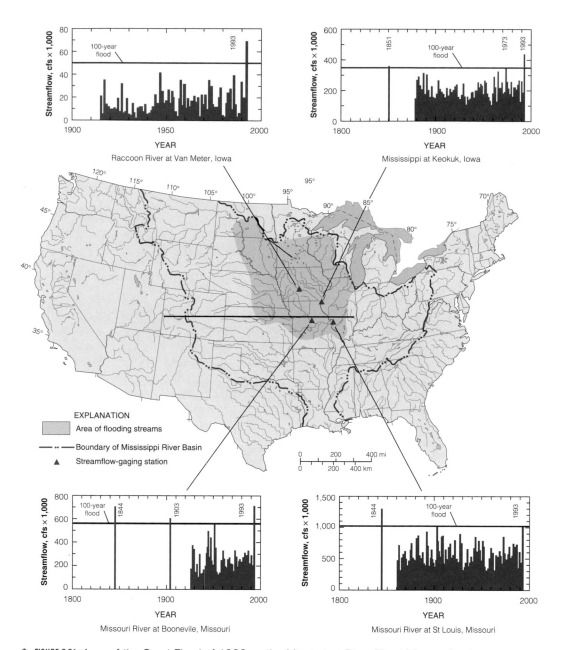

FIGURE 9.21 Area of the Great Flood of 1993 on the Mississippi River. The 100-year flood levels were met or exceeded at the four gauging stations shown. Note that the estimated flood of 1844 exceeded the 1993 event at St. Louis.

adjacent floodplains, so the danger of flooding is ever present.

Riverside land adjacent to the old French Quarter of New Orleans is extremely valuable tourist property. The natural levees here were about 2.1 meters (7 ft) above sea level (*a.s.l.*) in the 1700s, and natural flood stages probably never exceeded 3.0 meters a.s.l. However, confinement of the river upstream by artificial levees raised the flood level to 6.5 meters (21 ft) in modern times, and artificial levees were constructed.

Due to the real estate's historic and economic value, it was felt that additional protection should be provided as a safety factor—such as that provided by extending the levees to an elevation of 7.4 meters a.s.l. (24 ft). Because levees require a large land area (to provide a slope of 20 meters horizontal to 1 meter vertical), it was decided to construct a **floodwall** just landward of the existing levee's crest instead. The floodwall has gates that close during flooding events (see Figure 2 in Case Study 9.2). The 13-kilometer-long (8-mi) wall looks

July 4, 1988

(a)

July 18, 1993

(b)

FIGURE 9.22 Satellite photos of the confluence of the Mississippi, Illinois, and Missouri Rivers during (a) the drought of 1988 and (b) the flood of 1993. The top of the photos is north. The Illinois River flows south to the Mississippi, joining it at Grafton, Illinois. The Missouri River flows eastward across Missouri and empties into the Mississippi just north of St. Louis.

innocent enough, but it is built of 2-foot-thick reinforced concrete resting on a continuous wall of wooden piles driven 7.6 meters (25 ft) into the ground. The "sheet" piles in turn rest upon concrete piles that extend down another 15 meters (almost 50 ft). The sheet piles prevent underseepage of water during floods. New Orleans's solution to the flooding problem has served the city well for some time, but eventually the river will win the battle and reclaim its cypress swamps and floodplains, or it will shift its course to another distributary.

Insurance, Floodproofing, and Floodplain Management

The National Flood Insurance Act of 1968 provides federally subsidized insurance protection to property owners in flood-prone areas. The Federal Insurance Administration chose the 100-year-flood level as the **Regulatory Floodplain** on which to establish the limits of government management and set insurance rates. Maps have been developed that delineate Regulatory

Floodplains in literally thousands of communities in the United States. The subsidized flood insurance is not available on new homes within the regulatory floodway (Figure 9.24). Pre-1968 houses in the floodway were "grandfathered" (allowed to stay and to be insured under the government program), but they cannot be rebuilt at that site with flood-insurance money if they are more than 50 percent damaged by flooding.

Unfortunately, there are serious problems with the 1968 act. Because the insurance is government-subsidized, owners need not assume the full insurance risk for homes in flood-prone areas. This subsidy serves to encourage *more* building in flood-prone areas, because the land is cheap and the risk is subsidized. Also, as with much well-meaning federal legislation, enforcement is at the local-government level and thus varies considerably from place to place.

Three methods of flood-proofing are (1) raising structures above the 100-year-flood level by artificial filling with soil, (2) building walls and levees to resist floodwaters, and (3) using water-resistant building

Case Study | 9.3

The Colorado River—In Need of a Flood

The thought grew in my mind that the canyons of this region would be a Book of Revelations in the rock-leaved Bible of geology.

JOHN WESLEY POWELL, GEOLOGIST AND EXPLORER

The Colorado River begins as a mere rivulet on the west side of the drainage boundary called the **Continental Divide.** All water that falls on the continent west of the divide flows to the west, and only two large rivers in the contiguous 48 states flow to the Pacific Ocean—the Colorado and the Columbia. Although the Colorado's flow is puny compared to those of the Columbia and Mississippi Rivers, its saltiness and siltiness distinguish it among rivers in the United States. Its major tributaries, the Gunnison River of Colorado and the Green River of Utah and Wyoming, flow over sedimentary rock outcrops containing salt deposits. The river's steep gradient and predominantly sedimentary geology account for the large mass of sediment its transports, up to a half-million tons a day.

The Colorado's most celebrated explorer, John Wesley Powell, noted prophetically that "all waters of . . . arid lands will eventually be taken from their natural channels." Southern Utah's Lake Powell, impounded by the Glen Canyon Dam, bears his name. The dam is one of ten major dams that control the Colorado's flow and meter out its precious water to more than 20 million people.

That dams adversely alter downstream riverine environments is explained in this chapter, and the Glen Canyon Dam is no exception. Since its closing in 1963, the dam has restricted the river's natural cycle of springtime flooding and sedimentation in the Grand Canyon (◁▷ Figure 1). In 1972 river runners and scientists noticed that beaches were disappearing and that animal and plant life were changing on the river below the dam. The river has lost the sand bars and backwater sloughs that were home to so many riverine animals and plants. Because of this, the Glen Canyon Environmental Studies Program was established in 1982 and has been active since then in assessing the health of the river.

Before damming, the spring floods in the Grand Canyon carried 125,000 cubic feet of water per second (cfs) and 65 million tons of sediment. Today, only 6.5 million tons of sediment slip through the dam, and the major tributaries supply a meager 2.9 million tons. Flows through the dam today are controlled at 12,000 to 18,000 cfs. Cooperating agencies of the Department of the Interior decided to replicate the historic scouring action of spring flooding in 1996 by releasing an "artificial" flood into the Grand Canyon. The hydroelectric turbines in Glen Canyon Dam were turned off, releasing 177

◁▷ FIGURE 1 Annual floods on the Colorado River near the Grand Canyon before damming by the Glen Canyon Dam and after damming. Except for the unusual wet-year floods in 1983–1984, dam-controlled flows are about a fourth of natural flood flows.

billion gallons of water into the river. The flood's design spec-ified an initial flow of 8,000 cfs, a gradual increase to 45,000 cfs for about a week, and then a slow reduction to 8,000 cfs. The weeks of late March and early April were chosen be-cause they fall in the middle of the historic spring runoff period.

Postflood inspection showed that sandy beaches had increased by 30 percent (◀▶ Figure 2), Lake Powell had low-ered 1.1 meters (3.5 ft), and the river surface had raised 3.0 to 4.5 meters (10–15 ft) during the flood period. Sediment churned up from the river-channel bottom had been de-posited on banks, creating beaches and quiet backwater habitats—the desired end result. Not then known was how long the beaches would last and whether they would be worth the cost, about $4 million. Time will tell.

◀▶ FIGURE 2 Cross sections of the Colorado River channel at mile 68 above Tanner Rapids before the 1996 flood, labeled March 28, and after the flood, labeled April 1. At least 3 meters of sand were deposited on the bank.

(a)

(b)

(c)

◀▶ **FIGURE 9.23** Artificial levees protect land from the river and the sea and sometimes need protection themselves. (a) A levee with a road on the top protects an industrial area on the floodplain northeast of New Orleans. (b) A levee protects homes from the waters of Lake Pontchartrain. (c) A damaged levee on the Mississippi River was protected by snowfencing and sandbags.

tense storms. Nonetheless, they are extensively developed in resort areas of the Southwest such as Palm Springs, California, although they are subject to strict regulation. Fans present a special problem, because the chance of flooding near the fan's apex does not determine the probability or depth of flooding below that point. Flood intensities decrease downfan, from life-threatening channelized debris flows near the apex, to a zone where streams rapidly change channels in a "braided" drainage pattern, to the area of little hazard at the toe of the fan (◀▶ Figure 9.25). The building pads of homes built on alluvial fans near Palm Springs are required to be raised artificially several feet above the natural ground so as to locate the structure above possible floodwater levels. When built in this manner, they qualify for federally subsidized flood insurance.

Urban development presents special problems, because it alters the hydrology of a watershed. A conceptual hydrograph can be used to illustrate the fact that because of urbanization, floods peak sooner and their total runoff and peak flow are greater (◀▶ Figure 9.26). To geologists and hydrologists, urbanization represents the replacement of permeable soil with impermeable pavement, which produces the observed greater flooding and damage. It has been found that when 50 percent of a drainage basin is urbanized, the frequency of overbank flows (floods) is four times what it was before urbanization.

Development has literally poured onto floodplains in the United States, and annual flood losses have increased accordingly. Annual flood damage (in constant dollars) almost tripled between 1950 and the mid 1990s to an average of $3 billion per year. As a direct result of the great flood of 1993 in the Midwest, the Senate

materials. Good floodplain management limits the uses of floodplains in chronically flood-prone areas. As noted earlier, alluvial fans in southwestern deserts are subject to flash floods and debris flows from in-

Consider This...? You wish to build a house on an alluvial fan in the desert. In order to buy federally subsidized flood insurance you would have to raise the building pad several feet near the apex of the fan, but very little or not at all near the toe. What is it about the hydrology of alluvial fans that justifies these differing requirements? ■

Regulatory floodway—Kept open to carry flood water: no building or fill permitted.

Regulatory floodway fringe—Use permitted if protected by fill, flood-proofed, or otherwise protected.

- - - Regulatory flood limit—based on technical study; outer limit of the floodway fringe.

— · — Standard project flood (SPF) limit—Area subject to possible flooding by large floods.

◀▶ **FIGURE 9.24** Fluvial flood-hazard areas as defined by the U.S. government. No building is allowed in the regulatory floodway, and only protected or raised structures are allowed to be built in the fringe area.

Committee on Environment and Public Works stopped legislation that would have authorized dozens of new flood-control projects. The new belief is that flood policies must be changed, because traditional structural solutions—levees and dams—contribute to the problem. The new emphasis should include examination of upstream areas, where land use determines how water runs off and where some towns should be moved and the land reverted to floodplain.

Floods are natural and inevitable, as demonstrated in this chapter. A congressional report perhaps summed it up best in noting that "Floods are an act of God. Flood damages result from acts of men."

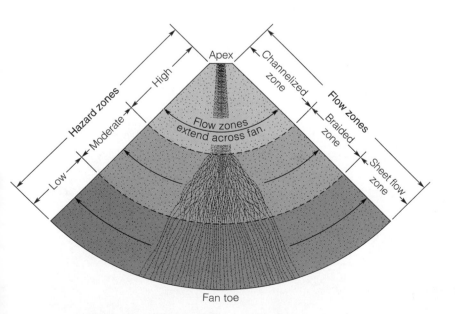

◀▶ **FIGURE 9.25** Flood hazard zones on an alluvial fan. Hazards exist across the full width of the fan.

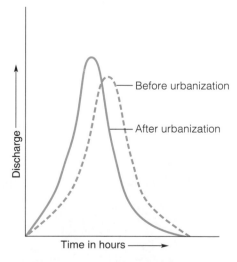

◀▶ **FIGURE 9.26** Hypothetical hydrographs of flooding patterns before and after urbanization. After urbanization the maximum flood stage occurs sooner and is higher.

Gallery

Going ... Going ... Gone!

Floods resulting from dam failures are a great threat to human life (see Table 9.1). The earth-filled Teton Dam in Idaho was built for water storage and to provide flood protection in the Snake River watershed. The dam collapsed on 5 June 1976, shortly after the reservoir reached its full capacity. The verdict was that poor-quality foundation rock led to a leak in the right abutment (on the right as one looks upstream toward the dam) and that the reservoir water pressure "piped," or tunneled, through the compacted fill to the face of the dam (◇ Figure 9.27a). Eventually the fill of the

◇ **FIGURE 9.27** The Teton Dam on the Teton River in Idaho, June 1976. The 95-meter (310-ft) high earth-filled dam sprung a leak on the right abutment and failed in a matter of minutes. (a) About 10:30 A.M. on 5 June 1976. The dam is going ..., (b) going ..., (c) gone!

(a)

(b)

(c)

(a)

(b)

(c)

◁▶ FIGURE 9.28 Baldwin Hills Dam, Los Angeles, December 1963. (a) A hole opened on the reservoir side of the dam about 15 minutes before failure. At the same time, water was being released through outlet pipes in order to reduce the water volume and level in the reservoir. (b) A residential street below the dam minutes after failure. Houses on the near side (lower right of photo) of the street have been totally swept away. At least 20 houses were lost. Not even their foundations remained. (c) A week later, the view through the breach toward the low Santa Monica Mountains and the high San Gabriel Mountains in the far background.

right abutment collapsed, releasing the stored water and flooding the small communities of Rexburg and St. Anthony, Idaho, with the loss of 14 lives (◁▶ Figure 9.27b and c).

In December 1963 the Baldwin Hills dam, a water-storage reservoir in Los Angeles, collapsed and drained into an urban environment. Again the verdict was water pressure

piping along an ancient fault beneath the earth-filled dam and reservoir. The fault is a strand of the active Newport–Inglewood fault system that caused the 1932 earthquake in Los Angeles (see Chapter 4). In less than eight hours from the time the leak was discovered, the dam breached and its water drained onto the streets below. Only by Herculean

◇ **FIGURE 9.29** (a) The St. Francis Dam in San Francisquito Canyon, southern California, early March 1928. The right abutment is in Pelona schist; the left abutment (center of photo), in the Sespe Formation, which is mainly sandstone. (b) All that remained of the dam on 13 March 1928.

(a)

(b)

efforts did the police and fire departments manage to evacuate the residential area below the dam (◇ Figure 9.28). Many people thought it was a joke when they were told, "The *dam* is collapsing! Get *out*!" Some residents had not even known there was a reservoir at the top of the hill.

To geologists the most famous U.S. dam failure was that of the St. Francis Dam in southern California in 1928. That failure took 450 lives. The dam was part of an aqueduct sys-

tem bringing water from the east slope of the Sierra Nevada to semiarid Los Angeles and its growing suburbs. The left abutment of the concrete dam was founded in sandstone, and the right abutment in mica schist (◇ Figure 9.29a). The two rock formations were in contact along a 2-meter-wide shattered fault zone at the left abutment. The dam was completed in 1926, and more than two years was required to fill the reservoir with 38,000 acre-feet (12.5 billion gallons) of water. On 12 March 1928, just one week after the reservoir had filled, the dam failed. Traditional theory has it that the dam failed along the fault zone in the left abutment, as the rocks there contained gypsum. Water percolating through the fault zone dissolved the gypsum, and eventually, full reservoir pressure blew out the weakened sandstone. Newer analysis, however, suggests that the right abutment was placed into an unrecognized landslide in the schist, and that the slide mass collapsed when the reservoir filled, leading to failure of the right abutment and the dam (◀▶ Figure 9.29b). Whatever the cause, the man who built the dam, William Mulholland, took the failure personally and died a short time later. He is said to be the four hundred fifty-first victim of the flood.

A dam just isn't supposed to mess up a beautiful trout stream, as the one in ◀▶ Figure 9.30 used to. Whether the hole was due to poor construction or the wrath of an angler or cattle rancher is not known.

◀▶ FIGURE 9.30 Pristine trout stream in Colorado and a useless dam that probably should never have been built.

FLOOD PLANNING

The following flood and flood-control information is useful in the planning process:

- Floods are the most consistently destructive natural hazard in the United States.
- Flooding cannot be controlled completely, but its damaging effects can be mitigated by restricting floodplain development or by government acquisition of floodplain lands.
- Flood damage costs have been increasing at a rate of 5 percent annually over the past several decades in real dollars.
- Urbanization increases flood peaks on small streams by two to six times.
- Use of the 100-year-flood standard for regulation is effective and should be continued, and the 500-year flood should prevail for such critical facilities as hospitals, water-supply facilities, and power plants.

Don Barnett, the mayor of Rapid City, South Dakota, at the time of its disastrous 1972 flood, said, "Elected public officials must give the same attention and priority to their drainage problems as they give to their police and fire problems. In the history of Rapid City, perhaps 35 people have died in fires and another 35 have been killed during the commission of crimes, but in just two hours, 238 died in a flood."

Summary

RIVER SYSTEMS

DRAINAGE BASIN
The fundamental geographic unit or tract of land that contributes water to a stream or stream system. Drainage basins are separated by divides.

DISCHARGE
The amount of water flowing in a stream channel depends upon the amount of runoff from the land, which in turn depends upon rainfall or snowmelt, the degree of urbanization, and vegetation.

EROSION
The erosive power of a stream is a function of its velocity; the greater the velocity, the greater the erosion. Stream velocity is determined by discharge, channel shape, and gradient.

BASE LEVEL
The lowest level to which a stream or stream system can erode. This is usually sea level, but there are also temporary base levels such as lakes, dams, and waterfalls.

GRADED STREAM
A stream that has reached a balance of erosion, transporting capacity, and the amount of material supplied to it. This condition is represented by a concave-upward profile of equilibrium.

STREAM FEATURES

ALLUVIUM
Sediment deposited by a stream, either in or outside the channel.

ALLUVIAL FAN
A buildup of alluvial sediment at the foot of a mountain stream in an arid or semiarid region.

DELTA
Forms where a sediment-laden stream flows into standing water. An example is the delta of the Nile River.

FLOODPLAIN
A low area adjacent to a stream that is subject to periodic flooding and sedimentation; the area covered by water during flood stage.

MEANDERS, OXBOW LAKES, AND CUTOFFS
Flowing water will assume a series of S-shaped curves known as meanders. The river may cut through the neck of a tight meander loop and form an oxbow lake.

FLOODING AND FLOOD FREQUENCY

FLOODS
Upland floods come on suddenly and move with great energy through narrow valleys. Upland floods are often "flash" floods; the water rises and falls in a matter of a few hours. Riverine floods, on the other hand, inundate broad adjacent floodplains and may take many days, weeks, or even months to complete the flood cycle.

HYDROGRAPH
A graph that plots measured water level (stage) or discharge over a period of time.

RECURRENCE INTERVAL
The length of time (T) between flood events of a given magnitude. Mathematically, $T = (N + 1) \div M$, where N is the number of years of record and M is the rank of the flood magnitude in comparison with other floods in the record. This enables engineers and planners to anticipate how often floods may occur and to what elevation the water may rise.

FLOOD PROBABILITY
The chance that a flood of a particular magnitude will occur within a given year based on historical flood data for the particular location. Using the calculation for T established for the recurrence interval, probability is calculated as $1/T$. Thus the 100-year flood has a 1 percent chance of occurring in any one year. It is the 1-in-100 *chance* flood.

MITIGATION

STRUCTURES
Dams, retaining basins, artificial levees, and raising structures on artificial fill are common means of flood protection. They also present problems.

FLOOD INSURANCE
The National Flood Insurance Act of 1968 provides insurance in flood-prone areas, provided certain building regulations and restrictions are met. There are many problems with this tax-supported program.

URBAN DEVELOPMENT

HYDROLOGY
Urbanization causes floods to peak sooner during a storm, results in greater peak runoff and total runoff, and increases the probability of flooding. Floods, including coastal flooding during hurricanes, are the greatest natural hazards facing humankind.

◆ Key Terms

alluvial fan	discharge	graded stream	point bar
alluvium	drainage basin	gradient	recurrence interval
artificial levee	drainage divide	hydrograph	riparian
base level (stream)	flood frequency	hydrology	Regulatory Floodplain
Continental Divide	floodplain	meander	
cutoff	floodwall	natural levee	
delta	geologic agents	oxbow lake	

◆ Study Questions

1. Distinguish between weathering and erosion.
2. How can we prevent floods or at least minimize the property damage caused by flooding?
3. How is the velocity of a stream related to its erosive power, discharge, and cross-sectional area?
4. What has been the impact of urbanization on flood frequency, peak discharge, and sediment production?
5. According to the local newspaper, your town on the Ohio River has just experienced a 50-year flood. What is the probability of a repeat of this event the following year? What is the weakness of such statistical forecasts?
6. How does the hydrograph record of a flash flood differ from that of a lowland riverine flood? Where do we find flash flooding, and what precautions should be observed in such areas?
7. What has the federal government done to alleviate people's suffering due to flooding? Do you believe this is a good thing? Does it promote more development in flood-prone areas? If so, does this development itself have possible negative impacts? Support your answers with examples and rationales.
8. Consider a stream whose headwaters are 1,500 meters (4,900 ft) above sea level and that flows 300 kilometers (180 mi) to the ocean. What is the average gradient of the stream in m/km and in ft/mi?
9. What benefits do river dams impart to humans? What are some negative impacts of dams upon human beings and the environment?
10. What is being done today to reverse some of the negative impacts of dams on the environment? Cite examples.

◆ For Further Information

BOOKS AND PERIODICALS

Baker, Victor. 1997. *Hydrological understanding and societal action.* Forest hydrology symposium presentations, 10 April 1997, Department of Geography, University of British Columbia, Vancouver, B.C.

Carrier, Jim. 1991. The Colorado: A river drained dry. *National geographic,* June.

Collier, Michael, and others. 1996. *Dams and rivers: A primer on the downstream effects of dams.* U.S. Geological Survey circular 1126.

Devine, Robert S. 1995. The trouble with dams. *Atlantic monthly,* August: 64–74.

Dinacola, Karen. 1996. The "100-year flood." U.S. Geological survey fact sheet FS-229-96.

EOS. 1996. *Controlled flood of Colorado River creates stream of data.* Transactions of the American Geophysical Union 77, no. 24 (11 June).

Hays, W. W., ed. 1981. *Facing geologic and hydrologic hazards: Earth science considerations.* U.S. Geological Survey professional paper 1240-B. Washington, D.C.

Holmes, Robert R., Jr., and Amit Kapadia. 1997. Floods in northern Illinois, July 1996. U.S. Geological Survey fact sheet FS-097-97.

Leopold, Luna B. 1968. *Hydrology for urban land planning: A guidebook on the hydrologic effects of urban land use.* U.S. Geological Survey circular 554.

Leopold, L.; M. G. Wolman; and John P. Miller. 1964. *Fluvial processes in geomorphology.* San Francisco: W. H. Freeman and Co.

Lucchitta, Ivo, and Luna B. Leopold. 1999. *Floods and sandbars in the Grand Canyon. GSA today* 9, no. 4 (April).

McPhee, John. 1989. *The control of nature.* New York: Farrar, Straus & Giroux.

National geographic. 1993. Water: the power, promise and turmoil of North America's fresh water. Special edition, October.

Rogers, J. David. 1992. Reassessment of the St. Francis dam failure. In Pipkin, B. W., and Richard Proctor, eds., *Engineering geology practice in southern California.* Belmont Calif.: Star Publishing Co.

Saarinen, T. F.; V. R. Baker; R. Durrenberger; and T. Maddock, Jr. 1984. *The Tucson, Arizona, flood of October 1983.* Washington D.C.: National Research Council, National Academy Press.

Teller, R. W., and M. J. Burr. 1998. Floods in north-central and eastern South Dakota, spring 1997. U.S. Geological Survey fact sheet FS-021-98.

U.S. Geological Survey. 1993. Mississippi River flood of 1993. *U.S. Geological Survey yearbook:* 37–40.

_____. 1999. *The quality of our nation's water.* Circular 1225.

Waananen, A. O., and others. 1977. *Flood-prone areas and land-use planning: Selected examples from the San Francisco Bay re-gion, California.* U.S. Geological Survey professional paper 942.

◆ World Wide Web Activities

USGS Long Baseline Real-Time Stream Flow Data for the United States
http://water.usgs.gov/dwc/
This Web page displays a map of the United States with real-time stream flow for specific gauging stations with 30 or more years of documented measurements. Colored dots tell you current dry-to-wet conditions on a 7-point scale (see map legend).
1. Which regions are at an all-time stream flow high? What can you conclude regarding the current potential for regional flooding?
2. Louisiana shows no gauging stations. What aspect of rivers is likely to be of more concern in this region than water availability?
3. Click on your own state. Describe the location and major river names for areas of relatively high stream flow.

Landsat Images Help Save Lives in the Great U.S. Flood of 1993
http://observe.ivv.nasa.gov/nasa/exhibits/flood/flood_1.html
"Just Doing My Job": This series of four linked pages with interactive graphics shows how Landsat images of the upper Mississippi basin south of St. Louis were used to swiftly evacuate people at risk from the cresting River Des Peres during the great flood of 1993.
1. Why is St. Louis, Missouri at significant risk from upper Mississippi Basin flooding?
2. On the false-color Landsat images, what do you think the pink color represents?
3. On page 4, what digital information was already in the St. Louis Community Development Agency computers, and what "real time" data did they add from the Army Corps of Engineers to guide rescue operations?

Benefits of Dam Removal and Danger from Aging Dams
http://www.glencanyon.org/damchronos.htm
"A Chronology of Dam Removal" mostly addresses dams in the United States. Included are both intentional dam removals and dam failures.
1. How many lives have been lost from the dam failures detailed here?
2. What specific advantages, both speculative and realized, occur as the result of dam removal?
3. Under 1998, link to "aging dams in the United States pose a serious risk." What are the specific problems associated with older existing dams in the United States?

Efforts to Maintain the Lower Mississippi in Its Current Channel
http://www.ce.utexas.edu/stu/mcbraymc/ce385proposal.html
This site explains in detail the history and current relationship of the lower Mississippi River to the adjoining Atchafalaya River, which is threatening to pirate Mississippi flow into its drainage outlet to the Gulf of Mexico.
1. In section 1.1: Why is stream piracy of the Mississippi flow into the Atchafalaya drainage favored by nature? What nineteenth-century human effort increased this potential piracy?
2. In section 2.4: What current natural and manmade structures are holding the Atchafalaya in place?
3. In section 3.0: What would be some of the major consequences to the lower Mississippi Basin and human activity if/when the Atchafalaya manages to capture much of the Mississippi flow?

 # Earth Systems Today CD Correlations

CHAPTER CONCEPT: Coastal Flooding Is the Most Deadly of Natural Disasters
This segment of the CD deals with changes in global weather patterns, both seasonally and over longer time periods.

SECTION MODULE: Weather and Climate
Examine the archived data for pressure systems and fronts for the United States.

QUESTION: *The Mississippi delta region, and other parts of the Gulf Coast of the United States, are subject to occasional severe storms. Describe the general pattern of pressure systems and fronts that brings these storms to the delta.*

CHAPTER CONCEPT: River Flooding and Dam Failures Pose Serious Threats to Human Populations Located within Floodplains and River Channels

This segment of the CD links to current events around the globe that are endangering humans and other segments of the biosphere.

SECTION MODULE: Earth Today: Earth Alert
Activate this CD segment, and scan the globe for any river-related disasters in progress.

QUESTIONS: *Do you find any floodplain inundation, or coastal flooding? Where is it? Is this a seasonal pattern? Is the event particularly severe historically? Is it linked to any other phenomenon such as a dam failure, El Niño climatic conditions, and so on?*

chapter | ten

The Coastal Ocean Environment

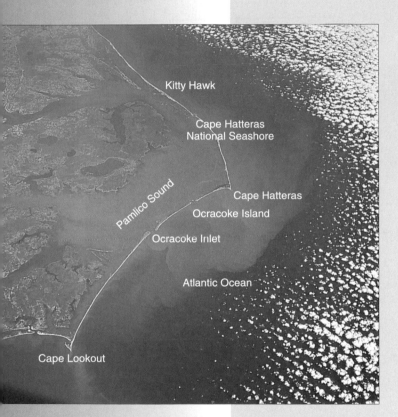

North Carolina's Outer Banks are a chain of islands composed of sand that form a barrier between lagoons and the ocean. Cape Hatteras, near the center of the photo, juts the farthest out into the Atlantic.

Q. How do you move a 200-foot, 420-ton, brick lighthouse?

A. Very carefully!

North Carolina's Outer Banks are a line of barrier islands stretching along the coast for about 200 kilometers (125 mi). With names like *Bodie, Hatteras, Ocracoke, Portsmouth, Smith, Roanoke,* and *Core Banks,* they form a barrier between lagoons and the Atlantic Ocean. The banks are known for their *capes* (seaward projections of land), especially Cape Hatteras, Cape Fear on Smith Island, and Cape Lookout on Core Banks, which have been the scene of more than 600 shipwrecks. Cape Hatteras, the easternmost point of North Carolina, has long been known as the "graveyard of the Atlantic" because of the many ship groundings on nearby shoals. The banks are battered by storm waves off and on all year long. Because of this hazard, particularly for sailing vessels, the islands were hideouts for eighteenth-century pirates, including the infamous Blackbeard.

Cape Hatteras Lighthouse, with its black and white candy-cane paint job, is the tallest brick lighthouse in the world. Built in 1870 with state-of-the-art lenses, the 63-meter (208-ft) structure is so distinctive that it is on the National Register of Historic Places. Besides saving lives, the lighthouse has hosted millions of visitors. It is one of the most enduring landmarks in North Carolina and the nation. This part of the Outer Banks, designated the Cape Hatteras National Seashore by Congress, is a protected environment.

Fixed structures built on barrier islands are inevitably at the mercy of beach erosion and, ultimately, direct wave action. This is because all U.S. barrier islands

Ready

Rolling

have been "migrating" shoreward due to the rise in sea level over the past few thousand years. When it was built in 1870 the lighthouse was 463 meters (1,500 ft) from the ocean. By 1919 the sea had come to within 100 meters (310 ft) of its base, and by 1935, to within 30 meters (100 ft). When it comes to a battle of man against the sea, the sea always wins. Illustrating the futility of shore-protection efforts during times of rising sea level, the National Park Service and the Corps of Engineers spent millions of dollars attempting to save the beloved lighthouse—constructing groins to trap sand, replenishing beaches and artificially depositing a barrier-sand-dune system in front of the structure, installing offshore wave-energy dissipaters, and finally, constructing hard structures such as seawalls that reflect wave energy. By the 1980s the sea was lapping at the base of the lighthouse, and direct measures were required. The National Park Service decided to move the lighthouse, and in 1990 the brick structure was reinforced to withstand the rigors of the proposed move.

Moving structures of great weight has become easier with the development of modern hydraulic systems. Between 1994 and 1999, three lighthouses along the New England coast had been moved to remove the threat of collapse into the sea. The massive Cape Hatteras Lighthouse was moved similarly to those lighthouses by using horizontally mounted "pusher" jacks that pushed it along a track

system in five-foot increments. The move took just 22 days, 19 of which were "pushing" days. A remarkable engineering feat, indeed.

The powerful light that had been out for nine months was relit on Saturday, 13 November 1999 at ceremonies on the original lighthouse site in Buxton, North Carolina. ■

Home safe

Coastal states in the United States experienced enormous population growth in the late twentieth century. Apparently, a lot of people want to live near a beach. Fifty percent of the U.S. population can now drive to a coastline in less than an hour. In spite of advances in technology and scientific understandings of coastal dynamics, Matthew's warning is largely unheeded. Nearshore environments are constantly being modified by natural and human intervention, usually with negative impacts. For example, Hurricanes Dennis and Floyd dumped huge amounts of rainfall on North Carolina in 1999, displacing thousands of residents; each spring and summer nitrates from agricultural fertilizers wash down the Mississippi River, creating algal blooms and a 7,000-square-mile "dead zone" in the Gulf of Mexico that threatens commercial fisheries; wetlands are being filled or made into marinas; and shoreline modifications continue to silt in harbors and erode beaches.

Wind-driven waves that arise in the open ocean eventually strike shorelines, generating dynamic processes that constantly alter beaches. Beach erosion is a serious, costly problem in most coastal areas. Armed with knowledge of how waves erode beaches and of the fate of eroded materials, we may be able to mitigate many of the human-induced coastal problems.

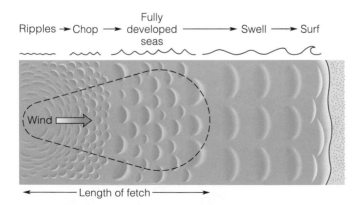

Ripples → Chop → Fully developed seas → Swell → Surf

Wind ⇒

← Length of fetch →

◁▷ **FIGURE 10.1** The evolution of wind-generated sea waves from a rippled water surface to fully developed seas (storm waves). Long waves with very regular spacing and height that outrun the storm area are called *swell*.

Wind Waves

WAVES IN DEEP WATER AND AT THE SHORE

When wind blows over smooth water, surface ripples are created that enlarge with time to form local "chop" (short, steep waves) and, eventually, wind waves. The factors that determine the size of waves at sea and ultimately the size of the surf that strikes the shoreline are

- the wind velocity,
- the length of time the wind blows over the water, and
- the **fetch,** the distance along open water over which the wind blows (◁▷ Figure 10.1).

For a given fetch, wind velocity, and wind duration, wave dimensions will grow until they reach the maximum size for the existing storm conditions. This condition, known as a *fully developed sea,* represents the maximum energy that the waves can absorb for wind of a given velocity. For example, theoretical wave heights of 32 meters (100 ft) are possible for a 50-knot (93 km/h, 58 mph) wind blowing for 3 days over a fetch of 2,700 kilometers (1,600 mi). Wind waves will move out of the area of strong atmospheric disturbance as regularly spaced, long waves known as *swell* and eventually become huge breakers on a distant shoreline.

Important dimensions of wind waves are their length, height, and *period*—the length of time between the passage of equivalent points on the wave. These dimensions are measures of the amount of *energy* the waves contain; that is, their ability to do work (◁▷ Figure 10.2). Although waves with long wavelengths and long periods are the most powerful, both long and short waves are capable of eroding beaches.

The actual water motion in a wind-generated wave describes a circular path whose diameter at the sea surface is equal to the wave height. This circular motion diminishes with depth and becomes essentially zero at a depth equal to half the wavelength ($L/2$), the depth referred to as **wave base.** Wave base is the maximum depth to which waves can disturb or erode the sea floor. (This explains why submersibles at depth are not tossed about during storms as surface ships are.) Although the waveform moves across the water surface with considerable velocity, there is no mass forward movement of water. Such waves are called **waves of oscillation.** Floating objects simply bob up and down with a slight back-and-forth motion as wave crests and troughs pass.

As waves approach a shoreline and move into shallow depths, the sea floor interferes with the oscil-

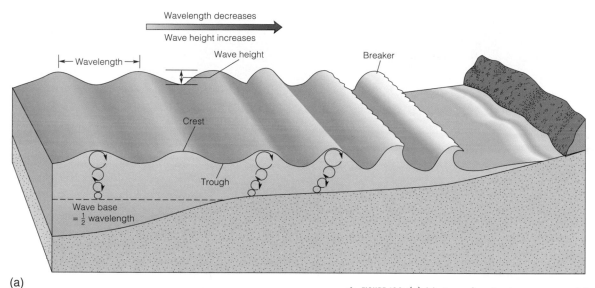

Wavelength decreases →

Wave height increases

Wavelength

Wave height

Breaker

Crest

Trough

Wave base = ½ wavelength

(a)

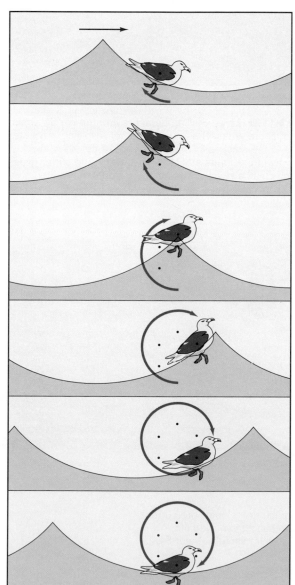

(b)

◆▶ FIGURE 10.2 (a) Motion of water in a wave and the terminology of wave geometry. Note that the orbital water motion decreases with depth until it is essentially nil at a depth of one-half the wave length ($L/2$), a depth called *wave base*. The height on land reached by the broken wave is called its *runup*. (b) A floating gull's movement is circular (see dots) as wave passes. This demonstrates that water particles do not advance forward with the wave form.

lating water particles. This friction causes wave velocity to decrease. This in turn causes the wave to steepen; that is, the wavelength shortens and the wave height increases, a phenomenon easily seen by surf watchers. Eventually the wave becomes so steep that the crest outruns the base of the wave, and the crest jets forward as a "breaker." Breakers are **waves of translation,** because the water in them physically moves landward in the surf zone. Where the immediate offshore area is steep, we find the classic tube-shaped *plunging breakers* typical of Sunset Beach and Waimea Bay on the north shore of Oahu (◆▶ Figure 10.3). This is because the wave peaks rapidly on a steep sea floor. On a flat nearshore slope, on the other hand, waves lose energy by friction, so they build up slowly, and the crest simply spills down the face of the wave. This latter type are called *spilling breakers*. They are typical of wide, flat nearshore bottoms such as that at Waikiki Beach on Oahu (◆▶ Figure 10.4). Regardless of the type of wave, *swash,* the water that rushes forward onto the foreshore slope of the beach, returns seaward as *backwash* to become part of the next breaker. A common misconception is the existence of a bottom current called *undertow* that pulls swimmers out to sea. The strong pull seaward one feels in the surf zone is simply backwash water returning to be recycled in the next breaker.

(a)

(b)

◆▶ **FIGURE 10.3** (a) Plunging (tubular) breakers form on steep offshore slopes. The front face of the wave steepens until the top of the wave "jets" forward and down onto the wave's front slope. (b) A tubular plunging wave; Waimea Bay, Oahu, Hawaii.

◆▶ **FIGURE 10.4** (a) Spilling breakers form on gently sloping offshore areas. The forward face of a spilling wave steepens, and the crest spills down the front of the wave. (b) A spilling-wave front forms a "wall" across which the surfer glides, much as a skier moves across a slope, rather than straight downhill.

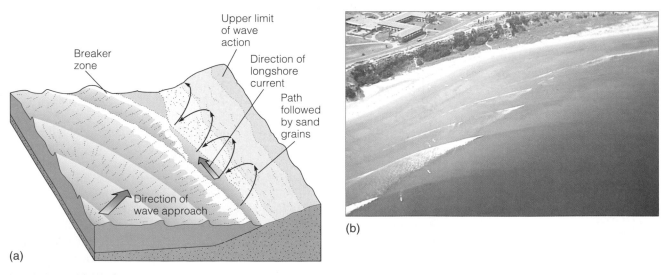

(a)

(b)

◆▶ **FIGURE 10.5** (a) Refraction, or bending, of wind waves and the generation of a longshore current parallel to the beach. The longshore current transports beach sand, resulting in a dynamic process called *beach drifting.* (b) Wave refraction along a relatively straight shoreline near Santa Barbara, California.

WAVE REFRACTION

Where waves approach a shoreline at an angle, they are subject to **wave refraction,** or bending. Waves may be seen to slow first at their shallow-water ends while their deeper-water ends move shoreward at a higher velocity. Thus a fixed point on a wave crest follows a curved path to the shoreline (◆ Figure 10.5).

Because waves seldom approach parallel to the shore, some water is discharged parallel to the beach as a weak current within the surf zone called a **longshore current.** The greater the angle at which the wave approaches the shoreline, the greater the volume of water that is discharged into the longshore-current stream. This current is capable of moving swimmers as well as sand grains down the beach. Sand becomes suspended in the turbulent surf and swash zone and moves in a zigzag pattern down the beach's foreshore slope (see Figure 10.5). Beaches are dynamic; the sand grains present on the foreshore today will not be the same ones we see there tomorrow. Beach-sand transport parallel to the shore in the surf zone is known as **littoral drift,** and its rate is usually expressed in thousands of cubic meters or yards of sand per year.

Another consequence of wave refraction is the straightening of irregular shorelines. This occurs because the concentration of wave energy at points causes shoreline erosion, and the dissipation of wave energy in embayments allows sand deposition (◆ Figure 10.6).

Longshore currents are likened to rivers on land, with the beach and the surf zone constituting the two banks. Just as damming a river causes sedimentation to occur upstream and erosion to occur downstream, so does building structures perpendicular to the beach. Such a structure slows or deflects longshore currents, causing sand accretion upstructure and erosion below

it (◆ Figure 10.7). **Groins** (French *groyne,* "snout") are constructed of sandbags, wood, stone, or concrete, ostensibly to trap sand and create a wider beach. For groins of equal length and design, the severity of the erosion-deposition pattern is a function of the strength of the waves and the angle at which they strike the shoreline. Most groins are short and are intended to preserve or widen a beach in front of a private home or resort. Unfortunately, because this causes neighbors' beaches below the groin to narrow, many of those neighbors build groins to preserve and widen *their* beaches, and so on. Like artificial levees on a river, groins tend to generate more groins, and we generally see fields of them along a given stretch of beach instead of just one (◆ Figure 10.8).

Just as rivers overflow their banks, longshore currents may overflow through the surf zone, particularly during periods of big surf. Overflow (really *out*flow to the open ocean) occurs because the water level within the surf zone is higher than the water level outside the breakers, making for a hydraulic imbalance. The higher water inshore flows seaward to lower water offshore as a **rip current** through a narrow gap, or *neck,* in the breakers (◆ Figure 10.9). Because the location of rip currents, erroneously called rip *tides,* is controlled by bottom topography, they usually occur at the same places along a beach. Rip currents can become a problem to swimmers under big-surf conditions, when velocities in the neck may reach 4–5 knots. Awareness can save your life, so be alert to a gap in the breakers, white water beyond the surf zone, and objects floating seaward—all of which are evidence of a strong rip current flowing out to sea (◆ Figure 10.10). Should you find yourself in a rip current, swim parallel to the beach until you are out of the narrow neck region of high velocity.

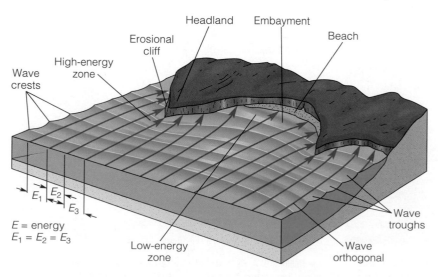

◆ **FIGURE 10.6** Wave refraction leads to concentration of energy on points and dissipation in embayments. The arrowed lines, called *orthogonals,* are constructed perpendicular to the wave crest and show the wave path. E_1, E_2, and so on, represent equal parcels of wave energy and, therefore, equal ability to erode. Note that energy is focused toward the points and spread out in the bays.

(a)

(b)

◁▷ **FIGURE 10.7** (a) Accretion and erosion of beach sand at jetties constructed to improve an inlet to a navigable lagoon. (b) This exact condition is seen at Boca Raton, Florida.

Consider This... ? After a few hours of frolicking in the surf, you find you have drifted along the shore to a spot where the waves aren't breaking. Suddenly, you find yourself being drawn seaward by an invisible hand, no matter how vigorously you swim toward the beach. What is happening? What should you do, and why? ■

Shorelines

Where the sea and land meet we find *shorelines,* some of which are shaped primarily by erosional geologic processes and others by depositional geological processes. Shorelines that are exposed to the open

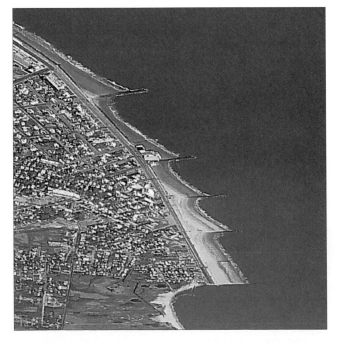

◁▷ **FIGURE 10.8** Groin field at Cape May, New Jersey. Note the bread-knife, or sawtooth, appearance that has resulted from sand accretion and erosion above and below the groins.

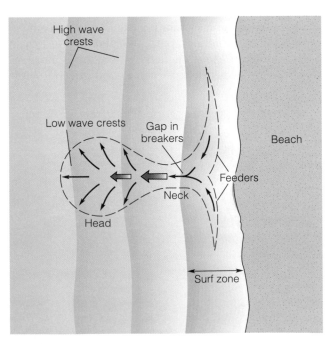

◁▷ **FIGURE 10.9** A rip current is characterized by relatively weak *feeder currents,* a narrow *neck* of strong currents, and a *head* outside the surf zone, where the rip dissipates.

(a)

(b)

◆ **FIGURE 10.10** (a) Small rip currents. Note the gap in the breakers and the white water seaward of the surf zone. (b) Signs warn swimmers about dangerous currents in the surf at some beaches.

ocean and high-energy wave action are likely to have *erosional* features such as sea cliffs or broad, wave-cut platforms. Shorelines that are protected from strong wave action, by offshore islands or by the way they trend relative to the direction of dominant wave attack, are characterized by *depositional* features such as sand beaches.

Technically, the shoreline is the distinct boundary between land and sea that changes with the tides, whereas the *coast* is the area that extends from the shoreline to the landward limit of features related to marine processes. Thus, a coast may extend inland some distance, encompassing estuaries and bays.

BEACHES

In addition to their usefulness as recreational areas, beaches also serve to protect the land from the erosive power of the sea. They are composed of whatever sediment rivers and waves deliver to them (◆ Figure 10.11). For example, in some places in Hawaii, black sand forms from weathered basalt and also from hot lava that disintegrates when it flows into the sea. In contrast, reef-fringed oceanic islands are characterized by beautiful white coral sand and shell beaches. Main-

land beach materials are largely quartz and feldspar grains derived from the breakdown of granitic rocks. *Shingle* beaches are composed of gravels; they typically occur where the beach is exposed to high-energy surf. Amusing or not, tin cans from a local dump once formed a beach at Fort Bragg, California.

Types of Beaches

Four types of beaches can be recognized, according to the geomorphic setting where they are found:

- coastal-plain, or mainland, beaches,
- pocket beaches,
- barrier islands, or barrier beaches, and
- sand spit beaches (◆ Figure 10.12).

Each of these has its own peculiar stability problems, but all are influenced by seasonal differences in wave action. As a rule, winter beaches are narrow or nonexistent, because high winter-storm waves erode sand from the beach berm and deposit it in offshore sandbars parallel to the beach (◆ Figure 10.13). You may have experienced this while wading in waist-deep water that abruptly shallowed as you walked onto a sandbar. With the onset of long, low summer waves, the sand in the offshore bar gradually moves back onto the beach, and the beach berm widens. Thus, the natural annual beach cycle is from narrower and gravelly in the winter, to wider and sandy in the summer. On the East Coast, beach narrowing is related to storm events that may occur at any time of the year.

Barrier islands form the most extensive and tenuous beach system in the United States (◆ Figure

Consider This...? Your favorite beach disappears during a large winter storm. Will you have to find a new beach for next summer? If not, will the summer beach look any different and, if so, how will it be different? Explain your answer. ■

(a)

(b)

(c)

(d)

◇ **FIGURE 10.11** Beach materials. (a) Black sand beach; American Samoa. (b) Coral sand; Caribbean. (c) Mineral-grain beach sand; Santa Barbara, California. (d) Cobble, or shingle, beach; Oceanside, California.

10.14). They extend southward from Long Island and Coney Island, New York, to Atlantic City, continuing as the Outer Banks of North Carolina (see the chapter opener) and then down the coast of Florida to Daytona Beach, Cape Canaveral, Palm Beach, and Miami Beach. The barrier system begins again on the west coast of Florida and continues to the Padre Islands in Texas, interrupted only by the Mississippi River delta. These barrier beaches are transient features; formed by wave action, they are reoriented by wave action and frequently overtopped by storm surges. They are separated from the mainland by shallow lagoons and marshland over which they migrate as large storms overwash the islands, transporting sand from their seaward sides to their landward sides.

Hog Island, Virginia, is a barrier island that was a popular hunting and fishing spot of the rich and famous in the late 1800s (◇ Figure 10.15). Its town of Broadwater had 50 houses, a school, a cemetery, and a lighthouse. After a 1933 hurricane inundated the island, killing its pine forest, Broadwater was abandoned. Today Broadwater is under water a half-kilometer offshore. Typical shoreward migration rates for an Atlantic-coast barrier island are 2 meters per year, but extraordinary displacements have been reported. Hatteras Island off North Carolina migrated 500 meters (1,550 ft) landward between 1870 and 1990 (4.2 m/y; more than 14 ft/y). The Cape Hatteras lighthouse, a notorious graveyard for ships, was built almost a half-kilometer in from the shoreline in 1870. Because of erosion, it was moved in 1999 (see chapter opener).

(a)

(b)

(c)

(d)

◀▶ **FIGURE 10.12** Examples of the four types of beaches. (a) Coastal plain beach; Huntington Beach, California. Note that the high angle of wave approach produces a strong longshore current along the beach from top to bottom in the photo. (b) Pocket beaches on a cliffed shoreline; Laguna Beach, California. (c) Barrier beach; New Jersey. Note the groin field and accretion of sand on the upcurrent side of the structures. (d) The sand spit that encloses San Diego Bay is the result of a huge volume of sand transported northward from the Tijuana River in Mexico. Naval Air Station Coronado is at the south entrance to the bay, and rock-supported Point Loma forms the north entrance.

◀▶ **FIGURE 10.13** Winter and summer beach profiles and some beach terminology.

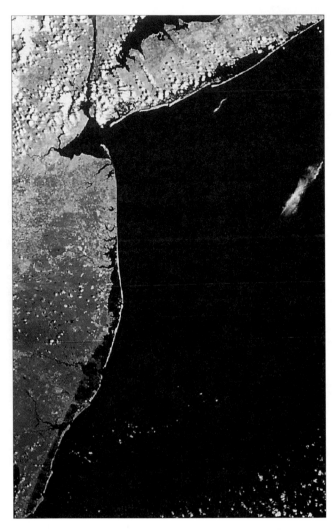

◆▶ **FIGURE 10.14** Satellite photo of the barrier island system off Long Island (top center) and New Jersey.

Spits, such as the one at Cape Cod, Massachusetts, are elongated masses of sand attached at one end to the mainland. Spits may be formed by longshore currents when they transport and deposit sand across the mouths of bays or estuaries. Newport Beach, California is a spit deposited across Balboa Bay, an estuary of an abandoned Pleistocene river channel. Coronado Island, a sand spit that protects San Diego Bay, formed from huge volumes of sand delivered from the south by Mexico's Tijuana River (Figure 10.12d).

Beach Erosion—The River of Sand Gone Dry

A beach is stable when the sand supplied to it by longshore currents replaces the amount of sand removed by waves. Where the amount of the supplied sand is greater than the available wave energy can remove—typically near the mouth of a river—the beach widens until wind action forms sand dunes. Significant areas of

coastal dunes occur in Oregon, on the south shore of Lake Michigan in Indiana, near the Nile River in Egypt, and on the Bay of Biscay on France's west coast, to name but a few (◆▶ Figure 10.16). The more usual case is for wave action and longshore currents to move more sand than is supplied, and here we find eroding beaches or no beaches at all. All 30 of the U.S. coastal states have beach erosion problems of varying magnitudes resulting from both human activities and natural causes (◆▶ Figure 10.17).

Inasmuch as sand comes to beaches by way of rivers, we should look to the land for causes of long-term beach erosion. Water-storage and flood-control dams trap sand that would normally be deposited on beaches. The need for upland flood protection is clear, but it can conflict with other needs, including the need for well-nourished beaches. Urban growth also increases the amount of paving, which seals sediment that might otherwise be eroded into local rivers and ultimately to the ocean.

Significant natural causes of beach erosion are protracted drought and rising sea level. Little runoff occurs during drought conditions, and therefore, little sand is delivered to beaches. This is true even for unmodified watersheds without dams. The impact of drought on beach nourishment is less where the watershed is behind a dam. Dams trap sediment so that even with normal rainfall sand is prevented from reaching the beach.

A rise in sea level could have catastrophic results, as indicated by the map of coastline changes in the eastern United States over the past 20,000 years (◆▶ Figure 10.18). On a shorter time scale, it has been estimated that as the earth warms over the next century, sea level will rise 1 to 3 meters (up to 10 ft) due to thermal expansion and melting of glacial ice. On flat coastal plains where slopes are as gradual as 0.2–0.4 meters per kilometer, a sea-level rise of 2.54 centimeters (1 in) would cause the shoreline to retreat between 62 and 125 meters (200–400 ft). If the coast is subsiding, this shoreline migration landward would be magnified. On the other hand, tectonic uplift along parts of the Alaska and California coasts is faster than sea-level rise, which makes for a net shoreline retreat seaward. Careful tidal gauge measurements at New York City, where the shoreline is tectonically stable, indicated a sea-level rise of 25 centimeters (almost 10 in) between 1900 and 1970. If that trend continues, plan to take a boat to the Statue of Liberty's big toe and to say goodbye to Long Island, Miami Beach, and much of Texas's Galveston County by the middle of the next century.

Local Erosion

Long-term beach erosion is aggravated locally by poorly conceived engineering works, such as *break-*

FIGURE 10.15 The south end of Hog Island, a Virginia barrier island, migrated almost 3 kilometers (1.9 mi) landward between 1852 and 1963.

1963
1955
1928
1852–71

Hog Island

N

Broadwater

Atlantic Ocean

0   1   2 mi
0   1   2 km

waters built to create quiet water for safe yacht anchorage, and long **jetties** built to prevent sedimentation in harbor or river inlets. Such structures cut off the littoral flow, causing accretion of sand in their immediate vicinity and erosion on their downcurrent side. Rates of erosion and accretion are calculated in cubic yards of sand per year, and it is a rule of thumb that a cubic yard of sand equals a square foot of beach. Thus, the littoral drift rates given in ◉ Table 10.1 can be roughly translated into gains and losses of beach area. However, you can bet that where beach sand has accreted and a beach has widened, it has occurred at the expense of some beach downcurrent.

Ocean City, Maryland, is at the south end of Fenwick Island, one of the chain of barrier islands extending down the U.S. east coast from New York to Florida. Long jetties were constructed there in the 1930s to maintain the inlet between Fenwick and Assateague Island to the south. As a result, the beach has widened at Ocean City—because the north jetty blocked the southward longshore transport of sand—and the Assateague Island beach has been displaced 500 meters (1,650 ft) shoreward due to sand starvation (◆ Figure 10.19). Even though the Ocean City beach had widened, a strong storm in March 1962 inundated all but the highest dune areas, causing $7.5 million worth of damage.

In the 1920s Santa Barbara's civic leaders decided to mold their community into the "French Riviera" of California. In spite of unfavorable reports by the Corps of Engineers, city officials authorized construction of a

⊙ TABLE 10.1	Rates of Littoral Drift at Selected U.S. Locations	
LOCATION	**DRIFT RATE, yds³/yr**	**DIRECTION**
West Coast (all California)		
Newport Beach	300,000	south
Port Hueneme	290,000	south
Santa Barbara	250,000	south
Santa Monica	160,000	south
Anaheim Bay	90,000	south
East Coast		
Fire Island Inlet, N.Y.	350,000	west
Rockaway Beach, N.Y.	260,000	west
Sandy Hook, N.J.	250,000	north
Ocean City, N.J.	230,000	south
Palm Beach, Fla.	130,000	south
Great Lakes		
Waukegan, Ill.	43,000	south
Racine Co., Wis.	23,000	south

SOURCES: U.S. Army Corps of Engineers and other sources.

(a)

(b)

◈ **FIGURE 10.16** Indiana Dunes National Lakeshore. (a) Glacially deposited sands were reworked by wave action and then by wind to form these impressive coastal dunes on the south shore of Lake Michigan. (b) Someone's beach home was buried when sand was blown to the lee side of the 25-meter-high (80-ft) coastal dunes. This is an example of dune migration (see Chapter 12).

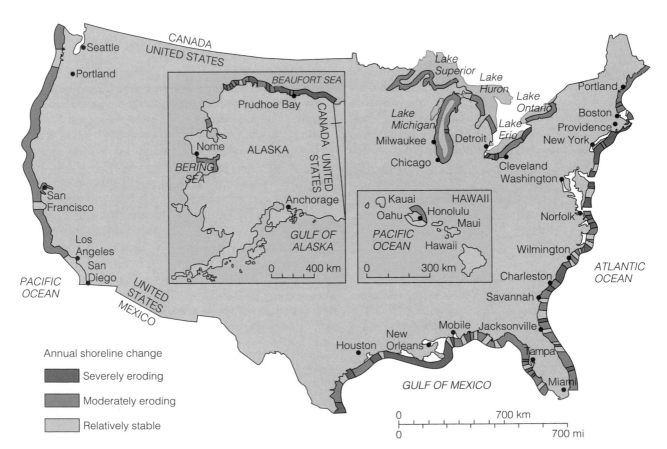

◈ **FIGURE 10.17** Present coastal erosion in the United States. Note that most of the shorelines are eroding. Uncolored coastal areas on the map represent shorelines for which data are not available.

drift rate here is about 250,000 cubic yards per year, which necessitates dredging sand from the harbor and pumping it to the beach about a mile downdrift. Summerland Beach, about 16 kilometers (10 mi) downdrift, retreated 75 meters (250 ft) in the decade following construction of the breakwater.

As a postscript, the City of Santa Monica to the south built a breakwater parallel to the shoreline in hopes of avoiding the kinds of problems experienced in Santa Barbara. The parallel breakwater created a *wave-energy shadow* on its landward side, causing sand to accumulate behind the structure but no serious downdrift erosion (◆ Figure 10.21). However, dredging is still required to keep the sand from building out to the breakwater and filling in the harbor.

Communities along the shores of the Great Lakes have erosion problems, also. Early in this century, protective structures were built on the Lake Michigan shoreline. Many of these have now deteriorated, particularly during the record high lake-level of 1984 (582.9 feet above mean sea level) (◆ Figure 10.22). Lake levels began dropping around 1960, and there was little concern at the time about shoreline erosion. Since 1964 (record low of 576.6 feet above mean sea level), levels have been rising, and there is increasing public pressure to put more protective structures on the shoreline (see Mitigation of Beach Erosion later in this section). A large volume of federal shoreline legislation affecting all coastal states is sponsored by U.S. congresspersons and senators who represent the Great Lakes area.

SEA CLIFFS

Most people enjoy ocean views, particularly from atop a 100-foot-high sea cliff. Unfortunately, such cliffs are eroded both by wave action at the toe and by mass-wasting processes on the face and the top. An **active sea cliff** is one whose erosion is dominated by wave action. This results in a steep cliff face and little rock debris at the base (◆ Figure 10.23a). The erosion of an **inactive sea cliff** is dominated by running water and mass-wasting processes. Its slopes are much gentler, and angular debris accumulates at the toe (◆ Figure 10.23, part b). Note that the nature of the debris at the foot of a sea cliff can aid in determining if it is active or inactive. Wave erosion produces rounded cobbles, whereas mass-wasting debris (talus) is angular. Rates of cliff erosion vary from a few centimeters to tens of centimeters per year, depending on the geology, climate, tectonic setting, and relative sea-level changes.

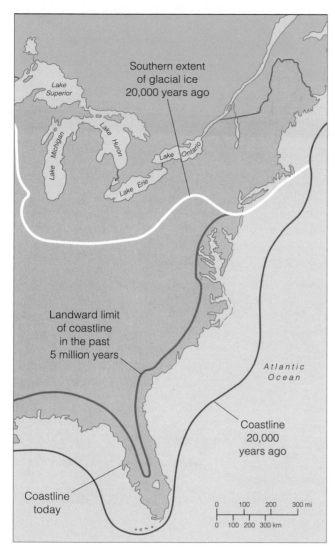

◆ **FIGURE 10.18** Position of the coastline in the eastern United States in late Pleistocene time (20,000 years ago), Miocene time (5 million years ago), and today.

long, L-shaped breakwater for a yacht anchorage in 1929 (◆ Figure 10.20). The predominant drift direction here is southeasterly, and as the beach west of the breakwater grew wider, near-record erosion occurred on the beaches east of the structure. Eventually, the southeast-moving littoral drift traveled around the breakwater, forming a sandbar in the harbor entrance. The bar built up, diminishing the size of the harbor and creating a navigational hazard that remains to this day. When the Santa Barbara city fathers asked Will Rogers, the noted American humorist, what he thought of the harbor in the 1930s, he is alleged to have replied, "It would grow great corn if you could irrigate it." The

(a)

(b)

(c)

◆▶ **FIGURE 10.19** Jetties built at Ocean City inlet in the 1930s cut off the supply of sand from the north. The north end of Assateague Island is now 500 meters shoreward of the south end of Fenwick Island. (a) Map of the islands, 1849 and 1980. (b) Recent aerial photo of the inlet, looking north. (c) Ocean City today.

It is standard practice to consider 50 years as the lifetime of a house or other structure. Therefore, a structure should be set back from the top of a cliff at least 50 times the annual rate of cliff retreat. In all too many cases, the rates are not known or people choose to ignore them, with inevitable results (◆▶ Figure 10.24). Orrin Pilkey of Duke University (see For Further Information) has developed a checklist for people to use when they are considering buying property on cliffs and bluffs. Pilkey's list includes such items as looking for evidence of failure at the toe of the cliff, determining rock hardness, and learning the history of cliff recession in the area from local land records and photographs.

◆▶ **FIGURE 10.20** The Santa Barbara, California, breakwater in 1988. Sand accumulation in the harbor is clearly visible, as is the dredge used to pump sand to the beaches downdrift.

◆▶ FIGURE 10.21 Beach at Santa Monica, California, about 1988. Note the sand accumulation in the "wave shadow" of the breakwater. The breakwater is the line of whitewater parallel to the shoreline.

MITIGATION OF BEACH EROSION

Most coastal engineering works are built to protect the land from the sea. This battle against the sea is older than the tale of the Danish king Canute who commanded his men to whip the ocean in order to keep back the rising tide. It is perhaps presumptuous to think we can build works that will stop the rise of sea level or withstand enormous storm surges, but that does not keep us from trying. Although the sea will ultimately win the battle, we should attempt to build pro

tective works that conform to nature as much as possible and that may outwit her for awhile.

Rock revetments and concrete seawalls are built landward of the shoreline and parallel to the beach not to protect the beach, but rather, to protect the land behind them (◆▶ Figure 10.25). Commonly, these "hard" structures reflect wave energy downward, removing sand and undermining the structure. There is a question among coastal engineers and geologists that goes something like "Do you want a revetment, or do you want a beach?" Because waves reflecting off hard shoreline structures remove the beach material in front of them, the best-designed rock revetments have sufficient permeability to absorb some of the wave energy. Another problem is that wave action erodes around the ends of revetments. When that happens, they must be extended to prevent damage to structures located there.

Groins are constructed to trap sand for the purpose of widening beaches. They interrupt sand flow on their upcurrent sides and deprive downcurrent beaches of sand. Once sand has built out to the end of a groin, it flows around the tip of it and begins filling in the eroded area downcurrent. Frequently another groin is built to widen *that* beach, and the cycle is repeated. The result is a beach with groins every few hundred feet and a shoreline with a scalloped or bread-knife appearance. The northern New Jersey shore and Miami Beach are classic textbook examples of multiple-groin shorelines. Permeable groins are often constructed today. They allow some sand to filter through them for nourishing downcurrent beaches.

◆▶ FIGURE 10.22 Heavily engineered shoreline of Lake Michigan at Chicago. Note that the groins have trapped sand, causing the beach to widen.

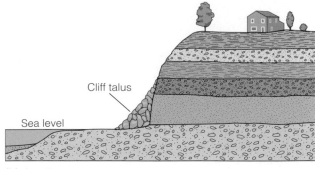

◆ **FIGURE 10.23** (a) Active and (b) inactive sea cliffs are distinguished by their dominant eroding agent and shape.

Jetties are the real culprits of beach erosion, because they are very long and they usually occur on both sides of a harbor entrance or where a river flows into the sea. They have caused entire beaches and whole communities to disappear. The jetties at Ocean City, Maryland, have already been cited.

Breakwaters intended to create quiet water for yacht sanctuary can create unintended environmental problems. We have already seen the results of breakwater construction at Santa Barbara and Santa Monica in California (Figures 10.20 and 10.21). Venice Beach, California, a community with a reputation for the unusual, built a parallel breakwater. It was constructed so close to shore that the sand deposited in the wave "shadow" behind the breakwater has connected the breakwater to the land. This sand feature is called a **tombolo** (◆ Figure 10.26).

At present, the designs of most long structures to be built perpendicular to the shoreline incorporate permanent *sand-bypassing works*. These features allow for the pumping of sand from the upcurrent side of the structure to nourish the beaches downcurrent. In addition, downcurrent beaches can be artificially nourished by pumping in sand from offshore areas or by transporting it to them from land sources (◆ Figure 10.27). Unfortunately, artificially nourished beaches are generally short-lived (see Case Study 10.1). One reason is that the imported sand is often finer than the native beach sand and is thus easily removed by the prevailing waves. The other reason is that artificial beaches usually have a steeper shoreface and are thus more subject to direct wave attack than is a gently sloping, natural shore, which dissipates wave energy. Artificial beach nourishment is expensive and it is most commonly done where a government project caused the erosion problem.

Because theory is lacking, thorough shoreline-impact studies using models should be performed for all large projects. Such a model is an exact replica of the project at a scale of about 1:100 (◆ Figure 10.28). The usual safeguard, where wave action is a threat, is to place the structure on wooden or steel piles that extend through the beach sand into firm bedrock or a hard layer. Thus if sand is removed, the building will remain standing (◆ Figure 10.29).

ESTUARIES

Estuaries (Latin *aestuarium,* "tidal") are semienclosed basins in which river water (fresh water) mixes with tidally influenced seawater. The river water flowing into the estuary "floats" on the denser seawater, which is constantly moving landward and seaward with the tide. The mixing of fresh water and saltwater generally occurs below the water surface, and the resulting mixture is dilute, or brackish, water. There are almost 900 estuaries along U.S. coasts, covering 68,000 square kilometers (27,000 mi^2). Based upon their geology, four major types of estuaries are recognized:

1. *Coastal-plain estuaries* form where the land subsides or sea level rises and submerges a river valley. Chesapeake Bay (◆ Figure 10.30) is an example of this type, as are Delaware Bay, the Hudson River, the Mississippi River, the Thames River in England, and the Seine River in France.

Consider This... ?
Several states have outlawed hard protective structures between beaches and the land. Why did they do this? Should other states, including those on the Great Lakes, enact similar legislation? ■

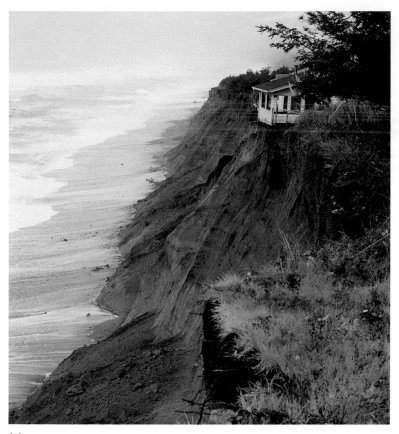

(a)

FIGURE 10.24 (a) Rapid sea-cliff retreat near Big Lagoon, Humboldt County, California. These cliffs near the Oregon border are composed of sandy and gravelly uplifted marine sediments. As shown by the photograph, they are subject to direct wave attack. (b) Drastic undercutting by wave action has left this house in San Diego County, California, in a precarious position. The near-vertical cliff and the rounded cobbles at the base indicate that this is an active sea cliff (see Figure 10.23a).

(b)

2. *Fiords* are drowned U-shaped valleys formed by glaciers (see Chapter 11). They are common in Scandinavia, Greenland, Alaska, British Columbia, and New Zealand.

3. *Fault-block estuaries* (grabens; see Chapter 3) form on tectonically active coasts such as the southern part of San Francisco Bay.

FIGURE 10.25 Seawall at Palm Beach, Florida. Large waves reflecting from the wall remove beach sand from in front of the wall.

◆ **FIGURE 10.26** This breakwater at Venice, California, was built so close to the beach that sand deposited behind it has connected the breakwater to the beach, forming a tombolo.

(a)

4. *Bar-formed estuaries* occur as lagoons between barrier islands and the mainland. Examples are Pamlico Sound within the Outer Banks of North Carolina and Laguna Madre in Texas.

Chesapeake Bay, a coastal-plain estuary, has the added distinction of having been impacted 35 million years ago by a spectacular body from the far reaches of space. A **bolide** (meteor or comet; pronounced bōʹlĭd) plopped into the coastal plain between Richmond and Norfolk, Virginia, leaving a crater 1.3 kilometers (4,250 ft) deep and 85 kilometers (53 mi) in diameter—deeper than the Grand Canyon and twice the size of Rhode Island. It is three times larger than any other U.S. crater and is the sixth-largest crater known on the planet today. Studies by C. Wylie Poag of the U.S. Geological Survey (U.S.G.S.) involving core drilling, mineral analysis, and seismic profiles identified

(b)

◆ **FIGURE 10.28** (a) King Harbor Marina, Redondo Beach, California, during a damaging storm in January 1988. (b) Model of King Harbor built after the storm to aid in assessing potential damage from waves of varying height and source direction in order to make design improvements.

◆ **FIGURE 10.27** Artificial beach nourishment by dredging from an offshore source; Redondo Beach, California. Typically, about 15 percent of dredged artificial beach fill is too fine-grained to be retained and is lost back to the sea.

the presence of the subsurface crater and the impact debris (◆ Figure 10.31). Consequences of the impact that affect humans around the bay today include:

- land subsidence due to water-saturated impact breccia at depth;
- accelerated relative sea-level rise (◆ Figure 10.32);
- disruption of the courses of the James and York Rivers to the northeast around the rim of the crater;
- alteration of coastal aquifers due to the briny water trapped in the breccia; and
- the location of the bay due to rivers, particularly the Susquehanna, having cut a channel down to a lower sea level, into which the sea has advanced during the past 10,000 years.

▶ **FIGURE 10.29** "Watch that first step!" would be a good warning to anyone exiting these stilt houses that once rested on the beach. Strong waves reduced the beach level almost four meters during one storm. Westhampton, New York.

(a)

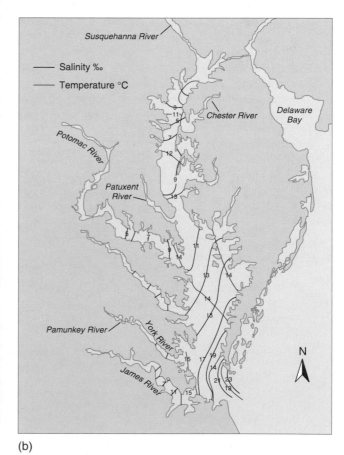

(b)

▶ **FIGURE 10.30** Chesapeake Bay, a coastal-plain estuary that formed as sea level rose and drowned the valleys of the Susquehanna, Potomac, and James Rivers and many smaller watersheds. The bay is 311 kilometers long and up to 40 kilometers wide (193 mi x 25 mi). (a) Satellite photo and (b) map of the area showing spring salinities in parts salt per thousand parts seawater. Normal seawater has a salinity of about 34 parts salt per thousand (3.4%).

Case Study

Case Study | 10.1

The Care and Feeding of Undernourished Beaches

Most shorelines in the world's developed nations are eroding. In the United States erosion control is managed by the U.S. Army Corps of Engineers as part of their mission to maintain navigable waters. They usually address the problem by installing "hard" shoreline-stabilization structures such as seawalls or by using the "soft" solution of artificial beach nourishment. A third solution, not advocated by the corps for a variety of engineering and political reasons, is to relocate any structures that are present on the beach and allow the shoreline to establish its own equilibrium. Because of the need to preserve recreational beaches, the use of hard shoreline protection is now being restricted, and New Jersey, North Carolina, South Carolina, and Maine have made hard stabilization illegal. This leaves artificial beach replenishment, which does not have a good longevity record, as the option of choice in many places. And this is the focus of controversy between the engineers who advocate it and the many geologists who question its effectiveness.

More than 90 beaches were replenished by 270 separate pumping contracts along East Coast barrier-island shores between the early 1960s and

(a)

(b)

◆ **FIGURE 1** (a) Miami Beach had little beach width in the early 1970s. (b) Miami Beach in 1981, after what has proven to be a successful artificial sand-replenishment project.

◀▶ FIGURE 2 Replenishment history of Wrightsville Beach, North Carolina, showing the cumulative amount of sand placed on the beach from 1965 to 1991. The steady rise of the curve indicates that regular additions of sand were required to maintain the beach and that no steady-state condition, or equilibrium, had been reached by 1991.

the early 1990s. Studies by Orrin Pilkey at Duke University from 1988 to 1992 show that the life spans of these replenished beaches are highly variable, but that most of them are very short lived. At Miami Beach, a 16-kilometer (10-mi) $60 million artificial beach was still largely in place after 15 years (◀▶ Figure 1). This is in sharp contrast to the experience at Ocean City, Maryland. In January 1989, Ocean City's mayor extolled the virtues of the community's new $12 million replenished beach. Two months later, the winter's first big storm had removed much of the new beach, some of it clear back to its prereplenishment location (see Figure 10.19). Essentially no replenished beaches north of Florida have lasted more than five years, because the newly placed beach fill is quickly removed by storm waves. North Carolina's Wrightsville Beach, one of the most extensively replenished beaches in the United States, has been replaced ten times (◀▶ Figure 2). ◀▶ Figure 3 shows Wrightsville Beach just after one of these replenishments and then three

years later. According to Pilkey, the main problem with artificial replenishment is many coastal engineers' strong belief in the theoretical "textbook" explanation of sand transport on a shoreline. That explanation is based on many questionable assumptions about the strength of future storms, the rate of sea-level rise, and coastal-zone development.

Artificial beach replenishment is expensive and temporary. A growing number of geologists believe the only feasible long-term solution is relocation, moving out whatever buildings are there—or simply letting them fall into the sea. This would preserve the beach, save shoreline stabilization costs, and preserve the buildings if moved (as the Cape Hatteras lighthouse was in 1999; see the chapter opener). By now it is clear that the shoreline-erosion crisis is human-made; if no one lived next to the shore there would be no problem. This alternative has merit and has many strong advocates, but it is politically unpopular.

(a)

(b)

◀▶ FIGURE 3 Wrightsville Beach (a) after one of its many replenishment projects (note width of the beach berm) and (b) three years later. The beach was replenished and destroyed again in 1996 by Hurricane Fran.

West   East

◆ FIGURE 10.31 Cross section showing main features of the 35-million-year-old Chesapeake Bay impact crater as disclosed by the three core holes. The central peak and peak ring formed similarity to how a central peak and ring form in standing water when a water drop strikes the water.

This discovery is an example of clever collaborative research using core holes drilled for water-resource evaluation; seismic lines taken by Texaco, the National Geographic, and the U.S.G.S.; and mineral sleuthing that found shocked quartz crystals and melted grains that confirm the impact.

Estuaries and the marsh wetlands on their shores are habitats for flourishing biological communities. The waters of estuaries have high nutrient contents that stimulate phytoplankton (plant) growth, which in turn supports a rich and diverse fauna. Estuaries are the habitat of a host of shellfish—including blue crab and common shrimp—and the breeding place of many fish species. The tidal marshes on their shores support a va-

riety of waterfowl. It is estimated that 95 to 98 percent of the important commercial fish species are spawned in estuaries.

Urbanization around estuaries has placed conflicting demands on their waters. For example, the fishing and shellfish industry that the Hudson River estuary once supported has drastically declined since 1960. The Hudson is now the emergency water supply for New York City and also the receptacle for the sewage of 16 million people. Because of the Clean Water and Ocean Dumping Acts of 1972, no raw sewage or sludge (solids) may be discharged into the river, but discharges of treated sewage have resulted in elevated bacteria levels, the presence of heavy metals and organic compounds

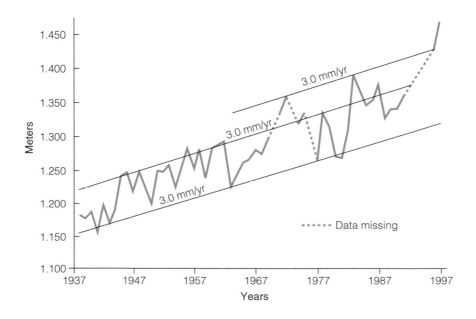

◆ FIGURE 10.32 The average annual sea-level rise recorded at the tide gage at Solomons Island, Maryland, was 3 millimeters per year between 1937 and 1997.

in bottom sediments, and a reduced dissolved-oxygen content in the water column. The aesthetic amenities and resources of this estuary have been severely degraded by human activities (see Case Study 10.2).

Hurricanes and Coastal Flooding

Andrew, Mitch, Camille, Opal, Gilbert, and *Floyd* are innocent-sounding names, but onshore surges of water and intense rainfall produced by hurricanes with such names are our greatest natural hazard. Hurricanes (called *typhoons* in the North Pacific Ocean and *cyclones* in the South Pacific and Indian Oceans) get their energy from the equatorial oceans in the summer and early fall months when seawater is warmest, usually above 25°C (about 80°F). Hurricanes begin as tropical depressions when air that has been heated by the sun and oceans rises, creating reduced atmospheric pressure, clouds, and rain. As atmospheric pressure drops, the warm air moves toward the center of low pressure, the eye of the impending hurricane. When the rising air reaches higher elevation it releases heat, causing it to rise even faster and creating even lower atmospheric pressure and greater wind velocities. Wind blowing toward the center of low pressure is given a counter-clockwise rotation by the Coriolis force in the Northern Hemisphere (◆ Figure 10.33).

Thus, what begins as a localized tropical disturbance can grow to be 800 kilometers (500 mi) across, with wind velocities far in excess of 120 kilometers per hour (74 mph), the lower velocity limit for hurricane

(a)

(b)

 FIGURE 10.33 (a) Cross section of a typical hurricane showing vertical air motions in red (warm air) and blue (cooler air), clouds (white), and precipitation (gray). (b) Satellite photograph of Hurricane Gilbert practically filling the Gulf of Mexico, 15 September 1988. Gilbert's rotation and central eye are clearly visible.

Case Study

Case Study | 10.2

Sea Sickness: Bleached Reefs and Oceanic Dead Zones

Pollution, overfishing, and overuse have put many of our unique reefs at risk. Their disappearance would destroy the habitat of countless species . . . [and have] . . . a devastating effect on coastal communities from Cairns to Key West, Florida.
WILLIAM JEFFERSON CLINTON, PRESIDENT OF THE UNITED STATES, REFLECTING ON THE ESTABLISHMENT OF THE INTERNATIONAL YEAR OF THE CORAL REEF, 1997

Charles Darwin may have been the first to recognize that coral reefs are the largest and most important ecosystems in the marine environment. Corals are animals related to jellyfish that secrete a hard skeleton of calcium carbonate (limestone). They grow only in warm water (above 20° C, which is approximately 70° F) within the zone of light. Corals require light because they live in symbiosis with the tiny marine algae *Zooxanthellae*. These plants live within the coral polyp and give the calcareous skeleton its vibrant color (◁▷ Figure 1). They provide food and oxygen to the coral, and the coral provides the carbon dioxide *Zooxanthellae* need for photosynthesis. In 1998 an unprecedented bleaching of corals over large areas in the world was noticed, with the most severe bleaching occurring in the Indian Ocean region. Bleaching is also found throughout the Gulf of Mexico and Caribbean Sea—in "the ABC Islands" (Aruba, Bonaire, and Curaçao) as well as off the coasts of Florida and the islands of the Antilles. When the water is too warm the algae are expelled, and the coral turns bone-white and eventually dies. Heated water from power-plant cooling systems is the demonstrated cause in some cases, but other possible causes include the exceptional El Niño event of 1998 and global climate change. Some scientists suggest that coral bleaching is the first widespread warning of more serious changes due to global warming. Whatever the cause, bleaching turns a beautiful underwater garden into a spiny wasteland (◁▷ Figure 2).

Although the Gulf of Mexico is still one of the world's great fisheries for shrimp and many fish species, an area off the delta of the Mississippi River no longer supports marine life such as shrimp, crabs, and other bottom-dwelling organisms. This "dead zone" is about the size of New Jersey and results from discharge of agricultural nutrients from the Mississippi River drainage basin (◁▷ Figure 3). The zone once formed each spring and summer, but it is now permanent. The river-borne nutrients, mostly nitrates, promote phytoplankton blooms (microscopic plants) that, in turn, promote

◁▷ FIGURE 1 Healthy coral reefs exhibit a variety of colors and diversity. A healthy reef off the Bahama Islands, with the staghorn coral, *Acropora palmata,* growing to the low-tide water level. This coral is diagnostic for determining past sea levels in drill cores, as it always grows to this surface when healthy.

◁▷ FIGURE 2 A bleached reef in the Bahamas that has lost most of its symbiotic algae. The limestone of the coral skeleton is clearly visible. If the environmental stress is not removed, the coral polyps will die. The coral here is elkhorn coral, *Acropora cervicornis.*

the growth of zooplankton (small animal floaters). When these organisms die, they sink to the bottom and are decomposed by bacteria. This process consumes oxygen, and eventually the bottom waters become *anoxic*. At dissolved-oxygen levels below 2 milligrams per liter, anything that can swim leaves the area, and sedentary animals such as clams, snails, and worms die for lack of oxygen. Such "dead zones" are consequences of human activities—deforestation, urbanization, and agriculture.

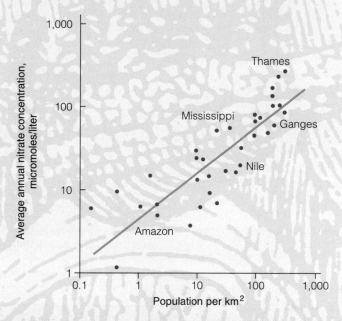

◆ FIGURE 3 Relationship between the population density of a river's watershed and the nitrate concentration of its waters. The data are from about 1990.

◇ **FIGURE 10.34** Regions where tropical storms form, their typical paths, and their terminology.

designation. Hurricanes travel as coherent storms with velocities of 10 to 35 knots (12–40 mph), but their exact course and where they will strike land is not completely predictable (◇ Figure 10.34). If we divide a storm system in the North Atlantic into four equal quadrants, we find the strongest winds in the northeast quadrant (designated as quadrant I), because the cyclonic winds and the storm system are moving in the same direction there (◇ Figure 10.35). When a storm makes a landfall, the greatest wind damage usually occurs in this sector.

A "hill" of water called a **storm surge** is pushed ahead of the hurricane that may cause sea level to rise many meters above the normal high-tide level. The reduction in atmospheric pressure associated with a hurricane also causes sea level to rise—like fluid rising by

Consider This… ?
"Why sweat a swamp?" is what some developers say about the fuss over preserving estuaries and marshlands. Is this attitude justified, or are these features valuable to the environment? ■

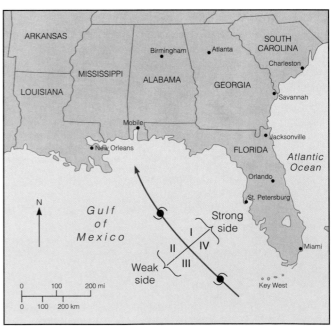

◇ **FIGURE 10.35** Northern Hemisphere hurricane quadrants. The highest winds occur on "the strong side."

suction in a soda straw. Average sea-level atmospheric pressure is 1,013 millibars (or 29.92 inches of mercury), and the rise in sea level is about 1 centimeter for every millibar of pressure drop. Hurricane Camille in 1969, the strongest storm to hit the U.S. mainland, had a near-record low barometric pressure of 905 millibars (26.61 inches of mercury). Hurricane surge and winds devastated the entire Gulf Coast from Florida to Texas.

Twenty-five residents of the Richelieu Apartments in Pass Christian, Mississippi, on the state's south coast were enjoying a hurricane party when the police chief came to alert them of the danger. They laughed when he requested the names of their next of kin because they did not appear to be willing to take his advice and evacuate. The storm surge at Pass Christian reached 7 meters (23 ft) above mean sea level, and only two people survived the party. One of them was rescued from a floating sofa 8 kilometers (5 mi) from the apartment complex.

Camille killed 130 people and flooded 40,000 homes. Three thousand homes simply disappeared, and property damage exceeded $1.3 billion. Yet real estate was resold, and new homes were built on the same sites where homes had been removed by the storm. Camille was rated as a category-5 hurricane, the highest category on the Saffir–Simpson Scale (◉ Table 10.2). Only two category-5 hurricanes were recorded in the United States in the twentieth century. However, neither Hurricane Camille nor any other natural disaster to date has caused as much dollar-damage as Hurricane Andrew caused in 1992—$25 billion (◈ Figure 10.36). Andrew was the third-largest hurricane to strike the United States.

Some hurricanes veer eastward and follow a track up the U.S. East Coast, causing flooding and beach erosion from the Carolinas to Maine. In 1989 Hurricane Hugo struck the coast just north of Charleston, South Carolina, with sustained winds of more than 215 kilometers per hour and a storm surge of 6 meters (20 ft; ◈ Figure 10.37). Hurricane Floyd was the 1999 storm in North Carolina that wouldn't go away. Although not a strong hurricane, it created flooding of biblical proportions in the eastern part of the state, aided by Hurricane Dennis that had dropped 25 centimeters (10 inches) of rain two weeks earlier. Thousands of square miles were underwater for weeks, and the flood area became a fetid swamp with the bodies of 10,000 hogs and 2.5 million chickens and turkeys. Forty-six Carolinians lost their lives, 6,000 were left homeless, and the monetary damage exceeded $6 billion. Although Hurricane Floyd did not make a direct landfall, it dropped 55 centimeters (20 inches) of rain, leading a resident of Rocky Mount, one of the cities most impacted by the storm, to say, "The last time it rained like this, Noah was alive."

The 1995 hurricane season in the North Atlantic Ocean was the worst in more than 60 years. Wet weather in northwest Africa spawned slow-moving disturbances, some of which grew into tropical storms and then became hurricanes over the ocean. Some of these hurricanes struck the U.S. Gulf and East Coasts. Opal was the worst of these and she made a direct hit on coastal Alabama and Florida in October. Hurricane predictors are discussed later.

◯ TABLE 10.2 Saffir–Simpson Hurricane Scale				
	WIND VELOCITY			
CATEGORY	km/h	mph	DAMAGE	EXAMPLES
1	119–153	74–95	minimal	Agnes, 1972 Juan, 1985
2	154–178	96–110	moderate	David, 1979 Bob, 1991
3	179–210	111–130	extensive	Betsy, 1965 Frederic, 1979
4	211–250	131–155	extreme	Hugo, 1989 Andrew, 1992 Opal, 1995
5	>250	>155	catastrophic	Labor Day, 1935 Camille, 1969

SOURCE: Coastal Weather Research Center, University of South Alabama, Mobile.

(a)

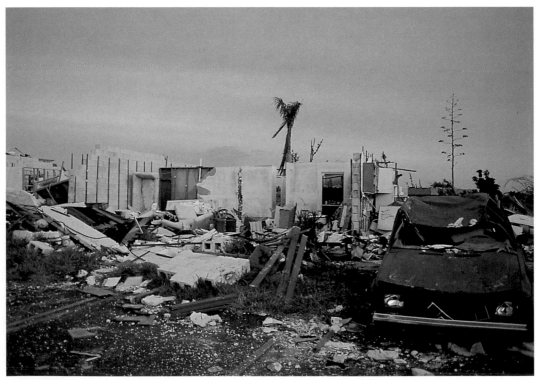

(b)

◆▶ **FIGURE 10.36** The path and result of Hurricane Andrew during the last 9 days of its life, 17–25 August 1992. (a) Positions as recorded at 10 P.M. CDT. Although the storm was spawned in North Africa during the second week of August, it did not reach hurricane proportions until it acquired heat and moisture from the Atlantic Ocean. (b) Wreckage of a condominium in southern Florida after Andrew passed through on 24 August 1992.

(a)

(b)

◀▶ **FIGURE 10.37** (a) Homes at beautiful Folley Beach, North Carolina, before Hurricane Hugo. Note the rock revetment facing the beach to prevent sand erosion and damage to structures. (b) The area had five fewer houses after Hurricane Hugo, but the palm trees and the rock revetment remained. Three houses in the background were damaged by battering-ram action of the demolished first-row structures. This is the hard way to become an owner of beachfront property.

Coastal flooding induced by hurricanes or typhoons is the greatest natural hazard to humans. Nowhere has this been shown more vividly than in flood-prone Bangladesh, where a third of the country lies less than 20 feet above sea level. In 1970 a half-million people lost their lives there in a massive storm surge on the low-lying delta of the Ganges River. In April and May 1991, the area was struck by another fast-moving wall of water more than 6 meters high. A string of islands along the delta margin were inundated, and a reported 100,000 people died this time. Winds of 250 kilometers per hour (155 mph) were reported on the low-lying islands of sand with names like *Chittagong, Cox's Bazar,* and *Sandwip.* Testimony to the storm's power was the recovery of a man's body 20 kilometers (12 mi) inland from his village. If the storm had hit the U.S. coast, it would have been classified a category-5 event, as Camille of 1969.

Natural disasters such as flooding, earthquakes, and massive debris flows are most devastating to people of developing countries. One explanation is that communication and warning systems are virtually nonexistent in many of these nations. Fortunately, Indian forecasters projected the 1991 storm path far enough in advance to evacuate three million people. In contrast, a weaker cyclone 21 years earlier killed five times as many people. Another explanation for the large death tolls is that overpopulation in many of these areas forces people to settle on terrain that is subject to life-threatening natural hazards. Bangladesh, occupying a land area the size of Wisconsin, has 115 million people, about half the U.S. population, and most of them are desperately poor. Even in areas where people were warned of the 1991 storm, many people were reluctant to leave their few possessions to looters.

EL NIÑO AND THE COASTAL ZONE

El Niño is a perturbation, or disruption, of the normal cold-water conditions that exist along the west coasts of North and South America. During El Niño years high-pressure cells replace the normally low atmospheric pressure of the western Pacific Ocean. Because the pressure difference that drives the trade winds between the eastern Pacific and the western Pacific lessens, the trade winds stop, or may even reverse direction. A bulge of warm surface water (as much as a meter high) called a *Kelvin wave* surges eastward toward the coast of the Americas and displaces the cooler water there (◀▶ Figure 10.38). First noticed by Peruvian fishermen around Christmastime in 1892, this invasion of warm water was dubbed the *Corriente del Niño* ("Current of the Christ Child"). Today, the atmospheric pressure reversal in the western Pacific region, called the *southern oscillation,* is known to be the trigger for El Niño events. We now call this connection and the ensuing warm-water events the **El Niño–Southern Oscillation,** usually shortened to its easily articulated acronym, **ENSO.**

Significant El Niño events of 1972–1973 and 1976–1977 and the super El Niño of 1982–1983 established it as a global phenomena. An almost continuous El Niño condition persisted from 1991 to 1994 (◀▶ Figure 10.39), and then another super El Niño occurred in 1997–1998. During El Niño years, because of atmospheric convection shifts, there have been drought in Africa, Australia, and Indonesia; flooding in California, Ecuador, Peru, and Tahiti; and impressive storm waves and beach erosion along both the east and west coasts of the United States. More bad news was the

(a)

(b)

FIGURE 10.38 Oceanography, meteorology, and geography of an El Niño. (a) The onset. Trade winds maintain warm equatorial currents toward the east. As El Niño begins, the trade winds weaken, and the equatorial countercurrent strengthens. This creates Kelvin waves, surges of warm water moving eastward. When a Kelvin wave reaches the Americas, it divided northward and southward, traveling along the west coasts. (b) Sea-surface temperature anomalies during a fully developed El Niño in 1998. Note the warm waters (yellow and red) progressing poleward along the west coasts of North and South America. The result was increased rainfall and storm activities in these regions.

◈▶ **FIGURE 10.39** The numbers on the left side of the graph are an ENSO index that incorporates these measured parameters: air temperature, atmospheric pressure, cloudiness, and wind speed and direction. El Niño events are shown in red (+1); La Niña events, in blue (−1).

unprecedented number of landslides and debris flows in California, Oregon, and Washington during the 1997–1998 event. According to Jim O'Brien at the Center for Ocean–Atmosphere Prediction Studies (COAPS) at Florida State University, there is also good news (Web site listed in For Further Information). El Niño:

- decreases Atlantic hurricane activity;
- reduces forest fires in Arizona, Texas, and New Mexico;
- reduces tornadoes;
- produces better citrus and vegetable crops; and
- improves the fishing off California because of warm water.

Interestingly, Hawaiian hurricanes occur only during El Niño years.

El Niño is a good example of the earth functioning as a system—that is, the atmosphere, hydrosphere, and lithosphere interacting to create natural events. El Niño's "sister," La Niña, appears to be a media concoction. La Niña is a cold-water condition, a condition that most experts call the normal condition for the west coast of the Americas. Even though it has been linked to drought and increased hurricane activity by the media, a direct scientific connection has yet to be demonstrated.

HURRICANE PREDICTORS

We know that hurricanes are part of the tropical severe weather pattern and that they will occur seasonally every year. According to meteorologist William Gray at Colorado State University, certain predictors can foretell

◈▶ **FIGURE 10.40** A series of 19 tropical storms, of which 11 were hurricanes and 5 were major hurricanes, occurred during the 1995 season in the Atlantic Ocean. Seen marching across the equatorial Atlantic on 24 August 1995 is Hurricane Luis (right center of the photo). Luis was preceded by Felix and followed by Hurricanes Marilyn, Opal, and Roxanne.

the severity of a hurricane season. Using these predictors, Gray has issued forecasts since 1984. For instance, in 1995 the hurricane "factory" of the eastern Atlantic pumped out 19 tropical storms, 11 of which were hurricanes. Five of those developed into major hurricanes with winds over 185 kilometers per hour (111 mph; ◈ Figure 10.40, p. 341). Gray had forecast a severe season with 9 hurricanes. Also, meterologists at the National Hurricane Center in Coral Gables, Florida, have made some impressive forecasts using sophisticated computer modeling. The predictors Gray uses require a massive data-collection effort, some of which are:

- Tropospheric winds up to an altitude of 12,000 meters (40,000 ft). Strong winds in the troposphere work against hurricane circulation by "shearing off" the tops of the vertical buildups that are required for storms to reach hurricane strength. Daily weather-balloon launches in the Caribbean Sea are used to measure wind shear at altitude. When wind shear is low, as it was in 1995, many hurricanes may develop.
- West African climate. Dry years promote westerly (from the west) winds at 12,000 meters above the tropical Atlantic. This promotes wind shear and reduction of hurricane-strength buildups. During wet years winds blow from the east at 12,000 meters, greatly reducing wind shear and thus promoting hurricane-strength storms.
- El Niño. The warming of the equatorial Pacific Ocean promotes westerly winds that blow into the Caribbean and Atlantic basins, where trade winds blow at lower levels from the east. This promotes wind shear that knocks down hurricane buildups. Between 1991 and 1994 repeated El Niño resulted in few hurricanes. In 1995 and 1999 El Niños gave way to a cooling of the equatorial Pacific Ocean (La Niña), and hurricane activity greatly increased.
- Atmospheric pressure. Low atmospheric pressure in the Atlantic Ocean and Caribbean Sea indicates areas of warm water, which fuels hurricanes, promotes convergence of air masses, and stimulates vertical buildups of moisture-laden clouds (◈ Figure 10.41).

The most difficult part of hurricane forecasting is predicting where and when a hurricane will make landfall. Unfortunately, this is the information that coastal-state residents need most of all. A few computer forecasts have been successful, but there is still much research to be done before landfall prediction is perfected.

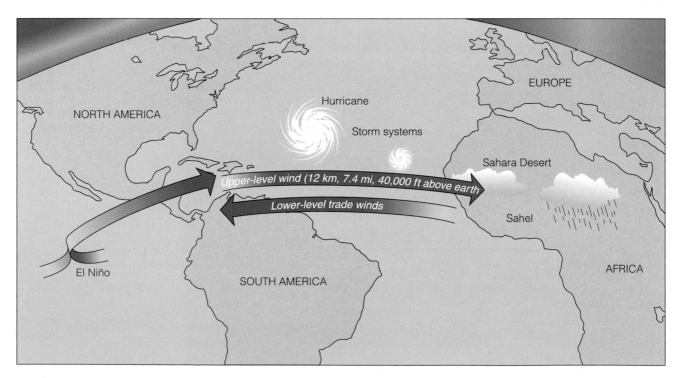

◈ **FIGURE 10.41** Predictors of a strong or mild hurricane season. West-flowing trade winds, El Niño, and upper-level winds create wind shear and prevent hurricanes from building. Conditions of high pressure and warm water in the Caribbean Sea fuel hurricanes and tropical storms and promote vertical buildup. Wet conditions in West Africa promote west-flowing winds and promote the development of tropical storms and hurricanes.

Gallery

Shoreline Wonders

The interaction of the ocean and the atmosphere at shorelines creates interesting geologic features, some of which become tourist attractions. **Stacks,** outlying remnants of retreating sea cliffs, are picturesque features along all cliffed shorelines. None are more picturesque than the "Twelve Apostles," a must for tourists who visit the south coast of Australia (◁▷ Figure 10.42) Sometimes a stack or small island close to a shore creates a wave-energy shadow—as a parallel breakwater does (see Figure 10.26)—and becomes connected to the mainland by a sandbar de-

posited in its lee. Such a sand bridge is known as a *tombolo,* and the system is called a *tombolo and tied island.* The Rock of Gibraltar on the south coast of Spain is such a system, but literally thousands of less famous and just as photogenic examples exist (◁▷ Figure 10.43). As sort of a defiant parting shot, Hurricane Hugo left one family a reminder of the weakness of human efforts to hold back the sea in comparison to nature's power to claim the land (◁▷ Figure 10.44). Hurricane Camille also contributed to our appreciation of nature's forces (◁▷ Figure 10.45).

◁▷ **FIGURE 10.42** The "Twelve Apostles" sea stacks on the south coast of Australia near Apollo Bay.

◁▷ **FIGURE 10.44** A revetment's half-ton to one-ton stone blocks were lifted and hurled through the walls and windows of a living space they were intended to protect; Hurricane Hugo, 1989.

◁▷ **FIGURE 10.43** A tombolo and tied island near Friday Harbor, San Juan Islands, Washington State.

◁▷ **FIGURE 10.45** Delta House Movers appeared to have added boat-moving services after Hurricane Camille in 1969; Gulfport, Mississippi.

Summary

COASTAL OCEAN

IMPORTANCE

1. Impact on local climate, weather, and shoreline structures
2. Estuaries
3. Beaches, an important natural resource

WAVES

CAUSE

Wind blowing over the water surface creates ripples, then stormy irregular seas, and finally regularly spaced swell that outruns the source area.

FACTORS OF SIZE

Wave size depends upon wind velocity, the length of time wind blows over the water, and fetch. A fully developed sea has the highest waves attainable for the three variables.

MOTION

Water parcels move in circular paths to a depth of half the wavelength called *wave base*. This wind causes movement of the waveform known as *waves of oscillation*. Water physically moves shoreward when waves break in the surf zone. These breaking waves are called *waves of translation*.

TYPES OF BREAKERS

Steep offshore slopes develop plunging breakers, and gentle slopes produce spilling breakers.

WAVE REFRACTION

Waves generally approach the shore at some prevailing angle. This results in longshore currents that move sand along the beach. The transported sand is known as *littoral drift*. Groins act as dams to littoral drift, causing sand to be deposited upcurrent and eroded downcurrent. Wave refraction concentrates wave energy on points and dissipates it in bays, thus tending to straighten an irregular shoreline.

BEACHES

DEFINED

Narrow strips of shore that are washed by the waves or tides, usually covered by sand or pebbles.

TYPES

1. Beaches backed by coastal plains
2. Pocket beaches along cliffed shorelines
3. Barrier islands (barrier beaches)
4. Spits

BEACH EROSION

DEFINED

Less sand is supplied by littoral currents than is removed by wave action.

NATURAL CAUSES

1. Drought
2. Rising sea level

HUMAN CAUSES

1. Dams that impound sediment
2. Groins, jetties, and breakwaters that impound sediment
3. Hard structures such as seawalls and revetments

MITIGATION

Design of groins, breakwaters, and jetties can incorporate features that allow sand to pass through or around them. Seawalls, revetments, and artificial nourishment are only temporary solutions. Model studies of existing structures and shorelines help us to better understand the problem and thus to effect remedies.

SEA CLIFFS

DEFINED

Cliffs formed by wave action.

TYPES

Active and inactive.

EROSION

Annual rates between a centimeter and tens of centimeters.

MITIGATION

Structures' optimal (safest) setback from the edge of a cliff can be determined when the erosion rate is known.

ESTUARIES

DEFINED

Basins open to the sea in which fresh water mixes with seawater.

TYPES

1. Coastal-plain estuaries, drowned river valleys (Chesapeake Bay, site of an ancient bolide impact)
2. Fiords, drowned glacial valleys (Alaska)
3. Fault-block estuaries (south San Francisco Bay)
4. Bar-formed estuaries landward of barrier beaches (Laguna Madre, Texas)

IMPORTANCE

Estuaries and their surrounding wetlands are areas of high biological productivity for marine life and waterfowl.

HUMAN IMPACT

Used for dumping sewage, commercial fishing, transportation, and recreation. Structures are built on filled marshland.

COASTAL FLOODING

HURRICANES

Storms generated at sea characterized by counterclockwise winds greater than 120 kilometers/hour (74 mph). They travel across the sea at velocities of 10–35 knots (12–40 mph).

FLOOD DANGER

A storm surge is a "hill" of water due to high winds and low atmospheric pressure. It may be many meters above normal high-tide levels. Storm surges on low-lying deltas such as the Ganges River delta in Bangladesh are the greatest natural hazard to human life.

CASE HISTORIES
Hurricanes Camille (1969), Hugo (1988), Andrew (1992), and Floyd (1999) are described.

EL NIÑO (ENSO)
Associated with floods, landslides, and accelerated beach erosion. Also associated with decreased hurricane activity in the Atlantic, fewer tornadoes in the Midwest and fewer forest fires in the Southwest, and better fruit and vegetable crops. Hawaii experiences hurricanes only in ENSO years.

HURRICANE PREDICTION
Utilizes data on tropospheric winds, West African climate, El Niño, sea-surface temperature, and atmospheric pressure in the Caribbean Sea. Heat is the fuel that drives hurricanes.

◆ Key Terms

active sea cliff	estuary	longshore current	tombolo
barrier island	fetch	rip current	wave base
bolide	groin	spit	wave of oscillation
El Niño–Southern Oscillation (ENSO)	jetty	stack	wave of translation
	littoral drift	storm surge	wave refraction

◆ Study Questions

1. Why do submarines and submersibles not experience severe storms at sea?
2. What shoreline structures impede littoral drift and cause beach erosion? How may these structures' impact be mitigated so as to minimize damage?
3. Where and when do hurricanes originate, and how do they obtain their energy? Distinguish a hurricane from a typhoon and a cyclone.
4. Name two types of breakers and explain what causes one or the other to occur along a given shoreline.
5. How does the width of a beach vary naturally over the year? Where is the sand stored that is removed from the beach in winter and returned to the foreshore in summer?
6. What is a storm (hurricane) surge, and why is it such a danger in low-lying coastal communities?
7. How do longshore currents develop, and how do they impact beaches and *swimmers* in the surf zone?
8. How can hazardous currents in the surf zone be recognized?
9. Why are there fewer hurricanes during El Niño years than during "La Niña" times?
10. Why do warm ocean temperatures and low atmospheric pressure stimulate hurricane development?

◆ For Further Information

BOOKS AND PERIODICALS
Bascom, Willard. 1964. *Waves and beaches.* Garden City, N.Y.: Doubleday.

Dean, Cornelia. 1999. *Against the tide: The battle for America's beaches.* New York: Columbia University Press.

Denison, Frank E., and Hugh S. Robertson. 1985. Assessment of 1982–83 winter storms damage, Malibu coastline. *California geology,* September.

Griggs, G., and L. Savoy, eds. 1985. *Living with the California coast.* Durham, N.C.: Duke University Press.

Kaufman, W., and Orrin Pilkey. 1979. *The beaches are moving.* Garden City, N.Y.: Anchor Press/Doubleday.

Larsen, Curtis E. 1998. *The Chesapeake Bay: Geologic product of rising sea level.* U.S. Geological survey fact sheet 102-98.

Peierls, B. L.; N. F. Caraco; M. L. Pace; and J. J. Cole. 1991. Human influence on river nitrogen. *Nature,* no. 350: 386–87.

Pendick, Daniel. 1996. Mean season. *Earth magazine: The science of our planet.* June.

Pilkey, Orrin H. 1991. Coastal erosion. *Episodes* 14, no. 1 (March).

Pilkey, Orrin H., and others. 1993. The concept of shoreface profile of equilibrium: A critical review. *Journal of coastal research* 9, no. 1: 255–278.

Pilkey, Orrin; David M. Bush; and William Neal. 1999. The Cape Hatteras lighthouse controversy: The opposing coastal management strategies of armoring, soft solutions, and relocation. *Abstracts of the Geological Society of America annual meeting.*

Pipkin, B.; H. Robertson; and R. S. Mills. 1992. Coastal erosion in Southern California: An overview. In Pipkin, B. W., and Richard Proctor, eds., *Engineering geology practice in Southern California.* Belmont, Calif.: Star Publishing Co., pp. 461–483.

Poag, C. Wylie. 1998. *The Chesapeake Bay bolide impact: A new view of coastal plain evolution.* U.S. Geological Survey fact sheet 049-98.

Scientific American. The oceans. 1998. *Scientific American quarterly.* Fall.

Shinn, E. A. 1993. Geology and human activity in the Florida Keys. *Public issues in energy and marine geology* (U.S. Geological Survey), October.

Williams, J. S.; K. Dodd; and Kathleen Gohn. 1990. *Coasts in crisis*. U.S. Geological Survey circular 1075.

Woods Hole Oceanographic Institute. 1981. The coast. *Oceanus* 23, no. 4.

_____. 1993. Coastal science and policy I. *Oceanus* 36, no. 1.

INTERNET SITES

Center for Ocean-Atmosphere Prediction Studies (COAPS) at Florida State Univerrsity. Excellent for El Niño information.

http://www.coaps.fsu.edu/lib/enso_sites.html

Colorado State University. Information on weather methodology.

http://www.cira.colostate.edu/ramm/advimgry/toc.htm

NOAA. Site for El Niño and La Niña.

http://www.elnino.noaa.gov/lanina.html

_____. National Weather Service.

http://www.nws.noaa.gov

◈ World Wide Web Activities

Predicted Coastal Land Loss from Global Warming by 2080
http://www.meto.govt.uk/sec5/CR_div/Brochure97/pics/p14f3-max.gif

The map and key show projected wetlands loss and risk to global populations due to sea-level rise from global warming by the year 2080. Assumptions in creating the map are a doubling of coastal populations and a global sea-level rise of 44 cm, between 2000 and 2080.

1. Why does the East Coast of North America show great vulnerability of wetlands, while the West Coast does not? (Think about the tectonic setting of each coast.)
2. It has been pointed out that developing nations will bear the greatest burden of global warming. Analyze the map and write a paragraph that specifically delineates the global regions affected, and what the specific effects are projected to be.

Animations of Wave Motion
http://www.kettering.edu/~drussell/Demos/waves/wavemotion.html

This dynamic site provides animations of particle motions in longitudinal waves, transverse waves, and the combined effect on water molecules in water-wave motion.

1. What is the movement path of "tagged" particles in longitudinal wave energy transfer? In transverse waves? What is the particle motion path in the combined water waveform?

2. In water waves, if individual water molecules have no net forward motion, what *is* moving directionally (here, from left to right)?

Wave Refraction at "Jaws" Surfing Beach, Hawaii:
A Computer Model
http://www.coastal.udel.edu/ngs/

Studies of the bottom profile and computer modeling of wave propagation demonstrate wave refraction and focusing of wave energies on a submerged ridge at the "Jaws" surfing site on Maui, Hawaii.

1. Describe the effect of the submerged ridge on incoming waves of period 15 seconds and height of 10 feet.
2. How does this effect differ for waves with a 5-second period at the same site?

Bleached Corals and Global Warming
http://www.marinebiology.org/coralbleaching.htm

This site gives an overview of the problem of bleached and bleaching coral reefs, including underwater photographs and real-time global sea-surface temperatures.

1. What specific factors have been identified as contributory to coral bleaching?
2. Coral reefs are limited to warmer-water regions of the world's oceans. Will continued global warming expand their range?

Earth Systems Today CD Correlations

CHAPTER CONCEPT: Fundamental Changes in Wave Propagation Occur when Waves Reach Shallow Water ($d \leq L/2$)

This part of the CD demonstrates the changes in wave propagation caused by waves dragging bottom in nearshore regions.

SECTION MODULE: The Ocean in Motion: Waves

Observe the animation of a water wave approaching the shore.

QUESTION: *Describe the changes in the wave in terms of height, shape, and velocity at depths ≤L/2.*

CHAPTER CONCEPT: Subsiding Coasts Have Retreating Shorelines, while Tectonically Uplifting Coasts Advance Seaward

This part of the CD shows the location of plate boundaries worldwide.

SECTION MODULE: Plate Tectonics Introduction: Global Map of Plate Boundaries by Type, and Location of Modern Volcanoes and Earthquakes

Click on the button that will display all plate boundary types on a global map.

QUESTION: *Characterize the East and West Coasts of the United States according to these parameters: relative distance to the nearest plate boundary, tectonic activity, retreating or advancing shorelines.*

chapter eleven

Glaciation and Long-Term Climate Change

Perhaps the most benign climate for modern humans began in the fifth century, about the time of the fall of the Roman Empire, and lasted until the beginning of the fourteenth century. During this period, a relatively storm-free Atlantic Ocean was warmer, and Europe experienced mild temperatures. About 1300, however, the climate began to deteriorate, and glaciers in Europe and elsewhere advanced, in the period now known as the Little Ice Age. Paintings and written records from that time reveal images of European glaciers whose waxings and wanings affected human activity.

Glacier activity during the Little Ice Age is known. In Norway, advances between 1700 and 1750 displaced people and overrode and destroyed several farms. Glaciers in Iceland increased greatly in size during the 1500s, and by 1700 a glacier had overriden farms at Breitha and Fjall. By the early 1900s, the ice had shrunk back, and today, once productive farmlands are barren glacial deposits. Paintings and photographs of the Unterer Grindelwald Glacier near Interlaken, Switzerland, dating as early as 1600, record that the glacier has retreated and advanced several times over about a third of a mile and has lost more than a mile of its length since 1860. It began a short-lived advance again in 1977 but has been retreating since then.[1] As the figures testify, land covered with ice during the Little Ice Age is now dotted with houses. ■

(a)

(b)

[1] Michael Hambrey and Jürg Alean, *Glaciers* (New York: Cambridge University Press, 1992); Bjørn G. Anderson and Harold W. Borns, Jr., *The Ice Age World*, (Oslo: Scandinavian University Press, 1994) and Thorleifur Einarsson, *Geology of Iceland* (Reykjavík; Mál og menning, 1994).

(a) *The Unterer Grindelwald*, painted by Samuel Birmann (1793–1847) in 1826. The Grindelwald Valley has two famous glaciers, the Upper and the Lower (German *Unterer*) Grindelwald. (b) The Unterer Grindelwald Glacier today. The village of Grindelwald has expanded into the area formerly covered by the glacier's toe.

Glaciers advance and retreat in response to cyclic global climate changes and consequently they may provide evidence of past climates. In order to better understand global climate changes over time, we will first examine glaciers and glaciation. Important in understanding climates are the interrelationships of the atmospheric, oceanic, and biologic systems with plate tectonics and the orbital elements of the earth–sun–moon system.

Glaciation

About 97 percent of the earth's water is in the oceans, and three fourths of the remainder (2.25%) is in **gla-**

ciers. A tenth of the earth's land area is covered by glaciers at present, the same amount that is cultivated globally. During the Pleistocene Epoch, which radiometric dating places as beginning 1.6 million years ago and lasting until about 10,000 years ago, as much as 30 percent of the earth was covered by glaciers (◀▶ Figure 11.1). Today's glaciers store enormous amounts of fresh water, more than exists in all of the lakes, ponds, reservoirs, rivers, and streams of North America. Farmers in the Midwest grow corn and soybeans in soils of glacial origin, Bunker Hill in Boston is a glacial feature, and the numerous smooth, polished, and grooved bedrock outcroppings in New York City's Central Park were eroded by a glacier that flowed out of Canada (as explained in Case Study 11.2 later in the chapter). Glaciers in the Himalayas, Norway, Switzerland, Alaska, Washington State, Alberta, and British Columbia exert a vital effect on regional water supplies during dry seasons. They do this by supplying a natural base flow of meltwater to rivers that helps to balance the annual and seasonal variations in precipitation. Meltwater from glaciers serves as the principal source of domestic water in Switzerland. The Arapahoe Glacier is an important water source for Boulder, Colorado, so much so that Boulder residents take pride in telling visitors that their water comes from a melting glacier. (Actually, the glacier is only one part of a drainage basin of several tributaries that together supply the city's water.)

The steady water supply provided by melting glaciers during long periods of summer drought is widely used as a water resource for generating hydroelectric power in many temperate regions of the world. Utilization of glacial meltwater is highly developed in Norway

(a)

(b)

◀▶ **FIGURE 11.1** Major centers of ice accumulation and the maximum extent of Pleistocene glaciation in (a) North America and (b) Europe.

and Switzerland, for example. Switzerland's hydropower stations, built mainly in the 1950s and 1960s, use water from melting snow and glacial ice. Dams store the meltwater in the summer, when electrical consumption is low in that country, and in winter, when demand is highest, half of the country's electrical production is generated by water released from the reservoirs.

ORIGIN AND DISTRIBUTION OF GLACIERS

Glaciers form on land where for a number of years more snow falls in the winter than melts in the summer. Such climatic conditions exist at high latitudes and at high altitudes. The pressure increases that occur as the thickness of the snowpack increases cause the familiar six-sided snowflakes to change into a coarse, granular snow called **firn** ("corn snow" to skiers). Continued accumulation increases the pressure and causes much of the trapped air to be expelled from the firn. Recrystallization into larger crystals occurs, and eventually the dense crystals of "blue" glacial ice form.

Once the ice attains a thickness of about 30 meters (100 ft), it begins to deform as a viscous fluid, flowing due to its own weight. It moves downslope if in mountains, or radially outward if on a relatively flat surface. This kind of behavior of an apparently solid material, described as *plastic flow,* takes place in the lower part of the glacier called the **zone of flowage** (◆ Figure 11.2). The upper, surficial layer of ice behaves in a more familiar manner as a brittle solid, often breaking into a jagged chaotic surface of *crevasses*—ominous deep cracks—and *seracs*—towering prominences of unstable ice. It is the deformation by plastic flow that distin-

guishes glaciers from the perennial snow fields that persist at higher elevations in many mountain ranges. Glaciers are more than just masses of frozen water. Although they are largely ice, they also contain large quantities of meltwater and vast amounts of rock debris acquired from the underlying bedrock or mountains where they originated and across which they have moved on their journey downslope. There are several types of glaciers, a classification of which appears in ⦿ Table 11.1.

GLACIER BUDGET

When the rate of glacial advance equals the rate of wasting, the front (terminus) of the glacier remains stationary, and the glacier is said to be in equilibrium. Glaciers **ablate,** or waste away, by melting or, if they terminate in the ocean, by calving. **Calving** is the breaking off of a block of ice from the front of a glacier that produces an iceberg. If the rate of glacier advance exceeds the rate of melting, the terminus advances. Conversely, if the rate of advance is less than the rate of wasting, the terminus retreats. It is important to realize that glacier ice always flows toward its terminal or lateral margins, irrespective of whether the glacier is advancing, is in equilibrium, or is retreating. Glaciers move at varying speeds. Where slopes are gentle, they

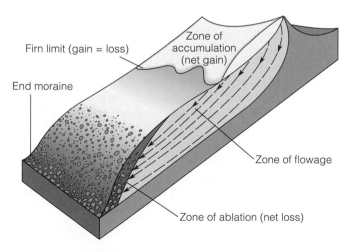

◆ **FIGURE 11.2** Typically, about two thirds of a glacier lies at elevations above the firn limit in the zone of accumulation, where more snow falls in winter than melts in summer. Below the firn limit is the zone of ablation, where glacial ice and the entrained rock debris are exposed by melting in the summer. The ridge of rock debris formed at the edge of the melting ice is an end moraine.

⦿ TABLE 11.1	**Classification of Glaciers**	
TYPE	**CHARACTERISTIC LOCATION**	**EXAMPLE**
Valley (alpine)	mountain valleys	Vaughan Lewis Glacier, Alaska
Ice-field	extensive mountain glaciers that bury many adjoining valleys, leaving only the highest peaks and ridges rising above the ice	Juneau Ice Field, Alaska
Ice-cap	relatively smooth land surfaces in high latitudes or mountain summits where ice flows radially outward	Vatnajökull, Iceland
Piedmont	at the foot of mountain ranges where several valley glaciers emerge, flow laterally, and coalesce	Malaspina Glacier, Alaska
Continental (ice-sheet)	vast land areas	Greenland and Antarctica

Consider This... ?   If an Alaskan glacier's ice accumulates at an average rate of 50 centimeters a year and the glacier is 500 meters thick, how many years will be recorded in a complete core drilled to the base of the glacier? ■

may creep along at rates of a few centimeters to a few meters per day; where slopes are steep, they may move eight to ten meters (26–32 ft) a day (◆ Figure 11.3). Some glaciers, under exceptional conditions, may episodically accelerate with speeds up to 100 meters (328 ft) a day, a condition known as *surging* (see Case Study 11.1).

Probably the best-documented and most dramatic retreat of a glacier is found at Glacier Bay National Park, Alaska. When George Vancouver arrived there in 1794, he found Icy Strait, at its entrance, choked with ice and Glacier Bay only a slight dent in the ice-cliffed shoreline. By 1879, when John Muir visited the area, the glacier, by then named the "Grand Pacific," had retreated nearly 80 kilometers (50 mi) up the bay. The

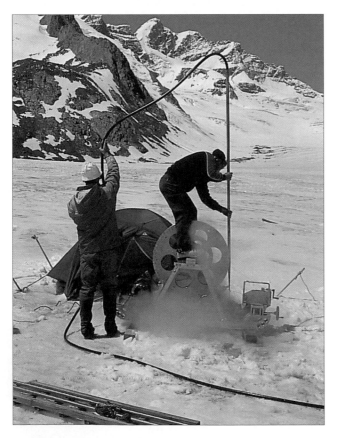

◆ **FIGURE 11.3** Scientists use a jet of hot water to drill a borehole in a glacier. A stake will be placed in the hole that will enable scientists to measure the glacier's movement and summer ablation.

Grand Pacific Glacier had retreated another 24 kilometers (15 mi) by 1916, and today a 104-kilometer (65-mi)-long fiord occupies the area that just 200 years ago held a valley glacier that was as much as 1,220 meters (4,000 ft) thick (◆ Figure 11.4).

GLACIAL FEATURES

The ice was here, the ice was there,
The ice was all around;
It cracked and growled and roared and howled,
Like noises in a swound!

SAMUEL T. COLERIDGE, "THE RIME OF THE ANCIENT MARINER" (1798)

Glaciers carry all manner of rock debris that is deposited directly by melting ice as unsorted rock debris, or **glacial till. Moraines** are landforms composed of till and named for their site of deposition. **End moraines** (also called *terminal moraines*) are deposited at the ends of the melting glaciers. **Recessional moraines** are series of nested end moraines that record the stepwise retreat, or meltback, at the end of an ice age (◆ Figure 11.5).

After melting, continental and ice-cap glaciers leave a subdued, rounded topography with a definite "grain" that indicates the direction of glacial movement. They may also leave behind numerous **kettle lakes,** water-filled, bowl-shaped depressions without surface drainage (◆ Figure 11.6). For example, there are about a dozen kettle lakes in northern Indiana's Valparaiso moraine. These lakes are believed to have formed when large blocks of stagnant ice that were wholly or partly buried in the deposits left behind by retreating glaciers melted (◆ Figures 11.6 and 11.7). End moraines now form the hilly areas extending across the Dakotas, Minnesota, Wisconsin, northern Indiana, Ohio, and all the way to the eastern tip of Long Island, New York (Case Study 11.2). Some Chicago suburbs have names that reflect their location on high areas of hilly moraines: Park Ridge, Palos Hills, and Arlington Heights, for example. In addition, poor drainage or deranged drainage characterizes many areas that have undergone continental glaciation.

Following a period of glaciation, a **fiord** is formed when the seaward end of a coastal U-shaped valley, eroded by a valley glacier, is drowned by an arm of the sea (◆ Figure 11.8).

Consider This... ?   How can a "retreating" glacier still be flowing downhill during a time of global warming? ■

◀▶ **FIGURE 11.4** Map of Glacier Bay National Park, Alaska, showing the terminal positions of the retreating Grand Pacific and other glaciers from about 1750 to the present.

The Effects of Glaciation

Although the great ice sheets of the Pleistocene Epoch seem remote in time and modern glaciers seem remote in distance, both have impacts on modern society that are far from trivial.

◀▶ **FIGURE 11.5** Recessional moraines; Coteau des Prairies, near Aberdeen, South Dakota.

Case Study | 11.1

When a Glacier Surges It's a Moving Experience

A rare but spectacular type of glacier movement is **surging.** A surge begins when ice in the upper part of the glacier reaches some critical thickness. Subsequently the ice surface breaks up into thousands of crevasses, which rapidly spread downglacier, and then the ice movement accelerates from perhaps only a few centimeters a day to as much as 100 meters (328 ft) a day. If the surge reaches the glacier front, the glacier advances dramatically. Surging may cause a glacier to advance several kilometers within a few weeks and be accompanied by a spectacular release of meltwater that causes great damage downvalley. Alaska's Bering Glacier—the earth's largest surging glacier—began surging in May 1993. After a minor retreat for seven months, it resumed surging in April 1995 and was still surging in March 1996. The surge advanced with maximum rates of 7–10 meters a day during the winter of 1993–94, pushing the glacier terminus forward by about 12 kilometers (7 mi) and producing a large number of icebergs, several of them more than a half-kilometer (1,640 ft) long. The U.S. Geological Survey is monitoring the surge using aerial and field observations, paying special attention to iceberg production. As with the icebergs produced by the Columbia Glacier in the 1980s, any icebergs entering the Pacific Ocean could pose a threat in the oil-tanker lanes leading to and from nearby Prince William Sound.

Surges are generally not considered climatically induced events, and it is important to understand them in or-der to predict surges that could endanger installations or people. Such is the case of another Alaskan glacier, the Black Rapids Glacier, which is being monitored because a surge by it could threaten a major highway and the Alaska Pipeline. That the concern is justified was demonstrated by its surge in 1936–37. The owners of a lodge on the Richard-son Highway correctly perceived that the glacier's 3-kilometer (1.8-mi)-wide front was moving toward them and notified authorities. The glacier was found to be advancing at rates of up to 65 meters (213 ft) a day. Fortunately, the surge halted a short distance from the highway; had it continued, it would have destroyed the lodge, dammed a major river, and cut through the highway.

Surges of the Plomo Glacier in the Argentinean Andes have led to the formation of ice-dammed lakes. When one of these lakes bursts through the ice dam, the resulting flood of water can create destruction downvalley. Several deaths were caused by the Plomo surge of 1934, and 13 kilometers (8 mi) of the Mendoza–Santiago railway and seven bridges were washed out. Similar events in the Karakoram Mountains on the border of Kashmir and China have caused extensive damage and loss of life.

Why glaciers surge is still uncertain, but research on the 1982–83 surge of Alaska's Variegated Glacier has provided new information on surging behavior. Located at the northern end of the Alaskan panhandle, the Variegated Glacier (◂▸ Figure 1) flows off a shoulder of the St. Elias Range. It is

DISTRIBUTION OF SOILS

Farming and forestry generally depend upon the distribution and quality of Pleistocene deposits. Some of the best agricultural soils in the world lie on sediments deposited in former glacial lakes and on glacio-eolian (loess) deposits (see Chapter 6). As Pleistocene glaciers expanded and moved over western Europe and North America, they scraped off soils from the underlying land and redeposited them as moraines of glacial till where the ice melted. Today some till-covered areas are excellent forestland and excellent-to-poor farmland (◂▸ Figure 11.9).

GROUND-WATER RESOURCES

Glacial outwash is sediment that was deposited by melt-water rivers originating at the melting edge of a glacier. The great Pleistocene ice sheet that extended out of Canada into the northern Mississippi River valley released copious amounts of sediment-laden melt-

20 kilometers (12 mi) long and is normally a well-behaved tongue of ice, flowing at a rate of about 0.1–0.2 meters (0.33–0.67 ft) a day. For some time before the onset of the surge, studies revealed considerable thickening in the upper reaches of the glacier and thinning in the lower part. The consequent steepening of the glacier increased the stress at the base of the ice in the upper area, which gradually closed the subglacial channels that carried away meltwater. Squeezed out of the channels, the water was forced to spread laterally as a layer across the entire glacier bed. This buoyed up the glacier slightly, which reduced the friction and allowed it to slide faster. Rapid movement began in the upper area, and as the faster-moving upper ice pressed on the slower-moving downglacier ice, stresses began transferring to the lower, thinner part of the glacier, and the surge spread downglacier as a wave. The surge's velocity tripled in its upper reaches and doubled in the middle. After the stresses were relieved, the subglacial channels began to reopen, the subglacial reservoir of water emptied in a concluding flood, and the glacier dropped down on its bed and stopped sliding.

SOURCES: Michael Hambrey and Jürg Alean, *Glaciers* (New York: Cambridge University Press, 1992); Barclay Kamb and others, "Glacier Surge Mechanism," *Science* 227 (1 February 1985), pp. 469–479; and Fleisher, P. J. and others, 1995, The surging advance of Bering Glacier, *Journal of Glaciology* 41: n. 137, pp. 207–213.

◈ FIGURE 1 During the surge of the Variegated Glacier, the glacier's accelerated motion was measured using surveying equipment. The umbrella was used to protect the heat-sensitive electronic equipment from the heat of the sun.

water, resulting in vast sheets of outwash sediments. Some of these glacial-outwash deposits serve as important ground-water aquifers in the end-moraine regions of the Midwest. Moraines associated with the outwash are poor aquifers due to the low porosity and permeability of the clays they contain. Thus geophysical exploration, test-drilling, and careful mapping are necessary in order to delineate the aquifers of porous and permeable sands within the buried stream channels. Glacial outwash can serve as an important source of sand and gravel (◈ Figure 11.10).

SEA-LEVEL CHANGES

The transfer from the oceans to the continents of the millions of cubic miles of water that became the enormous Pleistocene continental ice sheets caused sea level to lower about 100 meters (330 ft). This exposed extensive areas of continental shelves. Consider that a drop of this amount today would increase the area of Florida real estate by more than 30 percent, would leave New Orleans high and dry, and would cause the Panama Canal to drain. Conversely, if all the polar ice

◆ **FIGURE 11.6** Kettle lakes at Kewaskum, Wisconsin.

were to melt, sea level would rise on the order of 40 meters (130 ft), and London, New York, Tokyo, and Los Angeles would be mostly under water. As sea level dropped during the Ice Ages, the rivers in coastal regions began downcutting across the exposed continental shelves as they adjusted to the new conditions. When sea level rose to its present level, about 5,000 years ago, the mouths of the coastal rivers were drowned, forming what are now harbors and estuaries. This last rise of sea level began about 18,000 years ago, a mere twitch in the scale of geologic time.

ISOSTATIC REBOUND

The Ice Age continental glaciers and ice caps, with thicknesses of two miles or more, were so massive that the crust beneath them was depressed by their weight. In the time since the ice retreated at the end of the last ice age, 8,000 to 11,000 years ago, the land regions exposed by the retreating ice sheets in Scandinavia,

Canada, and Great Britain have risen due to the geologically rapid removal of the great load of ice. Uplift of the crust due to unweighting is caused by **isostatic rebound,** the slow transfer by flowage of mantle rock to accommodate uplift of the crust in one area that causes subsidence in another area. Evidence of isostatic rebound in deglaciated areas is seen in the raised beaches around Hudson Bay, the Baltic Sea, and the Great Lakes (◆ Figure 11.11). Careful surveying in Europe shows that as Scandinavia and Great Britain have been rising, a corresponding subsidence of Western Europe has occurred that is especially pronounced in the Netherlands. The uplift and subsidence in these regions remain active. Above the modern shorelines of the Gulf of Bothnia in Finland and Sweden are raised beaches that show a maximum uplift of 275 meters (900 ft). It is estimated that an additional 213 meters (700 ft) of rise will occur before equilibrium is reached. The rise is so rapid in some places in Scandinavia that docks used by ships are literally rising out of the sea. Uplift in Oslo Fjord, Norway, is about 6 mil-

End moraine

Subglacial river

Ice block left by retreating glacier

(a)

◆ **FIGURE 11.9** Rocky soil developed on glacial till; Adirondack Mountains, New York.

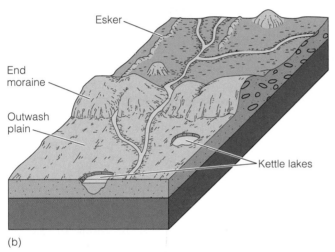

Esker

End moraine

Outwash plain

Kettle lakes

(b)

◆ **FIGURE 11.7** Formation of end moraines, kettle lakes, and eskers (a) during glaciation and (b) after glaciation.

limeters (¼ in) per year, which adds another 60 centimeters (2 ft) of elevation per century. Sea-level mooring rings the Vikings used for tying up their dragon boats a thousand years ago are now 6 meters (20 ft) above sea level. Not even the tallest Viking would be able to moor his boat to those rings today. Similar uplifting is recorded along the northern shores of Hudson Bay, where the average uplift is about a meter (3 ft) per century (◆ Figure 11.12).

◆ **FIGURE 11.8** A fiord, Gastineau Channel near Juneau, Alaska.

◆ **FIGURE 11.10** Part of a vast plain, underlain by Pleistocene glacial outwash near Elgin, Illinois. Outwash sediments, deposited along the margins of the melting continental ice sheets, are spread for thousands of square kilometers across the north-central United States. The ponds in the photograph are the excavations remaining from sand and gravel mining operations.

Case Study | 11.2

Continental Glaciers and the New York Transit Authority

A superb example of a major continental end moraine, the Harbor Hill moraine, extends along Long Island's north shore from Orient Point southwestward through Queens and into Brooklyn, where the rolling ground surface in Prospect Park betrays the glacial origin of the park grounds. The last pulse of the advancing Canadian ice sheet reached this point about 20,000 years ago, leaving the moraine as a kind of glacial footprint. South of the Harbor Hill moraine is the older and less obvious Ronkonkoma moraine. Much of the Harbor Hill moraine on Long Island is inaccessible because it is on private property, but from Port Jefferson to Orient Point, the moraine is easily visible, because its edge is hugged by state highways 25, 25A, and 27 (◆ Figure 1). Sagamore Hill National Historic Site, at Oyster Bay, Long Island, the family home of Theodore Roosevelt, twenty-sixth President of the United States, sits on a recessional moraine that is younger than the Harbor Hill. The cliffs along the bay at Garvies Point, next to the Nassau County Museum of Natural History at Glen Cove, expose a 1–2 meter (3–7 ft)–cover of glacial till overlying Cretaceous sands and clays. The most striking evidence of Pleistocene glaciation are the **erratic** boulders, large rock fragments on the beach and near the museum, carried here by the glacial ice and deposited with the till (◆ Figure 2). These erratics consist of granites and gneisses, unlike any bedrock that crops out nearby, which illustrates the variety of rocks eroded and carried here from distant areas far north of Long Island. At Prospect Park in Brooklyn, which is easily reached by sub-

way, you may stroll along the undulating ground surface of the moraine across terrain described as "knob-and-kettle topography," the *kettles* being depressions that mark former locations of chunks of glacial ice that were buried in the sand and gravel of the moraine and left behind when the ice retreated. Eventually the trapped ice blocks melted, and the ground surface collapsed into the rounded depressions. The moraine continues southwest from Prospect Park to Staten Island, where the moraine gravels form Todt Hill, the highest point in the region at 125 meters (410 ft), and then on into New Jersey and Pennsylvania.

When the Harbor Hill moraine extended across the area that is now the entrance to New York Harbor, it dammed the meltwaters coming down the valley of the modern Hudson River, forming a lake. Eventually the lake waters broke through the moraine dam and eroded the channel to the sea that is now called the Verrazano Narrows. Passing near the Verrazano Narrows Bridge on the New York Transit Authority's Staten Island Ferry or driving across the bridge, one might ponder whether the breakthrough of the moraine was a slow process or a catastrophic one some 20,000 years ago.

In addition to Pleistocene glaciation being responsible for the development of the Long Island landscape, it also set the stage for the region around Hempstead to become the cradle of American aviation. Before urbanization, the moraines and till of Long Island, like much of New York State, were heavily forested. However, the glacial outwash plain south of the moraines supported a mat of grasses so

◆ FIGURE 1 Locations of the Harbor Hill and Ronkonkoma moraines and other features mentioned in the case study.

(a)

◆ FIGURE 2 Glacial erratic boulders on the beach at Garvies Point, Glen Cove, Long Island, New York.

dense that it excluded trees. It was this treeless plain around Hempstead that Glenn Curtiss and other aviation pioneers recognized as the most suitable site for a landing field for their early flying exploits. Thus, the fledgling aircraft industry got its start here about 1910, and Mitchell and Roosevelt airfields were built soon afterward; it was from Roosevelt Field that Charles Lindbergh began his epic solo transatlantic flight in 1927.

Central Park in Manhattan, accessible by bus or subway, reveals its share of glacial footprints. There are numerous *roche moutonnée* (French, "rock sheep"), rounded knobs of bedrock so sculptured by a large glacier that their long axes are oriented in the direction of ice movement. Some of these show *glacial grooves* that were rasped out by boulders. An especially good set of grooves and glacial polish is found on a large roche moutonnée in the southwestern part of the park immediately south of Heckscher baseball fields numbers 3 and 4 (◆ Figure 3). The orientation of the grooves shows that the glacier scraped across Manhattan Island from the northwest to the southeast.

(b)

◆ FIGURE 3 (a) A smoothed roche moutonnée (rock knob) in Central Park, New York City. (b) The manner of ice flow and the influence of jointing on the development of a roche moutonnée. The glacier flows upward and over the unjointed part of the rock, smoothing and polishing it to form the gentle side, and then plucks out and removes blocks from the jointed part of the rock, forming the steep side.

◆▶ **FIGURE 11.11** Isostatically raised beach ridges at Richmond Gulf, Hudson Bay, Quebec, Canada. The highest ridge in the photo lies about 300 meters (984 ft) above sea level and is about 8,000 years old.

Uplifting in the Great Lakes region is slow but continuing and eventually it may cause problems for the residents of Chicago and elsewhere in Illinois. The rebound threatens to cause a drainage change that will force more of Lake Michigan to drain southward into the Mississippi River system.

The rapid unloading of the crust at Glacier Bay, Alaska, in the last 200 years has resulted in measurable rebound in the lower parts of the bay. Small islands are slowly emerging, and existing islands are rising at a rate of about four centimeters (1½ in) per year.

HUMAN TRANSPORTATION ROUTES

Beginning in 1980, the Columbia Glacier on Prince William Sound, Alaska, began a catastrophic retreat by calving thousands of icebergs each year, some of them as large as houses. The glacier had been in equilibrium

◀▶ **FIGURE 11.12** Locations of maximum postglacial (after glaciation) isostatic uplift in the Hudson Bay region and of land features resulting from the retreat of the late Pleistocene ice sheet about 10,000 years ago. Proglacial lakes (temporary, ice-marginal lakes) overflowed at different times through various spillways. One of the arrows east of the Finger Lakes represents the spillway pictured in Figure 11.17. The spillway draining Lake Agassiz in South Dakota is pictured in Figure 11.18. Windblown loess was deposited adjacent to major glacial meltwater systems. Waters from Lake Michigan drained southwestward into the Mississippi River at the time, which may occur again as isostatic uplift continues.

at much the same position for at least 200 years—it was first observed in the late eighteenth century—when it began to show evidence of retreat in the late 1970s. Fortunately, the moraine shoal at its terminus is sufficiently shallow to trap the larger icebergs and prevent them from drifting into Prince William Sound, where they would pose a major threat to the supertankers passing through these waters to and from the oil-loading terminal at Valdez. Because the terminus of the glacier had retreated some 6 kilometers (3.6 mi) or more by 1992, it appears that a new fiord is forming.

Juneau, Alaska, is the only state capital in the United States without highway access, even though it is only about 160 miles (by air) west of the nearest highway in Canada. A highway connecting Juneau with Canada would be desirable, but it is not to be, because the advancing Taku and Hole-in-Wall Glaciers threaten to close off the only possible route along the west shoreline of Taku Inlet (◆ Figure 11.13).

Whereas glaciers impede transportation in some areas, in other areas they have served to facilitate it. In wet boggy regions in the early days of New England, the crests of **eskers** provided

good travel routes. Eskers are long, winding, steep-sided ridges of stratified sand and gravel deposited by subglacial or englacial streams that flowed in ice tunnels in or beneath a retreating glacier. Hence, especially in Maine, the eskers were called *horsebacks* (Figure 11.7b and ◆ Figure 11.14).

PLEISTOCENE LAKES

The cooler climates of the Pleistocene brought increased precipitation, slower evaporation, and the runoff of glacial meltwater. Lakes dotted the basins of North America. Accumulations of water, mostly in valleys between the fault-block mountain ranges in the Great Basin, formed hundreds of lakes far from

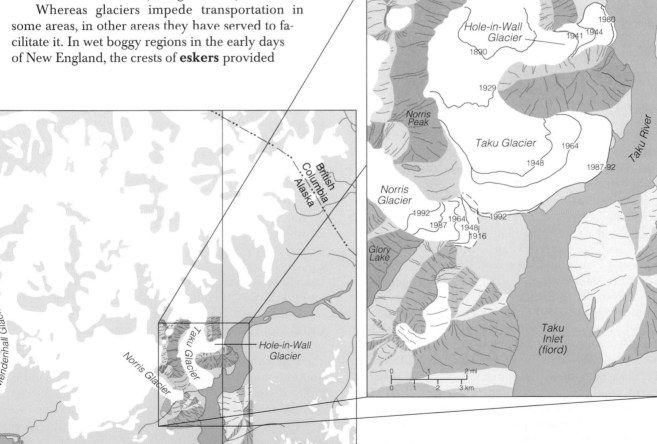

◆ **FIGURE 11.13** Map recording the behavior of the retreating Norris Glacier and the advancing Taku Glacier, Alaska, from 1890 to 1990. The reason for the different behavior of the neighboring glaciers is that the Norris Glacier has a much smaller and lower-elevation accumulation zone than the Taku. Only two percent of the Norris's accumulation zone is above 1,370 meters (4,500 ft), while about 40 percent of the Taku's is above that elevation.

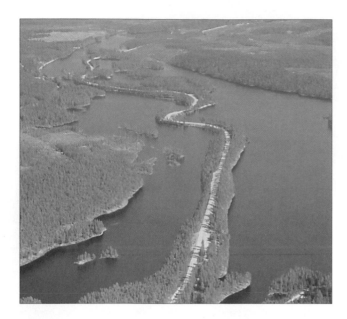

◆ **FIGURE 11.14** Esker with a road on its crest; Malingarna area, southern Sweden.

continental glaciers (◆ Figure 11.15). Today many of these valleys contain aquifers that were probably charged by waters of the Ice Age lakes. California had dozens of lakes, now mostly dried up; among them were Owens Lake, Lake Russell (the ancestor of Mono Lake, ◆ Figure 11.16), and China Lake. Lake Manley covered the floor of what is now Death Valley National Park, being named after the Manley party that suffered great hardships there in 1849, and Rogers Dry Lake, near Landcaster, California, is the location of Edwards Air Force Base and a landing site for the space shuttle. The Great Salt Lake in Utah is all that remains of the once-vast Pleistocene Lake Bonneville. The Bonneville Salt Flats, an area that is ideal for setting automobile speed records, formed by evaporation of the ancient Pleistocene lake. Further east, the Great Lakes and the Finger Lakes of New York State are of glacial origin. In the Pleistocene Epoch, when drainage to the north from the Great Lakes and the Finger Lakes was blocked by ice, these lakes drained through various spillways at different times (Figure 11.12 and ◆ Figure 11.17). Major disruptions in drainage produced large temporary ice-marginal lakes, called **proglacial lakes,** which were dammed partly by moraines and partly by glacial ice.

Ice Age lake sediments are the basis of the rich soils of the northern United States and southern Canada. The rich wheat lands of North Dakota and Manitoba, for example, formed as sediments in Pleistocene Lake Agassiz (Figure 11.12). Centered in present-day Manitoba, the lake was the largest known North American lake; its area of more than 518,000 square kilometers (200,000 mi^2) was more than twice

Consider This... ?   If there had been no large-scale Pleistocene continental glaciation in North America, what differences might we find in today's midwestern states? (*Hint:* Carefully examine Figures 11.1a and 11.2). ■

that of the five present-day Great Lakes combined. (It was named after Louis Agassiz, the nineteenth-century scientist who fostered the theory of the Ice Ages.) The lake formed when the great Canadian ice sheet blocked several rivers flowing northward toward Hudson Bay. For a time, a large spillway channel carried outflow southward from the lake, forming what is now the 200-foot-deep Browns Valley on the border between South Dakota and Minnesota (◆ Figure 11.18). The many remnants of Lake Agassiz include Lake Winnipeg and Lake of the Woods. Northwest of this area are Great Bear Lake, Great Slave Lake, and Lake Athabaska, descendants of other large ice-marginal lakes in the present-day Northwest Territories, Alberta, and Saskatchewan.

GLACIATION AND CLIMATE CHANGE

In the past century it has come to be recognized worldwide that Pleistocene—commonly referred to as *the Ice Age*—glacial deposits of different ages exist. The oldest deposits, weathered and gullied from erosion, have been found beneath younger ones, which are fresher-appearing, less weathered, and less eroded. By applying the Laws of Superposition and Cross-cutting Relationships to these observations, it is clear that multiple stages of glaciation occurred during the Pleistocene Epoch. It is now generally accepted that there have been at least four major glacial stages in North America and Eurasia. Independently, oceanographers studying fossil shells contained in cores of sea-floor sediments from three oceans have determined that the oceans experienced twenty or more periods of deep cooling with intervening warm periods during the Pleistocene. Because evidence of glaciations on land is easily destroyed by later ice advances, it is assumed that the continental record is probably incomplete.

Global temperatures during Pleistocene Ice Age averaged only about 4° to 10° C (7° to 18° F) cooler than the temperatures at present. The lowering of the average Ice Age temperature was not as significant as the range of seasonal extremes; winters were much colder, but summers were moderate. Consequently, climate zones shifted toward the equator, accompanied by similar shifts of plant and animal communities. The Rocky Mountains, the Sierra Nevada, the Alps, the Caucasus, and the Pyrenees were topped with ice fields

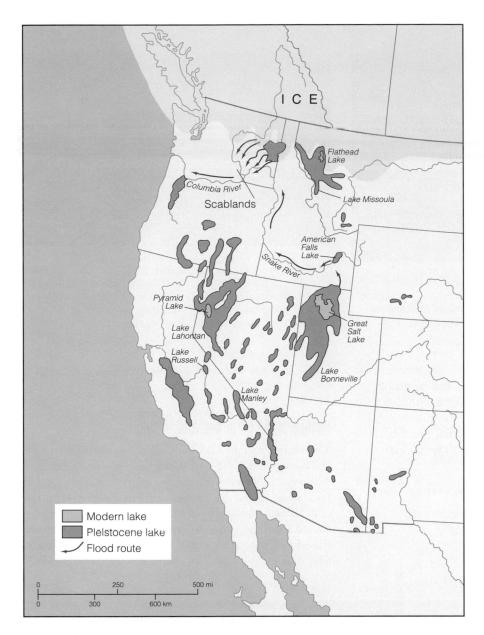

ICE

Flathead Lake

Columbia River
Scablands

Lake Missoula

American Falls Lake

Snake River

Pyramid Lake

Great Salt Lake

Lake Lahontan

Lake Russell

Lake Bonneville

Lake Manley

☐ Modern lake
■ Pleistocene lake
↙ Flood route

0		250		500 mi
0	300		600 km	

similar to the modern ice field in the mountains near Juneau, Alaska. Small mountain glaciers existed in high mountains of the midlatitudes much closer to the equator than any similar modern glaciers. Even the summit of Hawaii's Mauna Loa volcano, only 19° north of the equator, shows evidence of glaciation. It also was cooler south of the equator, but because the landmasses in the Southern Hemisphere's middle latitudes are small or narrow, extensive glaciation was limited to the South Island of New Zealand and the southernmost part of the Andes in South America. Antarctica and Greenland were ice-covered much the same as they are today.

The recognized relationships between cyclic global climate change and multiple stages of glaciation lead to speculation about common underlying causes. In order to better understand global climate changes over time, and why the earth experiences periodic ice ages, we next survey the factors that control climate.

Climate

FACTORS THAT CONTROL CLIMATE

Climate is the long-term average of day-to-day weather. Its definition also includes the annual extremes of weather elements within a specific region. Understanding climate requires recognizing that it is

◁▶ **FIGURE 11.16** Terraces marking the former shorelines of Pleistocene Lake Russell, east-central California (see Figure 11.15). The highest shoreline is approximately 185 meters (600 ft) above the level of the modern Mono Lake.

◁▶ **FIGURE 11.17** Typical subdued terrain resulting from continental glaciation at Lake George, New York. The lake is dammed by an end moraine. The break in the ridge on the right marks the channel of the ancient river that carried meltwater southward from the Pleistocene glacier. See Figure 11.12.

determined by the interactions of the geospheres that comprise the earth system—that is, the atmosphere, hydrosphere, biosphere, and lithosphere, coupled with the orbital elements of the earth–sun–moon system (see Chapter 2). The **climate controls,** the relatively constant factors that dictate a region's climate, are

1. the intensity of the sun, which heats the earth and the atmosphere,
2. the distribution of land and water masses,
3. winds and atmospheric circulation,
4. circulation of the oceans,
5. mountain barriers.

Some of these controls are treated in more detail in Chapter 12.

Heating of the Earth and Atmosphere

The earth's surface and atmosphere is warmed by radiant energy from the sun. Because part of the sun's radiation is reflected back to space by clouds and the atmosphere, about 51 percent of the solar energy that could reach the earth actually does. This energy received from the sun is mainly visible light and near-infrared radiation, particular wavelengths of solar radiation that do not "see" the atmosphere through which

◁▶ **FIGURE 11.18** Big Stone Lake now occupies part of Lake Agassiz's spillway channel. It is in Browns Valley on the boundary between western Minnesota and northeastern South Dakota. See Figure 11.12.

Oblique rays

Vertical rays

Oblique rays

FIGURE 11.19 The intensity of the sun's heat depends upon the angle at which the sun's rays hit the earth. The intensity and heating is greatest where the rays hit vertically.

they pass on their way to the earth, where they heat the ground and oceans. The intensity of the sun's heat at a particular place on the earth's surface varies with the angle at which the sun's rays hit the earth (❖ Figure 11.19). This incoming solar radiation is sometimes referred to as *short-wave radiation.* Because all warm objects radiate heat, the warmed surface of the earth radiates energy back to the atmosphere, but at longer wavelengths known as *infrared,* or *long-wave, radiation.* The atmosphere is warmed by this long-wave radiation due mainly to the absorption and reradiation of heat by atmospheric CO_2, water vapor, and other trace gases. Consequently, the atmosphere acts as a blanket that absorbs most of the long-wave radiation and keeps the earth warm. This process, called the **greenhouse effect** by analogy to the solar-heating method that warms greenhouses, serves to keep the earth's surface temperature from dropping excessively during winter and night.

Winds and Atmospheric Circulation

Wind is movement of air from a region of high pressure to a region of low pressure due to unequal heating of the earth's surface. Because the sun's rays strike the earth more directly at the equator than at other places, more radiation (heat) per unit area is received there (❖ Figure 11.20). Solar heating at the equator warms the air, and it rises, creating a zone of low atmospheric pressure known as the **intertropical convergence zone (ITCZ),** and the **doldrums.**

In polar regions, the sun rays strike the earth at a low angle, resulting in less-intense heating per unit area. There the colder, drier, denser air creates a zone of high pressure known as the **polar high.** The cold, dry, polar air sinks and moves laterally into areas of lower pressure at lower latitudes. Upper-level air flow from the equatorial regions moves toward the poles. When it reaches approximately 30 degrees north and south of the equator, it sinks back to the earth's surface as dense, dry air to form high-pressure belts known as the **subtropical highs,** or the **horse latitudes.** Winds, then, are driven by differences in air pressure caused by unequal heating, but as they move they do not follow straight paths. Instead, they are deflected by the Coriolis effect.

The **Coriolis effect** is caused by the rotation of the earth. A simple analogy of the Coriolis effect is to consider two people playing catch on a merry-go-round (❖ Figure 11.21). If the merry-go-round is stationary, the ball will follow a straight path. An apparent paradox occurs when the merry-go-round is turning, however. As seen by stationary observers not on the merry-go-round, the thrown ball still travels in a straight path, but to the people on the turning merry-go-round, the ball appears to follow a curved path. The reason for this paradox is that the merry-go-round and its passengers are turning below the ball as it travels from one person to the other. To the people on the merry-go-round, the path of the ball appears to be deflected as though some force is acting on it. The curve will be to the right if the merry-go-round is turning counterclockwise, the same direction the earth rotates if observed from above the North Pole. The deflection of freely moving objects also occurs on the earth. Rifle bullets, ballistic missiles, ocean currents, and winds that travel long distances all appear to follow curved paths because of the earth's rotation. The path is deflected to the right in the Northern Hemisphere and to the left in the Southern Hemisphere.

The rising warm air in the equatorial belt expands and cools as it rises. Cooling reduces the air's ability to hold moisture, resulting in the formation of clouds and heavy precipitation. For this reason, the intertropical convergence zone (ITCZ) is one of the rainiest parts of the world. We generally do not find deserts there. Air also rises at latitudes of approximately 60° north and south, where cold polar surface air and mild-temperature surface air converge but tend not to mix. At this boundary, called the **polar front,** surface air rises, forming a zone of low pressure called the **subpolar low.** It is along the polar front that rising air causes storms to develop. In the subtropics, in contrast, cool, dry air sinks, forming high-pressure regions. As the air sinks, it is compressed, which causes its temperature to increase. This warming enables the air to absorb water and precludes the formation of clouds. This is the reason that the climate belts of subtropical highs, from about 30° to

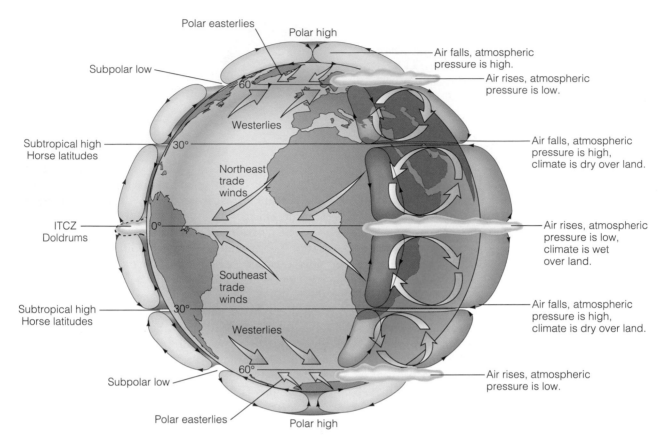

FIGURE 11.20 In the general scheme of global atmospheric circulation, heat transfer occurs in steps generating three cells between each pole and the equator. This results in persistent global air-circulation patterns that generate atmospheric pressure zones and surface wind belts. Descending air causes zones of high pressure (the polar highs and the subtropical highs), and rising air produces zones of low pressure (the ITCZ and the subpolar lows).

Consider This... ? The old expression "caught in the doldrums" is sometimes used to describe a person who lacks motivation and is listless. What is the relationship of this expression to the traditional term for the ITCZ? ∎

35° north and south of the equator, the so-called horse latitudes, experience warm, dry winds with generally clear skies and sunny days. It is here that we find most of the world's hot deserts. (Deserts are the topic of Chapter 12.)

FIGURE 11.21 The Coriolis effect in a game of catch. (a) The path of the ball is straight when the game is played on a stationary platform, and an observer perceives it as straight. (b) On a rotating platform the ball still follows a straight path, but to an observer on the platform, the path appears to be curved. This is the Coriolis effect.

(a)

(b)

Oceanic Circulation

Unequal heating of the earth's surface by the sun moves the atmosphere, and the resulting winds drive the surface currents of the oceans. Each major ocean basin is distinguished by currents that rotate clockwise in the Northern Hemisphere and counterclockwise in the Southern Hemisphere in accordance with the Coriolis effect (◆ Figure 11.22). In both hemispheres, these currents carry heat from the equator toward the poles along the west sides of the basins and, in turn, carry cooled water from polar latitudes back toward equatorial regions along the east sides of the basins. This transfer of heat occurs in all oceans. For example, the California Current cools the west coast of North America, imparting mild summer temperatures to San Diego. In contrast, the warm Gulf Stream on the east coast imparts hot, humid summers to Savannah, Georgia, at the same latitude.

A major oceanic process affecting coasts in the "trade-wind" belts north and south of the ITCZ (Figure 11.20) is **upwelling.** Upwelling occurs when winds blowing from the land to the sea *(offshore winds)* sweep surface water away from the shore, creating a condition of hydraulic instability. The displaced nearshore water is replaced by nutrient-rich, cold waters from depths in excess of 200 meters (650 ft; Figure 12.2, near the beginning of the next chapter, illustrates this). During upwelling, coastal waters nearshore are colder than offshore water, which further moderates local weather. Conditions for coastal upwelling are most commonly found along the western sides of continents in the trade-wind belts, as, for example, along the coasts of Peru and California. Upwelling brings dissolved phosphates and nitrates to the surface, which leads to the proliferation of marine plankton, which in turn promotes large fish populations and gives rise to commercial fisheries. However, if the trade winds break down or weaken and upwelling ceases, major climatic upsets such as El Niño may occur, causing floods and severe storms in some areas (see Chapter 10).

CAUSES OF CLIMATE CHANGE
Variations in the Earth's Orbit

Once scholars recognized that cyclic global climate changes were responsible for the multiple stages of glaciation, they began to seek answers to questions about underlying causes. An early answer was

◆ FIGURE 11.22 Major wind-driven oceanic surface currents. Warm currents are shown in red; cool currents in blue. Currents rotate clockwise in the Northern Hemisphere and counterclockwise in the Southern Hemisphere.

provided in the late nineteenth century by Scottish geologist John Croll. Serbian astronomer Milutan Milankovitch expanded upon Croll's work with elegant mathematics in 1941. Both of these scientists recognized that the cyclic pattern of the Pleistocene ice ages could result from periodic minor changes in the earth's rotation and revolution around the sun, with the cold periods of ice advance occurring when three variations in the earth's rotational and orbital elements were synchronous. The three variations are

1. the **obliquity,** or tilt, of the earth's axis;
2. the **eccentricity,** or shape, of the earth's orbit; and
3. the **precession,** or wobble, of the earth's axis.

See ◆ Figure 11.23. The earth's *obliquity* relative to the sun has been found to vary between 21.2° and 24.5° over a 41,000-year cycle. This slight change causes variations in the temperature range between summer and winter. When the earth is at the minimum tilt, sunlight hits the polar regions at a higher angle, and the global seasonal temperature variation decreases. Every 100,000 years, the orbit of the earth changes from nearly circular to more *eccentric*. Seasonal variation in temperature is greater when the orbit is more eccentric. The axis of the earth, which now points to Polaris, the North Star, *precesses* like a spinning top when its axis is disturbed. This precession causes the axis to trace out a cone in space, taking 23,000 years to trace out a complete cone. Consequently, the Northern and Southern Hemispheres "trade" their respective winter and summer seasons every 11,500 years.

   Each of these periodic factors alone could cause the limited global cooling that might lead to a "little ice age." More important is the interaction of the three cycles, however. When their cooling effects coincide, there is a sufficient change in the global heat budget to trigger glacial–interglacial episodes, though not enough to explain fully the 4° to 10° C cooling required for the major glacial cycles of the Pleistocene Epoch. In other words, the rotational and orbital factors adequately explain the *timing* of the glacial-interglacial cycles, but not the amount necessary to cause the *changes in global temperature*. There must be other factors.

Plate Tectonics and Variation in Atmospheric Carbon Dioxide

For glaciers to exist, the continents must be in the middle to higher latitudes, and high-standing mountains are "desirable." Plate tectonics can account for both the positioning of continents at "appropriate" latitudes and the uplifting of vast continental areas. Such repositioning of continents and higher elevations of land areas alters the circulation patterns of both the atmosphere and oceans,

(a)

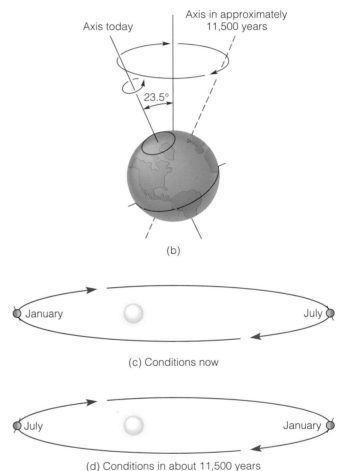

(b)

(c) Conditions now

(d) Conditions in about 11,500 years

◆ **FIGURE 11.23** Geometry of variation in the earth's orbital eccentricity, axial precession, and axial obliquity. (a) The orbital eccentricity varies from nearly circular *(dashed line)* to elliptical *(solid line)* to nearly circular and back again over a period of about 100,000 years. (b) The earth's axis slowly precesses, like a spinning top tracing out a cone in space, taking about 23,000 years for a complete cycle.
(c) Conditions today. The earth is closest to the sun in January while the Northern Hemisphere is experiencing winter. (d) Conditions in about 11,500 years. The earth will be closest to the sun in July while the Northern Hemisphere is experiencing summer.

which leads to long-term climate change. In addition, the higher the continental land masses, the greater is the rock surface area exposed to chemical weathering by atmospheric gases. As the rocks weather, they combine with water and atmospheric carbon dioxide (CO_2), which reduces the amount of the gas in the atmosphere. Further, input of CO_2 from volcanic activity is reduced when, at the end of a tectonic cycle, the volcanic fires at oceanic ridges cool down and the ridges become less active. Because CO_2 is the major greenhouse gas, these reductions promote global cooling. (In the earlier phases of the cycle, the ridges are active, which increases the CO_2 input to the atmosphere, and sea level is high, which reduces weathering. These two factors reinforce each other to increase atmospheric CO_2.) The relationship of CO_2 concentrations in the atmosphere, oceans, and biosphere is enormously complicated, and several feedback mechanisms exist that are not understood (Case Study 11.3). For example, diatoms, the microscopic oceanic plants that can be thought of as the "grass of the sea," may be part of a feedback loop in the biosphere that affects climate. Diatoms may step up productivity when there is excess atmospheric CO_2, thus drawing down the CO_2 content and cooling the climate.

Variations in Solar Radiation

Measurements by sophisticated instruments carried by satellites indicate that the output of solar radiation may vary considerably more than was once thought. Furthermore, it is known that the sun's energy output changes slightly with sunspot activity. *Sunspots*—enormous magnetic storms on the sun's surface that mark cooler regions surrounded by very bright areas radiating great bursts of energy—occur in cycles, with the maximum sunspot size and number occurring about every 11 years. During such times the sun produces about 0.1 percent more energy than it does at times of sunspot minimum. The 11-year cycle appears to collapse from time to time, as it did in the seventeenth century (at a time known as the *Maunder minimum*), when there were few if any sunspots. This "minimum" occurred during the coldest period of the Little Ice Age, when the average global temperature was about 0.5° C (1° F) lower than the long-term average.

Correlated with the 11-year sunspot cycle, is a 22-year cycle, the time required for completion of the sun's *magnetic cycle*—and an interval that closely corresponds to the 20-year cycle of drought on the Great Plains. The sun's general magnetic field increased 2.3 times between 1901 and 2000, and because the general magnetic field varies in time with the 11-year sunspot cycle, the varying magnetic field is also related to the sun's brightness. Furthermore, the impact of the sun's

magnetic field on the earth appears to be related to cloudiness. (The field influences the amount of cosmic rays that enter the earth's atmosphere, which in turn, may affect the formation of electrically charged particles, on which water condenses to form cloud droplets.)

Moreover, solar brightness also affects the stratospheric ozone layer. The 0.1 percent increase in solar energy produced during the 11-year maximum is ultraviolet radiation. Ultraviolet rays' wavelengths are absorbed by stratospheric ozone, the level of which also varies with the sunspot cycle. Because the time of maximum ozone content is also the time of maximum sunspot activity, additional ozone at such times absorbs the additional ultraviolet radiation.

If increased solar radiation is currently involved in global warming, it would be warming the upper and lower atmosphere as well as the earth's surface. This is not the case. The upper atmosphere is cooling, while the lower atmosphere and the earth surface are warming. Eventually it may be found that fluctuations in solar radiation account for some long-term climate changes, but much more research is needed. Correlation is not necessarily causation.

Greenhouse Gases

Scientists have been measuring and recording the level of CO_2 in the atmosphere since 1880. Their measurements over the years show that the concentrations of CO_2 and other greenhouse gases, as well as average global temperatures, have risen steadily (◆ Figure 11.24). This increase in atmospheric CO_2 is unprecedented in human history, and it is believed to be largely due to humankind's dependence upon fossil fuels. Other "greenhouse gases," include:

- methane (CH_4), which is expelled as flatulence from the world's cattle population, from rice paddies, and of all things, from termites in tropical rainforests;
- nitrogen oxide (NO_x) from fertilizers and automobile exhausts; and
- **chlorofluorocarbons (CFCs),** largely from air conditioners and spray-can propellents.

These substances combine to increase global warming.

In addition to their being a greenhouse gas, CFCs play another role in the atmosphere; they break down the upper-atmospheric (the stratospheric) ozone layer 10 to 48 kilometers (6–30 mi) above the earth surface. Ozone, a form of oxygen, serves as a vital shield that protects the earth's organisms from the harmful solar ultraviolet radiation that bombards the earth. The chlorine in one molecule of CFC has the potential to destroy as many as 100,000 ozone molecules before it

Case Study | 11.3

The Carbon Cycle

Complex interactions of the earth's geospheres—the pedosphere, the atmosphere, the hydrosphere, the biosphere, and the lithosphere—produce climate. The interactions are difficult to analyze and they are not well understood. By extending understandings about today's climate into the remote geological past, however, geologists have been able to shed some light on the workings of the climate system. They have reconstructed ancient climates by studying sedimentary layers and deducing their depositional environments, by studying fossil animals and plants, and by carrying out isotope studies that yield information on past temperatures of the earth's oceans and surface. Such *paleostudies* have laid important groundwork for understanding the earth's carbon cycle.

Presently, the earth's reserves of fossil fuels are being burned at a prodigious rate, and the large amount of carbon dioxide (CO_2) being released to the atmosphere by the burning appears to be impacting our planet and its inhabitants.

◆ FIGURE 1 The carbon cycle. (a) Carbon enters the atmosphere naturally by biological respiration, volcanism, and chemical weathering. Photosynthesis transfers it from the atmosphere to the biosphere. (b) Humans release carbon to the atmosphere when they burn fossil fuels and forests.

Carbon, the essential element that distinguishes living and once-living matter from other matter, is involved in the changing atmospheric composition that may be causing long-term climate change. Carbon exists in various forms in each of the geospheres:

- The biosphere is the system of all living and some decaying dead organisms, each cell of which contains carbon.
- The hydrosphere contains dissolved carbon dioxide.
- The atmosphere and pedosphere contain CO_2 gas.
- Carbon is stored in the lithosphere in deeply buried organic materials, in petroleum and coal, and as calcium carbonate in limestone and marble.

The transfer of carbon among these geospheres is termed the carbon cycle (◁▷ Figure 1). The carbon cycle is not just the cycling of CO_2; it incorporates all of the earth's carbon compounds, and there are many of them.

The biosphere is the key in the carbon cycle. Plants continuously remove CO_2 from the atmosphere and utilize it, by photosynthesis, to produce their food (sugar), much of which, in turn, is consumed by animals. Plant and animal waste materials contain carbon, which ultimately is carried to the oceans by surface runoff. Dead organisms decay by combining with atmospheric oxygen to form more CO_2, which is then cycled back to the atmosphere. So fast is the cycling of carbon through the biosphere that a complete recycling of atmospheric CO_2 occurs about every four years.

In the ocean, plankton, shellfish, and other marine life use CO_2 and calcium to form shells and skeletons. These materials eventually sink to the sea floor, where their carbon becomes sequestered in calcareous sediment. By the motion of the crustal plates, ocean-floor sediments eventually are cycled into the mantle by subduction. Heat drives the CO_2 out of the sediment, and CO_2 dissolved in magma eventually rises to the surface through volcanism. When a volcano erupts, large amounts of CO_2 are returned to the atmosphere to restart the cycle.

Not all dead organic matter in the biosphere decays to CO_2. Some is sequestered in sedimentary rock, forming reservoirs of carbon in petroleum and coal, the so-called fossil fuels. Some of this material naturally reenters the atmosphere when uplift and erosion expose the rock and its trapped carbon compounds (◁▷ Figure 2).

Atmospheric CO_2 is the smallest of all the carbon reservoirs, accounting for only about 0.03 percent of the atmosphere. Because of the atmosphere's relatively small carbon content, it is very sensitive to changes in the larger carbon reservoirs with which it exchanges. Of special concern is the recent acceleration of human-induced CO_2 production by the clearing of forests and the burning of fossil fuels. Both of these release CO_2 to the atmosphere from the biosphere and the lithosphere much faster than would occur naturally. The rate at which natural processes remove CO_2 from the atmosphere has not changed, however, and that is why there is a buildup of CO_2 in the atmosphere. Accordingly, this anthropogenic atmospheric-CO_2 imbalance to the greenhouse effect is the focus of the concern about global warming.

Despite the availability of high-speed computers and our ability to do sophisticated computer modeling, our knowledge about the carbon cycle is remarkably scant. Scientists have yet to account for about 25 percent of the sequestered carbon. Apparently, it is not in the ocean, but somewhere in the terrestrial biosphere, that is, in land-living plants and animals. Solving the puzzle of the "missing" carbon might be a big step toward solving the bigger global-climate-change problem. Such an effort will require the collaboration of paleoclimatologists, ecologists, ecosystem modelers, geochemists, and others. Such coordinated problem-solving is the key to understanding the carbon cycle and global change.

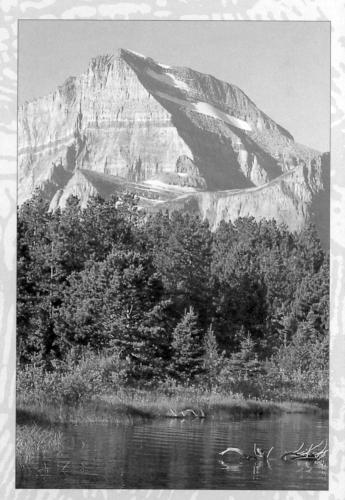

◁▷ FIGURE 2 These mountains in the Montana Rockies are composed largely of ancient carbonates that were formed as sea-floor sediments about 1.5 billion years ago. Such rock bodies constitute reservoirs of carbon that have been isolated from the carbon cycle. Combined uplift, erosion, and weathering processes are now slowly releasing the long-sequestered carbon back into the carbon cycle.

◀▶ **FIGURE 11.24** Atmospheric CO$_2$ and temperature changes since 1880. (a) Changes in concentration of CO$_2$ obtained from ice cores (smooth curve) and measured annual oscillations at Mauna Loa, Hawaii. (b) Average annual surface-air temperature at Mauna Loa, plotted as a deviation from the average, denoted as zero (0).

(a)

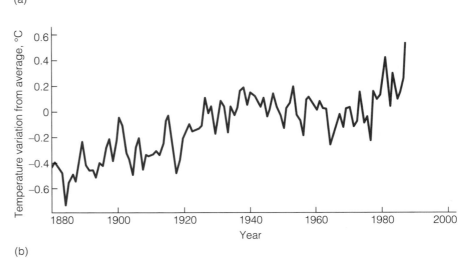

(b)

becomes inert. Any thinning of the ozone shield leads to global change.

Using data from satellites and airborne sensors, scientists discovered in the late 1970s and early 1980s that atmospheric ozone levels were decreasing at an alarming rate. By the spring of 1987, an "ozone hole" the size of the United States and Mexico combined had developed over Antarctica. The hole has continued to enlarge, appearing to document the continuing effects of human activity on earth. Ozone thinning is now recognized over the North Pole and the middle latitudes as well. Ozone loss in the stratosphere causes warming on the earth's surface as well as dangerous cooling in the stratosphere.

A thinning of the atmospheric ozone layer threatens humans with increased risks of skin cancer and cataracts and of weakening the immune system. The increased ultraviolet radiation reaching the earth also threatens to damage crops and phytoplankton, the microscopic plants that are the basis of ocean food chains. Some medical authorities reported that a person's chance of developing skin cancer by the 1990s was

about one in 75, up from one in 1,500 about 40 years earlier. Humans, of course, can protect themselves from ultraviolet radiation, but other animals cannot. For example, ultraviolet-radiation damage to DNA has been shown to account for severe losses of fertilized eggs in at least two species of Cascade Mountain frogs.

Also contributing to the depletion of the ozone layer is the burning of aviation fuel at altitudes above 9 kilometers (5.5 mi). Nitrogen oxides (NO$_x$) and water-vapor molecules released at altitudes between 9 and 13 kilometers (5.5 and 8 mi)—the usual cruising altitude for commercial jetliners—stay in the atmosphere about 100 times longer than those released near ground level. Although it seems contradictory, NO$_x$ emissions can either increase or decrease ozone concentrations in the atmosphere. In the lower atmosphere, where we live, NO$_x$, together with hydrocarbons, forms ozone (a major component of smog) in the presence of ultraviolet light. In the upper atmosphere, however, NO$_x$ breaks down ozone. CFC climate modeling, which does not take NO$_x$ emissions from air traffic into account, has underestimated the amount of ozone depletion. According to researchers, the amount of ozone loss over the Northern Hemisphere is about twice that calculated using only CFC climate modeling. Regular (subsonic) air traffic is believed to be responsible for a 10 percent increase in the nitric acid concentration at 21 kilometers (13 mi) above sea level, which may be the most critical altitude for ozone holes. Clearly, the chemistry of the atmosphere is complex, and it is the subject of continuing scientific research.

In the early 1990s, a few radio talk-show hosts and some government policymakers expressed doubts about the danger of CFCs and the depletion of the stratospheric ozone layer. They argued that chlorine from CFCs is trivial in comparison to the great quantities of chlorine released to the atmosphere from natural sources such as volcanic eruptions, which have been contributing chlorine to the atmosphere for many hundreds of millions of years. In spite of this, the ozone layer still exists, so, they argued, the concern about CFCs is overstated. Their argument was spurious, however, because natural chlorine is chemically reactive in the lower atmosphere and soluble in water; consequently, it gets rained out before it can reach the stratosphere. CFCs, in contrast, are absolutely inert and thus insoluble in the lower atmosphere. This allows them to rise through the lower atmosphere and into the stratosphere, where ultraviolet rays cause the release of chlorine. It was recognition of their chemical inactivity in the lower atmosphere—they react with *nothing*—that made them so appealing as coolants and propellants.[2]

Dust and Aerosols

Studies of glacial ice cores not only verify the Croll–Milankovitch theory of cyclic ice ages (Figure 11.23), but they also reveal that the atmosphere was dusty during glacial stages. Winds apparently lifted fine dust from glacial deposits and arid regions, causing the skies to become hazy. This haziness would have scattered incoming solar radiation back into space, causing further cooling of the earth.

Volcanic activity has been suggested as a contributor to global cooling, both because of the dust ejected by eruptions and due to volcanic releases of sulfur dioxide (SO_2), which is converted to sulfate-aerosol

droplets in the atmosphere. The sulfate aerosols, acting like tiny mirrors, cool the atmosphere by reflecting sunlight away from the earth. An illustration of the role played by volcanic eruptions in global climate is provided by the 1815 eruption of Tambora in Indonesia. After that event, with the spread of the eruption's aerosols and dust through the stratosphere, the amount of solar radiation reaching the earth was so reduced that the ensuing global cooling caused widespread crop failure. The starvation that followed caused the deaths of more than 90,000 people. Because the global cooling resulted in rain, late snows, and failed crops in Europe and in New England, 1816 became known as the "year without a summer."[3] Similar reductions in average global temperatures due to stratospheric sulfate aerosols and dust following volcanic eruptions in modern times are recognized, such as those after the 1982 eruptions of El Chichón in Mexico and (for two years following) the 1991 eruption of Mount Pinatubo in the Philippines (see Chapter 5).

Scientists once thought that all sulfate aerosols made clouds more reflective, thus helping to *cool* the earth. But that is true only of aerosols resulting from volcanic emissions, as they are blasted high into the stratosphere. In contrast, it is now recognized that sulfate aerosols released by coal-burning power plants and industry and by motor vehicles reside in the lower atmosphere, where they form droplets that make clouds dirtier and less reflective, thereby *increasing* global warming.

Change in Ocean Currents

Vagaries in ocean currents, too, seem to play a role in plunging the earth into glacial and interglacial climates. A giant, salty "conveyor-belt" current snakes through the world's oceans (◆ Figure 11.25). Originating in the North Atlantic by northward-flowing warm, salty (and thus dense) water because of evaporation, the belt is chilled as it moves northward, which further increases its density. It sinks into the depths flowing southward out of the Atlantic and into the other oceans, much of it rising in the Pacific, where it offsets the evaporation losses there. This conveyor is the main agent of oceanic heat transfer, with one part of it a deep, cold, salty (and thus dense) current (shown in green in Figure 11.25), and the other part of it a warm, less salty, surface current (shown in yellow). The northward-flowing surface water averages 8° C warmer than the cold, deep water flowing south.

[2]U.S. production of CFCs ended in December 1995. Now compounds that lack chlorine or that decompose in the lower atmosphere—the hydrofluorocarbons (HFCs) and the hydro*chloro*fluorocarbons (HCFCs)—are being used. In the rest of the world production of CFCs ended on 1 January 2000. The changeover requires that both refrigeration units and manufacturing facilities be modified to use the new refrigerants. Older refrigerators and air conditioners presently using CFCs are to be recharged only with recycled CFCs, but eventually they will be discarded or will be retrofitted to use the newer refrigerants.

[3]This cold summer made the *Old Farmer's Almanac* famous. When the printer was setting type for the 1816 issue, he found July's weather forecast to be incomplete. He decided to predict snow for the Fourth of July. Astoundingly, on that date, it *did* snow in New England, where a large proportion of Americans lived at the time, thereby establishing a solid reputation for that almanac's reliability.

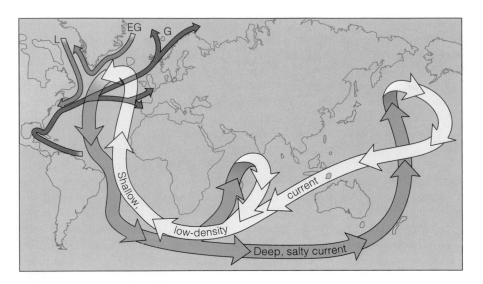

◁▶ **FIGURE 11.25** The giant "conveyor belt" current that snakes through the world's oceans. Two surface currents of major importance in determining the modern climates in Greenland, eastern North America, and western Europe are shown. The cold Labrador (*L*) and East Greenland (*EG*) currents (both shown in blue) cool the adjacent coasts. The Gulf Stream (*G*, red) transfers heat northward, influencing Arctic air currents and warming the coasts of Iceland, Greenland, and western Europe. This current was deflected southward toward North Africa during the coldest part of the last maximum glaciation. The importance of these currents, in combination with the atmospheric wind system, is illustrated by comparing the modern equitable temperature of northern Scandinavia, which is essentially unglaciated, with that of Greenland, which, at the same latitude, is covered by an enormous ice sheet.

When the conveyor is operating, the climate of the North Atlantic is mild, like that of today. It is vulnerable to abrupt disruptions and shutdowns, however, which can cause drastic shifts in global climate (see Case Study 11.4) and force the North Atlantic into bitter, ice-age cold.

When excessively large volumes of cold fresh water from melting icebergs and river runoff enter the North Atlantic, the conveyor shuts down and winter temperatures drop by five or more degrees.

GLACIER EVIDENCE OF CLIMATE CHANGE

Glacial ice provides evidence of cyclic changes in atmospheric chemistry, temperature, and amount of atmospheric dust that correlate with periods of cooling and warming during the Pleistocene Epoch. As snow accumulates and is packed down into ice, it carries information with it that gets locked into glaciers. The relative abundance of two slightly different oxygen atoms—the oxygen-18 to oxygen-16 isotope ratio, symbolized $^{18}O/^{16}O$—in ice molecules indicates what the temperature was when the snow from which it was formed fell. During a glacial period, when sea water evaporates from a cold ocean and precipitates on land

to form glaciers, water containing the lighter isotope, ^{16}O, is more readily evaporated than is water containing the heavier ^{18}O. Thus, the oceans come to contain more of the heavy isotope, and glaciers to contain more of the light isotope. Furthermore, each year's snowfall forms an easily recognized line in glacial ice. When core samples of glaciers are examined, these lines act like tree rings, allowing scientists to determine the years, and the $^{18}O/^{16}O$ ratios allow each year's temperature to be estimated. Together, these data allow a generalized overview and time scale of global climate change.

The bubbles of ancient atmosphere trapped in glacial ice obtained from 3,348-meter (10,981-ft)-deep cores drilled at Vostok Station in Antarctica, and the Greenland Ice Sheet Project 2 (GISP2) provide a 400,000-year record of variations in the atmosphere's temperature and composition (◁▶ Figure 11.26). The ice-core record and correlative deep-sea-sediment cores show that ice ages are marked by climatic jiggles—intricate cycling within the larger cycles of glacial advance and retreat due to orbital forcing by the Croll–Milankovitch effect. Peculiarities in the lower part of the GISP2 ice core, presumably representing atmospheric and climatic conditions existing about 120,000 years

Case Study | 11.4

Is There a New Beat to the Rhythm of the Ice Ages?

If you think today's weather (Table 11.4) seems weird, try to imagine weather conditions during the last ice age. Oceanographers' studies of sea-floor sediments have revealed that massive fleets of icebergs have moved across the North Atlantic Ocean every 2,000 to 3,000 years as the climate has flip-flopped back and forth between warm interglacial and glacial periods. The melting icebergs have chilled the North Atlantic and left trails of ground-up rock debris in layers on the ocean floor. After two or three such cycles there would be a larger flood of icebergs, which left a layer with a different composition. These greater floods, occurring every 7,000 to 12,000 years, have become known as *Heinrich events* (named for the German oceanographer who discovered them), and the smaller, more frequent events are informally called *flickers*.

It is now believed that the Heinrich events, recognized by layers of carbonate-rock grains in the sea-floor rock debris, resulted when the giant Keewatin ice sheet (Figure 11.1a) surged into Canada's Hudson Strait, causing enormous numbers of bergs to calve and drift into the North Atlantic. The more numerous flickers, in contrast, left layers of reddish, iron-coated sand on the ocean floor that has been traced to the present-day St. Lawrence Valley, which also was ice-covered at the time. The significance of the Heinrich and flicker events in the oceanic record is their implication that in the late stages of the Pleistocene ice ages, the Northern Hemisphere experienced numerous rapid climate changes unlike anything seen in historic times (◇ Figure 1).

Meanwhile in Greenland, glaciologists' studies of oxygen isotope ratios in ice cores have revealed wild flip-flops in climate from bitter cold to warm, between 115,000 and 140,000 years ago (an interval known as the *Eemian*), with the same timing as the Heinrich and flicker events recorded in the sea-floor rocks. What does this all mean? One hypothesis is that the St. Lawrence ice sheet accumulated and surged several times as the Keewatin ice sheet was growing, and that when the Keewatin ice sheet surged, the ice over the St. Lawrence did also. The two ice sheets, each acting independently and having different characteristics, must have surged as a consequence of climate warming. Variable currents in the North Atlantic—perhaps related to abrupt shutdowns in the oceanic conveyor system, brief ones of a century or less—may cause temperature swings every 2,000 to 3,000 years in a cycle that may run all the time, even today.

The debate continues on the explanations for the cycles of sediment deposition, the oceanic conveyor system, rapid swings in climate, and the discrepancies in the Greenland and Antarctic ice-core data. Conflicting data from different sources may drive the scientific community to develop more complicated, yet more realistic, views of the earth's climate and why it changes. With the earth expected to be bulging with people in 50 to 100 years, we should spare no effort in deciphering the seemingly chaotic behavior of the global climate system.

◇ FIGURE 1 The continuous GRIP temperature record (based on oxygen-18/oxygen-16 ratios) from the Greenland ice core. This record shows that during the Eemian interglacial interval (roughly 115,000–142,000 years ago), generally warm conditions suddenly gave way to drastic cold on several occasions. Compare this record with that of the Vostok core from Antarctica (Figure 11.27).

□ Warm interglaciation
□ Glaciation

(a)

(b)

◀▶ **FIGURE 11.26** The earth's two continental glaciers. (a) The largest glacier complex covers practically all of Antarctica, a continent that is more than half again as large as the 48 states. The ice sheet's thickness averages about 2,160 meters (7,085 ft), with a maximum of 5,000 meters (16,500 ft). (b) Greenland's ice sheet has a maximum thickness of 3,350 meters (10,988 ft). Ice thicknesses are contoured in meters. Vostok, GRIP, and GISP2 are the sites of core drilling projects discussed in the text.

ago, imply abrupt reversals every 2,000 or so years between glacial climate and warm interglacial climate, with the change occurring in as little as three years. Another study, the Greenland Ice-Core Project (GRIP), drilled about 11 kilometers (7 mi) away from GISP2, provided similar information. The climate interpretations of the upper part of the Vostok core and the two Greenland cores agree, but the wild changes portrayed by the Greenland cores are not seen in the Vostok core. The analysis is far from complete, and there is some evidence that the GISP2 and GRIP cores have been disturbed by intermittent melting and by layering distortions from nearby bedrock. Thus, it is believed that the Vostok record is the only undisturbed ice core that records a full glacial–interglacial cycle—that is, the conditions during the last interglacial period and the glacial period that preceded it (◀▶ Figure 11.27). The amounts of methane (CH_4) and CO_2, the principal greenhouse gases, in all three cores are lower in the layers accumulated during glacial stages and higher in the layers representing warm interglacial stages. The synchroneity of increasing and decreasing amounts of greenhouse gases with increasing and decreasing average temperature suggests that these changes are related. This correlation seems compelling, but a cause-and-effect relationship is still unknown: does the change in atmospheric gas content initiate a glacial stage, or does the global cooling of a glacial stage cause a change in the gas content? That greenhouse gases play a significant role in the glacial–interglacial cycle seems obvious, but it is still uncertain precisely what causes their percentages to change. On a human time scale, the ice reveals that since preindustrial times, the level of CO_2 has increased by about 40 percent and that CH_4 has increased more than 100 percent.

Today's Global Warming

I'll take ten-to-one-odds against anyone who says global warming is not well underway. And I'll take all the money they care to offer.

JERRY MAHLMAN, DIRECTOR, NOAA GEOPHYSICAL FLUID DYNAMICS LABORATORY, PRINCETON UNIVERSITY

EVIDENCE OF GLOBAL WARMING

Increasing Temperatures of the Earth's Surface and Lower Atmosphere

Temperature records from land areas and oceanic shipping lanes reveal that the average earth temperature increased about 1.1° F during the twentieth century, with half of that increase after 1970. The greatest amount of warming occurred at night over land in the higher latitudes. Some areas of Canada, Alaska, and Eurasia warmed as much as 5.5° C (10° F) between 1965 and 2000. These differences may seem small, but such changes in the past have had major consequences. As an example, the earth has warmed by only 3° to 5° C (5°–9° F) since the depths of the last ice age some 200,000 years ago.

The earth's present average temperature is the warmest it has been since 1400, which is as far back as we can go with certainty by direct and indirect evidence. Some authorities believe this may be the warmest period in 100,000 years. The warmest years of the six centuries were in the 1990s, with 1998 the warmest on record, even after correcting for the warming influence due to the active El Niño at the time.

Changes in Modern Glaciers

Modern glaciers provide relevant information because glaciers respond to climate changes by either advancing or retreating. Their behavior serves as an indicator of climate change and also as a "filter," smoothing out the record of seasonal and annual variations in temperature and precipitation. Because of their small size and ice volume, alpine glaciers are remarkably sensitive indicators of climate change: they respond quickly even to small perturbations in climatic elements. Data on the behavior of alpine glaciers (see ● Table 11.2) reveal that the number of advancing glaciers has decreased abruptly in Washington State, Italy, Switzerland, and Norway. For example, whereas all eight glaciers on Mount Baker in the North Cascades of Washington were advancing in 1976, all eight were retreating by 1990. Four of 47 alpine glaciers studied in the North Cascades have ceased to exist since 1984. The Palisade Glacier, the largest glacier in California's Sierra Nevada, retreated a third of a mile between 1940 and 1980 and is still retreating. Swiss glaciologists estimate that glaciers in the Alps have lost as much as half their mass since 1850. These glaciers are now responding to the higher average global temperature

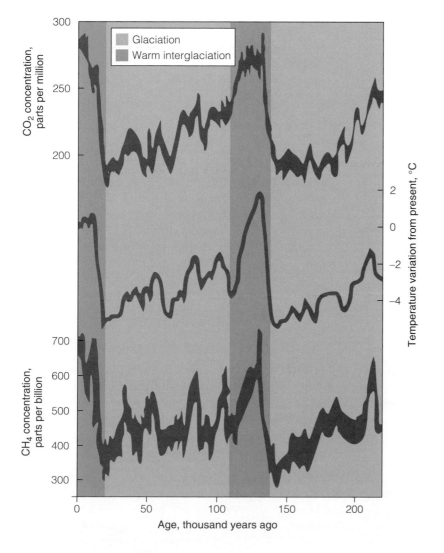

◀▶ **FIGURE 11.27** Variations in atmospheric concentrations of carbon dioxide (CO_2) and methane (CH_4) and air temperature over the past 220,000 years as determined from air bubbles trapped in the Vostok Station, Antarctica, ice core. The air temperature is plotted as deviation from present-day average temperature. Different line widths represent the ranges in estimates. The Holocene, the warm interglacial period of the past 9,000 to 10,000 years, was preceded by more than 150,000 years of generally colder temperatures. Compare this record with the GRIP (Greenland) record in Case Study 11.4.

TABLE 11.2 Changes in Alpine-Glacier Status, 1974–1995

	NUMBER OF GLACIERS		
	Advancing	*Stationary*	*Retreating*
Switzerland			
1980	70	3	28
1983	46	6	57
1986	42	9	13
1987	35	13	58
1989	19	3	83
1993	6	0	73
North Cascades			
1974	9	0	1
1984	5	11	31
1988	2	9	36
1990	1	5	41
1992	1	2	44
1995	0	0	47
Norway			
1980	5	0	7
1985	1	1	10
Alaska			
1976	15	18	18
1985	16	14	24
1994	15	12	27
Italy			
1981	25	10	10
1985	25	6	14
1988	26	13	92
1989	11	5	106
1990	9	9	123

SOURCES: Mauri Pelto, Director of the North Cascade Glacier Climate Project; Swiss Glacier Commission; Norwegian Water Resources and Electricity Board; U.S.G.S.; and *Ice* 98, no. 1, p. 3.

their much greater size, which results in a greater lag time in their response to climate variation.

Thawing in Boreal and Subarctic Regions

Climate changes in Alaska and the Bering Sea region have already had socioeconomic impacts on Alaskans, most of them negative. For example:

1. With less sea ice in the Bering Sea, weather events such as storm surges have increased in frequency and severity, causing increased coastal erosion, inundation, and threats to structures.
2. Subsistence lifestyles of the native peoples in the region have been adversely affected by such things as changes in sea ice conditions. Hunting on the ice, for example, is more dangerous.
3. Slope instability, landslides, and erosion are increasingly common in thawing permafrost terrain. This threatens roads and bridges and causes flooding.
4. The warmer climate has increased forest fire frequency and insect outbreaks, reducing timber yields.
5. Permafrost boundaries in some areas moved north by 130 kilometers (80 mi) in the 1900s.
6. The speed of permafrost thawing has caused major changes in forest, bog, grassland, and wetland ecosystems that have affected land use.

Rising Sea Level

Sea level was about 120 meters (393 ft) lower than the current level during the last glacial maximum, about 20,000 years ago, and its rate of rise has varied. In the nineteenth century it rose about 12 centimeters, and during the twentieth century the rise appears to have been about 25 centimeters (10 in). The accelerated rate is due to some combination of the thermal expansion of the oceans and the melting of mountain and Antarctic glaciers and sea ice. (Sea level rises about 24 centimeters with each 1° C increase in temperature, with a time lag of about 20 years.)

Biological Response

Observations in North America and Europe reveal that wildlife in various habitats are responding to warming climates. Case Study 11.5 provides summaries of some research studies.

Thinning of the Polar Ice Cap

Scientists believe that the Arctic Ocean ice cap is revealing a substantial climatic signal. Comparing ice

between 1977 and 1990, which was 0.4° C (0.7° F) above the 1940–1976 mean. No significant change in the behavior of Alaskan alpine glaciers has been noted in that time interval, but that is assumed to be due to

Case Study | 11.5

Global Warming Is for the Birds

With physical evidence of global warming continuing to accumulate, are there any indications of global warming in biologic systems? Yes. Late-twentieth century studies in Europe and North America show that some plants and animals are indeed responding to global warming.

In Britain, a 20-year study of populations of 101 bird species found that, coincidentally with the recent warming trend, most species had extended their northern range by an average of 19 kilometers (30 mi). Another study of the nesting habits of 20 species of British birds found that the species were laying eggs earlier in the spring. Similar results were found in Michigan, where some species were laying eggs as many as 21 days earlier than in the past.

Comparing data collected between 1992 and 1996 at more than 150 sites with historical records of the Edith's checkerspot butterfly (*Euphydryas editha*), a wide-ranging species of western North America, showed that population extinctions were four times higher at the southern edge of its range (in Mexico) than at the northern edge (in Canada). Furthermore, the rate of extinctions at lower elevations was more than double the rate at elevations above 2,440 meters (8,000 ft). Thus, the butterfly's range was found to be shifting northward and upward. A European study of the geographic ranges of 35 species of nonmigratory butterflies had similar findings. Two thirds of the species' habitats had shifted northward by 35 to 240 kilometers (21–149 mi) coincidentally with the warming of Europe and only 3 percent had shifted south.

In the mountain forests of Monteverde, Costa Rica, 20 of 50 species of frogs and toads disappeared following population "crashes" in 1987. These crashes probably resulted from ecosystem changes that altered the communities of reptiles, amphibians, and birds due to recent warming. The changes are associated with patterns of dry-season mist frequency, which are related to changes in sea-surface temperatures in the equatorial Pacific since the mid 1970s. The implication is that atmospheric warming has caused the base of the mountain-top cloud bank to raise in elevation and has negatively impacted the mountain forest ecosystem.

Temperature-limited environments in mountains, sub-arctic, and arctic regions also reveal evidence of warming. In the Swiss and Austrian Alps in the early 1990s, a study compared the plant life on mountain summits exceeding 3,000 meters (9,849 ft) to records collected 70 to 90 years earlier, when the annual temperature averaged 0.7° C (1.26° F) cooler. Altitudinal vegetation belts were found to have shifted to higher elevations by as much as 4 meters (13 ft) per decade, although many belts' shifts were less than 1 meter (3 ft) per decade. A study in Alaska found that **boreal** forests (forests in mountainous and northern regions of North America in which the mean temperature of the hottest season does not exceed 18° C [64.4° F]) expanded northward at a rate of 97 kilometers (60 mi) per 1° C (2° F) of temperature increase, and that the length of the growing season increased 20 percent in the last few decades.

SOURCES: Georg Grabherr and others, 1994, Climate effects on mountain plants, *Nature* 369: 448; J. Alan Pounds and others, 1999, Biological response to climate change on a tropical mountain, *Nature* 398: 611–15; Bruce F. Molina, 1998, Climate change workshop results begin to appear, *GSA Today* 8, September: 10–11; Camille Parmesan, 1996, Climate species' range, *Nature* 382: 765–6; Camille Parmesan, and others, 1999, Poleward shifts in geographical ranges of butterfly species associated with regional warming, *Nature* 399: 579–83; and Chris D. Thomas and Jack J. Lennon, 1999, Birds extend their ranges northwards, *Nature* 399: 213.

thicknesses measured by submarines between 1958 and 1976 with measurements made in the 1990s revealed that, on average, the ice cap had thinned by 4.3 feet and that it had lost 40 percent of its volume in that time (◆ Figure 11.28). Such a major change in the Arctic could have a significant impact on Northern Hemisphere weather. It is far from certain what has caused the thinning, whether it is due to human activities, natural global warming, a natural Arctic climatic fluctuation, or some combination of these factors.

Arctic research was stepped up in the 1990s through ice-camp-based, ship-based, and undersea explorations. Playing a major role in the research was the U.S. Navy's Scientific Ice Expeditions (SCICEX). It was SCICEX observations gathered between 1993 and 1997, compared with earlier measurements, that allowed scientists to recognize a 2° F temperature increase in a large part of the Arctic Ocean. At least part of the increase is due to incursions of warmer water from the Atlantic Ocean. Data that scientists believe would help them to understand trends and causes were still withheld as confidential by the U.S. and Russian navies in early 2000. Possible causes of the changes will remain speculative until the data are released to the civilian scientific community.

CAUSES OF TODAY'S GLOBAL WARMING

For every complex problem there is always a simple answer, and it's usually wrong.

H. L. MENCKEN (1880–1956), AMERICAN EDITOR

Of the five factors most likely responsible for the large-scale temperature changes presently being experienced, three vary naturally:

- stratospheric volcanic sulfate aerosols,
- climate variability, and
- solar radiation

and two, which have changed decisively, are due to human activity:

- greenhouse gases—mainly CO_2, but also CH_4, NO_x, and CFCs—and
- anthropogenic sulfate aerosols.

A decisive climatic change due to a change in one or more of these factors is called a *forcing*. Some scientists believe that solar forcing may have caused warming in the early 1900s, although natural climate variability and an anthropogenic factor (a human factor) cannot be ruled out. But warming from about 1950 to the end of the century, irrespective of any natural forcing, volcanic influence, or increasing solar activity, is most likely attributable to human activity. It was such an analysis that led the United Nations' Intergovernmental Panel on Climate Change (IPCC) in 1995 to recognize that, according to the best estimates from climate models, the climate changes due to human activity add up to just about the amount of expected change. The atmospheric patterns of temperatures match well the increased concentrations of greenhouse gases, sulfate aerosols (mainly from burning fossil fuels), and the reduction in stratospheric ozone from CFCs. In light of this combination of factors, the panel concluded that "the balance of evidence suggests that there is a discernible human influence on global climate."

Weather's Puzzling Signals

In August 1995 a record-breaking dry spell in the Northeast brought drought warnings to New York City,

◆ FIGURE 11.28 Walking on thin ice. Between 1958–1976 and the 1990s the Arctic Sea ice lost about 40 percent of its volume, thinning by 1.3 meters (4.3 ft) on average.

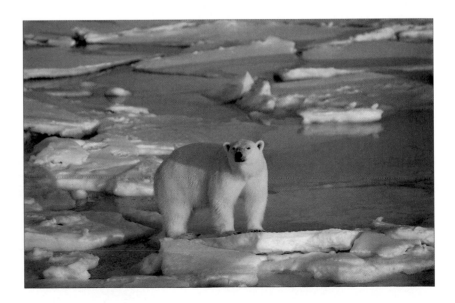

Westchester County, and the Catskills. The following winter brought record-breaking snowfall to the entire region. Puzzling weather anomalies have become almost commonplace (◉ Table 11.3). Since the mid 1980s these anomalies have included wide swings in the weather from cold winters (except for a peculiarly warm one in 1995) in New England to warm winters in Alaska, a six-year drought in California, a bitter cold snap for the 1994 Winter Olympics in Norway, the record-breaking flooding in the Mississippi River valley in 1993, and summer heat waves in central Europe.

Worldwide, the warmest ten years on record occurred in spite of cooling due to dust and aerosols emitted by the 1991 eruption of Mount Pinatubo.

Are such weather extremes likely to continue, and do they indicate global warming, or are they just coincidence? Scientifically, it is hard to tell, because climate is fickle. This episode of warming could be part of the natural changes that scientists have traced over the millennia as the earth has cooled, then warmed, then cooled again (Case Study 11.4). Weather patterns change naturally in the short term, and climate changes

◉ TABLE 11.3   Global Weather Extremes, 1995–2000

LOCATION	WEATHER/EVENTS
Worldwide	The 20th century was the warmest in at least six centuries. Warmest years were 1990, 1995, 1997, and 1998. Sea level rose 6+ inches in 1990s.
United States	Winter 1996–1997 one of the wettest in history. Annual rainfall increased 4% in 1900s; average temperature, 1° F.
Susquehanna River Basin (stretching from Maryland to New York)	Below-normal precipitation 9 of 11 months in 1994–1995 forced water-use restrictions.
Long Island, New York	Drought-associated brush fires, August 1995.
Arctic Alaska	Accelerated permafrost thawing caused major landscape changes 1987–1998. Growing season and growing days increased 20%. Subarctic forests expanding 60 miles northward per 2° F temperature increase.
Chicago, Illinois	Severe 8-day heat wave July 1995; 41° C (106° F) temperatures, 88% relative humidity.
California	Severe flooding January 1995; heaviest mountain snowpack in decades. Unseasonably cool summer 1999.
Southern California	Average temperature increased 4° F in 1900s.
Oregon and Washington	Severe flooding and mudslides winters of 1996 and 1998.
Southern Plains and southwestern U.S.	Drought during much of 1996.
Eastern U.S.	Unseasonably warm December 1994–January 1995. Record snowfall winter 1996. Worst drought of 1900s in Mid Atlantic and New England states summer 1995. Despite two hurricanes, drought returned 1998–2000. Water-use restrictions imposed in many areas 1995–2000.
Salt Lake City, Utah	Tornado hits downtown area in August 1999, the first since 1968; $150 million in damage, 1 death, 5 homes destroyed, 180 homes damaged.
Rocky Mountains	Unusually rainy summer 1999.
Bering Sea	Decreased sea ice allowed severe weather events such as storm surges to increase shoreline erosion 1987–1998.
Central America	Hurricane Mitch in 1998, the most destructive disaster in modern Central American history; 2 million people left homeless.
Tibet	Early snowmelt in spring 1995.
China, Tangtze River basin	Worst flooding in 40 years in August 1998; 3,000 people drowned, 14 million left homeless.
Himalaya Mountains	Forest fires summer 1995. Heavy rain, floods, mudslides, and rapid melting of glaciers (fastest in world) due to intense summer monsoon rains, August 1999.
Bangladesh	Rising Jamurna River caused worst flood in history summer 1998; 2/3 of country underwater 2 months. Indus River rose to unprecedented levels summer 1999, mainly from rapid melting of Himalayan glaciers.
United Kingdom	Record-breaking drought, summer 1995.

naturally in the long term. Thus, any human-induced contributions to climate change are superimposed on natural changes.

Clouds: A Wild Card in Global Warming

Clouds perform an important role in global temperature control, but scientists are unsure of their exact role. For instance, when the earth is warmer, there is more water vapor in the atmosphere, and consequently there are more clouds. Recognizing this relationship leads to the question, will clouds be the same as we know them today if the earth becomes warmer? Will they be at the same elevation, and will the different types of clouds exist in the same proportion as they do today?

To illustrate some of the uncertainty about the effect of cloud types, consider the global-warming feedback differences between low-altitude clouds and high clouds. Wispy, high cirrus clouds composed of tiny ice crystals (◆ Figure 11.29) are very effective at trapping infrared radiation that is emitted from the earth's surface and thus act to warm the atmosphere; they give positive feedback. On the other hand, the heat-trapping property of low clouds (◆ Figure 11.30) is very poor (negative feedback). In addition, low clouds reflect the sun's radiation back to space, which cools the earth; they give negative feedback in two ways.

The size of the water droplets in clouds is also decisive. Whereas clouds formed of large numbers of small droplets reflect more of the incoming solar radiation away from the earth (giving negative feedback), clouds with larger, fewer droplets give positive feedback, amplifying global warming.

Because of such uncertainties, many scientists have turned away from the debate about whether human-induced global warming is occurring and are now attempting to predict exactly what we can expect to happen in the next century.

LIVING WITH UNCERTAINTY

Atmospheric scientists are still uncertain about why climate changes, and they lack techniques for distinguishing between long-range climate trends and short-term weather events. "The temptation . . . is often to

◆ **FIGURE 11.29** Cirrus clouds, typical high-level clouds, normally do little to block solar radiation, but they are effective at trapping infrared radiation emitted from the earth surface. This warms the atmosphere (gives positive feedback to global warming). But when the clouds contain dust and sulfate aerosols that have been blasted into the stratosphere by volcanic eruptions, they reflect more solar radiation back into space, which serves to cool the atmosphere (negative feedback to global warming).

attribute any of these temporal and sometimes local variations to a wider and more pervasive change in climate," note several atmospheric scientists with the National Oceanic and Atmospheric Administration (NOAA). A respected senior climate scientist has shown that U.S. weather has become more extreme since 1980 and has estimated with 80 percent certainty

◆ **FIGURE 11.30** Stratocumulus and stratus clouds, typical low-elevation clouds. Their heat-trapping properties are low (negative feedback), and their qualities of blocking and reflecting solar radiation are high (negative feedback). When they contain sulfate aerosols, such as those released by industry and coal-fired power plants, however, the situation changes. The sulfate cloud droplets "dirty" the clouds, making them less reflective (positive feedback), thereby enhancing global warming.

Consider This... ?   With what you have learned about the increasing concentration of greenhouse gases in the atmosphere, what changes might you expect over the next 50 years where you live if global temperature should rise by 2° C? ■

that the extremes are due to human-induced greenhouse warming. In the case of greenhouse warming, built-in delays in the earth's climate system compound the potential dangers of waiting for 100 percent certainty; decades are required to warm or cool the atmosphere and oceans. If the world's policymakers wish to head off the potential for greater climate changes, they must act before the major effects of greenhouse-gas pollution are apparent. Regrettably, global warming has become political, and communication between scientists and policymakers is sometimes difficult (◆ Figure 11.31). If it *is* occurring, it presents a dilemma. If it is due to human activity, we must take measures now to slow or halt it, and enormous up-front costs will be required for mediating the impact. But the determination of certainty must be based on the uncertainty of the past record, and any actions taken may prove to be expensive mistakes if humans are not the primary cause for the apparent climate change.

Some climatologists are skeptical about global warming, and some of them doubt that it is occurring. Others feel that warming is indeed happening, but believe that the apocalyptic projections for the future are overblown. Nevertheless, most atmospheric scientists, basing their judgment on a variety of observed changes and studies, express growing confidence that human-induced global warming is occurring. Moreover, computer modeling of the world's climate at NOAA's laboratory throughout the 1990s successfully matched forecasts with actual conditions.

Implications for the Future

Prediction is difficult particularly when it concerns the future.
NEILS BOHR, NOBEL PRIZE LAUREAT FOR PHYSICS

Getting a group of scientists to agree on nearly anything, such as forecasting the long-term effects of global warming, is (according to a group of scientists) like trying to herd a bunch of cats. Scientists are especially argumentative when they are thrown a data curveball, as they were in 1999. The "curveball" was the finding that the rate of greenhouse-gas buildup in the atmosphere had decreased by about 25 percent since 1980. The amount of CO_2 from burning fossil fuels had increased only 0.4 percent in 1998 (despite the emissions from electric power plants increasing by 3.2 percent that year), down from a 1 percent per year increase in the previous few decades. This does not mean that greenhouse warming has stopped; only that it slowed, thereby giving humankind more time to cope with what many scientists think is inevitable global climate change. The decreasing rate, if it continues, means that the greenhouse-gas buildup will not double by 2050, as many climate forecasters had anticipated; the doubling time has merely been postponed until early in the twenty-second century. Make no mistake, there has been no reduction of greenhouse gas *emissions,* only in the atmospheric buildup. The reduction in build-up is caused by some unknown process, with the "missing" gases being sequestered in some mysterious manner in

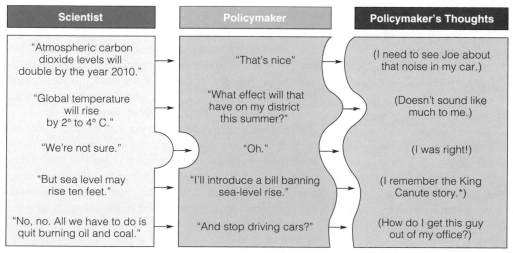

Scientist	Policymaker	Policymaker's Thoughts
"Atmospheric carbon dioxide levels will double by the year 2010."	"That's nice"	(I need to see Joe about that noise in my car.)
"Global temperature will rise by 2° to 4° C."	"What effect will that have on my district this summer?"	(Doesn't sound like much to me.)
"We're not sure."	"Oh."	(I was right!)
"But sea level may rise ten feet."	"I'll introduce a bill banning sea-level rise."	(I remember the King Canute story.*)
"No, no. All we have to do is quit burning oil and coal."	"And stop driving cars?"	(How do I get this guy out of my office?)

*Canute was a king of Denmark, England, and Norway, 1016–1035, who supposedly commanded the oceanic tide to stop rising.

◆ FIGURE 11.31 A hypothetical scientist briefs a hypothetical policymaker on the dangers of the greenhouse effect.

oceans, forests, lakes, or wetlands. Just where is still unknown (see Case Study 11.3).

In spite of these uncertainties, let us examine a scenario of anticipated effects of global warming developed by the Forum on Global Change Modeling in 1994. Most of the group's computer modeling assumes a continuing anthropogenic (human-caused) increase in greenhouse gases based on a 1 percent per year increase in CO_2.

Very Probable:

- Globally, the mean surface temperature will increase by the mid twenty-first century by about 0.5° to 2° C (0.9° to 3.6° F). Such warming will be due to the concentrations of greenhouse gases alone, assuming that no significant actions are taken to reduce these gases. If the CO_2 concentrations in the atmosphere should double, there is a range of estimates for the warming that will result. The best estimate is 1.5° to 4.5° C (2.7° to 8° F), and the most probable estimate is 2.5° C (4.5° F).

- Globally, mean precipitation will increase, but how it will be distributed is uncertain. A warming of the earth's surface temperatures will increase global mean precipitation because of the well-established relationship between evaporation rates and surface temperature.

- Northern Hemisphere sea ice will be reduced. The projected effects and timing of changes in Southern Hemisphere sea ice are uncertain.

- Arctic land areas will experience wintertime warming. Paleoclimate studies and computer modeling offer evidence for polar warming and reduction of snow cover on land areas. The magnitude of the warming is uncertain because of the complexity in understanding all of the variables.

- Globally, sea level will rise at an increasing rate, although the rate of rise may not be significantly greater than the last few decades of the twentieth century. The most reasonable estimate for the rate of sea level rise is 5–40 centimeters (2–16 in) by 2050, as compared to a rise of 5–12 centimeters (2–4.7 in) if the rates of the twentieth century continue. Incorporated in the modeling is a component due to seawater expansion, which is closely dependent on the amount of atmospheric warming, and an estimate of water derived from the retreat of mountain glaciers. The volume of water contributed by the melting of the polar ice sheets is less certain. If global temperatures increase by 3° C by the year 2050, as some climatologists believe, it is estimated that meltwater from mountain glaciers alone will raise sea level by another 20 centimeters (8 in).

Probable:

- Precipitation at high latitudes will increase, with potential feedback effects related to the influence of additional fresh water to the oceans causing changes in oceanic circulation. There will be increased precipitation on the polar ice caps.

- Summer dryness will increase in the mid latitudes of the Northern Hemisphere. Evaporation increases strongly with increasing temperatures. Soil moisture in the Northern Hemisphere could decrease by perhaps 40 percent if greenhouse gases continue to increase at predicted rates. This would make "dry" farming impossible or nearly so in much of North America and Europe.

- Episodic, explosive volcanic eruptions will cause short-term relative cooling of a few tenths of a degree lasting up to a few years.

Uncertain:

- Details of the climate changes over the next 25 years [for 1994–2019] are uncertain.

- Biosphere–climate feedbacks are expected, but whether these feedbacks will modify or amplify climate change is uncertain.

- Changes in climate variability will occur.

- Regional-scale climate changes will differ from the global averages. Unfortunately, there is limited capability to estimate how various regions will respond to climate change.

- Tropical-storm intensity may change. An increase in intensity is likely but uncertain because of the potential changes in poleward heat flow and other conditions.

Overall, warming will stress human activities because it will demand rapid adjustment of economic patterns and changes in much of society's infrastructure. There is also a growing indication that the potential effect on human health is no less serious.

In their worst-case scenarios, some global climate models point to the possibility of tens of millions more cases of infectious diseases, as mosquitoes and other pests expand their ranges, and of hundreds of thousands of additional deaths each year due to an increasing incidence of heat waves. Epidemiologists say the most deadly punch of global warming will be delivered to countries of the developing world. Their models predict an increase in such diseases as sleeping sickness, malaria, schistosomiasis, and yellow fever. These diseases already afflict more than 600 million people each year, killing more than two million. Some researchers estimate that global warming would increase the populations of the insects and snails that transmit these diseases, as they are cold-blooded insects and invertebrates that respond to subtle changes in temperature. Some scientific speculation links the 1° C increase in Rwanda's average temperature in 1987 with the country's 337 percent increase in the number of malaria cases that year. Medical researchers hope that monitoring vector (disease-carrying organism) populations will enable them to anticipate and combat such potential threats to human life. One global-warming model predicts that the annual mortality rate from heat-related deaths in New York City would increase from the current average of 320 per year (the highest in the U.S.) to 824, and that Montreal, Chicago, Cairo, and many other cities would have similar increases. Air-conditioning might lower the mortality rate in some cities somewhat, but it is unlikely that it would affect the numbers in Karachi or Cairo.

WHAT INDIVIDUALS CAN DO

What can we do about global warming? We can do nothing about any of its natural causes, but there are things we can do individually that may help to slow the pace of human-generated global warming that would hardly affect our life styles. More than 80 percent of the world's energy and most of the greenhouse-gas emissions comes from burning fossil fuels—oil, coal, and natural gas—and currently we are only getting close to cost-effective alternatives (see Chapter 14). But much can still be done. In the United States there are three main things individuals can do.

1. Buy products that reduce energy consumption.
2. Use energy more efficiently.
3. Actively recycle.

◉ Table 11.4 provides some more specific suggestions. The consequences of our choices and life styles are catching up with us, and even small steps will help.

WHAT INDUSTRY AND GOVERNMENTS CAN DO

I find myself increasingly persuaded that a climate effect may be occurring.

MARK MOODY-STUART, CEO, ROYAL DUTCH/SHELL OIL COMPANY

There is mounting evidence that the concentration of carbon dioxide in the atmosphere is rising and the temperature of the earth's surface is increasing.

JOHN BROWNE, CEO, BRITISH PETROLEUM

In debates about global warming, environmentalists and industrialists have commonly found themselves hollering at each other through clouds of scientific and economic jargon, respectively. This began to change in 1998, when several major corporations acknowledged that excess greenhouse gases could well be the cause of global climate change. Among the companies that have made the cross-over, warning that "climate change is serious business for all of us," are Boeing, British Petroleum (BP), 3M, United Technologies, Lockheed Martin, and Maytag. Royal Dutch/Shell, for example, announced that it was investing more than half a billion dollars on renewable-energy technology over five years. BP made a new commitment to investment in solar-energy research, and both of these companies backed out of the Gobal Climate Coalition, an industry lobbying group that opposes the 1997 Kyoto treaty. This treaty was the result of a 160-nation summit where an agreement was reached to reduce worldwide greenhouse emissions to below 1990 levels by government-imposed limits. The agreement has yet to be ratified by all nations involved.

Ford Motor Company, whose critics point out that it sells more sport utility vehicles (SUVs) and

○ TABLE 11.4 **Reducing Global Warming Begins at Home**	
ACTION TO TAKE	**ESTIMATED REDUCTION OF CO_2,** pounds per year
1. Insulate your home, install water-saver shower heads, clean and tune furnace.	2,480
2. Buy a fuel-efficient automobile with a rating of at least 32 mpg to replace your most-used car.	5,600
3. Do not drive your car two days a week.	1,590
4. Recycle all home waste metal, glass, newsprint, packaging, and cardboard.	850
5. Install solar heating to help provide hot water.	720
6. Replace washing machine with low-water-use, low-energy model.	420
7. Buy food and other products with recyclable or reusable packaging.	230
8. Replace refrigerator with a high-efficiency model.	220
9. Use a push lawn mower instead of a power mower.	80
10. Plant two trees.	20

SOURCE: Oregon Energy Office

Gallery

Wonders of Glaciation

Some of the world's most-visited scenic areas are popular because of the spectacular landscapes that remain as superlative legacies of the work of Pleistocene glaciers. ◈ Figures 11.32 to 11.36 are a collage of photographs exemplifying the beauties of glaciers and glacial landscapes.

◈ FIGURE 11.32 The Vaughan Lewis Glacier, a valley glacier, Alaska.

◈ FIGURE 11.33 A U-shaped glacial valley sculptured by an Ice Age valley glacier; Kern Canyon, Sequoia–Kings Canyon National Park, California.

◁▷ FIGURE 11.34 A glacial "erratic" that was transported by the Pleistocene valley glacier that occupied Lee Vining Canyon, California. This erratic boulder's location is more than 25 kilometers (15 mi) from the nearest outcropping of the particular bedrock type that could have yielded it.

◁▷ FIGURE 11.35 A cirque, an armchair-shaped depression at the head of a U-shaped valley, viewed from the summit of 4,313-meter (14,150-ft)-high Mount Sneffels, Colorado.

◁▷ FIGURE 11.36 Malaspina Glacier, a 2,200-square-kilometer (850-mi²) piedmont glacier, which was formed on Alaska's coastal plain by the discharge of about two dozen valley glaciers. It is the largest glacier in North America. Its contorted pattern is due to its entrained moraine rock debris being subjected to differential flowage when the valley glaciers that nurture the Malaspina have surged (episodic surging).

trucks—both major greenhouse-gas emitters—than its competitors, reported that its new minivans and SUVs would qualify as low-emission vehicles. Ford is also developing fuel cells that combine oxygen and hydrogen to produce electricity, and a fuel-cell-powered zero-emission vehicle (ZEV) has already been tested. But fuel cells use electricity to generate a "charge" of hydrogen and oxygen gases, as do the existing electric-powered vehicles (EVs). EV batteries require recharging, and the only economical source of that electricity is the power supplied by the electric utility industry, nearly 90 percent of which is derived from combustion of fossil-fuels (see Chapter 14). Worldwide, electric power generation is the number-one producer of air pollution. The use of EVs may improve air quality in large cities, but in the long run, with current technology, EVs will do little to reduce greenhouse-gas emissions without major changes in emission controls by the electric utility industry.

The new hybrid vehicles that combine EV technology with a small, low-emission, conventional internal combustion engine offer attractive possibilities for cutting greenhouse-gas emissions. They don't need to be "plugged in" for recharging because they are recharged by the car's engine itself. The batteries of the hybrid vehicles are smaller than those of EVs, however, and must be recharged more often. Reportedly, the Toyota hybrid gets about 52 miles per gallon.

Some researchers believe that most of the proposed global emission-reduction plans will do relatively little to slow greenhouse-gas emissions, because they consist mainly of modest, voluntary policies. Furthermore, the industrialized countries' CO_2 emissions may not be the major problem. The United Nations' economic and science advisors on global climate change predict that by 2025 the developing countries and those that are moving away from centrally planned economies (including the former Soviet-bloc countries) will be responsible for nearly 70 percent of all energy-related CO_2 emissions.

A U.S. trade group representing manufacturers, utilities, mining companies, coal and oil producers, and railroads insisted that reducing industrial emissions by 20 percent or more below the 1990 levels, as proposed at the 1995 Intergovernmental Panel on Climate Change (IPCC) conference, would be disastrous for the U.S. economy. They estimated that cuts of this magnitude could cost an average of 600,000 jobs a year every year until the year 2010.

Offering hope for mitigating CO_2 buildup is a chemical process that combines the gas with materials such as calcium oxide and magnesium oxide to form stable minerals. The process is viewed as a short-term insurance policy in case political pressure resulting from the threat of global warming forces industry to limit CO_2 emissions. The monetary costs of capturing CO_2 on an industrial scale would be very high. For example, it is estimated that removing the gas from coal-fired electric-power generating plants would double or triple the cost of electricity. Oceanographers have been cautiously exploring the feasibility of dumping finely ground iron particles into the ocean to stimulate the growth of phytoplankton, which, like all plants, metabolize CO_2 by photosynthesis. From initial experiments it is estimated that spreading a half-ton of iron across 100 square kilometers (39 mi^2) of a tropical ocean would stimulate enough plant growth to absorb some 350,000 kilograms (771,800 pounds) of CO_2 from seawater. Performed on a much larger scale, the iron fertilization of seawater could absorb billions of tons of CO_2, offsetting perhaps a third of global CO_2 emissions. The environmental side effects of such large-scale "iron fertilization" are difficult to foresee, but the enhanced bloom of plants could enrich the entire oceanic ecosystem.

In the short term—that is, on the scale of human lifetimes—it appears that there will be global warming. From a long-term, geological perspective, however, the warming is seen as a brief event and a nonrepeatable one—nonrepeatable because the earth's reserves of fossil fuels will be consumed within the next few hundred years. After they are consumed, atmospheric conditions should eventually return to normal. True, human-induced greenhouse warming may last a thousand years, but ultimately the variations in the earth's orbital geometry will change, and the earth will be thrust into yet another ice age.

Summary

GLACIERS

DESCRIPTION
Large masses of ice that form on land where, for a number of years, more snow falls in winter than melts in summer, and which deform and flow due to their own weight because of the force of gravity.

LOCATION
High latitudes and high altitudes.

CLASSIFICATION
Major types are continental (ice-sheet), ice-field, ice-cap, valley (alpine), and piedmont.

SOME IMPORTANT GLACIAL FEATURES
End moraine—Ridge formed at the melting end of a glacier composed of glacially transported rocky debris.
Kettle lakes—Water-filled, bowl-shaped depressions formed as glaciers retreated.
U-shaped canyon—Remains after a valley glacier melts.
Fiord—U-shaped canyon drowned by the sea.

EFFECTS OF PLEISTOCENE GLACIATION
1. Some soil conditions, including loess deposits transported by Ice Age winds and some areas of glacial till.
2. Local ground-water conditions.
3. Shore-line configuration due to sea-level changes and isostatic rebound.
4. Human transportation routes.

CLIMATIC CONTROLS

WIND
Movement of air due to unequal heating of the earth's surface. Caused by:
1. Variations in the heating of the earth at different latitudes because they receive varying amounts of solar heat energy due to the differing angles of the sun's rays.
2. Due to the earth's rotation, the Coriolis effect deflects winds to the right in the Northern Hemisphere and to the left in the Southern Hemisphere.

AIR PRESSURE, WIND, AND CLIMATE BELTS
Polar Highs—polar regions of high pressure; subsiding air produces variable winds and calms.
Westerlies—zones of generally consistent winds lying between 35° and 60° north and south latitudes.
Subtropical highs—zones of high pressure and subsiding air masses lying between 30° and 35° north and south latitudes; have variable winds and calms and, commonly, clear and sunny skies. Often called the *horse latitudes,* these zones are the regions in which most of the earth's deserts are located.
Trade winds—belts of generally consistent easterly winds lying between 5° and 30° north and south latitudes.
Equatorial low and intertropical convergence zone—zone of variable winds and calms lying between latitudes 5° north and 5° south. Sometimes called *the doldrums.*

HEATING OF THE EARTH

SOLAR RADIATION
Warming the earth is by ultraviolet radiation from sun.

RERADIATED ENERGY
Infrared energy from earth that warms the atmosphere.

GREENHOUSE EFFECT
The warming of the atmosphere by an increase in the amount of certain gases in the atmosphere, primarily CO_2 and CH_4, which increase the retention of heat that has been reradiated by the earth.

GREENHOUSE GASES
Trace amounts of H_2O, CO_2, CH_4, and CFCs that trap heat in the atmosphere.

OCEANIC CIRCULATION

DESCRIPTION
Each ocean has a characteristic current pattern, the general pattern being clockwise in the Northern Hemisphere and counterclockwise in the Southern Hemisphere. The net effect is the movement of cool water from polar regions toward the equator.

IMPORTANCE
Currents moderate coastal climate and weather, and coastal upwelling brings cool water and nutrients that are important for biologic productivity.

ICE-AGE CLIMATE

DESCRIPTION
Temperatures 4° to 10° C cooler than at present.

CAUSES
A number of interrelated factors contribute to climate changes in ways that are not clearly understood. They include:
1. Obliquity of earth's axis
2. Eccentricity of earth's orbit
3. Precession of earth's axis
4. Variations in content of atmospheric gases
5. Changes in "greenhouse gases"
6. Changes in landmass positions due to plate tectonics
7. Tectonic changes in the elevation of continents
8. Volume and temperature of the oceans
9. Changes in ocean currents
10. Changes in solar radiation
11. Dust and aerosols
12. Oceanic conveyor belt

SOME MODEL-BASED PREDICTIONS OF CLIMATIC CHANGE
1. Globally, average surface temperatures will increase.
2. Globally, average precipitation will increase, but its distribution is uncertain.
3. Northern Hemisphere ice will decrease; Southern Hemisphere ice may increase.

4. Arctic land areas will experience wintertime warming.
5. Sea level will rise at an increasing rate with drowning of low-level coastal plains.
6. Decreasing soil moisture in the Northern Hemisphere may make farming nearly impossible in much of North America and Europe.

7. Tropical-storm (hurricane) intensity and frequency may increase.

◆ Key Terms

ablate	esker	isostatic rebound	subpolar low
calving	firn	kettle lake	subtropical highs (horse
chlorofluorocarbons (CFCs)	fiord	moraine	latitudes)
climatic controls	glacial outwash	obliquity	surging, glacial
Coriolis effect	glacier	polar front	till, glacial
eccentricity	global warming	polar high	upwelling
end moraine	greenhouse effect	precession	wind
equatorial low	intertropical convergence	proglacial lake	zone of flowage
erratic	zone (ITCZ), "doldrums"	recessional moraine	

◆ Study Questions

1. What makes the wind blow? Why do winds in the Northern Hemisphere deflect to the right, and winds in the Southern Hemisphere to the left?
2. Explain the generalization "Glaciers are found at high latitudes and at high altitudes."
3. Glacial ice begins to flow and deform plastically when it reaches a thickness of about 30 meters (100 ft). What is the driving force that causes the ice to flow?
4. Describe the behavior of a glacier in terms of its *budget;* that is, the relationship between accumulation and ablation. How does a glacier behave when it has a balanced budget; a negative budget; a positive budget? What kind of a budget does a surging glacier have?
5. Areas far beyond the limits of the ice sheets were affected by Pleistocene continental glaciation. What are some of the effects? (*Hint:* Consider coastlines, today's arid regions, and modern agricultural regions.)
6. Describe isostatic rebound. What effect has the melting of the great ice sheets had on the crustal regions that were covered by thick masses of ice until 8,000 to 11,000 years ago? What are the effects of this rebound on society?
7. Describe the greenhouse effect. What causes greenhouse warming?
8. What can be done to modify (or even control) greenhouse warming?
9. What are some potential effects of greenhouse warming on human life?
10. What causes ice ages? Will there be more ice ages? Explain.
11. Using the "rule of 70" for determining the approximate number of years till doubling (70 *divided by* the rate of increase *equals* years for doubling), how long would it take for the amount of atmospheric CO_2 to double if its rate of increase is one percent per year; if the rate is 0.75 percent per year?

◆ For Further Information

BOOKS AND PERIODICALS

Ahrens, C. Donald. 2000. Possible causes of climatic change (pp. 510–525). *Meterology today.* Pacific Grove, Calif.: Brooks/Cole.

Andersen, Bjørn G., and Harold W. Borns. 1994. *The Ice Age world.* Oslo: Scandinavian University Press.

Broecker, Wallace S. 1995. Chaotic climate. *Scientific American* 273, no. 5:62–68.

Burger, Jack. 1992. New York: Take a geological field trip. *Earth: The science of our planet* 1, no. 2:60–67.

Cowan, Robert C. 1999. Arctic ice cap: It's thinner than we thought. *The Christian Science monitor,* 17 November:2.

Greenland Ice Core Project (GRIP). 1993. Climate instability during the last interglacial period recorded in the GRIP ice core. *Nature.* 364, no. 6434 (15 July):203–207.

Hambrey, Michael, and Jürg Alean. 1992. *Glaciers.* New York: Cambridge University Press.

Kirk, Ruth. 1983. Of time and ice. *Glacier Bay: Official National Park handbook.* Washington, D.C.: Division of Publications, National Park Service, U.S. Department of the Interior.

Knickerbocker, Brad. 1998. The global warming debate cuts two ways. *Christian Science monitor,* 11 May:5.

MacCracken, Michael, and Tom Karl. 1997. Is the climate changing? Indeed it is. *Bob Ryan's 1997 almanac & guide for the weatherwise.* Washington, D.C.: WRC-TV.

MacKenzie, Fred T. 1998. *Our changing planet: An introduction to earth system science and global environmental change.* Upper Saddle River, N.J.: Prentice Hall.

Monroe, James S., and Reed Wicander. 1997. *Physical geology: Exploring the earth,* 3d ed. Belmont, Calif.: Wadsworth.

Motavalli, Jim. 1996. Some like it hot; global warming is not just a scientific prediction—it's also a hot political football. *Environmental magazine* 7, no 1:28–35.

Singer, S. Fred. 1999. *Hot talk cold science; global warming's unfinished debate,* revised 2d ed. Oakland, Calif.: The Independent Institute.

Tett, Simon F. B., and others. 1999. Causes of twentieth-century temperature change near the earth's surface, *Nature* 399:569–572.

Trent, D. D. 1983. California's Ice Age lost: The Palisade Glacier: *California geology* 36, no. 12:264–267.

Turco, Richard P. 1997. The climate machine (pp. 320–364) and greenhouse warming (pp. 365–406). *Earth under siege: From air pollution to global change.* New York: Oxford University Press.

Van Diver, Bradford B. 1992. *Roadside geology of New York,* 5th ed. Missoula, Mont.: Mountain Press.

◀▶ World Wide Web Activities

Pleistocene Catastrophic Flooding in the Columbia River Basin
http://www.opb.org/ofg/1001/missoula/floods.htm
The story of the Late Pleistocene Lake Missoula floods, which repeatedly released hundreds of cubic miles of water into eastern Washington drainages as a result of ice dam failures, is told here via text, photos, and video.
1. A single great flood from Lake Missoula released what percentage of today's rivers worldwide?
2. What is the name of the area scoured by these great floods, where is it located, and what surface features characterize it?

History of Pluvial Lake Bonneville
http://daac.gsfc.nasa.gov/DAAC_DOCS/geomorphology/GEO_7/GEO_PLATE_KL-12.HTML
Pleistocene Lake Bonneville was the largest of the pluvial lakes in western North America. Its remnant footprint is the Great Salt Lake in Utah. This site explains with maps and text the complex history of this lake basin.

1. What surface features are researchers using to identify former lake level stand (ancient shorelines) for Lake Bonneville? Why aren't remote sensing techniques very helpful?
2. Describe the sequence of events that finally drained the Lake Bonneville high-water stand.

Scientific Approaches and Reasons for Studying Paleoclimates
http://www.ngdc.noaa.gov/paleo/primer_care.html
This NOAA site provides an overview of paleoclimatology, with a link to "natural proxy sources," techniques scientists use to determine climatic data beyond the 140 years of actual records.
1. Why are NOAA and other agencies so interested in paleoclimate data? How might their findings impact humans today?
2. Navigate to the "natural proxy sources" link. Pick one of these features—tree rings, ice cores, fossil pollen, ocean and lake sediments, and corals—and explain how it provides information on past climates.

Earth Systems Today CD Correlations

CHAPTER CONCEPT: Human-Generated CFCs Break Down the Ozone Layer in the Upper Atmosphere that Is a Vital Shield Protecting Earth's Organisms from Harmful Solar Ultraviolet Radiation
This part of the CD gives historical and current information on the status of the ozone layer, particularly at its most vulnerable point above Antarctica.

SECTION MODULE: Weather and Climate: The Ozone Hole
Observe historical and current changes in the Antarctic ozone hole via interactive exploration of this CD module.

QUESTION: *Describe the overall pattern of changes you see in the Antarctic ozone depletion site over time. Is the pattern systematically related to CFC emissions?*

CHAPTER CONCEPT: For Glaciers to Exist, the Continents Must Be in Higher Latitudes, and High-Standing Mountains Help, Too.
This CD segment displays interactively the changing positions of the continents over the phanerozoic.

SECTION MODULE: Geologic Time: Time and Plate Tectonics
Observe the changing positions of the continents over the last 750 million years using the interactive globe.

QUESTION: *Given that high latitude landmasses and high altitude mountain ranges (from plate collisions) promote glaciation, decide whether these particular times were **likely** or **not likely** to have had continental ice sheets: Earth at 450 ma; at 250 ma; at 150 ma.*

chapter | twelve

Arid Lands and Desertification

◆ **FIGURE 1** Northwest Egypt near El Alamein during the North Africa campaign, 1942. A British soldier is watching an approaching dust storm generated by *gibleh* winds.

Deflation Basins, Winds, and the North Africa Campaign of World War II

The Qattara Depression in northwestern Egypt is an immense, hot wasteland of salt marshes and drifting sand that covers an area of about 18,000 square kilometers (7,200 mi²) at an elevation nearly 135 meters (443 ft) below sea level (◆ Figures 1 and 2). Its origin is partly due to deflation by wind. Wind scour has lowered the land surface to the capillary fringe at the top of the water table. This has led to evaporation of the saline water and crystallization of the salt as a sparkling white crust on the depression floor.

This impressive and forbidding quagmire played an important role during World War II in a decisive Allied victory in 1942. The impassable trough served to block German Field Marshal Erwin Rommel's Afrika Korps' attempt to turn the flank of the British Eighth Army led by General Bernard Montgomery. The geography forced Rommel to attack the Eighth Army's heavily fortified positions on El Alamein, a 64-kilometer (40-mi)-wide strip of passable land between the Mediterranean on the north and the hills to the south that mark the edge of the impassable Qattara. Rommel outran his lines of supply, sustained heavy losses, and was forced into retreat. This pushed him out of Egypt, and eventually westward across Libya.

Throughout the two-year North Africa campaign, the Allied and Axis forces had a common enemy: the *gibleh,* the searing hot winds that blow out of the Sahara. These winds raised temperatures as much as 19° C (35° F) in a few hours and sent sand grains swirling across hundreds of square kilometers. The onslaught of a gibleh dictated the actions of men and machines, preempting military tactics and strategies. Airplanes had to be grounded, and battlefield action stopped for days, as the sun was blotted out and visibility dropped to zero. Soldiers wore tight-fitting goggles to protect their eyes from the clouds of stinging sand and gasped for air through gas masks or makeshift cloth masks. Winds of hurricane strength that tore up camps and turned over vehicles were common. Swirling clouds of sand generated static electricity that rendered useless the magnetic compasses needed for navigating across the desert. Storm-generated static electricity during one gibleh detonated and destroyed an ammunition dump. ■

◀▶ FIGURE 2 Locations of the Qattara Depression and El Alamein. The decisive battle between the forces of Montgomery and Rommel was fought north of the Qattara Depression.

Deserts are natural laboratories in which to study the interactions of wind and sometimes water on the arid surfaces of planets.

A. S. Walker, U.S.G.S. geologist

Deserts

The term *desert* means different things to different people. For some, a desert is a dry barren region of sand dunes; for others, it is a desolate region of bare, rocky outcrops and cactus; for still others, it is a treasure trove of scenic wonders, unique plants, and animals. Regardless of how they are perceived, deserts comprise about one-third of the earth's surface and are characterized by meager rainfall, scanty vegetation, a distinct landscape, and a limited population of people and animals. In deserts, the effects of running water, tectonic forces, and wind are clearly apparent. Because these factors combine in different ways in different places, the appearance of various desert landscapes differ as well. Deserts may be hot, or cold. They may be mountainous and rocky, or vast, flat sand sheets, and some are covered with gravel. But all deserts are dry. Although deserts appear harsh and rugged, they are enormously

fragile, and the misuse of arid lands is a serious problem in many parts of the world.

KINDS OF DESERTS

A **desert** (Latin *desertus,* "deserted, barren") is defined as a region with mean annual precipitation of less than 25 centimeters (10 in), with a potential to evaporate more water than falls as precipitation, and with so little vegetation that it is incapable of supporting abundant life. Notice that high temperatures are not necessary for a region to be called a desert; only low precipitation, a dry climate, and limited biological productivity. On this basis we can identify four kinds of desert regions: polar, subtropical, mid-latitude, and coastal (⊙ Table 12.1).

- **Polar deserts** are marked by perpetual snow cover, low precipitation, and intense cold. The most desertlike polar areas are the ice-free dry valleys of Antarctica. The Antarctic continent and the interior of Greenland are true deserts. They are very dry, even though most of the fresh water there is in the form of ice.
- **Subtropical deserts** (also called *trade-wind deserts*) are the earth's largest dry expanses. They lie in the regions of subsiding high-pressure air masses in the western and central portions of continents. They receive almost no precipitation and have large daily temperature variation. The Saharan, Arabian, Kalahari, and Australian deserts are of this type.
- **Mid-latitude deserts,** at higher latitudes than the subtropical deserts, exist primarily because they

⊙ TABLE 12.1	Classification of Deserts	
TYPE	CHARACTERISTIC LOCATION	EXAMPLES*
Polar	region of cold, dry, descending air with little precipitation	ice-free, dry valleys of Antarctica
Subtropical	a belt of dry, descending air at 20°–30° north or south latitude	Sahara, Arabian, Kalahari, Australia's Great Sandy and Simpson
Mid-latitude	deep within continental interior remote from the influence of an ocean	Gobi, Takla Makan, Turkestan
Coastal	coastal area in middle latitudes where a cold, upwelling oceanic current chills the shore (Figure 12.2)	Atacama, Namib
Rain-shadow	leeward of a mountain barrier that traps moist ocean air (Figure 12.1)	Mojave, Great Basin

*Figure 12.4 shows the locations of these deserts.

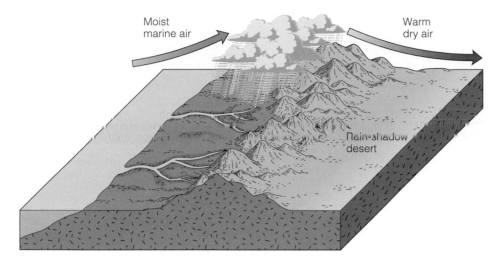

◁▷ **FIGURE 12.1** The origin of rain-shadow deserts in middle- and high-latitude regions. Moist marine air moving landward is forced upward on the windward side of mountains, where cooling causes clouds to form and precipitation to fall. The drier air descending on the leeward side warms, producing cloudless skies and a rain-shadow desert.

are positioned deep within the interior of continents, separated from the tempering influence of oceans by distance or a topographic barrier. The Gobi Desert of China, characterized by scant rainfall and high temperatures, is an example. Marginal to many middle-latitude deserts are **semiarid grasslands,** extensive treeless grassland regions. Drier than prairies, they are especially sensitive to human intervention. In North America and Asia the mid-latitude and subtropical arid regions merge to form nearly unbroken regions of moisture deficiency. Also in the mid latitudes are mountain ranges that act as barriers to the passage of moisture-laden winds from the ocean, a situation leading to the formation of **rain-shadow deserts.** As water-laden maritime air is drawn across a continent, it is forced to rise over a mountain barrier. This causes the air to cool and to lose its moisture as rain or snow on the windward side of the mountains. By the time the air descends on the mountains' leeward side, it has lost most of its moisture, and there we find desert conditions (◁▷ Figure 12.1). The Mojave Desert and the Great Basin, leeward of the high Sierra Nevada, are examples of rain-shadow deserts. Death Valley, the lowest and driest place in North America, is in the rain shadow of the Sierra Nevada. The Sierra Nevada receives heavy precipitation, and Mount Whitney, its highest point and the highest point in the conterminous 48 states, is a scant 100 kilometers (60 mi) from arid Death Valley.

■ **Coastal deserts** lie on the coastal side of a large land mass and are tempered by cold, **upwelling** oceanic currents (◁▷ Figure 12.2). Their seaward margins are perpetually enshrouded in gray coastal fog, but inland, where the air temperatures are higher, the fog evaporates in the warmer dry air. The Atacama Desert in Chile, which may be

the driest area on earth, is a rain-shadow desert as well as a coastal desert. The Andes on the east block the flow of easterly winds, and air from the Pacific Ocean on the west is chilled and stabilized by the Humboldt Current, the ocean's most prominent cold-water current.

Collectively, mid-latitude deserts and semiarid grasslands cover about 30 percent of the earth's land area and constitute the most widespread of the earth's geographic realms.

Desertification

Deluges of blowing sand, such as the one in the North African Desert described in the chapter opener, are common in many arid regions. In the past these winds have been rare in semiarid grasslands marginal to middle-latitude deserts, but this is changing. We now see a new kind of desert forming where such winds are

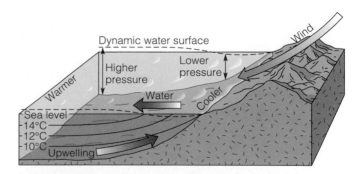

◁▷ **FIGURE 12.2** The mechanics of wind-driven upwelling of cool water from depth along a coastline. Upwelling delivers nutrients such as phosphates and nitrates to surface waters, where they promote plant plankton blooms and more marine life.

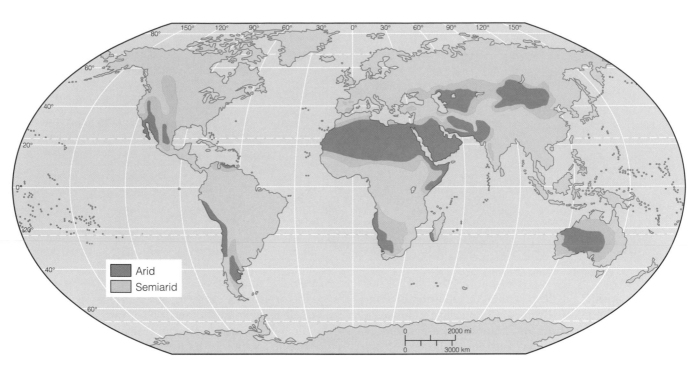

◆▶ FIGURE 12.3 Distribution of the earth's arid and semiarid regions.

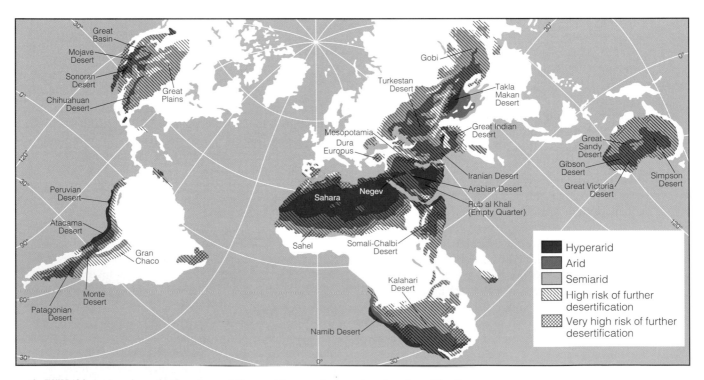

◆▶ FIGURE 12.4 Arid and semiarid regions of the world currently experiencing desertification.

◆▶ **FIGURE 12.5** Intensely overgrazed rangeland (left) and lightly grazed (right).

major causes of desertification. Other practices that lead to desertification include:

- clearing land of trees and brush for fuel without reforesting,
- surface mining without reclaiming the land,
- the rapid growth of large desert cities such as Phoenix and Las Vegas in the southwestern United States and other examples in countries such as Egypt and Saudi Arabia;
- depletion of ground water for desert irrigation and urban growth,
- replacement of native vegetation in semiarid regions with cultivated crops, and
- soil salinization due to evaporation of irrigation water (see Case Study 6.1).

Case Study 12.1 tells of the desertification in Africa's Sahel region.

For years the question has been asked, Does climate change cause desertification, or does desertification cause climate change? There has been no clear answer to this question until studies in the 1990s by the University of Arizona's Remote Sensing Center on the border between the United States and Mexico revealed a clear *signal*: higher temperatures were recorded on the Mexican side of the border, where the earth's resources were being exploited more heavily, than on the U.S. side.

Reversing desertification is a desirable goal, but disturbed desert and marginal grassland ecosystems are slow to recover. This is because those ecosystems are characterized by slow plant growth, little bacterial activity in the soils for recycling nutrients, little species diversity, and little water. Already-sparse vegetation that has been destroyed by overgrazing or off-road vehicles (◆▶ Figure 12.6) may require decades, perhaps centuries, to be restored to its natural state. The best program for reversing the trend is to halt exploitation immediately and to adopt measures such as planting trees, developing and using water resources properly, and establishing protective ecological screens that will help establish natural ecosystems.

A good example of natural recovery is seen in Kuwait. It is an unexpected consequence of the 1991 Gulf War. Thousands of unexploded land mines and bombs are scattered across the desert in the western part of the country. These have discouraged off-road driving by sport hunters of the region's animals. The native vegetation is reestablishing, and the desert now resembles a U.S. prairie. The number of birds in the area is much greater than it was before the war.

more common. These new deserts, which may be called *deserts of infertility,* are devouring more than 20,000 square miles of the earth's arid grasslands each year—an area about the size of the state of West Virginia—and putting another 70,000 square miles at risk. In fact, so much dust from some of these arid lands in Africa blows across the Atlantic Ocean during the summer that air quality in Florida and some other East coast states is seriously degraded by it. If the new air-quality standards proposed by the Environmental Protection Agency are adopted, some of these states may be found to be in violation of them.

Deserts of infertility are not controlled as much by climate as by human misuse. The expansion of deserts into formerly productive lands because of excessive human use is called **desertification,** and it is occurring at alarming rates in countries that can least afford to lose productive land. Understanding the causes of this class of desert is important, because much of humankind's future in many parts of the world depends on proper and careful use of arid and semiarid lands. In nearly 70 countries worldwide, deserts are expanding onto grazing lands and agriculturally productive lands (◆▶ Figures 12.3 and 12.4). The result is that millions of people are dying of starvation or having to migrate, in many cases to areas where more hunger and malnutrition await them.

Globally, overgrazing of sparse vegetation (◆▶ Figure 12.5) and trampling of the soil by livestock are the

Case Study | 12.1

The Sahel and Climate Change

For some time the world's deserts have been expanding (Figures 12.3 and 12.4), but the expansion of North Africa's huge Sahara Desert into the Sahel region has been especially severe since the late 1960s (◁▷ Figure 1). The Sahel is a belt of semiarid lands that stretches from the Atlantic Ocean on the west to the Red Sea on the east, 300 to 1,500 kilometers (180–940 mi) in width (◁▷ Figure 2). The region typically has dry winters and wet summers, when the intertropical convergence zone **(ITCZ)** moves into the region, bringing rain. Rainfall amounts vary from year to year, averaging 10 to 20 centimeters (4–8 in), but some years it

may be as much as 50 centimeters (20 in) in the southern part, where the Sahel blends into the belt of humid savannah grasslands.

Although the Sahel is inhabited mainly by pastoral nomads, who roam the region with their goats and cattle, there are some towns and there is some agriculture. Crops such as millet and peanuts can be grown in many areas.

In 1968 the summer rainfalls began decreasing, ushering in a period of severe drought conditions. Combined with the resultant overgrazing, the drought quickly turned thousands of square miles of natural pasture into desert. This led

◁▷ FIGURE 1 A sharp line marks the boundary between natural pasture and a dune encroaching from the Sahara in the west African country of Mali; a graphic example of desertification. The Sahara Desert is creeping southward into the semiarid Sahel region, and the Sahel, in turn, is expanding into the grasslands to the south of it.

to the deaths of 100,000 to 250,000 people by starvation and disease, the collapse of the agricultural development of five countries, and the disruption of millions of lives. Many more people would have perished had it not been for the massive influx of outside aid.

Under normal climatic conditions, such semiarid grasslands on the fringes of true deserts are parts of a delicately balanced ecosystem that is adapted to natural climate variations. The vegetation indigenous to these marginal lands can survive the rigors of a few dry years. Cultivated crops that replace the natural vegetation, however, lack drought tolerance, and thus, crop failure is common during drought. This leaves the soil bare and susceptible to erosion by wind and rain. In 1972, the Sahel received almost no rain. In 1973, at the climax of the drought, total rainfall was about half the long-term annual average. By then, the Sahara Desert had migrated as much as 70 kilometers (100 mi) southward into the Sahel, and 50 to 70 percent of the region's livestock had died.

It is still unclear just why the drought occurred. The wetter years of the mid twentieth century were due to the favorable positioning of the ITCZ, which brought intense summer rains, then the drier years experienced less intense rains. It might be that an initial decrease in rainfall and the consequent reduction in plant cover (due to the introduction of cultivated crops) modified the land surface sufficiently to reduce convective activity and create a positive-feedback mechanism that increased atmospheric drying even more. (Positive and negative feedback are explained in Chapter 11.)

Many questions remain. Is the current drought part of a long-term climatic fluctuation, or will the "normal" wetter climate of the 1950s and 1960s return? If we are entering a time of extended global warming, what will future precipitation patterns be, and what will become of the Sahel's people and their way of life?

▷ FIGURE 2 North Africa's semiarid Sahel is bordered by grasslands on the south and by the Sahara Desert on the north.

■ Sahara desert
■ Sahel
■ Grasslands
■ Tropical rain forest

◆ **FIGURE 12.6** Off-road vehicle damage to the desert. Jawbone Canyon in California is an example of how human activities contribute to desertification. Because the vehicles compact and alter the surface, they cause reductions in vegetation and permeability, which reduces soil moisture content and thus affects local biological systems. Desert rains have eroded the tracks down the slope, transforming them into gullies, which increases runoff and sediment yield.

Wind as a Geologic Agent in Deserts

Wind erodes by deflation and by abrasion. **Deflation** occurs when things are blown away. This produces deflation basins where loose, fine-grained materials are removed, sometimes down to the water table (◆ Figure 12.7). *Desert pavement,* an interlocking cobble "mo-

saic," results where winds have removed fine-grained sediments from the surface (◆ Figure 12.8). **Abrasion** occurs when mineral grains are blown against each other and into other objects.

Migrating dunes—small hills of migrating, wind-blown sand—and drifting sand, which are common in arid regions, can adversely affect desert highways and even communities (◆ Figure 12.9), but they are not

◆ **FIGURE 12.7** The Devil's Cornfield, a deflation basin that has been lowered to the capillary fringe at the top of the water table; Death Valley National Park, California. The rugged plants growing here—arrow weed, which resemble shocks of corn—are able to tolerate heat and saline soil, and their roots can withstand the effects of exposure to episodic sandblasting.

◆ **FIGURE 12.8** Desert pavement, an interlocking cobble mosaic that remains after winds have removed the surficial fine-grained sediments. The desert floor is left armored with a tightly interlocked mosaic of gravel. An unsorted mixture of coarse- and fine-grained sediments underlies the armor.

◆ FIGURE 12.9 Migrating sand dunes are encroaching on Mauritania's capital city, Novakchott, on northwest Africa's Atlantic coast.

limited to arid regions. Strong sea breezes in humid coastal regions and human activities also can cause dune migration (see ◆ Figures 10.16b and 12.10). The problems of migrating sand have been mitigated by building fences, by planting windbreaks of wind- and drought-tolerant trees such as tamarisk (◆ Figure 12.11), and by oiling or paving the areas of migrating sand. Dune stabilization in humid coastal areas of Europe, the eastern United States, and the California coast has been accomplished by planting such wind-tolerant plants as beach grass (*Elymus*) and ice plant "Hottentot fig" (*Carpobrotus*).

Arid regions of the world have some intriguing and bizarre natural features, as shown in ◆ Figure 12.12. Among the curiosities produced by wind are teetering rocks that have been nearly severed at the base by wind abrasion and playa rock scrapers, rocks that leave complicated tracks on mud-surfaced playas as they are pushed along by high winds.

◆ FIGURE 12.10 Coastal dune field near Pismo Beach, California. In the 1960s these dunes began migrating leeward onto agricultural lands due to the destruction of stabilizing plant cover by uncontrolled off-road vehicle use in the dune field.

◆ FIGURE 12.11 Windbreaks of salt-tolerant *tamarisk* (salt cedar) keep drifting sand from burying railroad tracks; Palm Springs area, California.

(b)

◆▶ **FIGURE 12.12** Desert oddities. (a) Mushroom rock; Death Valley National Park, California. Rigorous wind erosion in the abrasion zone just above the ground surface can produce unusual landforms such as this. (b) Remnants of a 1920s plank road, an early attempt to maintain a road in a region of drifting sand; Algodones dune field between El Centro, California, and Yuma, Arizona. (c) Playa lake scraper; Racetrack Playa, Death Valley National Park, California. The boulder's track is preserved on the mud-cracked surface. The boulder, 25 cm (11 in) at its widest, moved or was moved during a winter wet period due to just the right combination of rain-soaked mud and gusty high wind. (d) Devil's Golf Course, Death Valley National Park, California, the lowest elevation in the Western Hemisphere. The salt-encrusted surface has formed at the capillary fringe on the top of the underlying water table.

(c)

(a)

(d)

Summary

DESERTS
DESCRIPTION
Regions where annual precipitation averages less than 25 centimeters (10 in) and that are so lacking in vegetation as to be incapable of supporting abundant life.

CAUSES
1. High-pressure belts of subsiding, warming air that absorbs water and precludes cloud formation.
2. Isolation from moist maritime air masses by position in deep continental interior.
3. Windward mountain barrier that blocks passage of maritime air.

CLASSIFICATION
May be classed into four categories determined by climate belts and a fifth category due to human misuse.

TYPES
1. *Polar deserts*, regions of perpetual cold and low precipitation.
2. *Mid-latitude deserts* are found within the interior of continents in the middle latitudes, remote in distance from the influence of oceans.
3. *Subtropical deserts,* the earth's largest realm of arid regions, lie in and on the equatorial side of subtropical

zones of subsiding high-pressure air masses in the western and central portions of continents.
4. *Coastal deserts* occur on the coastal side of a land or mountain barrier in subtropical latitudes. Because they are bordered by ocean, they are cool, humid, and often foggy.
5. *Deserts of infertility,* generally variants of mid-latitude deserts, are not so much due to climate as to human misuse.

DESERTIFICATION
The expansion of deserts into formerly productive lands because of excessive human use or misuse.

WIND AS A GEOLOGIC AGENT
EROSION
1. *Deflation*–earth materials are lifted up and blown away.
2. *Abrasion*–mineral grains are blown against each other and into other objects.

CONTROL OF MIGRATING SAND
Accomplished with sand fences, paving, windbreaks, and stabilizing plants.

◆ Key Terms

abrasion	desertification	rain-shadow desert	upwelling
coastal desert	ITCZ	semi-arid grassland	wind
deflation	mid-latitude desert	subtropical desert	
desert	polar desert	sahel	

◆ Study Questions

1. Explain why the world's largest deserts are located on or near the horse latitudes.
2. What factors other than drought are responsible for the desertification that has resulted in starvation in Africa and elsewhere? (*Hint:* Consider carrying capacity, as discussed in Chapter 1.)
3. Explain why ground water can be found only a few centimeters to a meter (3 ft) or so below the salt-encrusted surface in some of the driest places in the world–for example, the Devil's Golf Course, at about 75 meters (250 ft) below sea level in Death Valley, and the Qattara Depression, at about 135 meters (443 ft) below sea level in the Sahara Desert.
4. List measures that can be taken to halt sand-dune migration.
5. Explain the interrelations between the biologic, atmospheric, and solid-earth systems in arid regions. You may use words, pictures, or both.
6. The gravel-covered desert surface called desert pavement plays an important role in arid regions. What is it?
7. List agricultural practices in arid and semiarid regions of the United States (such as California's Coachella Valley and western Nebraska) that increase the likelihood of desertification.
8. List measures that can be taken to halt desertification.

◆ For Further Information

BOOKS AND PERIODICALS

Ahrens, C. Donald. 2000. Global climate (pp. 470–502). *Meteorology today,* 5th ed. Pacific Grove, Calif.: Brooks/Cole.

Collier, Richard. 1977. *The war in the desert.* Chicago: Time-Life Books.

Cooke, R. A.; A. Warren; and A. Goudie. 1993. *Desert geomorphology.* London: UCL Press.

Ellis, W. S. 1987. Africa's Sahel: The stricken land. *National geographic* 172, no. 2:140–179.

Monroe, James S., and Reed Wicander. 1998. The work of wind and deserts (pp. 511–535). *Physical geology,* 3d ed. Belmont, Calif.: Wadsworth Publishing Co.

Sheridan, D. 1981. *Desertification of the United States.* Washington, D.C.: Council on Environmental Quality.

Scott, Ralph C. 1992. Climates of the world, (pp. 181–220). *Physical geography,* 2d ed. St. Paul, Minn.: West Publishing Co.

Sletto, B. 1997. Desert in disguise. *Earth* 6, no. 1:42–49.

Walker, A. S. 1982. Deserts of China. *American scientist* 70, no. 4:366–376.

_____. 1996. *Deserts: Geology and resources.* Denver: U.S. Geological Survey.

◆ World Wide Web Activities

Sand Dune Formation: A Link with Wet Climates?
http://www.geog.nottingham.ac.uk/~michele/research/geomorphology/sand.htm

This abstract with photo from the *Journal of Arid Lands* proposes that sand dune formation, rather than indicating hyperaridity, occurs in deserts during relatively wet times.

1. What evidence do the authors use to support their theory?
2. What specific weather event or condition would promote sand delivery and dune growth?

An Experimental Approach to Halting Desertification in Subsaharan Africa
http://www.kubota.co.jp/eng/problem/earth/earth3.html

The Sahil Greenbelt Plan Web page briefly describes an experimental plan by a Japanese consortium to stabilize a portion of sub-Saharan Africa from the threat of desertification.

1. What approach do scientists propose to increase water supplies in this region?
2. What energy source will be used to lift and distribute underground water supplies?

Salton Sea Ecology and Remedies to Hypersalinity
http://www-sci.sdsu.edu/salton/Salton%20Sea%20Description.html

This site explains the ecological value and water quality problems of the Salton Sea, a below-sea-level sag pond along the San Andreas Fault that is increasing in salinity as a result of high desert evaporation rates.

1. What is the current salinity of the Salton Sea in comparison to open ocean salinity? Why is it so salty?
2. What are the two main proposals for decreasing the salinity of the Salton Sea?

Scientific Studies in the Ice-Free Valleys of Antarctica
http://www.nrel.colostate.edu/soil/MCM.html

This Web site provides an introduction to the geochemistry and biology of soils in the Antarctic ice-free valleys as discovered by an ongoing research program of the National Science Foundation.

1. How do these ice-free valley soils compare geologically with soils in the rest of the world?
2. What is biologically unique about these soils? Why is the study of their biology so important?

Earth Systems Today CD Correlations

CHAPTER CONCEPT: The Atacama Desert Is a Coastal Desert, Created in Part by the Action of the Cold Upwelling Humboldt Current Offshore

This part of the CD shows general circulation of surface currents in the oceans.

SECTION MODULE: Currents

Observe the location of the Humboldt Current west of South America; examine the west coast of Africa for comparison.

QUESTION: *Given the similar geographic situation of Africa, is there also a coastal desert at this locality? If so, what is its name?*

CHAPTER CONCEPT: Rain-Shadow Deserts Are Found behind Active Mountain Ranges that Block Moisture-Bearing Winds from the Ocean

This part of the CD shows various plate boundary types on a global map.

SECTION MODULE: Plate Tectonics: Introduction, Interactive Map with Different Plate Boundary Types

Activate the **Convergent Boundaries** button to display plate collision sites that have created young, active mountain ranges.

QUESTION: *Which of these major world desert regions has a significant rain-shadow component to their origin (pick all that apply): Great Sandy (Australia), Gobi (Eurasia), Atacama (Chile), Namib (southwest Africa), Great Basin (United States), Sahara (North Africa), Antarctic dry valleys, Saudi Arabian peninsula desert?*

chapter thirteen

Mineral Resources and Society

The New Gold Rush

The biggest bonanza for Nevada mining interests since the 1860s boom days of the Comstock Lode was the discovery of the Carlin Trend, a 60-kilometer (37-mi) line of mostly low-grade but very large gold deposits in north-central Nevada. It is the largest gold district discovered in North America in the twentieth century. Typically the gold is submicron-sized, "invisible" discrete grains incorporated within sulfide minerals. So small are the grains that no colors can be obtained by the tried-and-true method of gold panning, even of high-grade ores. About a dozen major mines and several smaller ones, most of them open-pit operations, were mining the Trend by the mid 1990s.

The Barrick Goldstrike Meikle Mine on the Carlin Trend near Elko, Nevada.

Several processing techniques used to extract the gold include milling of the high-grade oxide ores, refractory treatment of the sulfide ores (roasting to burn off the sulfides and carbon), conventional cyanide heap-leaching, and bioleaching (using bacteria to loosen microscopic amounts of gold from problematic sulfide ores that would otherwise be mine waste). The cost of recovering gold from the open-pit mines is about $240 per ounce.

From the outset, the mining companies' reclamation staffs have worked closely with the mine-design staff. Reclamation bond money is set aside continuously, according to the conditions of the state permit, and it amounts to about $1 for each ounce of gold produced. Topsoil for future reclamation is stockpiled and seeded with grasses to protect it from erosion by rain and wind until it is needed. The mining permits require that "the land be returned to a reasonable state" when mining is completed. Specifically, this requires that the open pits be bermed and fenced and that waste-rock dumps be encapsulated with a 4-foot cover of topsoil and planted with native plants. A network of dispersion drainage ditches has been installed to collect and divert surface flow from the waste piles. Because Carlin has an arid climate (annual rainfall is about 10 inches), water penetration into the waste should be negligible. Thus, in spite of the ore bodies being rich in sulfides, acid drainage is not seen as a potential environmental problem.

The operating permit calls for a 3½-year review following completion of mining and obligates the mining companies to work on reclamation for 15 years. ∎

403

Every material used in modern industrial society is either grown in or derived from the earth's natural mineral resources. The mineral resources are usually classified as either metallic or nonmetallic, as shown in ◉ Table 13.1. Production and distribution of the thousands of manufactured products and the food we eat is dependent upon the utilization of metallic and nonmetallic mineral resources. The per capita U.S. consumption of mineral resources used directly or indirectly in providing shelter, transportation, energy, and clothing is enormous (◆ Figure 13.1). The availability and cost of mineral and rock products influences our nation's standard of living, gross national product, and position in the world.

With the value of nonfuel minerals produced in the United States running $40 billion per year and the value of materials processed from minerals in 1998 estimated at $415 billion (about 5 percent of the gross national product), it is amazing that the general public has relatively little knowledge of where these minerals naturally occur, the methods by which they are mined and processed, and the extent to which we depend on them. It is important that the public recognize that exploitable mineral resources occur only in particular places, having formed there due to unique geologic conditions, and that all deposits are exhaustible. Furthermore, development and exploitation of mineral resources is capital-intensive, requiring substantial long-term investment, and has environmental effects.

Understanding the origins, economics, and methods of mining and processing of mineral resources enables an individual to understand many of the difficult resource-related issues that are currently facing all levels of government. Increased land-use competition among the interests of mining, housing, wilderness preservation, agriculture, logging, recreation, wetlands preservation, and industrial development is forcing our government officials to make decisions that affect mineral-resource development and management and, thus, all of our lives.

Mineral Abundances and Distribution

ECONOMIC MINERAL CONCENTRATION

The elements forming the minerals that make up the earth's crust exist in many forms in a great variety of rocks. The higher the abundance, the richer the deposit and the more economically feasible it is to extract and process the desired element. Locally rich concentrations of minerals are called **mineral deposits.** If they are sufficiently enriched, they may be **mineral reserves,** deposits of earth materials from which useful commodities are economically and legally recoverable with existing technology. Reserves of metallic minerals are called **ores,** and the minerals comprising the deposits are referred to as **ore minerals.**

◯ TABLE 13.1 Classification of Mineral Resources	
METALLIC MINERAL RESOURCES	**NONMETALLIC MINERAL RESOURCES**
Abundant Metals	*Minerals for Industrial and Agricultural Use*
iron, aluminum, manganese, magnesium, titanium	phosphates, nitrates, carbonates, sodium chloride, fluorite, sulfur, borax
Scarce Metals	*Construction Materials*
copper, lead, zinc, tin, gold, silver, platinum-group metals, molybdenum, uranium, mercury, tungsten, bismuth, chromium, nickel, cobalt, columbium	sand, gravel, clay, gypsum, building stone, shale, and limestone (for cement)
	Ceramics and Abrasives
	feldspar, quartz, clay, corundum, garnet, pumice, diamond

SOURCE: James R. Craig, David L. Vaughan, and Brian J. Skinner, *Resources of the Earth* (Englewood Cliffs, N.J.: Prentice Hall, 1988).

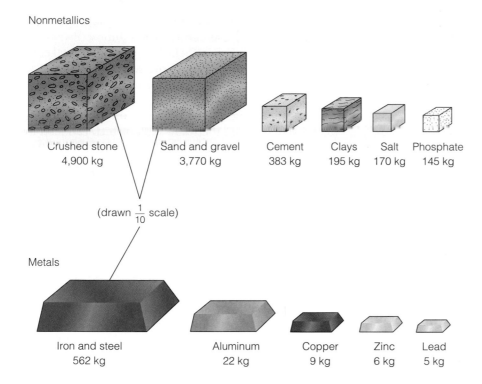

Nonmetallics

Crushed stone
4,900 kg

Sand and gravel
3,770 kg

Cement
383 kg

Clays
195 kg

Salt
170 kg

Phosphate
145 kg

(drawn $\frac{1}{10}$ scale)

Metals

Iron and steel
562 kg

Aluminum
22 kg

Copper
9 kg

Zinc
6 kg

Lead
5 kg

◀▶ **FIGURE 13.1** Approximate annual per capita consumption of nonfuel mineral resources in the United States. Nonmetallic resources constitute about 94 percent of the mineral use by weight. (A kilogram equals 2.2 pounds.)

A particular ore deposit may be described in terms of its **concentration factor,** its enrichment expressed as the ratio of the element's abundance in the deposit to its average continental-crustal abundance. ◉ Table 13.2 illustrates that the minimum concentration factors for profitable mining vary widely for eight metallic elements. Note in the table that aluminum, one of the most common elements in the earth's crust, has an average crustal abundance of 8 percent. A "good" aluminum-ore deposit is defined at this time as one that contains about 35 percent aluminum; thus, the present concentration factor for profitable mining of aluminum

ore is about 4. In contrast, deposits of some rare elements such as uranium, lead, gold, and mercury must have concentration factors in the thousands in order to be considered mineral *resources*.

Some elements are extremely common. In fact, 97 percent of the earth's crust is comprised of only eight elements: oxygen, silicon, aluminum, iron, sodium, calcium, potassium, and magnesium. For most elements, the **average crustal abundance,** the amount of a particular element present in the continental crust, is only a fraction of a percent (Table 13.2). Copper, for example, is rather rare; its average crustal abundance is only

◯ TABLE 13.2   Concentration Factors for Profitable Mining of Selected Metals

	PERCENTAGE ABUNDANCE		CONCENTRATION FACTOR
METAL	*Average in Earth's Crust*	*In Ore Deposit*	
Aluminum	8	35	4
Copper	0.0063	0.4–0.8	80–160
Gold	0.0000004	0.001	2,500
Iron	5	20–69	4–14
Lead	0.0015	4	2,500
Mercury	0.00001	0.1	10,000
Silicon	28.2	46.7	2
Titanium	0.57	32–60	56–105

SOURCE: U.S. Geological Survey Professional Paper 820, 1973.

0.0063 percent. In some localities, however, natural geochemical processes have concentrated it in mineral deposits that are 2 to 4 percent copper. A deposit with a high concentration of a desired element is called a **high-grade deposit.** A deposit in which the mineral content is minimal but still exploitable is called a **low-grade deposit.** For either grade of deposit, localities where desirable elements are concentrated sufficiently for economic extraction are relatively few.

FACTORS THAT CHANGE RESERVES

Reserves are not static, because they are defined by the current economics and technology as well as by the amount of a mineral that exists. Reserves fluctuate due to several factors: changing demand, discoveries of new deposits, and changing technology. In a free economy, a mineral deposit will not be developed at an economic loss, and prices will rise with demand. Consequently, low-grade deposits that are marginal or submarginal in today's economic climate may, if demand and prices rise, eventually become profitable to exploit. Advancements in technology also may increase reserves by lowering the cost of development or processing.

For example, the exploitable reserves of gold increased dramatically in the United States in the late 1960s due to a combination of changes in government policies and advances in technology. In 1968, when the price of gold was $35 per troy ounce, the Treasury Department suspended gold purchases to back the dollar and began allowing the metal to be traded on the open market. (A troy ounce equals 31.103 grams, whereas the avoirdupois ounce of U.S. daily life equals 28.330 grams.) Then in the 1970s the federal government removed restrictions prohibiting private ownership of gold bullion. In 1980 the gold price rose to more than $800 an ounce, and by 1999 it had settled at about $290 an ounce. Thus in a period of about 20 years the price of gold increased by 800 percent, meaning that many submarginal mining claims became economically profitable mines in that time. Furthermore, gold exploitation became more cost-efficient in the 1980s with the development of the new technology of cyanide heap-leaching, by which disseminated gold is dissolved from low-grade deposits and then recovered from the solution. A combination of changing economics and the new heap-leach technology initiated a new gold rush in the West beginning in the 1980s (see the chapter opener). But, by 1999, other factors had forced gold prices to record lows, in this instance due to changing market forces triggered by sales of gold bullion by the International Monetary fund, the Bank of England, and other central banks. Consequently, many gold mines were forced to close, including some in South Africa, the world's major gold-producing nation.

There are other factors that may reduce reserves, because extracting the earth's riches almost always requires the trade-off of aesthetic and environmental consequences (● Table 13.3 provides some examples). For example, large low-grade mineral deposits can be mined only by open-pit methods. Not only do these methods devastate the landscape, their excavations may reach nearly to the water table, which can lead to ground-water contamination, and they produce enormous amounts of waste rock and tailings that can pollute surface-water runoff. **Smelters,** large industrial plants that process ore concentrates and extract the desired elements, produce more air pollution, in the form of flue dust, than any other single industrial activity. Smelters also produce slag, a solid residue that can contain thousands of times the natural levels of lead, zinc, arsenic, copper, and cadmium.

The real and perceived negative environmental impacts that serve to rule out exploitation and development of known deposits also reduce reserves. In recent years there have been several widely publicized instances of wildlife and wilderness values being judged more important than exploitation of mineral or petroleum deposits. Examples include:

■ the extraordinary agreement reached between President Clinton and Crown Butte Mines in 1996 that halted proposed mining activity on the doorstep of Yellowstone National Park;
■ the two-year moratorium on hard-rock mining along a 100-mile-long region of Montana's Rocky Mountain Front, one of the nation's most important wildlife areas; and
■ the decision by the Office of Surface Mining and Reclamation not to allow surface mining of coal in the watershed of Fern Lake in Kentucky, because it would be incompatible with Cumberland National Historic Park's land-use plan, would destroy historic and natural systems, and would damage the region's water quality.

We can expect that controversies over land use will be increasingly common in the future as more and more lower-grade deposits will need to be exploited when higher-grade deposits become exhausted.

Political factors also may further limit or reduce resources. The United States imports substantial amounts of the mineral resources it needs from other countries. (This is discussed in detail later in the chapter.) As an illustration, consider that the Gulf War of 1991 was basically a resource war, and that the energy crisis of the 1970s was due to geopolitical events. Furthermore, resource availability may affect diplomatic policy. In spite of the general trade sanctions the United States enacted in the 1980s against South Africa to protest its

⊙ TABLE 13.3	Some Negative Environmental Consequences of Mineral Extraction and Processing Operations	
LOCATION	**OPERATION**	**ENVIRONMENTAL CONSEQUENCES**
Sudbury, Ontario, Canada	nickel-copper mining and smelting	This was one of the world's best-known environmental "dead zones." Little vegetation survived in 10,400-hectare (40-mi²) area around the smelter. Acid fallout destroyed fish populations in lakes within 65 km (40 mi). Conditions improved after completion of $530 million SO_2 smelter-abatement project and construction of the world's tallest smokestack; the city has a successful landscape reclamation program, but there is still significant SO_2 damage downwind of Sudbury.
Pará State, Brazil	Grande Carajas iron ore project	Wood requirements for smelting ore will require cutting 50,000 hectares (193 mi²) of tropical forest annually during the 250-year life of the project.
Amazon basin, Brazil	gold mining	The region has been invaded by hundreds of thousands of miners digging for gold, clogging rivers with sediment, and releasing some 100 tons of mercury into the ecosystem annually.
Ilo-Locumba area, Peru	copper mining and smelting	Each year, the Ilo smelter emits 600,000 tons of sulfur compounds, and nearly 40 million cubic meters (523 million yd³) of tailings containing lead, zinc, copper, aluminum, and traces of cyanide are dumped into the ocean, poisoning marine life in a 20,000-hectare (77-mi²) area. Nearly 800,000 tons of slag are dumped in the sea yearly.
Panguna mine, Bougainville, Papua New Guinea	copper sulfide ore mining	Before the mine closed in 1989, the operation had dumped 600 million tons of tailings into the Kawerong River. The wastes cover 1,800 hectares (7 mi²) of the Kawerong River system, including a 700-hectare (2.7-mi²) delta. No aquatic life survives in the river. Local anger generated by the destruction was a major cause of a civil war.
Nauru, South Pacific	phosphate mining	When mining is completed in 1998–2008, 80% of the 2,100-hectare (8-mi²) island will be uninhabitable. The people of Nauru have initiated legal action to control the mining.
Summitville, San Juan Mountains, Colorado	gold mining	Leaking leach pad, contaminated with cyanide and heavy metals, poisoned 27 km (17 mi) of the Alamosa River, on which the region's agriculture depends. When cleanup costs exceeded the posted bond, the Canadian mining company and its parent declared bankruptcy. Cleanup may cost U.S. taxpayers over $100 million.
Guyana, South America	gold mining	Cyanide spill from the Omai mine–backed by the same investors as Colorado's Summitville mine–released 3.2 million m³ (113 million ft³) of poisonous runoff into one of Guyana's largest rivers.

SOURCE: *Mining the Earth, State of the World* (Washington: World Watch Institute, 1992).

apartheid policy, ten minerals the United States imports from South Africa were exempted because of their importance to the United States.

DISTRIBUTION OF MINERAL RESOURCES

The global distribution of mineral resources has no relationship to the locations of political boundaries or technological capability. Some mineral deposits, like those of iron, lead, zinc, and copper, originate by such a variety of igneous, sedimentary, and metamorphic processes that they are widely distributed, though in varying abundance. On the other hand, bauxite, essentially the only ore mineral of aluminum (one of the most common elements in the earth's crust), is concentrated by only one geochemical process: deep chemical weathering in a humid tropical climate. For this reason, economically exploitable deposits of bauxite and other such minerals are unequally distributed and restricted to a few, highly localized geologic and geographic sites. Such localization of mineral deposits means that desired mineral resources do not occur in all countries. When this unequal supply pattern results in the scarcity of particular mineral resources that are crucial to a nation's economy, those minerals become critical or even strategic in the event of a national emergency. What constitutes "critical" minerals changes from time to time, mainly due to technological advances in mining or processing technologies.

Throughout history, nations' domestic mineral-resource bases and their needs for minerals unavailable within their borders have dictated international relations. Empires and kingdoms have risen and fallen due to the availability or scarcity of critical mineral resources. Two classic examples are those of the Greek and Roman empires. Silver mined by the Greeks near Athens financed the building of their naval fleet that defeated the Persians at the Bay of Salamis in 480 B.C. A century later, Greek gold supported the ambitious conquests of Alexander the Great. The Romans helped to finance their vast empire by mining tin in Britain, mercury in Spain, and copper in Cyprus. In the twentieth century, the efforts of Japan and Germany to secure mineral and fuel resources lacking within their borders were major factors leading to World War II, and the Soviet Union's annexation of the Karelian highlands of Finland following World War II provided the Soviets with new reserves of copper and nickel. More recently, Morocco acquired valuable phosphate deposits when it occupied part of the Spanish Sahara in 1975.

Origins of Mineral Deposits

Specific geological processes must occur in order to concentrate minerals or native elements in a mineral deposit. A simple classification of mineral deposits is based on their concentrating processes (● Table 13.4). The geologic processes responsible for many mineral deposits may be understood in terms of plate tectonics,

● TABLE 13.4   Genetic Classification of Mineral Deposits

GENETIC TYPE	EXAMPLE OF MINERAL(S) FORMED	LOCATION
Igneous Processes		
Pegmatite	tourmaline beryl	Oxford County, Maine; San Diego County, California Minas Gerais, Brazil
Crystal settling	chromite–magnetite–platinum	Bushveld complex, South Africa
Disseminated	copper–molybdenum porphyries diamonds	western North and South America kimberlite pipes in South Africa, Canada, and Australia
Hydrothermal	gold–quartz veins lead–zinc deposits in carbonate rocks	Mother Lode belt, California Mississippi River valley
Volcanogenic	massive copper sulfide deposits	Wrangell Mountains, Alaska
Sedimentary Processes		
Chemical precipitates in shallow marine basins	banded-iron formation	Lake Superior region
Marine evaporites	potassium salts	Carlsbad, New Mexico
Nonmarine evaporites	carbonates and borax minerals	Searles Lake, Trona, California
Deep-ocean precipitates	iron–manganese nodules	ocean floor, worldwide
Placer deposits	gold	Yukon Territory, Canada
Weathering Processes		
Lateritic weathering	bauxite	Weipa, Australia; Jamaica
Secondary enrichment	copper minerals	Miami, Arizona
Metamorphic Processes		
Contact-metamorphic	tungsten and molybdenum minerals	eastern Sierra Nevada, California
Regional metamorphic	asbestos	Quebec, Canada

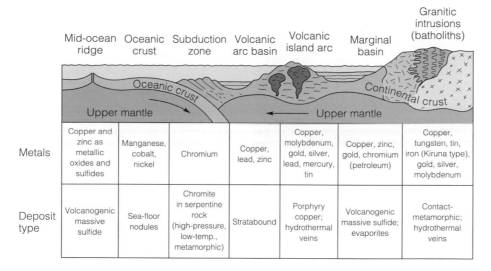

	Mid-ocean ridge	Oceanic crust	Subduction zone	Volcanic arc basin	Volcanic island arc	Marginal basin	Granitic intrusions (batholiths)
Metals	Copper and zinc as metallic oxides and sulfides	Manganese, cobalt, nickel	Chromium	Copper, lead, zinc	Copper, molybdenum, gold, silver, lead, mercury, tin	Copper, zinc, gold, chromium (petroleum)	Copper, tungsten, tin, iron (Kiruna type), gold, silver, molybdenum
Deposit type	Volcanogenic massive sulfide	Sea-floor nodules	Chromite in serpentine rock (high-pressure, low-temp., metamorphic)	Stratabound	Porphyry copper; hydrothermal veins	Volcanogenic massive sulfide; evaporites	Contact-metamorphic; hydrothermal veins

◆ FIGURE 13.2 Relationships between metallic ore deposits and tectonic processes.

as illustrated in ◆ Figure 13.2, with mineral concentrations occurring in distinct tectonic regimes. For example, the great porphyry copper deposits of the Western Hemisphere were formed by igneous processes at convergent plate margins at the sites of former volcanic island arcs. On the other hand, deposits of tungsten, tin, some iron ore, gold, silver, and molybdenum originated by hydrothermal activity or by contact metamorphism accompanying the emplacement of granitic plutons (batholiths) within continental lithospheric plates near convergent boundaries.

IGNEOUS PROCESSES
Intrusive Deposits

Pegmatites are small, tabular-shaped, very coarse grained, intrusive igneous bodies that may be impor-tant sources of mica, quartz, feldspar, beryllium, lithium, and gemstones. Reserves of pegmatites in the Western Hemisphere occur in the Black Hills of South Dakota; at Dunton, Maine; at Minas Gerais, Brazil; and in the tin-spodumene deposits of North Carolina (◆ Figures 13.3 and 13.4).

Crystal settling within a cooling magma chamber of ultramafic composition appears to be responsible for forming layers according to density, with the densest and earliest formed on the bottom and the less-dense and later-to-crystallize above or toward the top (◆ Figure 13.5), although alternate ideas for the layering have been proposed. Examples are the layered deposits of chromite at Red Mountain, Alaska, and the Bushveld Complex in the Republic of South Africa, the largest single igneous mass on earth (roughly the size of Maine). Chromium, nickel, vanadium, and platinum are obtained from this source.

◆ FIGURE 13.3 Granitic pegmatite, a very coarse grained igneous rock. The grains in this sample have lengths up to 3 centimeters.

◆ FIGURE 13.4 Tourmaline crystals. Tourmaline is commonly found as an accessory mineral in granitic pegmatites.

◆ **FIGURE 13.5** Layered mineral deposit; Red Mountain, Alaska. The black layers are chromite (FeCr$_2$O$_4$); the light layers are granitic rocks.

Some layered nickel-copper-cobalt deposits also are important sources of platinum-group elements.

Disseminated Deposits

Disseminated (scattered) **deposits** occur when a multitude of mineralized veinlets develop near or at the top of a large igneous intrusion (see Chapter 2). Probably the most common of these are the **porphyry copper** deposits (◆ Figure 13.6), in which the ore minerals are widely distributed throughout a large volume of granitic rocks that have been emplaced in regions of current or past plate convergence. The world's largest open-pit

mine, at Bingham Canyon, Utah, has extracted more than $6 billion worth of copper and associated minerals from a porphyry copper body (◆ Figure 13.7).

Hydrothermal Deposits

Hydrothermal (hot water) **deposits** originate from hot, mineral-rich fluids that are squeezed from cooling magma bodies during crystallization. Solidifying crystals force the liquids into cracks, fissures, and pores in both the magmatic rocks and the adjacent, "country" rocks (Figure 13.6). Hydrothermal fluids also may be of metamorphic origin, or they may form from subsurface waters that are heated when they circulate near a cooling magma. Regardless of their origin, the heated waters may dissolve, concentrate, and remove valuable minerals and redeposit them elsewhere. When the fluids cool, minerals crystallize and create hydrothermal vein deposits, which miners often refer to as **lodes** (◆ Figure 13.8). The specific minerals formed will vary with the composition and temperature of the particular hydrothermal fluids, but they commonly include such suites (associations) of elements as gold-quartz and lead-zinc-silver.

Volcanogenic Deposits

When volcanic activity vents fluids to the surface, sometimes associated with ocean-floor hot-spring **black**

◆ **FIGURE 13.6** A disseminated porphyry copper deposit.

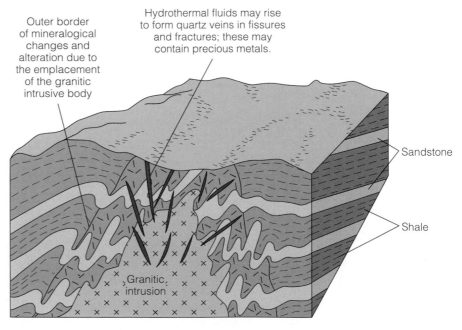

Outer border of mineralogical changes and alteration due to the emplacement of the granitic intrusive body

Hydrothermal fluids may rise to form quartz veins in fissures and fractures; these may contain precious metals.

Sandstone

Shale

Granitic intrusion

Disseminated deposits of copper form when mineralized solutions invade permeable zones and small cracks.

◁▶ **FIGURE 13.7** Kennecott Corporation's Bingham Canyon mine south of Salt Lake City, Utah, is the world's largest open-pit mine and largest human-made excavation. In the mid 1990s, the pit was more than 800 meters (½ mi) deep and nearly 4 kilometers (2½ miles) wide.

smoker activity, **volcanogenic deposits** are formed (◁▶ Figure 13.9). These deposits are so-named because they occur in marine sedimentary rocks that are associated with basalt flows or other volcanic rocks, and the ore bodies they produce are called **massive sulfides.** The rich copper deposits of the island of Cyprus, which have supplied all or part of the world's needs of copper for more than 3,000 years, are of this type. (The word *copper* is derived from Latin *Cyprium,* "Cyprian metal.") Cyprus's copper sulfide ore formed millions of years ago adjacent to hydrothermal vents near a sea-floor spreading center. Warping of the copper-rich sea floor due to convergence of the European and African plates brought the deposits to the surface when the island formed.

SEDIMENTARY PROCESSES

Surficial Precipitation

Very large, rich mineral deposits may result from evaporation and direct **precipitation** (the process of a solid forming from a solution) of salts in ocean water, usually in shallow marine basins. Minerals formed this way are called **evaporites.** Evaporites may be grouped into two types: marine evaporites, which are primarily salts of sodium and potassium, gypsum, anhydrite, and bedded phosphates; and nonmarine evaporites, which are mainly calcium and sodium carbonate, nitrate, sulfate, and borate compounds. Other important sedimentary mineral deposits are the *banded-iron formation* ores (◁▶ Figure 13.10). These deposits formed in Precambrian time, more than a billion years ago, when the earth's atmosphere lacked free oxygen. Without free oxygen, the iron that dissolved in surface water could be carried in solution by rivers from the continents to the oceans, where it precipitated with silica to form immense deposits of red chert and iron ore. Banded-iron deposits are found in the Great Lakes region, northwestern Australia, Brazil, and elsewhere, and they are enormous; they provide two hundred years or more of reserves even without substantial conservation measures.

Deep-Ocean Precipitation

Manganese nodules (◁▶ Figure 13.11) are formed by precipitation on the deep ocean floor. The nodules are

◆▶ **FIGURE 13.9** A "black smoker" hydrothermal vent on the Endeavor segment of the Juan de Fuca mid-oceanic ridge photographed during a dive of the submersible *Alvin*. The "smoke" is hot water saturated with dissolved metallic-sulfide minerals, which precipitate into black particles upon contact with the cold seawater.

◆▶ **FIGURE 13.8** Vein deposits of chalcopyrite (CuFeS$_2$) and pyrrhotite (Fe$_{1-x}$S) in a volcanic-rock-hosted massive sulfide deposit; Prince William Sound, Alaska.

mixtures of manganese and iron oxides and hydroxides, with small amounts of cobalt, copper, nickel, and zinc, that grow in onionlike concentric layers by direct precipitation from ocean waters. Any commercial recovery of this resource appears to be many decades away because of technological, economic, international political, and environmental limitations, but eventually it may be necessary to exploit the deposits.

Placer Deposits

Mineral deposits concentrated by moving water are called **placer deposits** (Spanish *placer,* "reef"; pronounced "plass-er"). Dense, erosion-resistant minerals such as gold, platinum, diamonds, and tin are readily concentrated in placers by the washing action of moving water. The less-dense grains of sand and clay are carried away, leaving gold or other heavy minerals concentrated at the bottom of the stream channel. Such deposits formed by the action of rivers are referred to as **alluvial placers;** ancient river deposits that are now elevated as stream terraces above the modern channels

◆▶ **FIGURE 13.10** Outcrop of banded-iron formation; northern Michigan.

◇ **FIGURE 13.11** Manganese nodules, an example of deep-ocean precipitation. The nodules consist mainly of manganese and iron, with lesser amounts of nickel, copper, cobalt and zinc.

are called **bench placers** (◇ Figures 13.12 and 13.13 and Case Study 13.1). Considerable uranium is mined from Precambrian placer deposits in southern Africa. The deposits formed under the unique conditions existing on the earth's surface before free oxygen was present in the atmosphere. Some placer deposits occur on beaches, the classic example being the gold beach placers at Nome, Alaska.

Sand and gravel concentrated by rivers into alluvial deposits are important sources of aggregate for concrete. River deposits formed by **glacial outwash**—that is, resulting from the runoff of glacial meltwater—are another type of placer deposit.

WEATHERING PROCESSES

The deep chemical weathering of rock in hot, humid, tropical climates promotes mineral enrichment, because the solution and removal of more soluble materials leaves a residual soil of less-soluble minerals. Because iron and aluminum are relatively insoluble under these conditions, they tend to remain behind in *laterite,* a highly weathered, red subsoil or material that is rich in oxides of iron and aluminum and lacking in silicates. When the iron content of the parent rock is low or absent, however, this lateritic weathering produces rich deposits of **bauxite,** the principal ore mineral of aluminum (see Chapter 6, Figure 6.15).

Ground water moving downward through a disseminated sulfide deposit may dissolve the dispersed

Consider This... ? How can a prospector use placer mineral deposits as a guide to finding ore deposits in bedrock? ■

metals from above the water table to produce an enriched deposit below the water table by **secondary enrichment** (◇ Figure 13.14). At Miami, Arizona, for example, the primary disseminated copper-ore body is of marginal to submarginal grade, containing less than one percent copper, but secondary enrichment has improved the grade to more than three percent.

METAMORPHIC PROCESSES

The high temperatures, high pressures, and ion-rich fluids that accompany the emplacement of intrusive igneous rocks produce a distinct metamorphic halo around the intrusive body. The result, in concert with the accompanying hydrothermal mineralization, is known as a **contact-metamorphic deposit.** If, for example, granite intrudes limestone, a diverse and colorful group of contact-metamorphic minerals may be produced, such as a tungsten-molybdenum deposit (◇ Figure 13.15). Asbestos and talc originate by **regional metamorphism,** metamorphism that affects an entire region.

Mineral Reserves

METALLIC MINERAL RESERVES

The metallic minerals are often grouped on the basis of their relative crustal abundance.

The Abundant Metals

The geochemically abundant metals—aluminum, iron, magnesium, manganese, and titanium—have abundances in excess of 0.1 percent by weight of the average continental crust. Economically valuable ore bodies of the abundant metals—such as iron and aluminum, for example—need only comparatively small concentration factors for profitable mining (Table 13.2) and are recovered mainly from ore minerals that are oxides and hydroxides. Even though they are abundant, these metals require large amounts of energy for production. Because of this, it is not surprising that the world's industrialized nations are the greatest consumers of these metals.

The Scarce Metals

The scarce metals are those whose crustal abundances are less than 0.1 percent by weight of average continental crust. Most of these metals are concentrated in sulfide deposits, but the rarest occur more commonly, or even solely, in rare rock types. Included within this

Case Study | 13.1

Mining the California Mother Lode

Although rumors of gold in Spanish-held California were circulating as early as 1816, the first rush for California's gold did not begin until 1842, following ranchero Francisco Lopez's discovery of gold in what is now Placerita Canyon on Rancho San Fernando near Pueblo de Los Angeles. The lack of water at the site resulted in mining by only the crudest of methods until the arrival of experienced miners from Sonora, Mexico, who introduced the method of dry washing. The spurt of interest quickly waned, however, because the small amount of gold available was soon extracted, and the lack of water for washing the gravels was discouraging to all but the most experienced miners. In 1848, however, James Marshall's recognition of gold at Sutter's Mill on the American River electrified the world and triggered the greatest gold rush in history, ultimately altering the development of western North America and perhaps of the United States (◇ Figure 1).

The crude early mining methods by which knives and spoons were used to extract gold from placer deposits soon gave way to the gold pan and "rocker," an open-topped wooden box mounted on curved boards. Using a rocker was a two-person operation: one miner shoveled sand and gravel and poured water into the box, while the second rocked the box back and forth. The dirt would wash through the box and the gold would be caught behind "riffles," slats placed across the bottom of the box perpendicular to the flow of water. Rockers in turn led to the development of the elongated rocker, and the wooden sluice box, with similar transverse riffles (◇ Figure 2). The sluice was positioned in a stream so that water would flow down the length of the box. Miners shoveled in goldbearing sand and gravel, and the gold was caught by the riffles. Cocoa matting was sometimes used to catch very small particles of gold, and oftentimes a copper plate coated with mercury was used to recover the very fine particles more efficiently by forming an *amalgam,* an alloy of mercury and gold—or silver; hence the reason that mercury is called *quicksilver.* When "cleaning up," the miners would heat the amalgam in a vessel to separate the dissolved gold from the amalgam by vaporizing and condensing the mercury, which could then be reused. The miners were careful to avoid breathing the fumes in recovering the vaporized mercury, because they knew of its toxicity.

◇ **FIGURE 1** The Mother Lode district, California's nineteenth-century gold-rush belt.

◇ **FIGURE 2** American and Chinese gold miners working a sluice box at Auburn, California, in 1852.

◆▶ **FIGURE 3** Hydraulicking a bench placer deposit.

◆▶ **FIGURE 4** Interior of a large stamp mill, like those used in the Mother Lode gold belt in the nineteenth century. Mills like this crushed ore 24 hours a day six days a week with a noise that must have been deafening. It is said that one could travel the length of the Mother Lode in the 1880s and never be out of earshot of the pounding stamp mills.

The easily obtainable gold was soon exhausted, and by 1853 the method of hydraulic mining—extracting gold by blasting a jet of water through a nozzle, called a *monitor,* against hillsides of ancient alluvial deposits—was developed (◆▶ Figure 3). Once the necessary ditches, reservoirs, penstocks (vertical pipes), and pipelines had been installed, it was possible to wash thousands of cubic yards of goldbearing gravels without hand labor. An attempt that same year to mine alluvial gold by dredging failed; it was not until 1898 that the first real success was achieved with a bucket-conveyor dredge (Figure 13.24).

Experienced miners immediately began searching for the *mother lode*—the bedrock source veins from which the placer deposits originated. The first gold-quartz lode vein was discovered in 1849 on Colonel John C. Fremont's grant near Mariposa. Discoveries of more gold-quartz veins of the lode system soon followed. In contrast to the simplicity of mining placer deposits, lode mining requires extensive underground workings (which require drilling, blasting, and tunneling) and a complex surface plant for hoisting men and ore from underground, for pumping water out of the mine, and for milling the raw ore. A few banks of old stamp mills

(◆▶ Figure 4), batteries of heavy weights that were alternately lifted and dropped to pulverize the ore, remain today in the gold country. The gold was recovered by gravity separation from a slurry of powdered rock and water as it flowed over riffles and by amalgamation with quicksilver. In the 1890s, cyanide extraction was introduced into the milling process to make recovery even more efficient. It had been learned that gold dissolves readily in a dilute solution of sodium or potassium cyanide. Hence the new method mixed crushed ore with cyanide and removed the gold later by chemical methods.

The influx of thousands of forty-niners, as the fortune-seekers were known, to the Mother Lode district led to widespread prospecting and to the rapid and complete exploration of the previously little-known, sparsely populated region. Within a year of James Marshall's gold discovery at Sutter's Mill, California had attained statehood and become, almost overnight, the most important Western state. California's gold rush also served as the impetus for widespread prospecting rushes in the 1850s that led to important gold strikes in Nevada, Colorado, Montana, Australia, and Canada. A conservative estimate placed California's annual gold production at $750,000,000 by 1865. The major world impacts of the California gold rush were the widespread redistribution of population, with its resulting problems and challenges, and the general increase of money in circulation.

(a) Ancient river pattern

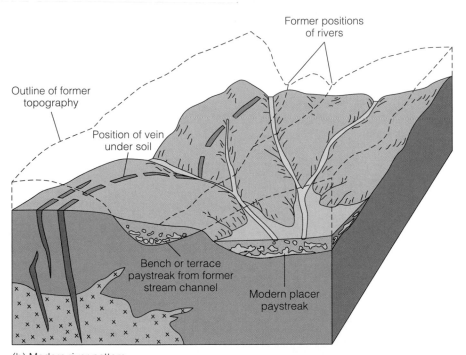

(b) Modern river pattern

◆▷ **FIGURE 13.12** The origin of gold placer deposits. (a) An ancient landscape with an eroding gold-quartz vein shedding small amounts of gold and other mineral grains that eventually become stream sediments. The gold particles, being heavier, settle to the bottom of the sediments in the channel. (b) The same region in modern time. Streams have eroded and changed the landscape and now follow new courses. The original placer deposits of the ancient stream channels are now elevated above the modern river valley as "bench" placers.

category are copper, lead, zinc, and nickel—which are widely known and used, but nevertheless scarce—along with the more obvious gold and platinum-group minerals. U.S. deposits of chromium, manganese, nickel, tin, and platinum-group metals are scarce or of submarginal grade. Important, less-well-known scarce metals include cobalt, columbium, tantalum, and cadmium, which are important in space-age ferroalloys; antimony, which is used as a pigment and as a fire re-

◆▶ **FIGURE 13.13** Gold nugget weighing 41.3 troy ounces, mined from a placer deposit in the Brooks Range, Alaska.

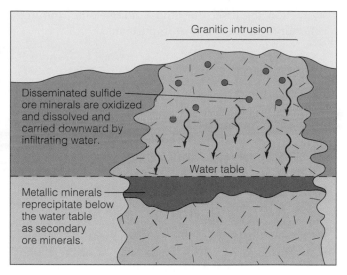

◆▶ **FIGURE 13.14** The development of ores by secondary enrichment. Descending ground water oxidizes and dissolves soluble sulfide minerals, carrying them downward and leaving a brightly colored residue of limonite and hematite. Below the water table new metallic minerals are precipitated as secondary ore minerals.

tardant; and gallium, which is essential in the manufacture of solid-state electronic components. The United States has all of these critical metals in the National Defense Stockpile, a supply of about 100 critical mineral materials valued at about $4 billion that could supply the nation's needs for three years in a national emergency.

NONMETALLIC MINERAL RESERVES

The yearly per capita consumption of nonfuel mineral resources in North America exceeds 10 tons, 94 per-

cent of which is nonmetallic, with the major bulk being construction materials (Figure 13.1). The nonmetallic minerals are subdivided for convenience into three groups based on their use in industry, agriculture, and construction. The number and diversity of nonmetallic mineral resources is so great that it is impossible to treat the entire subject in a book of this sort. Thus, this

◆▶ **FIGURE 13.15** The relationship of limestone and a granitic intrusion in the origin of contact-metamorphic mineral deposits.

subsection discusses only those nonmetallic resources generally considered most important.

Industrial Minerals

Industrial minerals include those that contain specific elements or compounds that are used in the chemical industry—sulfur and halite, for example—and those that have important physical properties—such as the materials used in ceramics manufacture and the abrasives, such as diamond, corundum, and garnet. Sulfur, one of the most widely used and most important industrial chemicals, is obtained as a by-product of petroleum refining (◊ Figure 13.16) and from the tops of salt domes in the central Gulf Coast states, where it is associated with anhydrite ($CaSO_4$) or gypsum ($CaSO_4 \cdot 2H_2O$). Salt domes originate from evaporite beds that are compressed by overlying sedimentary rock layers, causing the salt to be mobilized into a dome, which rises, piercing the overlying rocks and concentrating sulfur, gypsum, and anhydrite at the top. Superheated water is pumped into the sulfur-bearing caprock to melt the sulfur, which is then piped to a processing plant. More than 80 percent of domestic U.S. sulfur is used to make sulfuric acid, the most important use for which is the manufacture of phosphate fertilizers. Salt domes are also important sources of nearly pure halite (NaCl), common salt, and many domes have oil fields on their flanks (see Chapter 14).

In addition to being mined from salt domes, halite is mined from flat-lying evaporite beds in New Mexico, Kansas, and Michigan. The most important use of halite is for manufacturing sodium hydroxide and chlorine gas; about 15 percent of halite production is used for de-icing highways, and significant amounts are used in food products and in the manufacture of steel and aluminum.

Diamond is by far the most important industrial abrasive, although corundum, emery (an impure variety of corundum), and garnet also are used. Traditionally, natural diamonds have been important both as gemstones and as industrial abrasives, but artificial diamonds manufactured from graphite now hold about 70 percent of the market for diamond-grit abrasives. The major source of natural diamonds is kimberlite (named for Kimberley, South Africa), a rare and unique ultramafic rock that forms in pipelike bodies. Not only are kimberlite pipes rare, but most of them lack diamonds. The important diamond-bearing kimberlites are in South Africa, Ghana, the Republic of Zaire, Botswana, Australia, Belarus (formerly Belorussia of the USSR), Yakutia (the largest republic of the Russian Federation), and in Canada's Northwest Territories—where North America's first diamond mine began operations in 1998.

The metamorphic minerals corundum and emery, natural abrasives for which U.S. reserves are scarce or nonexistent, have largely been replaced by synthetic abrasives such as carbides, manufactured alumina, and nitrides. The United States still imports limited amounts of corundum from South Africa. Garnet, which is found in igneous and metamorphic rocks, is widely used in the United States for optical grinding, in sandblasting, and as a grit on finishing paper for wood. The main source of domestic industrial garnet is metamorphic rocks in the Adirondack Mountains in New York.

◊ **FIGURE 13.16** Sulfur awaiting shipment to Asia at Vancouver, British Columbia. The sulfur originated as a contaminant in natural gas produced from gas fields in Alberta. Processing plants have separated the sulfur from the natural gas in order to produce a gas whose combustion products meet environmental air-quality standards. The sulfur is a useful by-product of the process.

Agricultural Minerals

With the world's human population doubling every 40 years and the global population passing the 6 billion mark in September 1999, it seems obvious that food production and the corresponding need for fertilizers and agricultural chemicals also will continue to expand. Therefore, nitrate, potassium, and phosphate compounds will continue to be in great demand. Whereas nearly all agricultural nitrate is derived from the atmosphere, phosphate and potassium come only from the crust of the earth. Phosphate reserves occur in many parts of the world, but the major sources are in the United States, Morocco, Turkmenistan, and Buryat (the latter two are republics of the former Soviet Union). The primary sources of phosphate in the United States are marine sedimentary rocks in North Carolina and Florida (see Figure 13.28, later in the chapter), and there are other valuable deposits in Idaho, Montana, Wyoming, and Utah. The main U.S. supply of potassium comes from widespread non-marine evaporite beds beneath New Mexico, Oklahoma, Kansas, and Texas, with the richest beds being in New Mexico. Canada also has large reserves of potassium salts.

Construction Materials

Nonmetallic minerals used as construction materials include *aggregate,* the sand, gravel, and crushed stone used in concrete and for making roadbeds and asphalt road surfaces; limestone and shale, which are used in making cement; clay, used for tile and bricks; and gypsum, the primary component of plaster and wallboard. It is the mining of aggregate that is probably the most familiar to urban dwellers, because quarries customarily are sited near cities, the major market for the product, in order to minimize transportation costs. The quantity of aggregate needed to build a 1,500-square-foot house is impressive: 67 cubic yards are required, each cubic yard consisting of one ton of rock and gravel and 0.7 ton of sand. This amounts to 114 tons of rock, sand, and gravel per dwelling, including the garage, sidewalks, curbings, and gutters. Thus, it is no wonder that aggregate has the greatest commercial value of all mineral products mined in most states, ranking second only in those states that produce natural gas and petroleum. The annual per capita production of sand, gravel, and crushed stone in the United States amounts to about six tons, the total value of which exceeded $13 billion in 1998. The principal sources of aggregate are open-pit quarries in modern and ancient floodplains, river channels, and alluvial fans. In areas of former glaciation, aggregate is mined from glacial outwash and other deposits of sand and gravel that remained after the retreat of the great Pleistocene ice sheets in the northern United States, Canada, and Northern Europe (see Chapter 11).

Concrete, in addition to aggregate, contains cement. The major raw materials for manufacturing cement are large amounts of limestone or marble, from which is produced lime (CaO), and lesser amounts of gypsum or anhydrite, shale, clay or sand, and iron-bearing minerals such as hematite (Fe_2O_3) or magnetite ($FeO \cdot Fe_2O_3$). The principal raw materials are obtained by surface mining, and sufficient deposits of these materials occur worldwide.

Mineral Resources for the Future

Just as the world's fossil-fuel reserves are limited (Chapter 14), so are its mineral reserves. The earth has been well-explored, and fairly reliable estimates exist on the reserves of most important mineral resources. By dividing the known reserves of a resource by its projected rate of use, it is possible to calculate the number of years remaining until the resource is depleted. It was determined in the 1970s that reserves of many important elements would be exhausted by sometime early in the twenty-first century. Although more recent studies have found that the situation is not that serious, a few experts still view the exhaustion of some mineral deposits in the near future as likely.

U.S. consumption and production data for selected metallic and nonmetallic mineral resources appear in ◉ Tables 13.5 and 13.6, respectively. Notice in Table 13.5 that we consume greater amounts of many metallic mineral commodities than we produce. In the cases of chromium, cobalt, manganese, nickel, tin, and platinum-group metals, all or virtually all of those materials must be imported. The extent of U.S. import reliance for these and other mineral resources is shown in ◈ Figure 13.17. Perhaps more humbling is ◉ Table 13.7, the projected U.S. reserves of selected minerals, assuming 1994 rates of consumption, production, and imports without recycling. Thus, the United States is far from self-sustaining. In 1998, the United States imported an estimated $60 billion worth of raw and processed mineral materials. If nothing else, the data constitute a clear case against isolationism if we hope to sustain our present life-style.

Consider This...? How does the nation's need for mineral materials affect U.S. foreign policy? ■

TABLE 13.5 **U.S. Consumption and Production of Selected Metallic Minerals, 1995**

MINERAL	CONSUMPTION, thousands of metric tons	PRIMARY PRODUCTION, thousands of metric tons*	PRIMARY PRODUCTION AS PERCENTAGE OF CONSUMPTION**
Aluminum	7,300	3,300	45.2
Chromium	387	0	0
Cobalt	7.2	0	0
Copper	2,800	1,840	65.7
Iron ore	69,300	57,000	82.3
Lead	1,500	365	24.3
Manganese	695	0	0
Nickel	131	0	0
Tin	46	negligible	0
Zinc	1,220	540	44.3
Platinum-group metals, metric tons	127	8.3	16.3

*Primary production is production from new ore.
**Primary production divided by consumption and rounded to the appropriate number of significant figures.
SOURCE: *Mineral Commodity Summaries,* U.S. Bureau of Mines, 1995.

A brighter forecast is shown by the data in Table 13.6. Note in the table that the United States was self-sustaining in its consumption of domestic nonmetallic and rock resources as of 1995 and that it was producing a surplus of one mineral commodity. There are some apparent disagreements between these tables and Figure 13.17; the low need for imported lead reported in Figure 13.17, for example, does not seem to agree with the value shown in Table 13.5. The discrepancy results from extensive recycling of lead in the United States. Further, the high domestic production of aluminum reported in Table 13.5 results from a high re-

liance on imported bauxite (reported in Figure 13.17), because domestic sources of bauxite are insignificant.

As resources diminish in the coming decades, we can expect shortages of some minerals that presently are common. In many cases technological advances will develop substitute materials that will suffice, but in other cases, we will have to do without. Some minerals will become sufficiently scarce that deposits now classed as "submarginal" will be upgraded to "low-grade" and will be economically exploitable due to market forces. Projections of worldwide mineral reserves are shown in Table 13.8.

TABLE 13.6 **U.S. Consumption and Production of Selected Nonmetallic Minerals, 1995**

MINERAL	CONSUMPTION, thousands of metric tons	PRIMARY PRODUCTION, thousands of metric tons*	PRIMARY PRODUCTION AS PERCENTAGE OF CONSUMPTION**
Clay	37,900	42,300	112
Gypsum	26,000	17,300	66.5
Potash	5,390	1,425	26.4
Salt	47,400	39,500	83.3
Sulfur	13,450	11,300	84
Sand and gravel	922,000	922,000	100
Crushed stone	1,198,000	1,195,000	95.7

*Primary production is production from new ore.
**Primary production divided by consumption and rounded to the appropriate number of significant figures.
SOURCE: *Mineral Commodity Summaries,* U.S. Bureau of Mines, 1995.

Mineral Resources		Important Sources, 1994–1997
Graphite	100	Mexico, China, Brazil, Madagascar
Manganese	100	Rep. of South Africa, France, Australia, Gabon
Strontium (celestite)	100	Mexico, Germany
Bauxite and alumina	100	Australia, Guinea, Jamaica, Brazil
Diamond (industrial)	54	Ireland, China, Russia
Tungsten	78	China, Bolivia, Germany, Russia
Platinum-group metals	94	Rep. of South Africa, U.K., Russia, Germany
Tantalum	80	Germany, Brazil, Australia, Thailand, China
Tin	85	Brazil, China, Indonesia, Bolivia
Cobalt	77	Zambia, Canada, Norway, Finland
Chromium	75	Rep. of South Africa, Turkey, Zimbabawe, Kazakhstan
Potash	65	Canada, Russia, Belarus
Nickel	65	Canada, Norway, Australia, Russia
Antimony	84	China, Mexico, Rep. of South Africa
Cadmium	21	Canada, Mexico, Australia, Belgium
Zinc	70	Canada, Mexico, Spain, Peru
Magnesium compounds	28	China, Canada, Greece, Austria
Iron and steel	18	European Union (EU), Japan, Canada, Brazil, Mexico
Iron ore	17	Canada, Brazil, Venezuela, Australia
Lead	21	Canada, Mexico, Peru, Australia

◀▶ FIGURE 13.17 U.S. net import reliance for selected mineral materials as a percentage of apparent consumption, 1997. *Net import reliance* = imports − exports + adjustments for government and industry stock changes. *Apparent consumption* = U.S. primary + secondary production + net import reliance.

The need for mineral resources may require exploration and exploitation in more remote regions and more hostile environments. Coal has been mined for more than a half-century well north of the Arctic Circle on Norway's island of Spitsbergen (Svalbard), for example, and sources of fuels and minerals are currently being sought and exploited in the harsh climates of Canada's Arctic islands. Exploration by drilling cores has begun at a lead-zinc-silver prospect in western North Greenland. The Red Dog zinc-lead-silver mine, North America's largest zinc producer, on the lands of the Inupiat Eskimos north of the Arctic Circle in Alaska went into full production in 1991. Even environmentally sensitive Antarctica, which is known to have extensive deposits of coal and iron, may be the target of future exploitation as more is learned about that continent's mineral wealth. Deeper parts of the continents and the sea floor may also be targeted for exploration and exploitation as geophysical innovations and technological advances open up these presently inaccessible regions.

In the future we will experience more recycling of critical metals, such as has been done with gold for thousands of years. The gold in the crown on your tooth, for example, may have been in jewelry during the age of Pericles in Ancient Greece. Such metals as lead, aluminum, iron, and copper have been recycled for years. In many cities the large-scale recycling of

metals, especially lead and aluminum, as well as of paper, glass, and plastics, is being conducted successfully as part of municipal trash collection (see Chapter 15). Significant amounts of several metals consumed in the United States are now being obtained from recycled scrap. In 1994 these amounts were

METAL	AMOUNT OBTAINED BY RECYCLING
aluminum	37%
copper	40%
iron	59%
lead	65%
cobalt	22%
tin	27%
nickel	34%

Consider This...? As the geological sources of raw materials are exhausted, recycling of many metals will eventually be absolutely necessary. How would you educate the general public to the urgency of recycling? ■

TABLE 13.7   Estimated Lifetimes of U.S. Reserves of Selected Minerals Assuming 1994 Rates of Consumption, Primary Production, and Imports

MINERAL	PRODUCTION, thousands of metric tons	RESERVE BASE, thousands of metric tons*	ESTIMATED LIFETIME, years**
Bauxite	3,300	40,000	12
Chromium	0	0	0
Cobalt	0	860	negligible
Copper	1,840	90,000	49
Gold, metric tons	330	5,500	17
Gypsum	17,300	>700,000	>40
Iron ore	57,000	25,000,000	439
Lead	365	22,000	60
Manganese	0	0	0
Nickel	0	2,500	negligible
Phosphate	41,000	4,440,000	110
Platinum-group metals, metric tons	8.3	780	94
Potash	1,425	>6,000,000	>4,210
Silver, metric tons	1,400	72,000	51
Sulfur	11,300	230,000	20
Tin	negligible	40	negligible
Zinc	560	50,000	89

*The reserve base consists of the in-place demonstrated resources including presently economic mineral reserves, marginal reserves, and identified subeconomic resources.
**Estimated lifetimes were calculated by dividing the reserve base by the 1994 U.S. production (data from *Mineral Commodity Summaries*, U.S. Bureau of Mines, 1995) and rounding to the appropriate number of significant figures.
SOURCE: Data from *Mineral Commodity Summaries*, U.S. Bureau of Mines, 1995.

Collectively, the value of recycled metal and mineral scrap amounted to about 30 percent of the nation's mineral production.

Mining and its Environmental Impacts

The Mining Law of 1872 was important legislation that helped establish the mining industry as a fundamental element in the U.S. economy. At the same time, the law was the ultimate in *laissez-faire* regulation, as it allowed miners to exploit, to take profits at public expense, to pay little in return for the privilege of mining on federal land, and to walk away from the scars and waste materials remaining when they abandoned their mines. The original nineteenth-century law allowed miners to stake a mining claim on potentially profitable public land and, should the claim prove to contain valuable minerals, to obtain a patent (legal title) and reap the profits. This law helped "win" the West by enticing thousands of prospectors to seek their fortunes in gold, silver, and other valuable mineral commodities.

The 1872 law gave no consideration whatever to the environmental consequences of mining. The law is staunchly defended today by some public officials and the mining industry, who claim that it does a good job of providing the country with valuable minerals, employment, and economic development. The same law exists today, but it has been much modified over the years by more than 50 amendments, so that today's body of mining law fills six volumes. An example of the changes is the Mineral Lands Leasing Act of 1920, by which the government retains title to all federal lands possessing energy resources (oil, gas, and coal) and assesses royalties on the profits from the lands leased by developers. In addition, environmental restrictions added in the 1970s and 1980s dictate that modern mining must be conducted within a different policy framework than the simply drafted law of 1872.

Although interest in environmental protection and reclamation in coal mining districts began to develop

⊙ TABLE 13.8 **Estimated Lifetimes of Global Reserves of Selected Minerals Assuming 1994 Rates of Consumption and No Recycling**

MINERAL	PRODUCTION, thousands of metric tons	RESERVE BASE, thousands of metric tons*	ESTIMATED LIFETIME, years**
Bauxite	110,000	55,000,000 to 75,000,000	500 to 682
Chromium	9,400	11,000,000	1,170
Cobalt	22	8,800	400
Copper	9,300	1,600,000 (land), 700,000 (sea floor)	172 (land), 75 (sea floor)
Gold, metric tons	2,300	75,000	33
Gypsum	110,000	>2,400,000	>22
Iron ore	1,000,000	>800,000,000	>800
Lead	2,900	1,500,000	520
Manganese	7,100	4,900,000	690
Nickel	850	130,000	153
Phosphate	130,000	34,000,000	262
Platinum-group metals, metric tons	250	100,000	400
Potash	22,000	250,000,000	1,140
Silver, metric tons	14,000	420,000	30
Sulfur	52,000	180,000,000	350
Tin	180	10,000	56
Zinc	6,700	1,800,000	270

*The reserve base consists of the in-place demonstrated resources including presently economic mineral reserves, marginal reserves, and identified subeconomic resources.

**Estimated lifetimes were calculated by dividing the reserve base by the 1994 production (data from *Mineral Commodity Summaries*, U.S. Bureau of Mines, 1995) and rounding to the appropriate number of significant figures.

SOURCE: Data from *Mineral Commodity Summaries*, U.S. Bureau of Mines, 1995.

in the 1930s, serious efforts at regulation and reclamation did not begin until the late 1960s due to the increasing amounts of unreclaimed land, and water pollution and problems arising from the lack of uniform standards among state mining programs. Consequently, the Surface Mining Control and Reclamation Act (**SMCRA**) was passed in 1977. This act established coordination of federal and state efforts to regulate the coal industry in order to prevent the abuses that had prevailed in the past. SMCRA regulations apply to all lands—private, state, and federal. SMCRA established a special fee that is charged on coal production and generates funds for reclaiming abandoned coal-mine lands. SMCRA also provides for state primacy once a state's coal mine reclamation laws and policies have been approved by the Secretary of the Interior. This gives state regulators primary SMCRA enforcement jurisdiction as long as the state programs meet federal standards.

Federal reclamation standards and other safeguards established by SMCRA for the coal-mining industry have no counterpart for noncoal underground and surface mineral mining on state or private lands. Although there are federal regulations for noncoal mining on federal lands, it is left to the individual state governments to codify safeguards and enforce noncoal mine reclamation requirements.

Mining on federal lands must comply with the regulations of the *National Environmental Policy Act* (NEPA) of 1969 (see Chapter 10). A mining company must file an operating plan and an *Environmental Impact Report* (EIR), and conduct public hearings before the administrating agency can approve an operating permit. Even though an applicant may spend several thousand dollars in complying with NEPA permit application requirements, there is no assurance that an application will be approved.

There is no specific fund for cleaning up abandoned noncoal mines on state and private lands, nor are there uniform regulatory requirements for these lands. Action is left to the states and the mining companies. However, once any state with coal production

Case Study | 13.2

Reclamation of the Copper Basin

A well-known area of severe, long-lasting industrial degradation is the Copper Basin, near Ducktown, Tennessee, a region straddling the borders of Tennessee, North Carolina, and Georgia (◆ Figure 1). The area is also remembered as the site of the 1996 Summer Olympics white-water canoe and kayak events, which were held on the Ocoee River from south of Ducktown to Parksville Lake. Little did the contestants know of the area's history of degradation, which began in the 1850s with small-scale copper mining and the region's first smelter. Removal of trees to fuel the ore roasters and smelters quickly stripped the basin of its vegetation. Heap-roasting (open-pit smelting) of the copper ore yielded high concentrations of sulfur dioxide (SO_2) gas, which drifted over neighboring lands and extended into the neighboring states of Georgia and North Carolina, devastating the Ducktown area and causing severe plant and crop damage in neighboring areas. The basin's natural environment contributed to the problem. Its naturally acidic soils were highly leached and deficient in calcium; its meteorologic conditions (warm temperatures, high humidity, frequent fog, usually stagnant air, abundant and sometimes intense rainfall) in conjunction with its topographic setting produced atmospheric inversions that concentrated the acid and limited the lateral movement of the air. Consequently, the humid, unrefreshed, contaminated air lingered over the basin for extended periods of time.

Large-scale mining and smelting operations began in 1891. The fuelwood cutting for the smelters, mining activities, SO_2 emissions, and related disturbances killed all vegetation, eventually affecting about 130 square kilometers (50 mi²) of the mountain enclave. Subsequently, intense erosion of topsoil and subsoil occurred on about 9,290 hectares (23,000 acres). By 1905 the effects of the smelters' acid smoke and the acres of roasting ore piles on vegetation and agriculture were so devastating that the state of Georgia hauled Ducktown Sulphur Company and Tennessee Copper Company into court in an effort to get them to stop their polluting.

The mining companies began extending their smokestacks, one of them to 99 meters (325 ft) tall. High stacks have often served to satisfy complaints about industrial pollution, especially at copper smelters; the common solution for pollution being dilution. The tall stacks successfully removed fumes from ground level, but due to their higher release points, the fumes traveled even further and damaged vegetation over an even larger region. Court action dragged on for a decade, eventually reaching the U.S. Supreme Court. The Court ruled that the operations must "prevent the diffusing of sulphurous fumes over the Georgia border." In response to these problems, the mining companies constructed sulfuric acid plants to recover the SO_2 from the smelters, thereby converting an injurious polluting gas into a useful by-product needed by the fertilizer industry. So successful was the sulfuric acid production that by 1911 southeastern Tennessee had the two largest sulfuric acid plants in the world, and the sulfuric acid recovered from Ducktown's smelters accounted for 20 percent of U.S. production. There was a slow but encouraging improvement in the air quality, enough to satisfy Georgia and the Supreme Court, but the land in the Copper Basin by this time had become so devastated that it was a biological desert (◆ Figure 2). The barren landscape of deeply gullied red soil drew tourists out of their way to see it.

Erosion in the Copper Basin has now decreased the flood storage capacity of the

◆ FIGURE 1 The Copper Basin.

◇▶ FIGURE 2 The Tennessee Copper Company's smelter and sulfuric acid plant at Copperhill, Tennessee, 1912. The barren, denuded ground had been sterilized by acid fallout.

◇▶ FIGURE 3 Sediment deposits behind Ocoee Dam No. 2, a diversion dam for a hydroelectric generator. Sediment from the Copper Basin adversely impacts the Ocoee River's drainage, decreases dams' flood-storage capacities, and increases the cost of hydroelectric energy production.

Tennessee Valley Authority's (TVA's) three Ocoee River reservoirs and increased TVA's hydroelectric production costs due to sediment damage to turbines and power-generation facilities (◇▶ Figure 3). Heavy-metal pollution and siltation to surface waters draining the basin has become a high-priority water-resource problem in the Tennessee Valley. Measured annual erosion rates from the basin's denuded lands are as much as 195 tons per acre.

Efforts by the mining companies, the TVA, and other government agencies to revegetate the devastated 9,290 hectares began in the 1930s. The Tennessee Copper Company (TCC), in cooperation with the TVA, began research on restoration of the lands and instituted an extensive tree-planting program in 1939. The work was continued by the company and its successors, Cities Service, and BIT Manufacturing. By the late 1970s, aerial applications of fertilizer were enhancing tree growth and survival, and the increased vegetative cover had contributed to a marked slowing of erosion.

By 1984, driven by the need to improve the Ocoee's water quality, reduce the rate of sedimentation behind the

Ocoee River dams (Dam No. 3 had lost nearly 90% of its water-storage capacity), and reduce the adverse impact of sediment loads on hydropower operations, new techniques were implemented by the TVA, TCC, and others. These included chemically treating more than 1,200 hectares (3,000 acres) of land and planting more trees. By 1990, more than 16 million trees, most of them genetically improved strains of loblolly pine that can tolerate the poor soil, had been planted (◇▶ Figure 4). Additional projects planned for the 1990s and beyond included containing contaminated surface waters in critical drainage basins, operating wastewater treatment plants in critical areas, installing ground-water-monitoring wells in six towns, cleaning up the aniline-chemical and petroleum contamination on the property of a chemical plant, and continuing the reforestation of the TCC property. By combining these projects with the ongoing reclamation effort, major environmental improvements can be expected in the Copper Basin.

(a)

(b)

◇▶ FIGURE 4 The Copper Basin (a) before reclamation, March 1986, and (b) after reclamation, July 1991.

◆▶ FIGURE 13.18 A large underground mining operation; the Homestake Mine near Lead, South Dakota. The shaft headframe and the hoist building are the large structures on the ridge crest. The mills for processing the ore are the large "stair-stepped" buildings at the center left and the lower right of the photograph. This mine began operations in 1876 and is still producing gold.

has implemented safeguards and standards for coal mining, it may use the SMCRA abandoned mine fund to address reclamation of abandoned mines that exploited other commodities. For example, Utah's abandoned mine reclamation program uses SMCRA funds to reclaim metals mines.

Environmental pollution abatement has become a major concern of domestic gold producers in some states, especially in Alaska, where there are many small placer mines. Reclamation of mined land has become an integral part of an increasing number of gold mine operating plans. Abandoned mine sites, as well as other contaminated industrial sites that are deemed to pose serious threats to health or safety, may be cleaned up under the Comprehensive Environmental Response, Compensation and Liability Act (**CERCLA**) of 1980, which is overseen by the *Environmental Protection Agency* (EPA). For example, the Forest Service, a branch of the United States Department of Agriculture, has CERCLA authority to clean up contaminated mines on National Forest System lands. It was CERCLA that established the so-called *Superfund* to clean up the worst toxic or hazardous waste sites that have been assigned to the National Priorities List (see Chapter 15).

Some states have enacted *State Environmental Policy Acts* (SEPAs) that are patterned after NEPA, while other states may have little regulation. The differences in the various states' permit procedures, their mining and reclamation standards, and their enforcement of underground and surface mining laws can be illustrated by comparing the laws of three states, each noted for over a century of mining activity: Colorado, Arizona, and Michigan. Colorado has strong regulatory

provisions relating to mine operation, environmental protection, reclamation bonding, inspections, emergency response by the state, and even the authority over mine operation. In contrast, Arizona, the only state in the country without a mine-reclamation law until 1994, allows companies to self-bond their reclamation fund, to wait as long as 17 years after mining has stopped to begin reclamation, and to set their own reclamation standards. Furthermore, there is an escape clause that may allow companies a variance from any of the law's provisions. Michigan, which in 1994 ranked sixth among all states in the production value of nonfuel minerals, has no comprehensive mining law. No financial bond is required to ensure that a mine site will be properly closed and reclaimed if the company abandons the project or declares bankruptcy. No permit is needed to open a mine except a local zoning permit and a federal water-discharge permit—permits that are required of *every* industrial operation. Reportedly, there is no enforcement of the state's voluntary reclamation law.

As we consider the environmental impacts of mining in the United States in this section, it is important to distinguish between (1) the legacy of careless exploitation that is obvious at long-abandoned mines with their residual scars and toxic wastes and (2) the current mining scene, in which, depending upon the particular state and whether the site is on land administered by the U.S. Forest Service or the Bureau of Land Management, more than 30 different permits may by needed in order to develop and operate a mine legally. Many mine owners consider the current permitting requirements to be excessive, and they question whether they will be able to continue to operate in the United States at all if the regulations become more restrictive.

Mining in the past was conducted largely by underground methods with a surface plant for milling and processing and for hoisting workers, ore, and equipment (◆ Figure 13.18). Surface mining methods, typified by vast open-pit excavations such as the Bingham Canyon mine (see Figure 13.7), have now largely replaced underground methods. The surface plant remains much the same, however, although it usually operates on a much larger scale than in the past. Both surface and underground mining create significant environmental impacts on the land and air and on biological and water resources. In addition, the needs for housing and services in mining areas have social impacts.

IMPACTS OF COAL MINING

Much coal now mined in the United States is extracted by surface operations, which involve removing rock and soil that overlie the coal beds. Compared to underground mining (◆ Figure 13.19), surface mining is generally less expensive, safer for miners, and facilitates more complete recovery of coal. Surface mining, however, causes more extensive disturbances to the land surface and has the potential for serious environmental consequences unless the land is carefully reclaimed. The three major methods of surface mining of coal are contour mining, area mining, and mountain-top mining. Each of these is explained here.

Contour mining is typical in the hilly areas of the eastern United States, where coal beds occur in outcrops along hillsides. Mining is accomplished by cutting into the hillside to expose the coal and then following the coal seam around the perimeter of the hill. Successive, roughly horizontal strips are cut, enlarging each strip around the hillside until the thickness of the overburden is so great that further exposure of the coal bed would not be cost-effective. At each level of mining is a *highwall,* a clifflike, excavated face of exposed overburden and coal that remains after mining is completed. Augers (giant drill bits) are used to extract coal from the areas beneath the highwall. The SMCRA requires that the sites must be reclaimed, that is, returned to the original contour and highwalls must be covered and stabilized after contour mining is completed. This is accomplished by backfilling **spoil,** the broken fragments of waste rock removed in order to mine the coal, against the highwall, and spreading and compacting as necessary to stabilize the reclaimed hillside. *Hydroseeding,* spraying seeds mixed with water, mulch, fertilizer, and lime onto regraded soil, is used on steep reclaimed slopes to aid in establishing vegetation that will help prevent soil erosion (Case Study 13.3).

Area mining is commonly used to mine coal in flat and gently rolling terrain, principally in the Midwestern and Western states (◆ Figure 13.20). Because the pits of active area mines may be several kilometers long, enormous equipment is used to remove the overburden, mine the coal, and reclaim the land. Topsoil is stockpiled in special areas and put back in place when mining is completed. After replacing, it is tilled with traditional farming methods to reestablish it as crop or pasture land. Commonly, the land is more productive after reclamation than it was before mining.

Mountain-top mining is used predominantly in the eastern United States to recover coal that underlies the tops of mountains. After the coal is removed, the

◆ **FIGURE 13.19** The steps in conventional large-scale underground bituminous-coal mining.

Case Study | 13.3

Steps in Surface Mine Reclamation

There are five steps in reclaiming abandoned coal mine lands:

1. drainage control to eliminate acid mine drainage,
2. stabilization of landforms (◇▶ Figure 1),
3. revegetation (◇▶ Figures 2 and 3),
4. ongoing monitoring, and
5. returning the land to use (◇▶ Figure 4).

The reclamation work pictured in Figures 1, 2, and 3 was carried out near Beckley, West Virginia. The reclaimed site in Figure 4 is near Harding, West Virginia, where reclamation of this site eliminated more than a mile of highwall and sealed four hazardous abandoned mine openings. An underdrain was placed along the length of the highwall to collect drainage from the site. Today the site bears no resemblance to its appearance before reclamation.

◇▶ FIGURE 1 Drainage control and diversion at the disturbed area. Spreading and compacting of waste rock, subsoil, and soil against the highwall stabilizes slopes and buries reactive sulfide minerals.

◇▶ FIGURE 2 Immediate seeding, here with a hydroseeder, follows emplacement of topsoil so as to establish vegetation quickly. Biodegradable green dye is added to the seed to help the operator gauge coverage.

◇▶ FIGURE 3 A fast-growing, temporary cover crop such as rye grass prevents soil erosion and adds organic matter to the soil. The cover crop is selected carefully, as some plants are undesirably dominant or persistent.

◇▶ FIGURE 4 After the cover crop has died back, permanent legumes and grasses take over. They eventually restore the disturbed land to meadow or pasture so that it can be used for wildlife habitat, recreation, or livestock grazing. Trees eventually may return the site to forest.

FIGURE 13.20 Area (open-pit) coal mining. As the ground is opened up to expose more of the coal and the pit enlarges, bulldozers backfill the mined-out area to prepare it for reclamation.

mined area is returned to its approximate original shape or left as relatively flat terrain. Mountain-top mining allows for nearly complete removal of the coal bed. In many instances, removal of the mountain top results in terrain that is suitable for agriculture, grazing, or residential development after mining ceases.

IMPACTS OF UNDERGROUND MINING

The impacts of underground mining range from moderate to critically damaging, but they are often less severe than those of surface mining. Underground workings usually follow the ore body closely, which minimizes the amount of barren rock that must be removed and piled on the surface near the mine entrance. Some waste materials at currently active underground mines, such as the Homestake Mine in South Dakota (Figure 13.18), are used to backfill abandoned tunnels. This reduces the costs of energy for hoisting as well as minimizing unsightly surface waste-rock piles.

The most far-reaching effects of underground mining are ground subsidence (Chapter 7), the collapse of the overburden into mined-out areas (◆ Figure 13.21), and **acid mine drainage (AMD),** the drainage of acidic water from mine sites. AMD is generally considered to be the most serious environmental problem facing the mining industry today, as some acid drainage may continue for hundreds of years. AMD can result

when high-sulfur coal and metallic-sulfide ore bodies are mined. In both cases, pyrite (FeS_2) and other metallic-sulfide minerals are prevalent in the walls of underground mines and open pits; in the **tailings,** finely ground, sand-sized waste material from the milling process that remains after the desired minerals have been extracted; and in other mine waste. The reaction of pyrite or other sulfide minerals with oxygen-rich water produces sulfur dioxide (SO_2), which, facilitated by bacterial decomposition, reacts with water to form sulfuric acid (◆ Figure 13.22; Case Study 13.4).

◆ FIGURE 13.21 Store building subsiding into an abandoned underground coal mine south of Pittsburgh, Pennsylvania.

Case Study | 13.4

Acid Mine Drainage and Earth Systems

The origin of acid mine drainage (AMD) is an excellent example of the interaction of four of the spheres of the earth system. When sulfide ore bodies (components of the *lithosphere*) are mined, the surface area of the sulfide minerals in the underground workings, ore piles, waste heaps, and mill tailings is increased (due to their being broken). This increased exposure to chemical processes speeds up the minerals' reactions with both oxygen (an *atmosphere* component) and underground and surface water (components of the *hydrosphere*). The chemical interaction of components of these three earth spheres produces sulfuric acid and liberates free metallic ions such as those of iron, lead, zinc, copper, and manganese (Figure 13.22). Additionally,

certain bacteria (components of the *biosphere*) thrive in this acidic environment, deriving energy from the conversion of one type of iron (ferrous, Fe^{+2}) to another ferric, Fe^{+3}). In so doing, these bacteria cause a "cascading" effect in the conversion of Fe^{+2} to Fe^{+3}, increasing the rate by as much as a million times what it would be if the bacteria were absent. Heat generated by the chemical and biochemical activity provides positive feedback to the entire process, further speeding up the reactions. The acidic water that is produced by the reaction interacts with other components of the biosphere—aquatic life in streams and rivers (see Case Study 13.5) and land plants and animals, including humans, that utilize the water—often with injurious effects.

This not only acidifies surface and underground waters, but it also expedites the leaching, release, and dispersal of iron, zinc, copper, and other toxic metals into the environment. Such substances kill aquatic life and erode human-made structures such as concrete drains, bridge piers, sewer pipes, and well casings. (Zinc concentrations as low as 0.06 mg/L and copper concentrations as low as 0.0015 mg/L are lethal to some species of fish.) Estimates are that between 8,000 and 16,000 kilometers

(5,000 and 10,000 mi) of U.S. streams have been ruined by acid drainage. The oxidized iron in AMD colors some polluted streams rust-red ($\blacklozenge$ Figure 13.23). Where polluted water is used for livestock or irrigation water, it may diminish the productive value of the affected land. Although expensive to implement, technologies do exist for preventing and controlling AMD (see Case Study 13.5). The subject of surface- and ground-water protection is addressed later in the chapter.

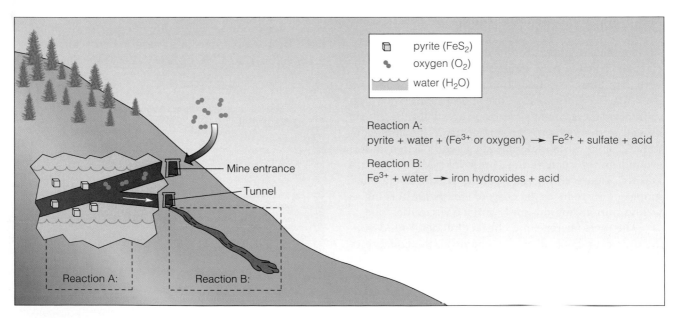

pyrite (FeS_2)

oxygen (O_2)

water (H_2O)

Reaction A:
pyrite + water + (Fe^{3+} or oxygen) $\longrightarrow$ Fe^{2+} + sulfate + acid

Reaction B:
Fe^{3+} + water $\longrightarrow$ iron hydroxides + acid

Mine entrance

Tunnel

Reaction A:

Reaction B:

$\blacklozenge$ **FIGURE 13.22** Pyrite reacts with air and water to produce sulfuric acid.

◆ FIGURE 13.23 Rust-red sludge of acid mine drainage from West Virginia coal mines that were abandoned in the 1960s. The acidic water has eaten away at the Portland cement–based concrete retaining wall, and the bridge supports will soon share that fate. Problems such as this are rarely caused by modern surface mines, because current mining and reclamation practices eliminate or minimize acid mine drainage.

Consider This... ? Pyrite (fool's gold) is common in deposits of gold, lead, zinc, silver, and copper. Why is a conspicuous rusty-colored outcropping often a clue to the existence of buried ore? ■

IMPACTS OF SURFACE MINING

A variety of surface-mining operations contribute to environmental problems. Hydraulic mining is largely an activity of the past, but **dredging**–scooping up earth material below a body of water from a barge or

◆ FIGURE 13.25 Placer gold mining with a bucket-conveyor dredge on the Yuba River, California, in 1993. Surrounding the dredge operation is the scarified landscape of waste-rock piles.

raft equipped to process or transport materials—continues in the 1990s (◆ Figure 13.24). Much of the current U.S. production of sand and gravel is accomplished by dredging rivers, and large dredges are used to mine placer tin deposits in Southeast Asia. Dredging causes significant disruption to the landscape; it washes away soil, leaves a trail of boulders, and severely damages biological systems. Scarification remains from former dredging in river bottoms in many Western states and elsewhere in the world (◆ Figure 13.25). Few gold-dredge areas have been reclaimed, but an outstanding example of what can be accomplished appears in ◆ Figure 13.26. Another example is a 73-square-kilometer (28-mi^2) dredge-scarified area near Folsom,

◆ FIGURE 13.24 Placer mining with a bucket-conveyer dredge near Platinum, Alaska, 1958. The dredge operated from the early 1930s until the late 1970s, during which time it was the major producer of platinum-group metals in North America.

◆ FIGURE 13.26 Reclaimed dredge-mined site, a former placer gold mine; Fox Creek, Fairbanks District, Alaska. The State of Alaska awarded a certificate of commendation for the reclamation work.

Case Study

Case Study | 13.5

Cleaning Up Historic Mining Sites in Colorado

Challenges of cleaning up abandoned mining regions are well-illustrated by ongoing activities in Colorado, where mine drainage is a very serious environmental problem. An estimated 1,300 miles of the state's streams are contaminated with metals due to contaminated drainage in mining areas. The water emitted from the mines, acid mine drainage (AMD), is commonly more acidic than lemon juice (citric acid) or vinegar (acetic acid)—often with pHs less than 4 (see Case Study 13.4)—and generally contains high concentrations of dissolved metals. The high metal concentrations are due to acid dissolution of the country rock and to oxidation of metallic sulfide minerals. These chemical reactions produce harmful concentrations of such elements as iron, manganese, zinc, lead, copper, arsenic, and cadmium in the waters (see Case Study 13.4). As water moves away from a mine site, additional chemical reactions may generate more acid, causing red, orange, and yellow iron-rich precipitates to settle on aquatic plant life and stream bottoms.

The Argo Tunnel and Clear Creek

The Argo Tunnel was excavated between 1893 and 1910 to support the mining industry near Idaho Springs (◆ Figure 1). Its purposes were to drain waters from several mines and provide access for ore from those mines to the Argo processing mill adjacent to the tunnel opening. Ore last moved through the tunnel in the 1940s, but water laced with acid (sometimes with a pH as low as 2.5) and metals still drains from the tunnel continuously. Until the late 1990s, Argo water discharged directly into Clear Creek, the source of drinking water for more than 250,000 people north of Denver. Argo's drainage has sometimes carried as much as a ton of metals into Clear Creek per day. Argo's AMD legacy includes manganese, iron, copper, zinc, arsenic, aluminum, and cadmium, with manganese and aluminum posing the greatest threats to aquatic life in Clear Creek. Until recently, drainage from the tunnel contaminated the water downstream for 35 miles. At that point, the drainage joined the waters of the South Platte river and became diluted.

In the 1980s the Argo Tunnel was identified as a Superfund site, and in 1998 a chemical water treatment plant began operating that will, continuously and forever, treat the metal-laden water issuing from the Tunnel (◆ Figure 2). The treatment removes 99.89 percent of the metals (about a ton of metals per day from the Argo discharge), which are collected as a sludge and trucked to a landfill site. The cost of constructing the plant was $4.7 million, 90 percent of which was paid by the EPA, and the remainder by the State of Colorado. Annual operating costs are about $1 million. Similar Colorado AMD-water treatment plants operate at the Eagle Mine near Minturn, in California gulch near Leadville, and at Summitville in Del Norte County.

Remediation Studies for the Animas River Watershed

In the area of Silverton, in southwest Colorado, many millions of dollars' worth of lead, zinc, silver, and gold were produced between 1874

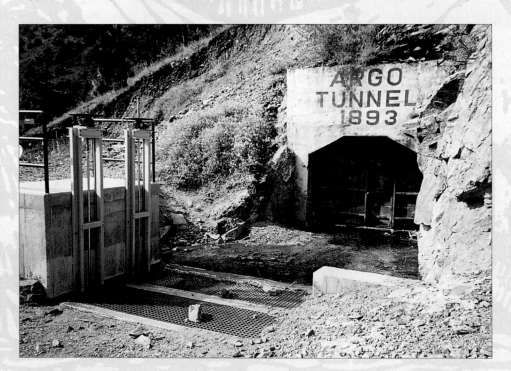

◆ FIGURE 1 AMD draining from the Argo Tunnel flows into a drain that carries the water to the Argo Water Treatment Plant. With a pH of about 3, the water is almost as acidic as vinegar. Its yellow-orange color is due to precipitates of the metallic ions carried in the water.

◇ FIGURE 2 The Argo Water Treatment Plant at Idaho Springs, Colorado, continuously treats AMD from the Argo Tunnel before its discharge into Clear Creek. Operation of the plant began in April 1998, and the State of Colorado anticipates operating the plant for perpetuity. Idaho Springs and the Argo Mill Historical Mining Site are a short distance west of Denver on I-70.

◇ FIGURE 3 Longfellow Mine, at Red Mountain Pass, one of an estimated 2,000 abandoned mines in the Animas River watershed near Silverton, Colorado

and 1991. By the late 1990s, most of the 2,000 mine sites within the Animas River watershed had been abandoned (◇ Figure 3), and metal-laden AMD issuing from the old mines had affected the river's aquatic life and water quality for more than 100 miles downstream.

One of the first objectives in cleaning up mining sites is *characterization;* that is, determining the relative "metal-loading" contributions to the watershed from various sources and identifying the pathways of the toxic metal contaminants. Such information is needed for developing a hydrologic model for the area and determining the watershed's baseline conditions. For the Animas river watershed, these objectives were met by:

- studying surface-water quality;
- inventorying abandoned mine sites;
- collecting water samples from each subwatershed at low-flow and high-flow stages;
- developing an overview of point sources, transport pathways, and partitioning of toxic metals within the watershed; and
- determining specific sites of contamination by various means.

Collectively, these measurements provide toxic-loading data for specific sites. With this information, technical experts from the EPA, the Colorado Division of Minerals and Geology, the U.S. Forest Service, the Bureau of Land Management, the Colorado Department of Public Health and Environment, and the U.S. Geological Survey established priorities for site remediation. Prioritization was based on:

- contaminating sites' levels of environmental impact,
- amounts of dissolved metals in the drainage waters,
- existing and potential aquatic life,
- feasibility of treatment,
- cost-effectiveness, and
- preservation of the area's historic and cultural values.

Of the 300 or so abandoned mines in the watershed recognized as contributors of AMD, 25 were identified as "major" sources of AMD and toxic metals, and four as "critical."

The first stage of the remediation efforts began before the list of priority sites was finalized. It consisted of constructing simple barriers between water bodies and mine wastes in order to minimize or eliminate the production of AMD. The second stage is evaluating various remediation alternatives for the high-priority sites. From the least expensive to the most expensive, the options include cementing and sealing mine openings, in-place bioremediation using metal-reducing bacteria, and constructing a water-treatment facility such as the one at Idaho Springs. As of early 2000, no final solution had been reached for the clean up of abandoned mines in the Silverton area. Colorado's effort to eliminate toxic AMD from old mines, the costs incurred, and the number of government agencies involved illustrates well the challenges faced in reclamation of abandoned mines.

SOURCES: U.S.G.S. Circular 1097, 1994; Dunn, Russell, and Morrissey, Remediating Historic Mining Sites in Colorado, *Mining Engineering,* August 1999; and *Argo Water Treatment Plant,* Colorado Department of Public Health and Environment, August 1999.

California. The area was reclaimed by reshaping the waste piles, covering the surface with topsoil, and then building a subdivision of homes on it. Landscaped with trees and grasses, no evidence remains of the once scarified surface.

In **hydraulic mining,** or *hydraulicking,* a high-pressure jet of water is blasted through a nozzle, called a *monitor,* against hillsides of ancient alluvial deposits (Case Study 13.1). Hydraulicking requires the construction of ditches, reservoirs, a *penstock* (vertical pipe), and pipelines. Once constructed, it is possible to wash thousands of cubic yards of gold-bearing gravel per day. The gold is recovered in sluice boxes, where mercury (the liquid metal that dissolves gold and silver to form an amalgam; it's explained in Case Study 13.1) usually is added to aid recovery at a low cost. Hydraulicking is efficient but highly destructive to the land (◆ Figure 13.27). Because nineteenth-century hydraulicking was found to create river sediment that increased down-

stream flooding, clogged irrigation systems, and ruined farmlands, court injunctions stopped most hydraulic mining in the United States before 1900 (although it still continues in Alaska on a limited scale). Hydraulicking is still being used in Russia's Baltic region for mining amber, and in 1979 thousands of gold seekers began invading remote areas of Brazil's Amazon basin and using it to extract placer gold. The danger of using mercury is that when it escapes from the mining activity, it can accumulate in the food chain; consumption of mercury-tainted food causes birth defects and neurological problems in humans and animals. Its current use in gold recovery in Brazil as well as in Indonesia and the Philippines is causing extensive contamination. It has been estimated that 100 tons of mercury is working its way into the ecosystem of the Amazon basin each year.

Strip-mining is used most commonly when the resource lies parallel and close to the surface. The phos-

◆ **FIGURE 13.27** A monument to the destructive force of hydraulic mining of bench placer deposits; Malakoff Diggins State Park, Nevada County, California. When the hydraulic mining operations here ended in 1884, Malakoff Diggins was the largest and richest hydraulic gold mine in the world with a total output of about $3.5 million. More than 41 million cubic yards of earth had been excavated to obtain gold. The site is now marked by colorful, eroded cliffs along the sides of an open pit that is some 7,000 feet long, 300 feet wide, and as much as 600 feet deep.

FIGURE 13.28 A rich phosphate bed in Florida is strip-mined by a mammoth drag-line dredge. The bucket, swinging from a 100-meter boom (that's longer than a football field), can scoop up nearly 100 tons of phosphate at a time. The phosphate is then piped as a slurry to a plant for processing. Fine-grained gypsum, a by-product of processing, is pumped to a settling pond *(upper left),* where it becomes concentrated by evaporation.

phate deposits of North Carolina and Florida are strip-mined by excavating the shallow, horizontal beds to a depth of about 8 meters (26 ft) (◆ Figure 13.28). After the phosphate beds are removed, the excavated area is backfilled to return the surface to its original form.

Open-pit mining is the only practical way to extract many minerals when they occur in a very large low-grade-ore body near the surface. The process requires processing enormous amounts of material and is devastating to the landscape. The epitome of open-pit mining is the Bingham Canyon copper mine in Utah, where about 3.3 billion tons of material—seven times the volume moved in constructing the Panama Canal—have been removed since 1906 (Figure 13.7). Now a half-mile deep, the pit is the largest human excavation in the world, and the waste-rock piles literally form mountains.

The environmental consequences of open-pit mining are several. The mine itself disrupts the landscape, and the increased surface area of the broken and crushed rocks from mining and milling sets the stage for erosion and the leaching of toxic metals to the environment. This is especially true of sulfide-ore bodies. They produce AMD, because the waste rocks and tailings are highly susceptible to chemical weathering.

IMPACTS OF MINERAL PROCESSING

Except for some industrial minerals, excavating and removing raw ore are only the first steps in producing a marketable product. Once metallic ores are removed

from the ground, they are processed at a mill to produce an enriched ore, referred to as a *concentrate.* The concentrate is then sent to a smelter for refining into a valuable commodity.

Concentration and smelting are complex processes, and a thorough discussion of them is well beyond the scope of this book. Briefly, the concentration process requires:

1. crushing the ore to a fine powder,
2. classifying the crushed materials by particle size by passing them through various mechanical devices and passing on those particles of a certain size to the next step, and
3. separating the desired mineral components from the noneconomic minerals by a flotation, gravity, or chemical method.

These means of separation are discussed individually here.

Flotation

The **flotation** separation process is widely used, especially for recovering sulfide-ore minerals such as lead, zinc, and copper sulfides from host rock. The process is based on the principles of wettability of mineral particles and surface tension of fluids. After crushing and concentration, the wettability of the *undesired* mineral particles is increased by treating the crushed ore

◁▶ FIGURE 13.29 The major components of cyanide heap-leach gold recovery.

1. Spray cyanide solution on heap.

2. Collect gold-bearing solution.

4. Add cyanide and adjust pH.

3. Recover gold.

Pumps

Ore

Carbon-filled columns

Impermeable base (clay) with drainage blanket (gravel) on top

Pregnant pond

Barren pond

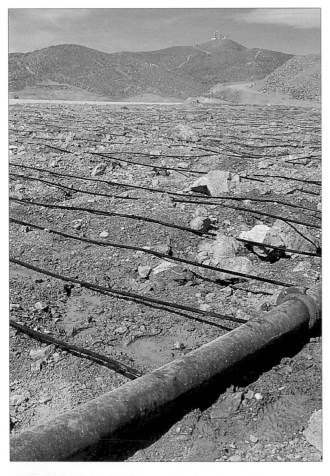

◁▶ FIGURE 13.30 Cyanide drip lines on a leach heap. The cyanide concentration is 200 parts per million (one five-thousandth of one percent), and the solution is kept highly alkaline to prohibit cyanide gas from forming.

chemically—usually with liquid hydrocarbons—to ensure that the undesired minerals will sink. Air is then bubbled into the slurry of crushed ore and water, forming a froth that collects the *desired* mineral particles of low wettability. The froth, with the attached desirable mineral particles, is skimmed off the top of the flotation tank and dried; this is the concentrate. The undesired mineral particles, the tailings, sink to the bottom of the flotation tank. They are drawn off and piped to the tailings pond. Although they are usually environmentally undesirable, tailings are an unavoidable waste product of mining.

Gravity Separation

Gravity separation methods are used in recovering high-density ore minerals such as gold, platinum-group metals, tungsten, and tin. By this process, mineral particles mixed with water are caused to flow across a series of riffles placed in a trough. The riffles trap the desired high-density particles, and water carries away the undesired low-density minerals, the tailings. The rockers, sluice boxes, and dredges used in gold placer mining, explained in Case Study 13.1, are all examples of gravity separation methods.

Chemical Methods

For minerals whose physical properties make them unsuitable for separation by flotation or gravity methods, chemical processes are used, the major ones being leaching and cyanidation. *Leaching* is often used in treating copper-oxide ores. Sulfuric acid is added to crushed ore to dissolve the copper and produce a solution of copper sulfate. The dissolved copper is then recovered by placing scrap iron in the copper sulfate solution; the copper plates out onto the iron. The acidic waste materials are chemically neutralized by treating them with lime.

Cyanidation

Cyanidation, used to recover gold and silver since 1890, makes use of the special property of cyanide to dissolve gold and silver. An innovation of cyanide recovery, **cyanide heap-leaching,** began to be widely used in the United States in the 1980s. During the 1990s there have been about 150 heap-leach operations active in Alaska, Montana, Idaho, Colorado, Nevada, California, Washington, Utah, South Dakota, and South Carolina. Although heap-leaching is efficient, it is controversial in environmental circles because the open-pit mining, waste-rock dumps, and tailings piles are destructive to the landscape. Furthermore, cyanide is perceived as a hazard to wildlife and a contaminant to ground and surface waters. In Montana alone, 51 cyanide releases were documented between 1982 and 1999. The use of cyanide is illustrated in ◈ Figure 13.29. Ore from an open-pit excavation is pulverized, spread out in piles over an impervious clay or plastic liner, and sprayed with a dilute cyanide solution, commonly about 200 parts per million (0.02 %) (◈ Figure 13.30). The solution dissolves gold and silver (and several other metals) present in small amounts in the ore as it works its way through the heap to the pregnant pond (◈ Figure 13.31). The gold and silver are recovered from the resulting "pregnant" solution by adsorption on activated charcoal, and the barren cyanide solution is recycled to the leach heap. The precious metals are removed from the charcoal by chemical and electrical techniques, melted in a furnace, and poured into molds to form ingots.

Monitoring wells are placed downslope from leach pads and ponds containing cyanide for detecting possible leakage. Maintaining the low concentrations of cyanide required by state regulations requires regular monitoring of the solution. The cyanide solution is kept highly alkaline by additions of sodium hydroxide, a strong base, in order to inhibit the formation of lethal cyanide gas. No measurable cyanide gas has been detected above leach heaps where instrumental testing for escaping gas has been carried out. Unforeseen peculiarities of the ore chemistry and unusual weather have caused unexpected increases in cyanide values at some mines. This has required an occasional shutdown of operations until the chemistry of the solution could be corrected to the established standards. In a few cases, stiff fines have been assessed where cyanide levels exceeded

◈ **FIGURE 13.31** A typical "pregnant pond" for collecting the cyanide solution containing dissolved gold from a leach heap. The face of the leach heap is at the upper right of the photograph. The pipes on the far side of the pond drain the "pregnant" solution from beneath the leach heap. The pond is lined with two layers of high-density polyethylene underlain by a layer of sand and a leakage detection system of perforated PVC pipe. In the rare chance that leakage should occur, it is trapped in the pregnant pond, and appropriate repairs are made. The pond is covered with fine-mesh netting to keep birds from the toxic solution.

the limits of the operator's permit. Upon abandonment of a leach-extraction operation, federal and state regulations require flushing and detoxification of any residual cyanide from the leach pile. Regular sampling and testing at ground-water-monitoring wells also may be required for several years following abandonment.

Securing the open cyanide ponds to keep wildlife from drinking the lethal poison is an environmental concern at several operations. After 900 birds died on the cyanide-tainted tailings pond at a mine in Nevada in 1989–1990, the company pleaded guilty to misdemeanor charges, paid a $250,000 fine, and contributed an additional $250,000 to the Nature Conservancy for preservation of a migratory bird habitat. Reported wildlife mortalities at cyanide-extraction gold mines in Arizona, California, and Nevada from 1980 to 1989 appear in ◉ Table 13.9. Of major concern are a number of endangered, threatened, or rare mammal species whose geographic ranges include cyanide-extraction mines. About 34 percent of the mammal mortalities reported in the table are bat species that are on the federal government's list of rare and endangered species.

Various techniques for discouraging wildlife from visiting cyanide-extraction operations have been tried: stringing lines of flags across leach ponds in an effort to scare birds away, covering pregnant ponds with plastic

⦿ TABLE 13.9	**Mammal, Bird, Reptile, and Amphibian Mortalities Reported at Cyanide-Extraction Gold Mines in Arizona, California, and Nevada, 1980–1989**					
STATE	NO. OF MINES	MAMMALS	BIRDS	REPTILES	AMPHIBIANS	TOTAL
Arizona	1	52 (12.7%)	357 (87.3%)	0	0	409
California	11	34 (5.2%)	606 (92.7%)	14 (2.1%)	0	654
Nevada	63	433 (6.6%)	6,034 (92.2%)	24 (0.4%)	55 (0.8%)	6,546
Totals	75	519 (6.8%)	6,997 (92.0%)	38 (0.5%)	55 (0.7%)	7,609

SOURCE: D. R. Clark and R. L. Hothem, "Mammal Mortality at Arizona, California, and Nevada Gold Mines Using Cyanide Extraction," *California Fish and Game* 77 (1991), pp. 66–67.

sheeting, and blasting recorded heavy-metal rock music from loudspeakers. At one mine, the heavy-metal recordings were effective in keeping migratory waterfowl away, but resident birdlife adjusted to the din and remained in the area. One wonders what effect the blasting of recordings by Limp Bizkit or Nine Inch Nails might have. The current standard practice is to enclose the operations with chain-link fencing to keep out larger animals and to stretch netting completely across the ponds to protect birds (Figure 13.31). The fences and netting have been successful; California's 12 to 15 active heap-leach operations, for example, have reduced bird losses to only 50 to 60 a year in the cyanide ponds. This number is less than trivial when compared to the thousands of birds killed each year in the state when they fly into roof-mounted television antennas and the paths of automobiles.

Smelting

Historically, smelters have had bad reputations for causing extensive damage to the environment (see Case Study 13.3). Sulfurous fumes emitted as byproducts of smelting processes have polluted the air, and toxic substances from smelting operations have contaminated soils and destroyed vegetation. Because water was necessary for operating early-day smelters, they were located near streams. The accepted practice at the time was to discharge mill wastes and tailings into streams or settling ponds, which commonly spilled over into streams. Improved smelting technologies are now eliminating these problems. For example, Kennecott's former smelter at the Bingham Canyon, Utah, copper mine released 2,136 kilograms (4,700 pounds) of sulfur dioxide (SO_2) per hour to the atmosphere. Kennecott's new smelter, designed to meet or exceed all existing and anticipated future federal and state emission standards, was put into service in 1995. SO_2 emissions were cut by 96 percent to only 91 kilograms (200 pounds) per hour, less than the rate for the world's cleanest smelters now operating in Japan.

Mine Land Reclamation

The volume of coal-mine spoil and hardrock-mining waste rock is usually greater than the volume occupied by the rock before mining. SMCRA requires that coal-mine spoil be disposed of in fills, usually in the upper reaches of valleys near the mine site. Because some settlement of this material can be expected over time, construction on these reclaimed sites may be risky. Waste rock from hardrock mining is usually deposited in great mesa-like piles near the mines. In recent years, many of these piles have been contoured to break up their unnatural, flat-topped appearance. In both cases the features are engineered and structured for stability with terraces and diversion ditches for controlling surface-water flow and preventing erosion, and then they are landscaped. Ground water within the fills is commonly channeled through subsurface drains.

SURFACE- AND GROUND-WATER PROTECTION

High-quality, reliable water is critical for domestic, industrial, and agricultural activities, and it is especially precious in arid and semiarid regions. SMCRA requires that a surface coal-mining operation be conducted in a way that will maintain hydrologic balance and ensure the availability of an adequate water supply for postmining use. In the early 1990s, nearly 1,500 abandoned U.S. coal mines had been identified as having water problems. Of these, about 500 have been reclaimed or have been funded to begin reclamation.

During surface coal mining all of the runoff water that collects in the pit is required to be collected and treated. Because surface mining destroys the original plant cover and exposes the soil to erosion, mitigation of erosion and sediment loss is needed during mining and reclamation. AMD problems have arisen from coal mining in Pennsylvania, West Virginia, Maryland, and Ohio and from the hardrock mining of sulfide ores in the Western states and in many areas of Canada. More than 100,000 abandoned hardrock mines in the West may pose AMD problems (see Case Studies 13.4 and 13.5).

SMCRA requires coal mine operators to treat water from their mines before releasing it into streams and rivers. Control measures for AMD include:

- Holding mine drainage water in entrapment ponds and neutralizing it by adding alkaline materials from other industries such as kiln dust, slag, alkaline fly ash, and limestone before releasing it from the mine-permit area.
- Grading and covering acid-forming materials to promote surface-water runoff and inhibit infiltration.
- Backfilling underground mines with alkaline fly ash, which absorbs water and turns into weak concrete, in order to minimize the flow of water from the mine and reduce the amount of oxygen in the mine. Filling is done by pumping a slurry down several boreholes. This process has been carried out by the U.S. Bureau of Mines to reduce AMD and stabilize the ground surface of many communities built over mined-out areas.
- Using wetlands to treat AMD issuing from both operating and abandoned coal mines. Cattails and other wetlands plants have been found to be effective in removing toxic metals and other substances from water. One such Pennsylvania wetland has saved an electric power company $50,000 in water treatment costs each year since it was built in 1985.

Wastewater from hardrock mines and from processed ore tailings is another source of contamination. These waters are usually treated by chemical or physical processes before disposal. When those processes proved unsuitable at the Homestake Gold Mine at Lead, South Dakota (Figure 13.18), the company developed a biological treatment. The bacterial bioxidation process breaks down residual cyanide from the mill's wastewater into harmless components that are environmentally compatible with the disposal creek's ecosystem. The creek now supports a viable trout fishery for the first time in a century. A secondary benefit of the bacterial treatment is that precious metals are also recovered by the process.

◆ **FIGURE 13.32** A placer gold mining operation near Fairbanks, Alaska. A downstream settling pond traps placer tailings from the mine.

At some unprotected mine properties, siltation clogs streams and increases the threat of flooding. This problem may be mitigated by constructing settling ponds for trapping the sediment downstream from the source (◆ Figure 13.32).

Of all mining operations, sand and gravel mining are the most widespread and the most obvious to urban dwellers—especially when the operations are along or within rivers. Such modification to rivers results in a variety of consequences: increased erosion to channels and river banks, degradation of water quality, disconnection of links between terrestrial and aquatic ecosystems, and creation of unacceptable habitats for spawning fish such as anadromous (migrating) salmonids (◆ Figure 13.33). Fortunately, disrupted river channels can be restored, and more states are requiring restoration elements in mines' operating plans.

(a)

(b)

◆ **FIGURE 13.33** (a) Site of a sand and gravel mining operation along the Molalla River in Oregon before restoration, and (b) after reclamation. Not only is the site more attractive, the pit has been modified to decrease erosion, shallow ponds have been created to provide habitat suitable for migrating and spawning salmon, and the overall setting is improved for the general benefit of wildlife.

MILL AND SMELTER WASTE CONTAMINATION

As ore is milled to produce concentrate, about 90 percent of the mined material is discarded as tailings. Smelters that process metallic sulfide concentrates release waste products that can contain thousands of times the natural levels of heavy metals such as lead, arsenic, zinc, cadmium, and beryllium. These contaminants remain in smelter slag (rendered inert by the smelting process so that it is not bioavailable) and in flue dust that commonly remains around former smelter sites long after smelting has ended. The primary concerns with such contamination are the environmental impact of stormwater runoff into nearby streams or lakes and the risks to human health that result when people unwittingly inhale or ingest the heavy

metals. If the site is on the EPA's Superfund list, the cleanup will take one of two directions—either removal or in-place remediation such as enclosing or encapsulating the heavy-metal sources—determined by which one is considered appropriate for preventing human contact. At one Superfund site the discovery of elevated levels of arsenic in the hair and blood of children in a small community built on contaminated soil resulted in the EPA's moving the entire community of about 30 families and destroying the houses. If a contaminated Superfund area is populated, the EPA may require an ongoing blood-testing program to monitor residents' blood levels of heavy metals. Contaminated stormwater runoff may require construction of a stormwater management system with engineered drainage channels, ponds, and wetlands.

◆ **FIGURE 13.34** Removal of mill tailings at a Superfund site at Butte, Montana. Immense amounts of material must be moved at some Superfund sites.

◆ **FIGURE 13.35** Preparation of a repository site for contaminated smelter-flue dust near Anaconda, Montana. The pit will have a heavy polyethelene liner.

The EPA procedure for remediating contaminated smelter slag, flue dust, and tailings that pose a threat to surface or ground water is to remove the material and deposit it in specially constructed repository sites (◆ Figure 13.34). Commonly these are pits underlain with heavy plastic liners (◆ Figure 13.35) into which the contaminated waste is placed and covered with crushed limestone, sealed with topsoil, and revegetated. To eliminate leaching of the material by rainwater or snowmelt, a subsoil drainage system collects excess water and disposes of it in such a way that it poses no threat to human health or the environment. (Disposal of toxic wastes is discussed thoroughly in Chapter 15.)

REVEGETATION AND WILDLIFE RESTORATION

SMCRA regulations for coal mining and most state reclamation regulations for hardrock mining require healthy vegetation to be reestablished once mining has ended. Permanent vegetation is the principal means of minimizing erosion and reducing stream siltation. The types of vegetation that are to be used in reclamation are stipulated in the original mine permit, based on premining vegetation and intended postmining uses. Commonly, a straw mulch or chemical soil stabilizer is applied after seeding to inhibit erosion and retain moisture. Mine operators are responsible for maintaining the new plant cover until it is successfully reestablished; the SMCRA minimum is five years in the East and Midwest and ten years in the semiarid West.

Wildlife habitat is one of the most common uses of postmined land. Among the techniques the mining industry uses to meet the EPA, SMCRA, and state regulations for attracting and supporting desirable species of wildlife are contouring the land, introducing plant species that will support browsing and foraging, creating wetland habitats, and stocking ponds for sport fishing.

The Future of Mining

Concern about the future reserves of mineral resources has led to the obvious question, Are we running out? Numerous studies have been conducted over the years to determine reserves, and the answer to the question seems to be "Not yet." The question of availability is probably not the important one, however. Increasing needs for resources will require mining lower and lower grades of minerals. This will require improved technologies and larger, more powerful machines, which in turn will produce greater quantities of waste material. This will impose more stress on environmental systems. Because of these factors, the question that *should* be asked is: Can we afford the environmental and human costs required to satisfy our increasing need for minerals?

Today, we residents of industrialized countries enjoy living in comfortable homes, traveling by automobile, and having labor-saving appliances, all of which provide us with life-styles that only 100 years ago would have been considered luxurious. These life-styles are possible only because of the availability of inexpensive raw materials for manufacturing the products to which we have grown accustomed. Few of the earth's people realize that today's prices of mineral commodities typically reflect only the short-term, tangible costs of wages, equipment, fuel, financing, and transportation—just as they have since mining began in ancient times. Much of the *real cost* of exploiting mineral and fossil-fuel resources is intangible, and it has been externalized. Historically, the environment has borne a great deal of the cost of extracting raw materials; the consumer has paid only part of the real cost. The low prices of raw materials today do not cover the costs of polluted surface waters, dammed rivers, squalid mining towns, and devastated landscapes. Even in the United States, where regulations designed to protect the environment are imposed, mining and processing still cause substantial damage.

Worldwide, poor mining and mineral-processing practices contribute significantly to soil erosion, water contamination, air pollution, and deforestation, especially in developing nations, as illustrated by the examples in Table 13.3. Fortunately, the negative environmental impacts of mining and mineral processing are gaining more attention in the developing nations. For example, Mexico adopted new mining and environmental laws in 1992 before encouraging mining by privatizing that industry. And in 1994, Tanzania and Guinea began requiring environmental impact assessments to be submitted along with applications for mining licenses.

For over a century, the world's indigenous peoples unfortunate enough to live on mineral deposits have been pushed aside as miners removed the riches. This also is changing. A 1992 decision by Australia's High Court overturned the doctrine of *terra nullis*, the principle that the country's land belonged to no one when the first European settlers arrived. This decision is expected to result in mining companies having to pay compensation fees to the Aborigines and being prevented from extracting minerals in locations established as traditional Aboriginal lands or sacred sites.

In this country, the Crandon Mine Project in Wisconsin, which could become one of the world's largest zinc-copper mines, would border some of the state's most productive wild rice areas in the waters of the Wolf River. The Native Americans on the Menominee Reservation downstream from the mine site and other Native Americans have joined with the local Chippewa Band in opposing the project, because it threatens the livelihoods of the region's Native Americans who depend on the river's annual wild rice harvest for much of their income.

Mining generates twice as much solid waste as all other industries and cities combined, and most mining wastes in the United States are currently unregulated by federal law. In addition, the mineral industry is one of the greatest consumers of energy and a contributor to air pollution and to global warming. It is estimated, for example, that the processing of bauxite into aluminum alone consumes about one percent of the world's total energy budget. Miners admit that mines leave holes in the ground, but they go on to point out that when they abandon their currently active mines in the United States, they will be restoring the site to as near a natural appearance as is reasonable, except for the open pits.

In evaluating mining and its impact on the environment, we must recognize that the only way not to disturb the landscape is *not to mine*. This would mean not having the raw materials to build automobiles, houses, farm machinery, airplanes, radios, television sets—virtually all of the objects we view as essential to a modern society. The question thus becomes, What must be done to minimize the impacts?

U.S. MINING LAWS

In the 1970s some genuinely important environmental concerns related to mining and processing mineral materials became obvious in the United States and many other countries, and laws began being written to ad-

dress these concerns. Consequently, in the United States, the EPA now administers a number of acts regulating water quality and toxic wastes, and the Fish and Wildlife Service requires noninterference with rare and endangered species. Further constraints are imposed by numerous state and local laws requiring that mining operations meet certain air-pollution standards. Many states will not issue a permit for an operation to begin without prior approval of an adequately funded reclamation plan specifying that the waste and tailings piles will be restored to resemble the surrounding topography to the extent that it is possible, that the disturbed ground and the tailings will be replanted with native species, that all buildings and equipment will be removed, and that the open pits will be fenced and posted with warning signs.

It is claimed by some that enforcement of federal regulations pertaining to mining has been weak. Initially, the EPA did little to regulate mining wastes. Further, Congress specifically exempted hardrock-mining wastes and tailings from regulation as hazardous wastes in the Resource Conservation and Recovery Act (1976). Complicating the issues is the fact that instead of the federal government, it is the individual states that play the role of regulator on state and private lands.

Further reform of the Mining Law of 1872 may occur. Beginning in the 1980s public concern grew over the problems of protecting nonmineral values on public lands, the lack of meaningful federal reclamation standards, limited environmental protection, and the lack of royalty collection for exploited public land—issues of public-resource management that are not addressed in the current mining law. These issues were once again addressed in Congress in the 1990s, and a mining law reform bill that had passed in the House of Representatives was killed in the Senate in the final days of the 103rd Congress. A point of contention has been the issue of imposing royalties on mine income. This royalty income could be used to finance the reclamation of some 100 thousand abandoned hardrock mines in the Western states, much as is already done by SMCRA with abandoned coal mines using the Abandoned Mine Fund. Some critics of mining law reform believe that charging royalties for hardrock operations on federal lands will close down U.S. mining and drive more mineral production to foreign countries. Nevertheless, efforts at reform are expected to continue.

The future may also bring some changes in the basic attitudes and assumptions that underlie our capitalistic society. The economic assumption that prosperity is synonymous with mineral production is now being questioned. Environmental deterioration from today's unprecedented rate of mineral production will, if con-

> **Consider This... ?**
> Is it likely that regional devastation such as that in Tennessee's Copper Basin (Case Study 13.2) could develop in the United States today? Why or why not? What is the likelihood of such problems occurring in the world's developing countries? ■

tinued, eventually overwhelm the benefits gained from increased mineral supplies. As the world's developing nations strive to achieve the economies and life-styles currently enjoyed by the developed countries, there will be severe demands on, and competition for, the world's remaining mineral and fuel resources. By the middle of the twenty-first century when world population has reached 10–15 billion (see Chapter 1), new technologies and economic strategies must be in place. The goals of protecting and managing the environment while at the same time exploiting and expanding the mineral resource base in an environmentally responsible manner are not necessarily mutually exclusive. But how can both goals be achieved? Society will need to replace the current materials-intensive, high-volume, planned-obsolescent manufacturing processes—and consumption patterns—with those that use raw materials and fuels more efficiently, that generate little or no waste, and that recycle most of the waste that is generated. Whether technology can modernize fast enough to conserve natural resources at the same time that the world's wealth is increasing is the big question. This is the formidable challenge that must be faced jointly by industry, governments, and society.

Gallery

The Legacy of Mines and Mining

The collage of pictures in this Gallery illustrates the good, the bad, and the ugly of past and present mining in the western United States and Germany (◁▷ Figures 13.36–13.41).

◁▷ **FIGURE 13.36** Toxic legacies left for perpetuity by turn-of-the-century Colorado silver mining; waste-rock piles from abandoned mines above Silver Plume, Colorado. Although the mines have been shut down for decades, the barren piles still lack vegetation because of the accumulation of acid salts (indicated by the white coloration) due to evaporation of acidic seepage water that has percolated through the material. There are thousands of such abandoned mines in Colorado and perhaps hundreds of thousands in the Western states.

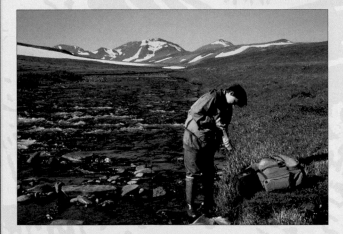

◁▷ **FIGURE 13.37** Humans can be exposed to waters carrying high concentrations of toxic metals even in "undisturbed," natural settings. Several thousand people living in and near the Brooks Range of northern Alaska rely on subsistence fishing and hunting for food. Massive metallic sulfide deposits in the region pose a potential hazard to all life there if drainage from these deposits enters the rivers and streams. In this photo a scientist is collecting water samples downstream from the Drenchwater massive sulfide deposit in the northwestern part of the Range. Water bodies' pH and concentrations of various metals are monitored in order to evaluate the health hazard of this lead-zinc-silver deposit.

◁▷ **FIGURE 13.38** The legendary "Mother Lode" gold-quartz vein is exposed in the wall of an open-pit mine; Jamestown Mine, Jamestown, California.

◁▷ FIGURE 13.39 Creek near an open-pit copper mine at Morenci, Arizona. The water's distinct blue color reveals a high concentration of dissolved copper leached from mine wastes.

◁▷ FIGURE 13.40 The *entire* landscape visible in this photo is productive reclaimed land; near Cologne, Germany. Extensive surface coal mining was carried out here in the 1960s and 1970s.

◁▷ FIGURE 13.41 Plattberg Hill, an accumulation of about 40 years' spoil from underground coal mining, near Kamp-Lintford, Westphalia, Germany. Landscaped hills of mine spoil such as this provide welcome relief to the otherwise flat terrane of the lower Rhine Valley. They provide refuges for wildlife and year-round recreation areas for humans. Mine headworks are at the lower right.

Summary

MINERAL ABUNDANCES, DEFINITIONS
MINERAL DEPOSIT
Locally rich concentration of minerals.

ORE
Metallic mineral resources that can be economically and legally extracted at the time.

MINERAL RESERVES
Known deposits of earth materials from which useful commodities are economically and legally recoverable with existing technology.

FACTORS THAT CHANGE RESERVES
TECHNOLOGY CHANGES
affect costs of extraction and processing.

ECONOMIC CHANGES
affect price of the commodity and costs of extraction.

POLITICAL CHANGES
cut off or open up sources of important mineral materials.

AESTHETIC AND ENVIRONMENTAL FACTORS
may reduce reserves.

DISTRIBUTION OF MINERAL RESOURCES
DESCRIPTION
Mineral deposits are highly localized; they are neither uniformly nor randomly distributed.

CAUSE
Concentrations of valuable mineral deposits are due to special, sometimes unique, geochemical processes.

ORIGINS OF MINERAL DEPOSITS
IGNEOUS DEPOSITS
1. *Intrusive*–occur (a) as a pegmatite, an exceptionally coarse-grained igneous rock with interlocking crystals, usually found as irregular dikes, lenses, or veins, especially at the margins of batholiths; and (b) by crystal settling, the sinking of crystals in a magma due to their greater density, sometimes aided by magmatic convection.
2. *Disseminated*–a mineral deposit, especially of a metal, in which the minerals occur as scattered particles in the rock but in sufficient quantity to make the deposit a worthwhile ore.
3. *Hydrothermal*–a mineral deposit precipitated from a hot aqueous solution with or without evidence of igneous processes.
4. *Volcanogenic*–of volcanic origin, e.g., volcanogenic sediments.

SEDIMENTARY DEPOSITS
Mineral deposits resulting from the accumulation or precipitation of sediment. These include

1. Surficial marine and nonmarine precipitation.
2. Deep-ocean precipitation.
3. Placer deposits.

WEATHERING DEPOSITS
1. *Lateritic weathering*–concentrations of minerals due to the gradual chemical and physical breakdown of rocks in response to exposure at or near the earth's surface.
2. *Secondary enrichment*–mineral development that occurred later than that of the enclosing rock, usually at the expense of earlier primary minerals by chemical weathering.

METAMORPHIC DEPOSITS
Concentrations of minerals, such as gem or ore minerals, in metamorphic rocks or from metamorphic processes.

CATEGORIES OF MINERAL RESERVES
ABUNDANT METALS
Average continental crustal abundances are in excess of 0.1 percent.

SCARCE METALS
Average continental crustal abundances are less than 0.1 percent.

NONMETALLIC MINERALS
Minerals that do not have metallic properties; include industrial minerals, agricultural minerals, and construction materials.

ENVIRONMENTAL IMPACTS OF MINING
1. Surface coal mining and land reclamation is regulated by federal law, the Surface Mining Control and Reclamation Act (SMCRA). No similar federal law regulates hardrock mining and land reclamation.
2. SMCRA requires surface coal mine pits to be backfilled with the mine waste (spoil) and then covered with topsoil and landscaped. Some states in which hardrock mining is conducted require similar handling of waste rock and tailings at currently operating mines.
3. Ground subsidence due to underground coal mining is less common than in the past because of the required backfilling. Surface collapse from hardrock (underground) mining is not common, and stabilization may be expensive.
4. Acid mine drainage (AMD) may be a problem at active and abandoned coal and sulfide-ore mines. Pit water in surface coal mines is required to be held and treated before it is released to streams or rivers. Sealing abandoned mines, covering waste rock with soil and landscaping, and installing drainage courses to direct surface drainage off tailings and waste-rock piles are some methods of controlling AMD formation. Wells are installed below waste-rock and tailings dumps to monitor water quality.
5. Scarified ground, despoiled landscapes, and disrupted drainage that remain after dredging and hydraulic mining can be reclaimed by reshaping, adding topsoil, and landscaping.

6. Abandoned open pits with oversteepened side walls despoil the landscape and may pollute hydrologic systems. They are difficult or impossible to reclaim. The public is protected by surrounding the pits with chain-link fencing and posting warning signs. The pit floor may be partially backfilled and sealed with impermeable clay to protect ground water from pollution.

7. Cyanide heap-leach gold-extraction methods normally cause minimal environmental problems because they are strictly regulated by state and federal agencies. There are occasional leaks of cyanide into the environment, however. Wildlife losses have been mitigated by placing nets over the lethal ponds. Surface and ground waters are monitored by sampling and monitoring wells, respectively.

8. The main concerns associated with heavy-metal contaminated mill and smelter waste are the environmental impacts resulting from stormwater runoff into nearby surface waters and the human health problems caused by inhaling or ingesting the heavy metals in dust. Superfund cleanups require either removal or in-place remediation by encapsulating the heavy-metal sources. Wetlands with appropriate plantlife may be constructed, or water treatment plants constructed, to clean toxins from surface runoff.

THE FUTURE OF MINING
WORLDWIDE
Mining and processing of mineral commodities are contributing to soil erosion, water contamination, air pollution, and deforestation.

THE UNITED STATES
Mining and processing of mineral commodities must meet standards of air and water quality set by the EPA and other government agencies. The Mining Law of 1872, although much amended, is being criticized and evaluated in light of current economics, domestic needs, and conflicting land uses.

◆ Key Terms

acid mine drainage (AMD)
alluvial placer
area mining
average crustal abundance
banded-iron formation
bauxite
bench placer
black smoker
CERCLA
concentration factor
contact-metamorphic deposit

contour mining
crystal settling
cyanide heap-leaching
disseminated deposit
dredging
evaporite
flotation
glacial outwash
gravity separation
high-grade deposit
hydraulic mining
hydrothermal deposit

lode
low-grade deposit
manganese nodule
massive sulfide deposit
mineral deposit
mineral reserves
mountain-top mining
open-pit mining
ore
ore mineral
pegmatite
placer deposit

porphyry copper
precipitation
regional metamorphism
secondary enrichment
SMCRA
smelter
spoil
strip-mining
tailings
vein deposit
volcanogenic deposit

◆ Study Questions

1. Distinguish between mineral deposits, ores, and reserves.
2. Why do some metals, such as gold, require concentration factors in the thousands, while others, such as iron, require only single-digit concentration factors?
3. Of which metals does the United States have ample reserves? How do we obtain the scarce mineral resources we need but lack within our borders?
4. How do you expect domestic and foreign supplies of critical mineral resources to change in the next 25 years?
5. What would cause currently marginal or submarginal mineral deposits to become important mineral reserves?
6. Where are the major deposits of porphyry copper and porphyry copper-molybdenum deposits located? How do these deposits relate to plate tectonic theory?
7. What steps are involved in the origin of a placer gold deposit? Of a bench placer deposit?
8. What minerals may eventually be harvested from the deep-ocean floor? What might be the environmental consequences of such mining?
9. Explain why you wouldn't go to the Hawaiian Islands to prospect for gold.
10. What is hydrothermal activity? What metallic mineral deposits are commonly associated with hydrothermal action?
11. Contrast the gold mining and extraction techniques of the nineteenth century with those of today. What are the environmental legacies of the nineteenth-century mining methods? What environmental concerns accompany modern gold-extraction technology?
12. Contrast the environmental hazards and mitigation procedures of underground, open-pit, and surface mining.
13. Why are mercury and cyanide, very dangerous poisons if they are mishandled, used in gold extraction?

14. Briefly describe the cyanide heap-leach gold-extraction process.
15. It has been suggested that as society exhausts the earth's reserves of critical minerals, lower-grade mineral deposits will be mined and supply-and-demand economics will dictate prices and costs. This rationale could be extended to a scenario in which average rock would eventually be mined for critical materials. What is the fallacy of such a rationale?

◆ For Further Information

BOOKS AND PERIODICALS

Baum, Dan, and Margaret L. Knox. 1992. In Butte, Montana, A is for arsenic, Z is for zinc. *Smithsonian* 23, no. 8:46–56.

Craig, James R., David J. Vaughan; and Brian J. Skinner. 1988. *Resources of the earth*. Englewood Cliffs, N.J.: Prentice-Hall.

Cutter, D. C. 1948. The discovery of gold in California. In Jenkins, O. P., ed. Geologic guidebook along Highway 49: Sierran gold belt, the Mother Lode country. *California Division of Mines bulletin* 141, (centennial ed.): 13–17.

Drew, Lawrence, and Brian K. Fowler. 1999. The megaquarry: A conversation on the state of the aggregate industry. *Geotimes* 44, no. 6:17–22.

Dunn, James; Carol Russell; and Art Morrissey. 1999. Remediating historic mine sites in Colorado. *Mining engineering* 51, no. 8:32–35.

Henning, Robert A., ed. 1982. Alaska's oil/gas and minerals industry. *Alaska geographic* 9, no. 4:216.

James, Patrick M. 1999. The miner and sustainable development. *Mining engineering* 51, no. 6:89–92.

King, Trude V. V., ed. 1995. *Environmental considerations of active and abandoned mine lands: Lessons from Summitville, Colorado*. U.S. Geological Survey bulletin 2220. Washington, D.C.: Government Printing Office.

Muncy, Jack A. 1991. *A plan for cooperatively completing revegetation of Tennessee's Copper Basin by the year 2000*. Technical note TVA/LR/NRM-91/3. Norris, Tenn.: Tennessee Valley Authority.

Newmont Gold Company. 1995. *Newmont*. Newmont Gold Company, P. O. Box 669, Carlin, NV 89822-0669.

Office of Surface Mining Reclamation and Enforcement, Department of the Interior. 1992. *Surface coal mining reclamation: 15 years of progress, 1977–1992, Part 1*. Washington, D.C.: Government Printing Office.

Quinn, M. L. 1991. The Appalachian Mountains' Copper Basin and the concept of environmental susceptibility. *Environmental management* 15, no. 2:179–194.

Silva, Michael A. 1988. Cyanide heap leaching in California. *California geology* 41, no. 7:147–156.

Stewart, K. C., and R. C. Severson, eds. 1994. *Guidebook on the geology, history, and surface-water contamination and remediation in the area from Denver to Idaho Springs, Colorado*. U.S. Geological Survey circular 1097. Washington, D.C.: Government Printing Office.

Wheeler, Gregory R. 1999. Teaching mineral resources as an aid to understanding international policy issues. *Journal of geoscience education* 47:464–468.

VIDEOS

Hydraulic gold mining. California Department of Parks and Recreation, Publications Section, P.O. Box 942896, Sacramento, CA 94296-0001.

The magic of Homestake gold from the discovery of gold to mining today. Homestake Mining Tours, Lead, SD 57754

Mining: Discoveries for progress. American Mining Congress, Communications and Education Department, 1920 N Street, NW, Suite 300, Washington, DC 20036-1662.

Poison in the Rockies. 1992. NOVA. WGBH, P. O. Box 2284, South Burlington, VT 05407-2284; 800-255-9424.

Public lands, private profits. 1994. Overview of the controversy surrounding the Mining Law of 1872. Frontline, 24 May. PBS Video, 1320 Braddock Place, Alexandria, VA 22314; 800-328-PBS1.

◆ World Wide Web Activities

Museum Scientists Collect a Massive Sulfide Chimney from the Ocean Floor
http://www.amnhonline.org/expeditions/blacksmokers/black_smokers.html
This American Museum of Natural History site describes an expedition to collect a "black smoker" chimney from the seafloor. These massive sulfide deposits are thought to model the origin of chemically similar ore deposits now on the continents.

1. Besides the production of massive sulfide ore deposits, what other aspects of black smokers are interesting to scientists?

2. Link to **black smoker sulfide chimney.** Where are the sulfide-rich waters coming from, and what forces their precipitation on the chimneys? Link to **Video and Image Gallery** for video images of black smokers in action.

Formation and Production of Borates in the United States
http://borax.com/nonflashsite/index.html
U.S. Borax maintains this master page and links regarding the mining, refining, and uses of borate minerals.

1. Link to **Mining.** Where is the largest U.S. deposit of borates? In what geological environment (igneous,

metamorphic, or sedimentary) were these salts crystallized? What is the specific environment of deposition for the South American (and Mojave) deposits?

2. Link to **Products and Applications.** What are boron salts used for?

High-Tech Approaches to Identifying and Cleaning Up Superfund Sites
http://borworld.usbr.gov/geo/calglch1.htm
This site summarizes the application of AVIRIS technology to the California Gulch Superfund clean-up site at Leadville, Colorado, where heavy metal contamination and acid mine drainage released from 130 years of hardrock mining are polluting downstream waterways.

1. What are the bedrock units mined here, and how and when did they become mineralized with metallic ore deposits?

2. Why is AVIRIS technology so important in the efforts to identify and mitigate hazardous waste sites like this one?

USGS Gateway to Mineral Resources in the United States
http://mrdata.usgs.gov/av/usa1.html
This United States Geological Survey metasite provides many layered databases relating to U.S. mineral resources.

1. To see relative mineral productivity by state, select these boxes: **USA** and **Value of Mineral Production.** What are the top four mineral-producing states in the coterminous United States?

2. Use the zoom tool to focus on your state. Select **Active Mines and Mineral Plants** and **Cities** in the left-hand window. Locate major mining operations in your state by nearest city.

 # Earth Systems Today CD Correlations

CHAPTER CONCEPT: Many Massive Sulfide Deposits Are Products of "Black Smoker" Hydrothermal Vents Found on Spreading Ridges
In this part of the CD, plate boundary types are delineated on a base map and geologic processes associated with them are explained.

SECTION MODULE: Plate Tectonics: Divergent Boundaries
Observe the development of a diverging ridge.

QUESTION: *If the black smoker metal-rich fluids are sea water thermally circulated through ocean floor basalts, which stage of seafloor development—1 (rifting), 2 (narrow ocean), or 3 (mature ocean)—would likely be most productive for these massive sulfide deposits?*

CHAPTER CONCEPT: Ore Deposits Originate in All Three Rock Families
The rock cycle segment of the CD explains conditions of formation for igneous, metamorphic, and sedimentary rocks.

SECTION MODULE: The Rock Cycle and Rock Laboratory
Observe the physical conditions (pressure, temperature) and chemical changes that take place as earth materials are cycled from one form to the next.

QUESTION: *Borates and other mined salts are what specific category of sedimentary rock? Gold veins in a quartz dike would belong to what subset of the igneous family? The dimension stone marble belongs to which rock family?*

Energy Resources

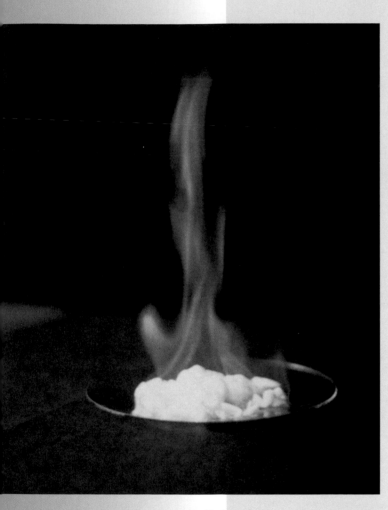

Flaming Ice, Energy, and the Greenhouse!

Setting ice balls on fire is U.S. Geological Survey oceanographer Bill Dillon's idea of fun, and it never fails to amaze his visitors. It is not ordinary ice that flames. Dillon uses hydrated methane (CH_4) recovered from sea-floor sediments deep below the ocean surface. This flame has become the focus of the energy-hungry nations of the world (◆ Figure 1). What actually burns is methane gas released from a crystalline solid of methane molecules encased in a "cage" of ice (◆ Figure 2). Methane is the gas that constitutes the bulk of the "natural gas" that is piped to your home. Methane hydrates are so extensive in deep-sea sediments that they have the potential to supply a share of the world's future fuel needs, *when and if* practical mining techniques can be developed. India and Japan, at present energy poor, are most interested in developing methane hydrates as an energy source. Experts agree that marine gas hydrates harbor twice the carbon contained in all known natural gas, petroleum, and coal deposits on earth—about 10,000 gigatons, or 10^{13} tons, of carbon. Extensive gas-hydrate deposits are currently known in hundreds of localities. Particularly interesting for potential development are those of offshore Oregon, Costa Rica, New Jersey, Japan, and Florida (◆ Figure 3). Also intriguing are the significant deposits that have been encountered in oil drilling through Arctic permafrost areas in Canada, Alaska, and Siberia (see Chapter 6).

Gas hydrates in marine sediments have been known for several decades, but their full extent and potential were not realized until the 1990s (◆ Figure 4). Conditions necessary for their formation are high water pressure (water depth >530 m in low latitudes and >250 m in high latitudes) and low ambient temperatures (<7° C) at the water–sediment interface. These conditions are met on the sea floor's continental slope and rise. The exact origin of the gas is not

FIGURE 2 Crystalline structure of methane hydrate. Methane gas (CH_4) generated by bacterial digestion of organic matter in sea-floor sediments is sometimes trapped in "cages" of frozen H_2O (the large polygons) to form methane hydrate.

FIGURE 3 Location of the world's known methane hydrate deposits, 1999. Dots are marine deposits; squares are permafrost deposits.

known, but it is strongly suspected that bottom bacteria consume organic-rich detritus and generate methane as a waste product. Under particular conditions, the gas becomes trapped in ice in the sediment instead of being released to the overlying water column.

The bad news is that these hydrated molecular structures can release their caged methane into the atmosphere if there is a slight decrease in pressure (if sea level is lowered, for instance) or a slight warming of the surrounding sea water. Because methane is a powerful "greenhouse" gas linked to global warming (it is about 10 times as strong as carbon dioxide), melting of the hydrates and release of methane could have a drastic impact on the earth's surface temperature (see Chapter 11). One model suggests that as sea levels rose toward the end of the last ice age, the waters covered land that was underlain by permafrost containing the methane hydrates. As the warmer water temperature melted the permafrost ice, the caged greenhouse gas was released, increasing the global warming and the melting of glaciers.

There is also a potential for gas hydrates to be a geologic hazard. Hydrates may occupy much of the pore space in bottom sediments. When the hydrate-bearing sediment becomes deeply buried, the temperature may increase to the point at which gas hydrates are no longer stable. When a solid–gas mixture becomes a liquid–gas mixture, high pore pressure and unstable slopes result (see Chapter 7). In other words, the gas hydrate dissociates at depth and fluidizes a layer of sediment at the base of the methane hydrate zone. This presents a danger to offshore drilling platforms, pipelines, communication cables, and other offshore engineering structures.

In conclusion, current interest in methane hydrate is threefold: as an energy source, as a factor in global climate change, and as a geologic hazard. Experts agree that, at present, methane hydrate is a hazard to human welfare as the search for offshore oil goes into deeper waters and as work there requires pipelines and engineered infrastructures.

FIGURE 4 Methane hydrate nodule in marine sediment from the Blake Ridge off the Florida coast.

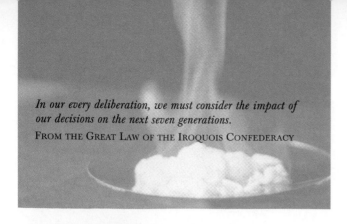

From the simplest pond scum to the most complex ecosystem, energy is essential to all life. Derived from the Greek word *energia* meaning "in work," *energy* is defined as the capacity to do work. The units of energy are the same as those for work, and the energy of a system is diminished only by the amount of work that it does.

Prosperity and quality of life in an industrialized society such as ours depend in large part on the society's energy resources and its ability to use them productively. We may illustrate this in a semiquantitative fashion with the equation:

$$L = \frac{R + E + I}{\text{population}}$$

where L represents quality of life, or "standard of living," R represents the raw materials that are consumed, E represents the energy that is consumed, and I represents an intangible we shall call *ingenuity*. As the equation expresses, when high levels of raw materials, energy, and ingenuity are shared by a small population, a high material quality of life results. If, on the other hand, a large population must share low levels of resources and energy, a low standard of living would be expected. Some highly ingenious societies with few natural resources and little energy can and do enjoy a high quality of life. Japan is a prime example. Some other countries that are self-sufficient in resources and energy—such as Argentina—are having difficult times. Thus we see that ingenuity, which is reflected in a country's political system, technologies, skills, and education, is heavily weighted in the equation and it can cancel out a lack of resources and a large population. ◆ Figure 14.1 shows annual world and U.S. energy consumption between 1925 and 1985.

To a physicist there is no energy shortage, because she or he knows that energy is neither created nor destroyed; it is simply converted from one form to another, such as from nuclear energy to heat energy. Fuels of all kinds are warehouses of energy, which can be tapped by some means and applied in some way to do work. Coal and oil, for example, are fossil fuels that have been storing solar energy in the lithosphere for millions of years.

Some forms of energy are *renewable;* that is, they are replenished at a rate equal to or greater than the rate at which they are used. Examples include solar, water, wood, wind, ocean and lake thermal gradients, geothermal, and tidal energy. The energy in all of these resources except for geothermal and tidal (gravitational) energy was originally derived from the sun. Renewable resources are dependable only if they are consumed at a rate less than or equal to their rate of renewal. If they are overexploited, some period of time will be required to replenish them. If geothermal heat is overexploited, for example, subsurface rocks or magma will cool to temperatures below the temperature at which steam is produced, and they will be reheated only after some time has passed. Peat, a fuel used extensively for space heating and cooking in Ireland and Russia, is estimated to accumulate at a remarkable 3 metric tons per hectare per year (1.3 tons/acre/year). Nonetheless, the conversion from plant litter to peat may take a hundred years. Wood energy may renew in a matter of a few decades, and water and wind are renewed continuously.

◆▶ FIGURE 14.1 World energy consumption, 1925–1995. Consumption increased more than 500 percent between 1925 and 1985, and increased 300 percent between 1950 and 1985. The increase beginning about 1950 was mostly in developing countries. The United States is the world's largest consumer of energy, and its rate of consumption has increased very little between the early 1970s and 1995.

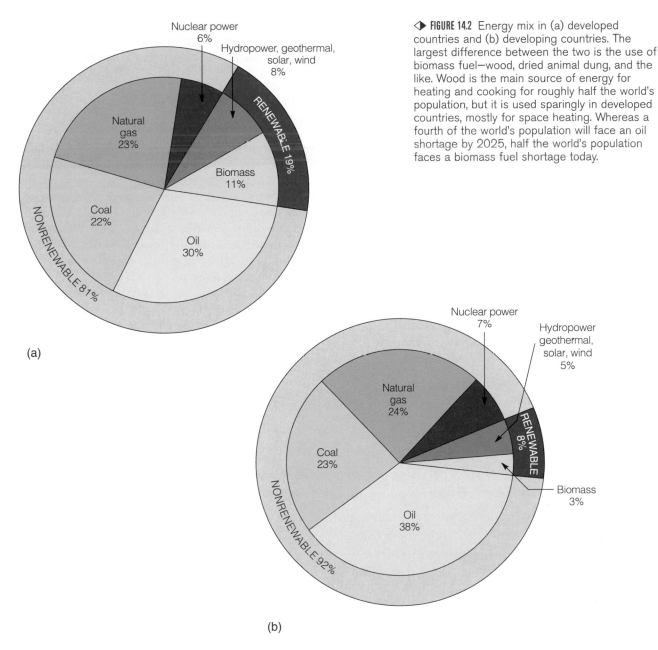

(a)

(b)

◆► **FIGURE 14.2** Energy mix in (a) developed countries and (b) developing countries. The largest difference between the two is the use of biomass fuel—wood, dried animal dung, and the like. Wood is the main source of energy for heating and cooking for roughly half the world's population, but it is used sparingly in developed countries, mostly for space heating. Whereas a fourth of the world's population will face an oil shortage by 2025, half the world's population faces a biomass fuel shortage today.

Nonrenewable resources are not replenished as fast as they are utilized, and once consumed, they are gone forever. Crude oil, oil shales, tar sands, coal, and fissionable elements are nonrenewable energy resources. Their quantities are finite. Supplies of crude oil, for example, are within a few human generations of exhaustion. The problem is that, owing to the use of gasoline-powered engines for transportation, demand far exceeds replenishment. Oil underground was discovered almost fifty years before the first automobile was operational. Prior to then, gasoline was a minor byproduct of refining oil for kerosene lamps. In 1885 Germans Gottlieb Daimler and Karl Benz independently developed gasoline engines, and in 1893 Massachusetts bicycle makers Charles and Frank Duryea

built the first successful U.S. gasoline-powered automobile. By the turn of the century, production automobiles were hitting the roads, horses were being put out to pasture, and refineries were stepping up production to satisfy an increasing demand for gasoline. The transportation revolution to automobiles spawned the largest private enterprise on earth: the exploration and production of petroleum. Today this industry employs half the world's geologists and hundreds of thousands of engineers, technicians, and managers. More than half the developed world's energy needs are met by oil and gas (◆► Figure 14.2). This has led to major geopolitical and economic problems, because the nations that consume the most do not produce in comparable amounts.

Petroleum

Although the carbon content of the earth's crust is less than 0.1 percent by weight, carbon is one of the most important elements to the earth's people. It is indispensable to life, and it is the principal source of energy and the principal raw material of many manufactured products. Crude oil, or petroleum (Latin *petra,* "rock," and *oleum,* "oil"), is composed of many **hydrocarbon compounds,** simple and complex combinations of hydrogen and carbon (⊙ Table 14.1). Petroleum occurs beneath the earth's surface in liquid and gaseous forms and at the surface as oil seeps, tar sands, solid bitumen (gilsonite), and oil shales. In addition to its use as a fuel, hydrocarbon compounds derived from petroleum are used in producing paints, plastics, fertilizers, insecticides, soaps, synthetic fibers (nylon and acrylics, for example), and synthetic rubber. Carbon combines chemically with itself and with hydrogen in an infinite variety of bonding schemes; some two million hydro-carbon molecules have been identified to date. The manufacturing process of separating crude oil into its various components is known as *refining,* or *cracking.*

ORIGIN AND ACCUMULATION OF HYDROCARBON DEPOSITS

Carbon and hydrogen did not combine directly to form petroleum; they were chemical components of living organisms before their transformation to complex hydrocarbons in crude oil. Porphyrin compounds found in petroleum are derived either from chlorophyll, the green coloring in plants, or from hemin, the red coloring matter in blood, and their presence is solid evidence for an organic origin for crude oil. The fact that large quantities of oil are not found in igneous or metamorphic rocks also rules out an inorganic source for oil.

Four conditions are necessary for the formation and accumulation of an exploitable petroleum deposit in nature:

⊙ TABLE 14.1 **Hydrocarbon Compounds Typically Found in Crude Oil**

MOLECULAR TYPE		HYDROCARBON COMPOUND		PERCENTAGE OF WEIGHT IN MEDIUM-GRADE CRUDE OIL
Name	*General Formula*			
Paraffins	C_nH_{2n+2}	methane, CH_4	lighter	25
		ethane, C_2H_6	↑	
		propane, C_3H_8		
		butane, C_4H_{10}*		
		pentane, C_5H_{12}**		
Aromatics	C_nH_{2n-6}***	benzene, C_6H_6	↓	17
Naphthenes	C_nH_{2n}	asphalt		50
Asphaltenes	(solid hydrocarbons)	gilsonite	heavier	8
				100

* Butane gives gasoline quick-starting capability.
** Pentane gives smooth engine warm-up.
*** Aromatics improve mileage and "knock" resistance.

- a source rock for oil,
- a reservoir rock in which it can be stored,
- a caprock to confine it, and
- a geologic structure or favorable strata to "trap" the oil.

Even where these geologic conditions are met, the human elements of exploration, location, and discovery still remain. Oil companies utilize the skills of geologists to interpret surface and subsurface geology, to locate the potential oil-bearing structure or stratum, and to specify the optimal location for a discovery well. Until a well is drilled, the geologists' interpretation remains in doubt, much like a medical doctor's diagnosis of an ailment subject to surgery; the diagnosis is tentative until the patient is opened up.

A **source rock** is any volume of rock that is capable of generating and expelling commercial quantities of oil or gas. Source rocks are sedimentary rocks, mostly shales or limestones, usually of marine (ocean) origin and sometimes of *lacustrine* (lake) origin. The biological productivity (biomass) of surface waters must have been high enough to generate a settling "rain" of dead organisms, and the bottom waters must have been low enough in oxygen to prevent the deposited organic matter from being oxidized or consumed by scavengers. Almost all source rocks are dark-colored, indicative of high organic-matter content, and some of them carry a fetid or rotten-egg odor.

Favorable marine environments are rich in microscopic single-celled plants known as *diatoms* (phytoplankton), which form the largest biomass in the sea. Where there are diatoms, we also find zooplankton—tiny protozoans and larvae of large animals. These together with diatoms provide the molecules that make up crude oil. As the rocks are buried, they are heated. The conversion from organic matter to petroleum takes place mostly between 50° C and 200° C. Thus a proper thermal history, ideally between 100° C and 120° C, is necessary to form liquid petroleum—too cool, and oil does not form; too hot, and the hydrocarbons "boil" away.

After petroleum has formed in a source bed, it is squeezed out and migrates through a simple or complex plumbing system into a **reservoir rock.** This migration is a critical element in the formation of an economically exploitable accumulation of oil. Reservoir rocks are porous and permeable (see Chapter 8). Commonly they are sandstones, porous limestones, or in some cases, fractured shales. Reservoir porosities range from 20 to 50 percent, meaning that for each cubic foot of reservoir rock we will find 1 to 4 gallons of oil. The unit of oil volume is the barrel (equal to 42 gallons), and a so-called giant field, such as the north slope of

Alaska, will yield a billion barrels of oil in its lifetime. Using standard recovery techniques, however, as much as 40 to 80 percent of the oil may be left in pore spaces and as films on mineral grains.

An impermeable **caprock** prevents oil from seeping upward to form tar pits at the surface or dissipate into other rocks. Such seals are analogous to aquicludes in ground-water systems and are most commonly clay shales or limestones of low permeability. Our discussion of the fourth requirement, a suitable geologic structure for trapping the oil, requires a separate subsection.

GEOLOGIC TRAPS—OIL AND GAS STOP HERE
Structural Traps

An **anticline** is an ideal structure for trapping gas and oil. It is a convex-upward fold in stratified rock (◆ Figure 14.3a). Analogous to a teacup inverted in a pan of water that traps a layer of air inside it, an anticline holds a reservoir of gas and oil. This occurs because crude oil floats on water and natural gas rises to the top of the reservoir. The anticlinal theory of oil accumulation was not developed until 1900, forty-one years after oil was discovered. Strata dip away from the central axis of an anticline at the ground surface, and most anticlines with surface geologic expression have been drilled. Today the search for oil is more difficult, because less obvious and geologically more complex traps need to be discovered. Many times faults form impermeable barriers to hydrocarbon migration, and oil becomes trapped against them (see Figure 14.3a). Faulted and folded stratified rocks in a single oil field may thus contain many isolated oil reservoirs.

Salt domes are of much interest to geologists (◆ Figure 14.3b). Not only do they create oil traps, but they are valuable sources of salt and sulfur and may be potential underground storage sites for petroleum and hazardous waste. More than 500 salt domes have been located along the U.S. Gulf Coast, both offshore and on land (◆ Figure 14.4). They rise as flowing fingers of salt, literally puncturing their way through the overlying rocks and buckling the overlying shallow strata into a dome (◆ Figure 14.5). Some fingers rise as high as 13 kilometers (8 mi) above the "mother" salt bed, the Louann salt, and would be taller than Mount Everest if they were at the surface of the earth. Because the salt is considerably less dense than the overlying rock, buoyancy forces drive the salt upward—much as a blob of oil will rise through water. Brittle solids, such as salt or ice, will flow over long periods of time, and many salt domes are still rising measurably. The "mother" salt

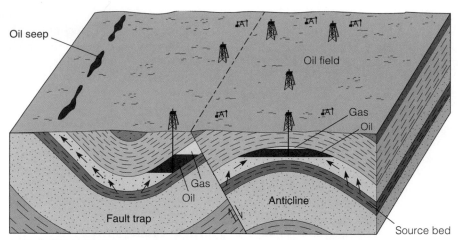

(a) Anticline and fault trap

(b) Salt dome

(c) Stratigraphic traps

◆▶ **FIGURE 14.3** (a and b) Common structural oil traps. (c) Stratigraphic oil traps.

FIGURE 14.5 Ancient salt deposits are buried deeply, and because salt is less dense than the overlying sediments, it rises buoyantly as pillars of salt, creating salt domes.

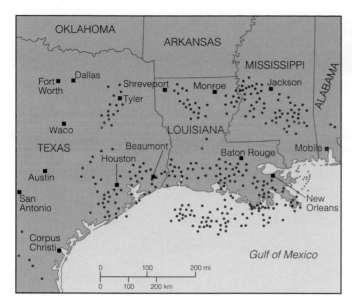

FIGURE 14.4 Locations of salt domes of the U.S. Gulf Coast. More than 500 domes have been discovered on land and in the shallow parts of the Gulf of Mexico, and more are known to be in deep water offshore.

was deposited in Jurassic time by evaporation of seawater when the embryonic Gulf Coast was connected to the open ocean by a shallow opening. The opening allowed seawater to enter the basin but did not allow the denser salt brines at the bottom to escape. Thus the brine became concentrated to the point of saturation, and the great thickness of salt that now underlies the entire Gulf Coast was deposited. Oil accumulates against the salt column and also in the dome overlying the salt. "Old Spindletop," a salt dome near Beaumont, Texas, that was discovered in 1901, is perhaps the most famous U.S. oil field. Soon after its discovery it was producing more oil than the rest of the world combined, and the price of oil plummeted to two cents a barrel.

Stratigraphic Traps

Any change in sedimentary rock lithology (its physical character) that causes oil to accumulate is known as a **stratigraphic trap** (in contrast to anticlines and salt domes, which are structural traps). Thus, if a stratum changes laterally from a permeable sandstone to an impermeable shale or mudstone, oil may be trapped in the stratum (Figure 14.3c).

Ancient coral reefs are ideal reservoirs, because they are porous and were biologically productive when they were living. Oil may be trapped in the porous, permeable debris on the flanks of the reef, and production from such fields can be measured in thousands of barrels per day (bbl/d). Oil fields of the Middle East are of this type, and their production potential is

tremendous. A comparison of reef production to that of sandstone reservoirs such as those of California or Texas, which typically yield only a few hundred barrels per day, explains why the Middle East can control oil production and therefore price.

OIL PRODUCTION

The first successful oil well was drilled in Titusville, Pennsylvania, in 1859 (Case Study 14.1). In modern jargon, this well would be called a **wildcat well,** because it was the discovery well of a new field. There is 1 chance in 50 of a wildcat well being successful, less favorable odds than those of winning at roulette. The probability improves when we consider all wells drilled, including those in known oil fields. Between 1970 and 1999 1 well in every 4 drilled was successful. Independent oil entrepreneurs take considerable risks; it is not a business for the faint of heart.

Most successful wells require pumping. If gas and water pressure are sufficient, however, oil may simply flow to the ground surface. High reservoir pressures develop from water pressing upward (buoyancy) beneath the oil and gas pressure pushing downward on the oil (Figure 14.6). In some cases dissolved gases "drag" the oil along with them as they spew forth, as though from a bottle of champagne. If high pressures are not controlled, rocks, gas, oil, and even drill pipe may shoot into the air as a "gusher." High-pressure wells— 749 in all—were set afire by Iraqi soldiers in Kuwait during Operation Desert Storm (Figure 14.7). Professional oil-fire-fighting firms from all over the world were summoned to squelch the fires, among them the well-known Red Adair Company and others with such interesting names as Boots and Coots and Joe Bowden's Wild Well. With the help of Kuwaiti *roughnecks* (the nickname applied to drilling-rig workers), the fires

Consider This...? An acquaintance has offered you a share in a sure-shot wildcat oil-drilling venture in the Sierra Nevada. How should you respond to the offer? ■

Case Study | 14.1

Rock Oil and the Colonel

The history of the petroleum industry is full of interesting characters, and probably none was more charismatic than the self-styled Colonel of Titusville, Pennsylvania, Edwin L. Drake (◁▷ Figure 1). In the 1840s the demand for whale oil and lard had outstripped the supply and was being supplemented by "rock oil" from seeps and tar pits. The oil on northwest Pennsylvania's Oil Creek, so-named because oil bubbled to its surface, had long been skimmed by the Seneca Indians and other Native Americans and used to waterproof their homes and boats. Later, Europeans skimmed it for use as lamp oil, paraffin, liniments, and lubrication.

Drake was a railway conductor in New Haven, Connecticut, who had an interest in petroleum exploration but no experience. There are several versions of why Drake went to Titusville. The most credible one is that he had bought some shares in the Seneca Oil Company in New Haven and was sent by one of its founders, a friend of his, to stimulate the seepage at Oil Creek, which the company had leased from the Pennsylvania Rock Oil Company. Drake had a "Lincolnesque" appearance and he frequently wore a knee-length topcoat and stovepipe hat. The company bestowed upon him the honorary title of *Colonel* to enhance his credibility in Pennsylvania.

Colonel Drake tried digging trenches to stimulate the flow, and he installed a series of skimmers to collect the oil. The process was so inefficient that he persuaded the company to lease the land and attempt drilling. Drake enlisted the help of local "salt borers," who drilled for brine from which salt was extracted. Often these drillers had struck oil instead of water and they had been run out of town because of the terrible mess it created. The Colonel finally found an experienced driller named Uncle Billy Smith, who used percussion tools featuring a chisel on a cable to chip the rock and a bailer to remove the cuttings. Trouble with cave-ins ensued, and Drake ingeniously drove a large-diameter pipe into the caving zone and drilled the hole through it. On Friday, 27 August 1859, they struck oil, although they didn't know it until they checked the well on Sunday. A black goo had filled the 21-meter (69-ft)-deep hole, and the world had its first successful oil well.

Drake's well produced 10–35 barrels per day, almost doubling the previous world output of rock oil. The oil price jumped to $20/bbl, and a thousand oil pits sprang up around Titusville. Within a year the price had dropped to $0.20/bbl, and the Seneca Oil Company sold the Drake Well. Nearby towns such as Pithole City flourished for a few years and then became ghost towns. At Oil City a few miles south of Titusville, a Cleveland hay and grain dealer named John D. Rockefeller built the first commercial oil refinery in 1863. Seven years later he established the Standard (now Exxon/Mobil) Oil Company.

Because Edwin Drake had filed no patents and established no lease claims, he did not benefit from his vision and labor. He later lost his money speculating on oil stocks and became impoverished. He was forced to live on the generosity of friends. Learning of his dire need in 1873, Titusville citizens took up a collection for him. Three years later the Pennsylvania legislature voted to give him an annual income of $1,500. Drake was buried in Bethlehem, Pennsylvania, in 1880, but his body was reinterred in Titusville in 1901, ironically, the same year that a gusher in Texas, "Old Spindletop," drastically changed the oil supply of the world.

Today, Oil Creek between Titusville and Oil City is the site of the Oil Creek State Park. The ghost town of Pithole is just east of the park, and Drake Well Memorial Park is at the park's northern boundary.

◁▷ **FIGURE 1** Colonel Edwin L. Drake (right) in front of his well in 1861 with his friend, Titusville druggist Peter Wilson. The men in the background are probably drillers.

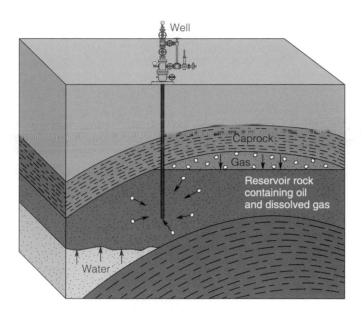

◆ FIGURE 14.6 Oil in an anticline is driven by gas pressure from above and by buoyant water pressure from below.

were extinguished in nine months, three months earlier than the most optimistic estimate.

It is possible to drill a well so that the drill hole slants from the vertical to penetrate reservoir rocks far from the drilling site. This is desirable where the oil structure is offshore and must be "slant-drilled" from land or from a drilling platform. It is also employed for tapping reservoirs beneath developed land, such as in Beverly Hills, where a large number of oil wells are slant-drilled under fashionable commercial areas. Modern methods use "smart" drill bits that are remotely controlled. Producers can thus withdraw oil from below a large area more economically and at the same time minimize the visual blight of drilling towers.

The ultimate drilling technology of the 1990s was horizontal drilling, a technique by which the drill bit and pipe follow gently inclined reservoir sands and limestones. By this means a single borehole can provide access to reservoirs beneath a much larger area than provided by vertical drilling or slant drilling. This method was superceded in the late 1990s by multilateral drilling, which enables operators to reach multiple oil-bearing reservoirs by means of lateral extensions from one well (◆ Figure 14.8). It is worth noting that some offshore drilling platforms are in water more than a mile deep and have strings of drill pipe dangling below them 8.5 kilometers (5 mi), about the height of Mount Everest. There are plans to drill exploratory holes in water depths greater than 3,100 meters (10,000 ft).

Secondary Recovery

Secondary recovery methods extract oil that remains in the reservoir rock after normal withdrawal methods have ceased to be productive. As much as 75 percent of the total oil may remain. Secondary recovery methods can be grouped into three categories: thermal, chemical, and fluid-mixing (miscible) methods (◆ Figure 14.9). All of these methods require *injection* wells for injecting a fluid or gas and *extraction* wells for removing the remobilized oil.

Thermal methods include steam injection, which makes the adhering oil less viscous and thus more free to flow, and fire flooding—in which air is injected into the reservoir in order to set fire to the oil and thus produce gases and heat that will increase the flow of oil. Water injection is a chemical method that utilizes large-molecule compounds, which when added to water, thicken it and increase its ability to wash or sweep the adhering oil films and globules toward an extraction

◆ FIGURE 14.7 The Kuwaiti night skies were illuminated by 749 blazing oil wells set afire by Iraqi troops during the 1991 Gulf War. The fires continued for nine months.

◆ **FIGURE 14.8** Multilateral drilling from an offshore drilling platform allows many oil-producing zones to be tapped from one platform. In this illustration three zones are tapped by lateral pipes, and the bottom zone is tapped by a horizontal pipe.

◆ **FIGURE 14.9** Secondary recovery. Steam, air, carbon dioxide, or chemicals dissolved in water are injected into a sluggishly producing formation in order to stimulate the flow of oil to extraction wells.

well. Fluids that mix with oil, that are **miscible,** are very effective in removing "stuck" oil from the reservoir. Miscible recovery methods use mixtures of water with propane or ethane extracted from natural gas or mixtures of CO_2 and water. Even after the use of these secondary recovery methods, as much as a fifth of the total oil may remain in the reservoir.

QUALITY AND PRICE

The price of a barrel of crude (unrefined) oil varies with its grade—its quality—and with market demand at the time. Light oils bring higher prices, because they contain large proportions of paraffins and aromatic hydrocarbons, which are desirable for gasoline and bottled gases. Heavy crudes contain lower proportions of those components and greater proportions of the heavy, less valuable asphalts and tars (see Table 14.1). A scale of crude-oil quality has been established by the American Petroleum Institute based upon its weight. Light crude oil is very fluid and yields a high percentage of gasoline and diesel fuel. Heavy crude, on the other hand, is about the consistency of molasses. The percentages of fuels and lubricating oils yielded by a barrel of medium-weight crude oil are shown in ◉ Table 14.2.

THE FUTURE

Our way of life is possible only if we have abundant energy. In the case of nonrenewable energy forms, these supplies are referred to as **reserves,** the amount of an identified resource that can be extracted *economically*. A general awareness of this began in 1980, when the U.S.

Geological Survey (USGS) predicted that the world's known oil reserves would be exhausted by the year 2006 if oil continued to be consumed at the 1980 consumption rate. In 1989, the USGS advanced the projected year of "the last drop of oil on earth" to 2049 and projected a "lifetime" of 120 years for natural gas reserves at 1988 consumption rates. ◆ Figure 14.10 summarizes the amounts of oil reserves by country in 1990. Oil and gas exploration and discovery are dynamic enterprises, however, and much can happen in a short period of time. In 1995 the amount of technically recoverable U.S. oil—that is, the sum of proven reserves, inferred reserves in existing fields, and undiscovered accumulations—had grown to 110 billion

◉ TABLE 14.2	**Typical Composition of an A.P.I. Medium-Grade Crude Oil**	
COMMON NAME	**NUMBER OF CARBON ATOMS***	**PERCENTAGE OF CRUDE-OIL WEIGHT**
Gasoline	5–10	27
Kerosene	11–13	13
Diesel fuel	14–18	12
Heavy gas oil	19–25	10
Lubricating oil	26–40	20
Heavy fractions	>40	18
		100

* Volatility decreases as the number of carbon atoms increases, C_3 to C_{40}.

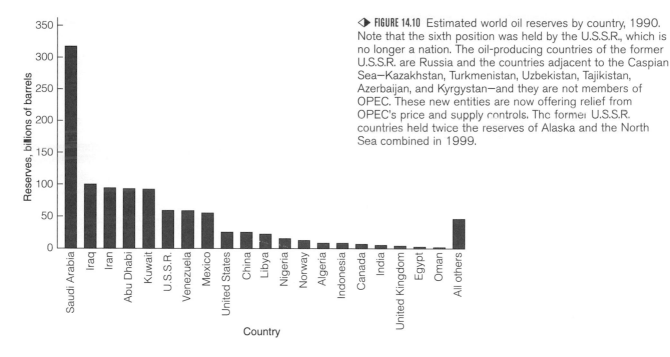

◆ FIGURE 14.10 Estimated world oil reserves by country, 1990. Note that the sixth position was held by the U.S.S.R., which is no longer a nation. The oil-producing countries of the former U.S.S.R. are Russia and the countries adjacent to the Caspian Sea—Kazakhstan, Turkmenistan, Uzbekistan, Tajikistan, Azerbaijan, and Kyrgystan—and they are not members of OPEC. These new entities are now offering relief from OPEC's price and supply controls. The former U.S.S.R. countries held twice the reserves of Alaska and the North Sea combined in 1999.

Consider This... ? Historically, Pennsylvania crude oil (such brands as Quaker State and Pennzoil) has been known for its superior lubricating properties compared to, say, heavier California or Texas crude oils. Why do you suppose this is so? (Table 14.1 can help you deduce a reasonable answer.) ■

barrels (BB), an increase of almost 41 percent over the similarly derived 1989 estimate of 78 BB. World reserves estimates were increased slightly in 1995 from those of 1989 to about 1.1 trillion barrels, yielding a lifetime of slightly more than 50 years. However, in an astounding 1995 report, USGS geologists estimated the world's "ultimate resources" of conventional oil at double that amount, 2.1 trillion barrels, of which 700 BB had been used by 1995. "The last drop" is now forecast to occur at the end of the twenty-first century (◆ Figure 14.11).

The Organization of Petroleum Exporting Countries (**OPEC**) includes all the major oil-producing countries of the equatorial climatic zone with the exception of Mexico. Other significant nonmembers are countries of the former Soviet Union. OPEC has the ability to control oil prices, as the oil "shock" of the 1970s revealed (◆ Figure 14.12). The United States and other nations responded to OPEC's price increases by instituting conservation measures and accelerating exploration—actions that eventually drove prices down from a high of $38/bbl to a low of $12/bbl. Persian Gulf nations have more than half of the world's known reserves, so it is likely they will

dominate oil production and marketing well into the twenty-first century.

The United States is the most energy-deficient of the 20 largest oil-producing countries. The difference between the U.S. consumption of 16 million bbl/d and its production of 8.4 million bbl/d represents its dependency on imported oil. This dependency has a major impact on the U.S. trade deficit and foreign policy. Another consequence of having to import oil is the risk of environmental damage due to oil spills. Tanker-transported oil is always subject to accidental spillage or catastrophic spills due to collisions or groundings (Case Study 14.2).

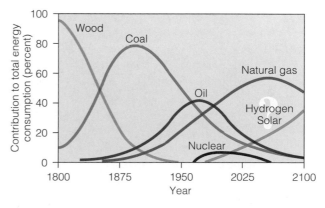

◆ FIGURE 14.11 Use of energy resources in the United States since 1800, with projections to 2100. The shifts from wood to coal and then from coal to oil and natural gas each took about 50 years. Many analysts believe a new shift, to increased use of solar energy and hydrogen gas, will occur over the next 50 years.

Case Study | 14.2

The Tragedy in Prince William Sound

Late in the evening of 24 March 1989, the supertanker *Exxon Valdez* left Valdez Fiord—in the area called the "Switzerland" of Alaska—and sailed into the designated outbound shipping lane on a southwest heading. The seas were calm and the winds light, and visibility was 10 miles. Everything was normal as the pilot left the ship outside the entrance to the fiord.

A short time later the officer in command radioed the Coast Guard, requesting permission to move over to the inbound lane so as to avoid a small iceberg that had calved off the nearby Columbia Glacier (see Chapter 11). The request was granted, and the ship altered its course 25° to port to change lanes. The ship continued on this more southerly heading for 30 minutes, sailing *past* the inbound lanes and into the shallow waters near Bligh Island (named by James Cook for Captain William Bligh of *Mutiny on the Bounty* fame). Realizing his error, the officer on the bridge gave urgent commands to turn starboard in order to return to the

shipping lanes, but it was too late. At 4 minutes past midnight, the *Exxon Valdez* drove up onto Bligh Reef, tearing a gash in her hull and coming to a stop balanced on a pinnacle of rock (◁▷ Figure 1).

Within the next two days, 10.1 million gallons of oil spilled into a bay teeming with marine mammals, fish, waterfowl, and eagles. Exxon and Alyeska Pipeline Service Company, which jointly operate the terminal at Valdez, responded to the emergency, but their efforts to mobilize containment booms were slowed because the booms were buried under snow. By the time the booms could be put into place to keep the oil from spreading and skimmers could begin removing the oil from the sea surface, much damage had already been done to the shoreline and its inhabitants (◁▷ Figure 2).

It was estimated that 1,600 otters and 37,000 marine birds died from the oil. Some biologists believe this bird count represents only 10–30 percent of the actual number killed, that the losses were 100,000–350,000 birds. About

◁▷ **FIGURE 1** Prince William Sound showing the tanker channel leaving Valdez. The *Exxon Valdez* went aground on Bligh Island and leaked millions of gallons of oil. Compare with Figure 3 to get an idea of the enormity of the spill.

150 bald eagles died after scavenging oil-covered corpses, and more than 80 percent of the area's nesting pairs failed to produce young that year, further affecting the eagle population. Compared to other oil-tanker spills, such as those of the *Amoco Cadiz* off Spain that killed 20,000 birds and the *Torrey Canyon* off England that killed 30,000 birds, the *Valdez* spill was a much larger catastrophe (◆ Figure 3).

Fish kill is very difficult to estimate. After a few weeks on the sea surface, the oil thickened and sank to the bottom, where it affected such fish as snapper and halibut. The impact on invertebrate marine populations and the ecological damage cannot be quantified. There is some evidence that the shoreline cleanup involving steam cleaning did as much damage to the environment as the oil spill did. It was learned in the California oil spill of 1969 that oil-eating bacteria do an efficient job removing oil from bottom sediments and rocks. Steam cleaning the shores of Prince William Sound killed the oil-consuming bacteria in the tidal zone, worsening the problem.

The lesson learned there is that oil spills of this magnitude are difficult if not impossible to contain. "Throwing money at it" will not change the results if weather and currents hamper the cleanup. As they say, "It doesn't do any good to close the barn door after the horse is out." In these days of satellite positioning systems, it would seem that restricted areas could easily be programmed into the navigation systems, and that audible warnings could be sounded when a vessel strays from the approved shipping lane. The double hulls on newer supertankers offer assurance that at least some groundings will not result in oil spills. The final cost to Exxon of this "mistake" will probably be in excess of $4 billion ($1.025 billion of that in fines). As long as humans are at the controls, however, incidents such as this one can be expected.

◆ FIGURE 3 Comparison of the area of the 24 March 1989 oil spill with the Eastern Seaboard.

◆ FIGURE 2 An oil-soaked bird is rescued and prepared for clean-up after the spill. Thousands of other marine birds were not so lucky.

Consider This... ? It has been forecast that the last drop of oil will not be pumped until some time beyond your life expectancy. In light of this, do you think we should go on consuming energy as usual, or should the world's peoples and their governments do something to delay the day when that last drop will be used? ■

The future is not so bleak when we factor U.S. ingenuity into the energy equation. Much oil remains to be discovered using current exploration techniques, although a crude-oil price of at least $25/barrel is required to make the expensive searches feasible. In addition, there are nonconventional sources of petroleum such as oil shale, tar sands, and coal from which liquid fuel can be derived. Coal is also a source of fuel-grade natural gas at present.

ENERGY GASES AND THE FUTURE

Some experts predict that the decline in the supply of *affordable* oil will begin as early as 2010, others say the crunch will not start until 2040 or 2050. It will happen, however, and the speculation about when the world will run completely out of oil is not relevant. What does matter is when production will fall off as demand inexorably continues to rise. In other words, price depends on the well-known supply and demand curves of the economists. Figure 14.11 reflects this. It shows the predicted decline in the use of oil in the twenty-first century and the increasing use of natural energy gases (methane and related combustible gases; see Table

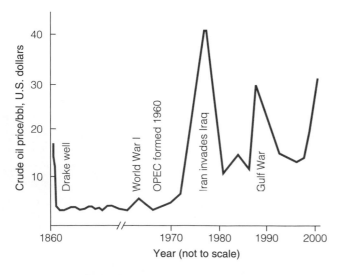

◆▶ FIGURE 14.12 Generalized history of crude-oil prices, since 1859. Since the founding of OPEC in 1960, prices have risen and have fluctuated with political stability and harmony in the Middle East.

14.1) and alternative energy sources. Natural gas is currently plentiful and burns more cleanly than coal, gasoline, or heating oil. It currently accounts for 75 percent of the total U.S. energy used in space heating and cooking. According to the U.S. Geological Survey:

> The issue, stripped of all the complexities that attend it, is this: Is it realistic to hope that increased use of energy gas in our energy mix will be the solution we need to fulfill our industrial needs and maintain our standard of living but also decrease the potential damaging effects of energy use on the natural environment?

Natural gas is measured as cubic feet of gas at an atmospheric pressure of 14.7 pounds per square inch and a temperature of 60° F. A trillion cubic feet (Tcf) of gas would fill a cube 2 miles on a side (8 mi³), and the United States consumes about 17 Tcf per year. We have proven reserves of 160 Tcf, which suggests that we will "run out" in 9.4 years. However, the *undiscovered* gas resource in the United States is estimated at between 300 and 500 Tcf. Unconventional sources of natural gas such as coal-bed gas and shale gas (discussed later in the chapter) add considerably to the total resource. And don't forget the methane hydrates.

Hydrogen has a bad reputation because it was the gas that filled the German dirigible *Hindenburg* that exploded at Lakehurst, New Jersey, in 1937. Also, the testing of the first hydrogen bomb in 1952 is still remembered by many with fear and awe. Many experts believe it is the energy of the future, however, as it burns without pollutants, emitting just water vapor. Unfortunately, there are no large reservoirs of hydrogen, and at present it must be manufactured using fossil fuels. It offers promise as a fuel for airplanes and autos, and will be economically viable when it can be produced in large quantities using solar or hydroelectric energy. (Fuel cells that burn hydrogen gas are discussed later in the chapter.)

Coal

Coal is the carbonaceous residue of plant matter that has been preserved and altered by heat and pressure. Next to oil and oil shales, coal is the earth's most abundant reservoir of stored energy. Deposits are known from every geologic period since Devonian time and the appearance of widespread terrestrial plant life some 390 million years ago. Permian coal is found in Antarctica, Australia, and India—pre-continental drift Gondwanaland (Chapter 3). The large fields of North America, England, and Europe were deposited during the Carboniferous Period, so-named because of the extensive coal deposits in rocks of that age. In the United States the Carboniferous Period is divided into the Mississippian and Pennsylvanian Periods, named after the rocks found in those states. Tertiary coals are found in such diverse locations as Spitsbergen Island in the Arctic Ocean, the western United States, Japan, India, Germany, and Russia.

COALIFICATION AND RANK

The first stage in the process of *coalification* is the accumulation of large amounts of plant debris under conditions that will preserve it. This requires high plant production in a low-oxygen depositional environment, the conditions usually found in nonmarine brackish-water swamps. The accumulated plant matter must then be buried to a depth sufficient that heat and pressure expel water and volatile matter. The degree of metamorphism (conversion) of plant material to coal is denoted by its **rank.** From lowest to highest rank, the metamorphism of coal follows the sequence *peat,* to *lignite,* to *subbituminous,* to *bituminous,* to *anthracite.* The sequence is accompanied by increasing amounts of fixed carbon and heat (Btu) content, and a decreasing amount of quickly burned volatile material. Once ignited, it is the carbon that burns (oxidizes) and gives off heat, just as wood charcoal does in a barbecue pit.

Bituminous, or soft coal, usually occurs in flat-lying beds at shallow depths that are amenable to surface mining techniques. Anthracite, the highest rank of coal, is formed when coal-bearing rocks are subjected to intense heat and pressure—a situation that sometimes occurs in areas of plate convergence. In the Appalachian coal basin, high-volatile bituminous coals occur in the western part of the basin and increase in rank to low-volatile bituminous coals to the east. Anthracite formed close to the ancient plate-collision zone of North America and Africa (pre-Pangaea), where the coals were more intensely folded and subject to higher levels of heat. Traveling east from the Appalachian Plateau to the folded Ridge and Valley Province, one goes from flat-lying bituminous terrain to folded anthracite terrain. West Virginia, Kentucky, and Pennsylvania are the leading coal producers of the eastern states. The interior coal basins of the Midwest are characterized by high-sulfur bituminous coal, whereas the younger Western deposits are low-sulfur lignite and subbituminous coal. ◉ Table 14.3 projects the development of coal-mining activity in the United States. The expansion of mining activity will occur mainly in the Western coal basins, because the coal seams there are very thick and easily mined and the coals are low in sulfur. Wyoming, with its low-sulfur coal, was the leading coal producer in 1993 (190 million short tons), followed in order by West Virginia, Kentucky, Pennsylvania, and Illinois (◆ Figure 14.13).

RESERVES AND PRODUCTION

The future of coal-derived energy in the United States is far more optimistic than the outlook for petroleum. Coal constitutes 80 percent of U.S. energy stores but only 18 percent of present usage. In 1989, electric utilities accounted for 86 percent of the coal consumed, with residential and industrial use accounting for the remainder. It is estimated that there are 84 trillion (84×10^{12}) recoverable tons of coal in the world, the equivalent of 34 trillion barrels of oil. The United States has 283 billion tons of recoverable reserves, which could last 200–300 years at the current rates of production and use. These reserves could meet only 50 years of total U.S. energy demand, however.

It has been estimated that energy demand in the early twenty-first century will be between 100 and 150 quads (a quad is a quadrillion British thermal units, 10^{15} Btu), compared to 73 quads in 1973 and about 80 quads in 1980 (◉ Table 14.4). Coal production will double to meet the demand, and this increase will be mostly in the Western coal basins, where the relatively inexpensive surface-mining techniques can be used.

	SURFACE MINES		UNDERGROUND MINES		TOTAL	
REGION	1977	2000	1977	2000	1977	2000
Appalachia*	185	130–175	205	380–500	390	510–675
Midwest	91	95–135	54	120–180	145	215–315
West	141	700–1,005	13	80–110	154	780–1,115

◉ **TABLE 14.3** **Active U.S. Coal Mines by Region, 1977 and 2000 (Projected)**

*Encompasses parts of Pennsylvania, eastern Kentucky, West Virginia, Virginia, Maryland, Tennessee, and Ohio.
SOURCE: U.S. Geological Survey.

◆▶ FIGURE 14.13 Production in short tons for each coal-producing region and the total for the United States, 1998. Note the shift in production from the Appalachian region to the Western coal states.

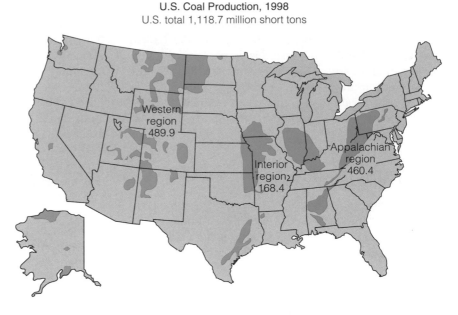

U.S. Coal Production, 1998
U.S. total 1,118.7 million short tons

Western region 489.9

Interior region 168.4

Appalachian region 460.4

Underground mining costs more, requires a greater capital investment, and takes more time to get into production. Also, miners who work underground face greater risks than do those who work at the surface.

The economic future of coal lies in its use as a substitute fuel for oil and gas. The technology is in place to make methane from coal—the product is called *synthesis gas*—and it has been used on a large scale to manufacture petrochemicals, including methanol (methyl alcohol). Methanol can be converted to gasoline using a mineral catalyst. Methanol is an interesting fuel: its combustion generates little nitric oxide (NO) and ozone, and it has a high octane number—about 110. Its high octane has made it attractive for years as a fuel for

Consider This… ? Your Christmas stocking contains nothing but a lump of anthracite coal. Being an optimist, you rejoice in this gift. Why? ■

race cars and high-compression engines. Although it has a lower heat (Btu) content than gasoline, it burns more efficiently, which evens things out. On the negative side, methanol releases greater emissions of carbon dioxide. **Biomass** alcohol (ethanol) made from sugar cane has been used in Brazil for years as an auto fuel, and at one time almost 90 percent of the cars there burned it. As the use of lead in gasoline is phased out, methanol can be added as an octane enhancer.

Direct liquefaction of coal shows great promise. Using coal- or petroleum-derived solvents, coal is dissolved and hydrogenated, resulting in about 75 percent gasoline and the remainder propane and butane. About 5.5 barrels of liquid are derived from a ton of coal at the approximate cost of the most expensive oil during the 1970s energy crunch, $35–$40/bbl. As long as we continue to rely on hydrocarbon-fueled engines, a major facet of our energy future lies in converting coal to liquid or gaseous **synthetic fuels,** often called simply *synfuels*.

Nonconventional Fossil Fuels

TAR SANDS

Tar sands contain oil that is too thick and viscous to flow at normal temperatures. These deposits are found

⊙ TABLE 14.4 Energy (Power) Units	
UNIT	EXPLANATION
British thermal unit (Btu)	The amount of heat required to raise the temperature of 1 pound of water 1° F; ≈ energy released by a burning match
Quad (quadrillion)	10^{15} Btu = 172 million bbl/oil
Barrel of oil	5.8 million Btu = 42 gallons
Bituminous coal (average)	25 million Btu/ton
Natural gas	Variable Btu content, measured in cubic feet (cf)
Megawatt (MW)	1,000 kilowatts (kw), or a million watts
Gigawatt (GW)	1,000 megawatts, or a billion watts

(b)

◆▶ **FIGURE 14.14** (a) Locations of Alberta's tar sands—Athabasca, Cold Lake, and Peace River. (b) After extraction by huge machinery, the sands are washed with hot water to extract the oil adhering to the grains. Cleaned sand is then returned to the open pit, and the land is restored to its original contours.

(a)

in Canada (◆▶ Figure 14.14), Venezuela, Madagascar, and the United States and they are surface-mined or drilled, using steam to mobilize and recover the oil. Canada's Athabasca sands (see Figure 14.14a) constitute the largest oil field in the world; they are estimated to contain more than 1.7 trillion barrels of oil, more than all the reserves of Saudi Arabia. At Athabasca the sand is extracted by open-pit mining techniques, and the oil is separated from the sand using steam or other thermal methods. The amount of material moved in these excavations is greater than the amount workers had to move to build the Great Wall of China, the Great Pyramid of Cheops, the Suez Canal, and the world's ten largest dams *combined*. After oil has been removed from the sand, the sand is returned to the excavation. The area is then leveled and relandscaped. Oil production at Athabasca was 205,000 bbl/day in 1998, and this supplied 12 percent of Canada's needs.

Tar sands that are too deep (500 m) for surface exploitation underlie Alberta's Cold Lake area, also shown in Figure 14.14a. Steam at 300° C (575° F) is injected at high pressure and allowed to "soak" for a month. The oil released from the sand is then brought to the land surface by the familiar "rocking-horse" pumps. More than 2,000 steam-injection and oil-production wells deliver 130,000 barrels of oil per day. This method is known as "huff and puff," because the same wells that are used to inject the steam are used to pump the oil to the surface. The area is estimated to contain 220 billion barrels of recoverable oil.

OIL SHALES

Oil shales are sedimentary rocks that yield petroleum when they are heated. Found on all continents, including Antarctica, they were originally deposited in lakes, marshes, or the ocean. The original rock oil used in kerosene lamps was produced from black oil shales. The most extensive U.S. deposits, the Green River shales, were formed during Eocene time in huge freshwater lakes in the present-day states of Wyoming, Colorado, and Utah (◆▶ Figure 14.15). They are not oily like tar sands, but when heated to 500° C (900° F) they yield oil. The source of oil in the rock is **kerogen,** a solid bituminous substance, and a ton of oil shale may yield 10–150 gallons of good-quality oil. Potential deposits of economic interest yield 25–50 gallons per ton, and the U.S. Geological Survey includes 3 trillion barrels of oil from shale in their estimates of reserves. Kerogen-rich lake deposits are also found in China, Yugoslavia, and Brazil.

Water is the limiting factor in oil-shale utilization, because processing the shale requires water volumes 3–4 times greater than that of the oil extracted. Furthermore, the extraction process destroys the natural landscape, and some of the richest shales are in scenic wilderness areas (◆▶ Figure 14.16). Because oil shales expand like popcorn when they are heated, disposal of the waste rock is a problem. At this time the oil shale industry cannot compete economically because of the high cost of transporting unprocessed oil shale

(a)

◆▷ **FIGURE 14.15** (a) Oil shale and the heavy oil that it produces when heated. (b) Distribution of oil shales in the Green River Formation of Wyoming, Colorado, and Utah.

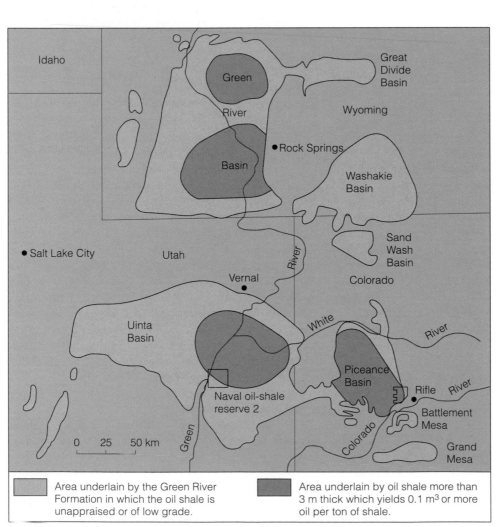

Area underlain by the Green River Formation in which the oil shale is unappraised or of low grade.

Area underlain by oil shale more than 3 m thick which yields 0.1 m³ or more oil per ton of shale.

(b)

◁▶ FIGURE 14.16 Green River oil shale exposed in Book Cliffs near Grand Junction, Colorado.

to consumers. The future for the trillions of barrels of oil in shales rests in developing *in situ* ("in place") extraction methods that will not disrupt the surface environment or create waste-rock disposal problems.

Problems of Fossil Fuel Combustion

AIR POLLUTION

Obtaining energy by burning fossil fuels creates environmental problems of global proportions. It produces oxides of carbon, sulfur, and nitrogen and fine-particulate ash. Carbon monoxide (CO), an oxide produced by combustion of all fossil and plant fuels, is converted to carbon dioxide (CO_2), which contributes to global warming (see Chapter 11). Burning coal also releases sulfur oxides (SO_x) to the atmosphere, where they form environmentally deleterious compounds. Nitrogen oxides (NO_x), mostly NO and NO_2, are products of combustion in auto engines and are the precursors of the photochemical oxidants, ozone and peroxyacetyl nitrate (PAN), which we associate with smog. Water and oxygen in the atmosphere combine with SO_2 and NO_2 to form sulfuric acid (H_2SO_4) and nitric acid (HNO_3), the main components of **acid rain**—rain with increased acidity due to environmental factors such as atmospheric pollutants (Case Study 14.3).

The Clean Air Act of 1963 as amended in 1970 and 1990 specifies standards for pollutant oxides and hydrocarbon emissions. By 1980 new cars were 90 percent cleaner than their 1970 counterparts. The Clean Air Act Amendments of 1990 (CAAA) set goals and timetables that are affecting how and what we drive. Beginning in 1992, fuel suppliers were required to sell only reformulated gasoline in 39 areas where winter air quality is a problem and to sell only reformulated gasolines in the high-ozone cities of Baltimore, Chicago, Hartford, Houston, Los Angeles, Milwaukee, New York, Philadelphia, and San Diego. Reformulated gasolines that enhance burning and reduce emissions will be the required auto fuel in the late 1990s. Reformulations using methyl tertiary butyl ether (MTBE) add as much as 2.5 percent oxygen to the fuel, causing it to burn cleaner and create less ozone, at a cost of about 10 cents more per gallon. Also, the CAAA mandated that automakers significantly reduce emissions leading to ozone formation in all new vehicles sold in the United States beginning in 1996.

Sulfur Emissions and Acid Rain

Sulfur occurs in coal as tiny particles of iron sulfide, most of which is the mineral pyrite, or "fool's gold" (◁▶ Figure 14.17). Coal also contains organic sulfur originally contained in the coal-forming vegetation. Upon combustion, mostly in coal-burning electrical generating plants, the sulfide is oxidized to sulfur dioxide (SO_2) and carried out through smokestacks into the environment. Emissions of SO_2 are known to have a negative impact on human health, particularly lung problems. Although most coal-burning in the United States occurs in the Midwest and the East, significant volumes of SO_x can travel 1,000 kilometers (600 mi) or more from their source. In 1980, 51 power plants in six Eastern states emitted 25 percent of the total U.S. sulfur dioxide. ◁▶ Figure 14.18 shows typical U.S. SO_2 emission densities, which reflect the pattern of coal usage very closely. Mobile sources using petroleum fuels emit

Case Study | 14.3

Baking Soda, Vinegar, and Acid Rain

The term *acid rain* refers to the atmospheric deposition of acidic substances including rain, snow, fog, dew, particles, and certain gases. Although volcanic activity (see Chapter 5) is the greatest source of materials that may form acids—sulfur, carbon dioxide, and chlorine—sulfur and nitrogen oxides introduced by human activities are approaching the amount of nature's contributions. Combustion of fossil fuels and the refining of sulfide ores are the major human sources of these contaminants. Acids form when these gases come in contact with water in the atmosphere or on the ground. Whereas carbon dioxide forms carbonic acid, a weak acid (see Chapter 6), chloride ion and the oxides of sulfur and nitrogen form the strong hydrochloric, sulfuric, and nitric acids, respectively.

The acidity or basicity (the chemical opposite of acidity) of an aqueous solution is referenced to the pH scale (◆ Figure 1), a measure of the solutions' hydrogen-ion activity. A solution containing no acid and no base will have a pH value of 7.0. The more hydrogen ions floating around in the solution, the more acidic it is, and the lower its pH will be. Note that the scale is reversed, so to speak; a solution with a pH below 7.0 is acidic, and one with a pH above 7.0 is basic, or "alkaline." Acid-rain events with acidities below pH 2.8, the pH of vinegar, have been reported in large cities and heavily industrialized areas. In contrast, neutral waters may become quite basic. For example, lakes without outlets, such as Lake Natron in Africa, may become extremely alkaline and have pHs greater than 11.0—well in excess of the pH of a concentrated solution of baking soda and water.

◆ FIGURE 1 The pH scale. A neutral solution has a pH of 7.0. A solution whose pH is less than 7.0 is acidic, and a solution whose pH is above 7.0 is basic, or alkaline.

less than 4 percent of the total. Nevertheless, these emissions amounted to almost a million tons of SO_2 in 1980 (◉ Table 14.5).

Sulfur dioxide emitted from a smokestack may be deposited dry near the facility, where it will damage plant life and pose a health threat to animals. If the gas plume from the smokestack is transported away from the facility, its SO_2 will react with the atmosphere and eventually fall to the earth surface as acid rain. Acid rain can damage crops; acidify soils; corrode rocks, buildings, and monuments; and contaminate streams, lakes, and drinking water. Scientists and government agencies disagree on the impact of acid rain on forests. The 1990 report of the National Acid Precipitation Assessment Program shows the distribution of acid rain to include southern and midwestern states and minimizes acid rain's impact on forests. The Environmental Protection Agency, on the other hand, believes it has evidence that acid rain does extensive forest damage. Meanwhile, it has been estimated that annual U.S. emissions of sulfur gases will have increased almost 30 percent from 1990 levels by 2010.

Sulfur oxides can be removed from combustion gases in smokestacks by using a device called a *scrubber*

◁▷ **FIGURE 14.17** Small pyrite masses in a bituminous coal matrix (gray).The light-gray material is the cell walls of the original plant material, and the darker gray is the solid material that was contained within the cell cavities. Earth pressures compacted the plant material, squeezed out any water, and collapsed the cavities. That pyrite formed within the coal before completion of compaction is evidenced by plant materials bent around solid pyrite masses. The pyrite "blebs" are about 50 microns across.

that selectively reacts with SO_2 and absorbs or neutralizes it. Many coal-burning facilities are switching to a process known as fluidized-bed combustion. Coal and pulverized limestone are burned together, and the heat of combustion converts the limestone into calcium oxide (CaO) and CO_2. CO_2 is emitted, and the CaO combines with sulfur and oxygen to form calcium sulfate $(CaSO_4$, a gypsumlike substance) as a solid waste product. Sulfur emissions are reduced, but CO_2 emissions increase. It is thus a tradeoff between the health benefits of reduced atmospheric sulfur and the environmental effects of increased contributions to global warming.

Nitrogen Oxides and Smog

Nitrogen oxide emissions (NO_x) are significant from both stationary (coal-burning) and mobile (vehicular) sources. Mobile sources are the most difficult to control, and their contribution to total NO_x is increasing as the number of automobiles increases. NO_x forms when engine combustion causes nitrogen in air to oxidize to NO and NO_2, which then react with oxygen, free radicals (incomplete molecules that are highly reactive), and unburned hydrocarbons to produce **photochemical smog.** Photochemical smog requires sunlight for its formation. Although the word smog is a contraction of

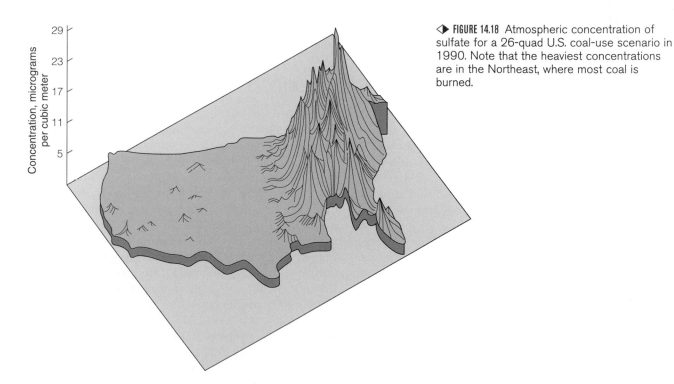

◁▷ **FIGURE 14.18** Atmospheric concentration of sulfate for a 26-quad U.S. coal-use scenario in 1990. Note that the heaviest concentrations are in the Northeast, where most coal is burned.

Concentration, micrograms per cubic meter

29
23
17
11
5

○ TABLE 14.5 U.S. Sources of Sulfur and Nitrogen Oxide Emissions, 1980

| | PERCENTAGE OF TOTAL EMISSIONS | |
SOURCE	Sulfur Oxides*	Nitrogen Oxides
Mobile sources**	3.8	44.0
Stationary sources***	80.0	51.0
Industrial processes	16.0	2.9
Other (agriculture, etc.)	0.2	2.1

* SO_2 emissions (tons/year) decreased 33% between 1975 and 1990.
** All transportation including recreational vehicles.
*** Utilities, industrial, institutional, and residential sources.
SOURCE: Council on Environmental Quality, *Twelfth Annual Report*, 1981, and various other sources.

the words *smoke* and *fog,* the air pollution in Los Angeles, Mexico City, and Denver has little direct relationship to either smoke or fog (Case Study 14.4).

Both NO_2 and ozone (O_3) are reactive oxidants and cause respiratory problems. A level of only 6 parts per million (ppm) of ozone can kill laboratory animals by pulmonary edema (water in the lungs) and hemorrhage within four hours. On 17 March 1992, Mexico City, the largest city in the world, had such high ozone concentrations (four times the normal amount) that it was necessary to close schools, restrict traffic, and have factories cut back production 50 percent. In humans, alcohol consumption, exercise, and high temperatures have been found to increase adverse reaction to ozone, whereas vitamin C and previous exposure seem to lessen the reaction. One investigator noted that the person most likely to succumb to ozone would be "an alcoholic with a 'snootful' arriving at LAX for the first time on a hot, smoggy day and jogging all the way to Beverly Hills." Such a scenario may not be all that unlikely in Southern California. A frightening aspect of ozone is that it is so reactive with human cells that when it is inhaled, only a fourth of it gets past the nasal and bronchial passages to the lungs, and that appears to be more than enough to cause lung damage. NO_2 has the same effect on humans as ozone, except at dosages that are 15 times greater.

Energy-efficient cars and new fuels have gone a long way to decreasing NO_x emissions. Exhaust-system catalytic conversion of nitrogen oxides, carbon monoxide, and hydrocarbons into carbon dioxide, nitrogen, and water are mandatory automotive equipment on new cars in most states.

Domestic Coal Burning

China is the world's largest producer of coal, accounting for 25 percent of global output. It currently produces 10 percent of the global emissions of CO_2, and it is estimated that by 2025 it will emit more CO_2 than the United States, Canada, and Japan combined. Coal provides 76 percent of the huge country's commercial energy, and it is the only fuel millions of people use.

In a typical home coal is burned in unvented stoves for cooking, space heating, and heating water. In many areas, foods are brought indoors in the fall for drying with coal fires, as the climate is too cool and damp to dry them outdoors. At least 3,000 people in Guizhou Province in southwest China suffer from severe arsenic poisoning caused by consuming chili peppers dried over high-arsenic coal fires (◆ Figure 14.19a). Although fresh chili peppers have less than 1 ppm arsenic, chili peppers dried over high-arsenic coal have up to 500 ppm arsenic! More than ten million people in the same province suffer dental and skeletal fluorosis (see Chapter 6) due to eating corn that has been cooked over coal containing large amounts of fluorine. Gathering about their coal fires at night exposes families to aromatic hydrocarbons from incomplete combustion of coal. This is thought to explain the high incidence of esophageal and lung cancers in some parts of China (◆ Figure 14.19b). About 3.5 billion people worldwide are exposed to toxic fumes indoors due to coal burning. Each year, an estimated 178,000 people die prematurely due to indoor air pollution in China alone. To combat these health hazards Chinese and U.S. Geological Survey geologists are working together to identify coal deposits whose high toxic-element concentrations make them unsuitable for domestic use.

MINE COLLAPSE

Whenever coal is extracted from shallow underground seams, a collapse of the overburden into the mine opening is risked. Lignite and bituminous (soft) coals are most susceptible to collapse because they are exploited from near-horizontal beds at relatively shallow depths. The "room and pillar" method recovers about 50 percent of the coal, leaving the remainder as pillars for supporting the "roof" of the mine. As a mine is being abandoned, the pillars are sometimes removed or reduced in size, increasing the yield from the mine, but also increasing the possibility of subsidence at some time in the future. Even when the pillars are left, failure is possible. The weight of the overlying strata can cause the pillars to rupture, and sometimes it physically drives the pillars into the

(b)

◇ **FIGURE 14.19** Burning coal in unvented stoves exposes millions of Chinese families to serious health risks. (a) Drying chili peppers over a coal fire that emits arsenic fumes increases the arsenic content of the chilies up to 500 times, and eating them causes arsenic poisoning. (b) As families gather about coal fires at night, they are exposed to polycyclic aromatic hydrocarbons from incomplete combustion of the coal. Breathing these substances can lead to lung and esophageal cancer.

(a)

underlying shale beds. Subsidence has been documented in Pittsburgh, Scranton, Wilkes-Barre, and many other areas in Pennsylvania. In 1982 an entire parking lot disappeared into a hole above a mined-out coal seam 88 meters (290 ft) below the surface. Wyoming, Montana, South Dakota, and other Western "lignite" states are pockmarked with subsidence features, but human activities and human lives are less likely to be affected by subsidence, or collapse, in those areas (◇ Figure 14.20).

Energy for the Future

The development of the oil-based economy that began at the turn of the twentieth century was driven by many scientific and engineering advances and by the needs of a growing industrial economy. Our oil-based economy is now "mature," however. About 80 percent of the oil produced today comes from oil fields dis-covered before 1970, and most of them are on the decline. For certain, the energy scenario of the twenty-first century will not be at all like that of the twentieth century, or even like today's. Undoubtedly, the primary energy source will be the sun, and the fuel may very well be hydrogen, the most common element in the universe.

DIRECT SOLAR ENERGY

Although the amount of solar energy that reaches the earth far exceeds all human energy needs, it is very diffuse. For example, the amount that strikes the atmosphere above the British Isles is 80 times Great Britain's energy needs; it averages about 1 kilowatt per square meter. Cloud cover reduces the amount by 80 percent, but there is still more than enough. So why aren't there solar-electrical plants in every city and hamlet? The reason is cost. The current technology requires a great deal of space, which makes installations expensive. There are essentially two ways to put the sun to work for us:

Case Study | 14.4

Make Smog while the Sun Shines

Hell is a city much like London,
A populous and smoky city.
SHELLEY, PETER BELL THE THIRD

Air pollution is the blight that results from burning fossil fuels. It was probably first recognized as a serious problem in Elizabethan London, filthy with smoke from thousands of domestic and workshop coal fires. In 1661 John Evelyn, one of the founders of the Royal Society and a noted do-gooder, published by royal decree a paper entitled "Fumifugium, or the Inconvenience of Aer and Smoke of London Dissapated; together with Some Remedies Humbly Proposed." It is said that Londoners' lungs have been black since that time, and that for many years when the city's medical students saw pink lungs in the anatomy laboratory, they suspected the deceased was from Ireland.

Many catastrophic air-pollution incidents have occurred since the industrial revolution. The few listed here give an idea of the magnitude of the problem.

YEAR	PLACE	FATALITIES
1872	London	thousands (estimated)
1930	Meuse Valley, France	60
1948	Donora, Pennsylvania	20 (and many ill)
1950	Poza Rica, Mexico	22
1952	London	4,000 (estimated)

The word *smog*, a contraction of the words *smoke* and *fog*, was probably coined in the course of explaining the smoke-fog deaths that occurred in Scotland in 1909.

All health-endangering smog events begin with an atmospheric temperature inversion, which "puts a lid on" polluted air, holding it close to the earth surface. Temperature normally decreases with altitude, about 2° C (3.6° F) per 1,000 feet, which causes air that has been warmed and expanded at the surface to rise and denser cold air from above to sink. When an inversion forms, a layer of warm air at, say, 1,000 feet above ground level prevents vertical circulation and keeps the bad vapors at the surface (◆ Figure 1).

The worst air-pollution disaster in history was the infamous "black fog" event in London in December 1952. The fog was so thick that the racing dogs at a greyhound track lost sight of the rabbit and became disoriented, and a large duck flying blind crashed through the glass roof of a train station. In two days the fog turned black when coal-derived soot mixed with sulfur oxides, and people, most of them elderly, began dying. Many stricken people could not get to hospitals, and 50 bodies were recovered from a small park in the city.

Photochemical smog (as opposed to true smog) is produced when automobile emissions are activated by ultraviolet radiation to produce ozone. In simplest terms, the process is this:

1. The high temperatures and pressures involved in fuel–air combustion in the engine form nitric oxide (NO), which, with unburned hydrocarbons, is emitted as exhaust.

2. NO reacts with the air to form nitrogen dioxide (NO_2).

3. Under ultraviolet radiation, NO_2 reacts reversibly with oxygen to yield ozone (O_3) and nitric oxide.

1. Solar-thermal methods utilize panels or "collectors" to warm a fluid, which is then used for heating or for generating electricity.

2. Photovoltaic cells use the second most common element on earth, silicon, to convert light directly into electricity.

In 1995, the electricity produced from photovoltaic cells cost about 16 cents per kilowatt hour (kwh), many times that produced by fossil fuels (◉ Table 14.6). However, it is estimated that photovoltaic electricity will soon cost less than 10 cents per kilowatt hour, and will compete with conventional electrical generation in this century.

Solar-thermal facilities called *power towers* illustrate a creative use of solar energy. The "Solar One" station in the Mojave Desert near Barstow, California, generated electricity between 1982 and 1988 for the Southern California Edison Company. It employed 1,818 sun-tracking mirrors focused on a liquid-filled receiver atop a tower where temperatures reached 900° C (◆ Figure 14.21). The super-heated liquid was transferred to a plant where it flashed to steam and drove turbines that operated an electrical generator. Solar One occupied 40 hectares (100 acres) and its receiver was visible for miles, looking like a miniature sun. Power towers are practical only in areas that offer more than 300 sunny days per year and inexpensive land. Because of the suc-

◆ FIGURE 1 Temperature and altitude factors of an inversion layer, which traps polluted air. If the base of the inversion layer lowers, the mixing depth decreases and air quality worsens.

◆ FIGURE 2 Typical daily variation of smog components in Los Angeles.

In chemical notation, steps 2 and 3 look like this:

$$NO_2 + O_2 \underset{sunlight}{\overset{\leftarrow}{\rightarrow}} O_3 + NO$$

nitrogen   oxygen *sunlight* ozone   nitric
dioxide   oxide

A "tossed salad" of reactions occur between the bits and pieces of unburned hydrocarbons that remain. One of the reactions forms PAN (peroxyacetyl nitrate), which makes your eyes tear and which itself decomposes to form more NO_2 (a feedback mechanism), which then forms more ozone. So once the smog begins, it is self-perpetuating as long as there is ultraviolet radiation to drive the reactions.

We can see that photochemical smog is a daytime blight; it abates at night, only to build up the next morning (◆ Figure 2). It's not uncommon for the Los Angeles Basin to be clear at 5:00 A.M. on a summer day and for visibility to be restricted to a few miles five hours later. Topography is an influence, and smog is chronic in cities such as Denver, Los Angeles, and Mexico City that have mountain barriers. However, any city with a plethora of cars and a temperature inversion can have air pollution problems.

cess of Solar One, a 10-megawatt facility was built on the same site and began operating in mid 1996. Solar Two uses molten nitrate salt as the heat transfer medium. Molten salt has excellent heat storing properties; the heat can be used immediately to generate steam, or it can be stored and used later, during cloudy periods or after sundown. Solar two generated enough power for 10,000 homes and operated from 1996 to 2000, when government funding ended.

Photovoltaic cells (PV cells) absorb pulses of light energy on semiconductor materials and turn that energy into an electric current. They are used widely as a power source for satellites, calculators, watches, and remote communication and instrumentation systems. In recent years the coast of PV cells has been reduced by 80 percent, and it is now possible to apply thin films of these materials to shingles, tiles, and window glass, enabling buildings to generate their own electricity. Costs need to fall another 50–75 percent to make them competitive with fossil-fuel generating plants.

Several U.S. utility companies have installed PV systems to supply small users in out-of-the-way places without building costly power-line extensions. In India, which lacks an extensive electrical transmission network, PV power is more economical than diesel power in remote locations. India has 4–5 million water pumps that could be powered by PV cells during daylight hours. In Sacramento, California, the electric company offered to

◁▷ **FIGURE 14.20** Surface effects of underground mining along the Tongue River in Wyoming. The pits, troughs, depressions, and cracks have formed since the mine was abandoned in 1914.

TABLE 14.6 Cost of Electricity by Energy Source	
ENERGY SOURCE	COST, cents/kwh
Sun	15.8*
Geothermal	13.3*
Biomass	13.0*
Wind	7.5**
Oil	4.4†
Gas	2.9†
Coal	1.2†
Nuclear	0.93†

* The rates the Southern California Edison Company was paying for purchased and manufactured energy in December 1995.
** 1998 value.
† The rates the Southern California Edison Company was paying for purchased and manufactured energy in December 1991 (more recent data not available).

install PV panels on the roofs of those customers willing to pay an extra $4 per month. Thousands of homeowners requested the panels, but since the panels cost about $20,000 each, the company had set up only about 500 units by 1999. Ongoing research is devoted to developing thermophotovoltaic cells that can generate electricity using waste heat from manufacturing plants.

INDIRECT SOLAR ENERGY

Wind Energy

Wind power was the fastest growing energy segment in the world between 1994 and 1998. Within that time the world wind industry added 2,000 megawatts (MW) of capacity, reaching a total of 9,600 MW. That capacity is sufficient for generating 21 billion kilowatts hours (kwh) of electricity, enough to supply 3.5 million U.S. homes. California leads the nation in wind power, producing enough electricity to light a city the size of San Francisco, about 1.5 percent of the state's total electric-

◀▶ **FIGURE 14.21** Solar One, a direct solar-energy power conversion facility in the Mojave Desert near Barstow, California. The heliostat (reflector) field directs sunlight to the central tower. Solar Two now exists on this site using a new molten salt heat-transfer technology.

by 1998, as the world's wind capacity had grown dramatically. According to the American Wind Energy Association (AWEA), the total U.S. wind capacity grew by more than 230 MW in 1998, with major additions in states outside of California. Most of the new growth was in Minnesota, Oregon, Wyoming, and Iowa. By the 2000s, AWEA projects between 120 and 250 MW additional U.S. wind-energy capacity. Internationally, Germany, Spain, and Denmark are leading wind-energy producers.

At the three California localities narrow passes through mountain ranges separate hot, desertlike valleys from areas with more maritime climates. As hot air rises in the valleys, cool air is drawn through the topographically narrow passes to replace the risen air. Areas of strong winds are favored, because winds' ability to do work *increases as the cube of the increase in wind velocity;* that is, with a doubling of wind velocity (2^3), 8 times as much power is available (◀▶ Figure 14.23). The original wind generators were plagued by design flaws and maintenance problems, but with modern composite materials and programmed "smart" turbines, more-reliable 300-kilowatt windmills are now mass-produced. Costs have been reduced to the point that wind-generated electricity is now price-competitive (Table 14.6). ◀▶ Figure 14.24 shows the wind-electric potential of the 48 conterminous states expressed as a percentage of each state's total energy need. Note in the figure that practically all of the electricity needed in the United States could be provided by winds over 12 states. Wind can provide electricity during peak demand periods, which typically occur late in the afternoon on hot summer days and early in the evening in the winter. Peak-demand power is most costly to the consumer, and an immediate gain can be realized where wind electricity is available to satisfy peak demand. Of course, windmills can be "becalmed." In areas where electricity is critical, they are used in conjunction with diesel electricity or other conventional generators.

Environmental problems that accompany wind generation of electricity include noise, interference with TV reception by steel blades, land acquisition difficulties, and most importantly, visual blight. Modern generators are quiet, and their wood or composite blades do not interrupt television signals. Although land disruption is minimal with windmills, they are hard to ignore, and many people feel they destroy the visual esthetics of a region. On the other hand, ranchers at Altamont love the windmills; they can still graze their cattle on the land and they receive land-use royalties that have increased the value of their land—to five times its prewindmill value.

ity. More than 13,000 wind turbines—95 percent of California's wind-generation capacity and output—are in three regions: Altamont Pass east of San Jose, Tehachapi Pass southeast of Bakersfield, and Banning Pass near Palm Springs (◀▶ Figure 14.22). In 1995 these areas produced 30 percent of the world's wind-generated electricity, but they had lost their standing

Hydroelectric Energy

Falling water, our largest renewable resource next to wood, has been used as an energy source for thousands of years. First used to generate electricity about 100 years ago on the Fox River near Appleton, Wisconsin, today it provides a fourth of the world's electricity. In Norway, 99 percent of the country's electricity and 50 percent of its total energy is produced by falling water. The principle is relatively simple: impound water with a dam and then cause the water to fall through a system of turbines and generators to produce electricity. Because of this simplicity, electricity coming from existing facilities is also the cheapest source of this power. Aitupu Dam on the Parana River between Paraguay and Brazil, completed in 1982, is the largest-producing electrical complex in the world; its potential is 12,600 megawatts. It provides much more electricity than the present demand in this part of South America. When completed, the Three Gorges Dam on the Yangtze River in China will be the largest hydroelectric project in the world. The dam will be 2.5 kilometers wide (1.5

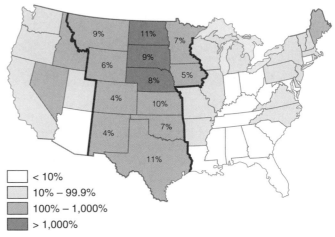

◀▶ **FIGURE 14.23** Wind's ability to do work increases exponentially with velocity. For this reason, wind "farms" are located in areas where winds are strong.

◀▶ **FIGURE 14.24** Wind-electric potential of the 48 contiguous states as a percentage of each state's energy need. Twelve states in the central part of the country could provide about 90 percent of the total U.S. need.

mi) and rise almost 180 meters (600 ft) above the world's third-largest river. Scheduled for completion in 2009, its construction has raised a great deal of controversy within China. Although it will help to control disastrous flooding on the river, its 600-kilometer-long reservoir (411 mi) has forced the relocation of 1.2 million people and innumerable historical treasures.

Because of land costs and environmental considerations, it is doubtful that any more large dams will be built in the United States. Therefore, dams that were built for other purposes have become attractive for retrofitting with generators. Hydroelectrical facilities are particularly useful for providing power during times of peak demand in areas where coal, oil, or nuclear generation provides the base load. Hydroelectricity can be turned on or off at will to provide peak power, and many utilities have built pumped-water-storage facilities for just that reason. During off-peak hours, when plenty of power is available, water is pumped from an aqueduct or other source to a reservoir at a higher level. Then during peak demand the water is allowed to fall to its original level, fulfilling the temporary need for added electricity.

Although hydroelectric energy is clean, dams and reservoirs change natural ecological systems into ones that require extensive management. One of the social consequences of dam building is displaced persons; for example, 80,000 people were forced to move when Lake Nasser was created after construction of the Aswan High Dam. This displacement is unacceptable in most societies today. Also unacceptable are recurring attempts to dam areas of great aesthetic value, such as the Grand Canyon. Finally, dam failures—most often geological failures of dams' foundations—result in floods that can take enormous human and economic tolls.

GEOTHERMAL ENERGY

The interior of the earth is an enormous reservoir of heat produced by the decay of small amounts of naturally radioactive elements that occur in all rocks. When the deep heat rises to shallower depths, this geothermal energy can be tapped for human use (*geo-*, "earth"; *-therm*, "heat"). Furthermore, this natural heat is cleaner than many other energy sources, and there are fewer negative environmental consequences of using it. Not surprisingly, the prospects for using geothermal energy are best at or near plate boundaries where active volcanoes and high heat flow are found. The Pacific Rim (Ring of Fire), Iceland on the Mid-Atlantic Ridge, and the Mediterranean belt offer the most promise. The

first facility to utilize geothermal energy was built in 1905 at Larderello, Italy. Today, the United States, Japan, New Zealand, Mexico, and countries of the former USSR are utilizing energy from earth heat. Iceland uses geothermal energy directly for space heating (one geothermal well there actually began spewing lava), and many other countries have geothermal potential. The versatility of earth heat ranges from using it to grow mushrooms to driving steam-turbine generators with it (◆ Figure 14.25).

Geothermal energy fields may be found and exploited where magma exists at shallow depths and

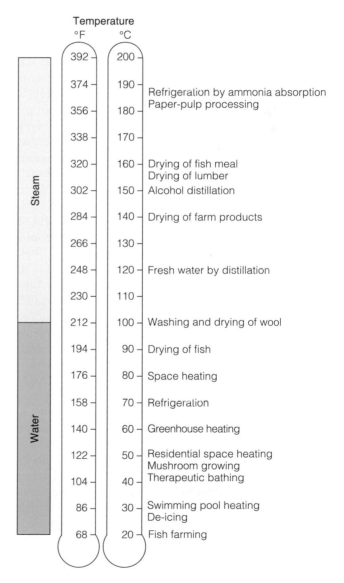

◆ FIGURE 14.25 Some established commercial uses of geothermal energy.

where there is sufficient underground water to form steam. Geothermal fields are classified as either *steam-dominated* systems or *hot-water-dominated* systems. Whereas deep, insulated reservoirs may produce live steam, shallower reservoirs contain hot water, some of which flashes to steam at the surface. Yellowstone National Park and Wairaki, New Zealand, are examples of hot-water reservoirs (◆ Figure 14.26). The two kinds of fields require different power-plant designs to generate electricity, as illustrated in ◆ Figure 14.27. A third kind of geothermal electrical generation utilizes hot water at temperatures below 100° C. The water is used to heat a low-boiling-point liquid, typically isopropane, and under pressure, the "steam" that is produced drives the generators. These *binary* plants are so named because they use two operating liquids.

Larderello, Italy, and The Geysers near Napa Valley in northern California (◆ Figure 14.28) are examples of steam-dominated fields. Such fields are the rarest and most efficient energy producers. They occur where water temperatures are high and discharge is low so that steam forms. They are much like an oil reservoir in that porous rocks that hold steam or very hot water are overlain by an impermeable layer that prevents the upward escape of the steam or water. ◉ Table 14.7 shows the electrical generating and direct-use capacities of selected U.S. geothermal fields in 1990. The Geysers is the largest producer of geothermal power in the world; its production is sufficient to supply electricity to a U.S. city of 1.3 million people. More than 600 wells had been drilled there by 1994, some of them as deep as 3.2 kilometers (10,400 ft). To illustrate geothermal energy's potential, consider this: whereas a nuclear power plant generates about 1,000 MW, The Geysers produced 1,300 MW in 1992.

Geothermal energy is clean. Coal-burning plants typically generate 5.5 metric tons of CO_2 for each megawatt of electricity produced, and relatively clean-burning natural gas (methane) releases 3 tons of CO_2 per megawatt. Geothermal energy, in contrast, releases from 0.5 metric ton to as little as 10 kilograms of CO_2 per megawatt. Similar contrasts exist for emissions of sulfur and sulfur oxides.

The problems associated with geothermal energy production are related to water withdrawal and water quality. Some geothermal waters contain toxic elements such as arsenic and selenium and heavy metals such as silver, gold, and copper. Technology is in place to transfer the heat of the toxic brine deep within the wells to a clean, working fluid that can be brought safely to the surface. In some geothermal areas removal of underground water can cause surface subsidence and perhaps even earthquakes. At The Geysers, subsi-

◆ **FIGURE 14.26** Wairaki, one of several large volcanic centers on the North Island of New Zealand. Significant amounts of geothermal energy are tapped here in a country that is not energy-rich.

dence of 13 centimeters (5 in) has been measured without noticeable impact. Reinjecting the cooled water back into the geothermal reservoir decreases the subsidence risk and helps to ensure a continued supply of steam to the electrical generators. All geothermal fields in California are being monitored for possible seismic activity related to steam production. Small quakes have been reported from The Geysers, but none of significant magnitude.

The future for geothermal energy looks good. About 20 geothermal fields in the United States are generating electricity, and direct-use hydrothermal systems are being developed at more than 30 sites. In spite of some technical and environmental problems, it is anticipated that geothermal energy will supply 1 percent of U.S. energy needs early in this century.

Fuel Cells

Fuel cells burn hydrogen gas (H_2) to generate electricity and do work. Discovered in 1829, they attracted a great deal of interest in the early twentieth century. However, modern applications required improvements that were not developed until the 1960s brought space-age technologies. The cells produce no air pollution, have no moving parts, are nearly silent, and are roughly twice as efficient as conventional engines. Cars powered by fuel cells could be refueled in a matter of minutes, compared to the several hours needed for recharging conventional electric cars. Yielding 2.5 times the energy per weight of gasoline, hydrogen is also attractive as a potential aircraft fuel. Current fuel

(a) Hot-water system

(b) Steam system

◆ FIGURE 14.27 (a) Generation of electricity with a hot-water geothermal system. Steam used to drive the generator is diverted at the separator, and wastewater from the separator and condenser are reinjected underground to keep the system going. (b) A steam-dominated electric-power generation system. Steam from the well is injected directly onto the turbine generator. Wastewater from the condenser is reinjected underground to help extend the life of the system.

◆ FIGURE 14.28 The Geysers geothermal field in Sonoma County, California, is steam-dominated. It is the largest geothermal-energy producer in the world.

cells are expensive, and producing hydrogen energy costs the equivalent of $2.50-per-gallon gasoline.

Since there are no large reservoirs of hydrogen gas, where and how are we going to get it? There is plenty of hydrogen in the world; it is in *water*. The challenge is to separate it from water economically (◆ Figure 14.29). Most proponents of fuel cells believe the electrical energy needed for separating hydrogen from water (called electrolysis) must come from the sun in the form of hydroelectricity, direct or indirect solar energy, or wind energy. Chemists developed a "water splitter" in the 1990s that almost doubles the efficiency of converting solar energy into hydrogen. Much hope is attached to this potential new fuel. It is hoped that a *solar–hydrogen revolution* will solve the world's fuel and air-pollution problems within the next 50 years. Some scientists believe that finding a cheap way to electrolyze water could make hydrogen the dominant fuel of the twenty-first century, just as hydrocarbons were in the twentieth century. In Germany there are plans for a solar–hydrogen system that would provide for all the heating, cooking, and electricity needs of a home.

⦿ TABLE 14.7   Electrical and Direct-Use Capacities of Selected U.S. Geothermal Fields, 1990

STATE	FIELD	ELECTRICAL CAPACITY, MW	DIRECT-USE CAPACITY, MW (THERMAL)
California	The Geysers*	1,970[1]	–
	Coso	256[2]	–
	Salton Sea area*	214[2]	–
	East Mesa	10[3]	–
	Long Valley (Mammoth Lakes)	43[3]	–
Colorado	Pagosa Springs	–	14
Hawaii	Puna	25[2]	–
Idaho	Boise	–	57
Nevada	Dixie Valley	50[2]	–
	Steamboat Springs	18[2,3]	–
New Mexico	Animas	–	8
New York	Auburn	–	1.2
Oregon	Klamath Falls	–	37.8
Utah	Cove Fort	13[2,3]	–
Wyoming	Thermopolis	–	9.0

* 1993 data.
[1] Steam is piped directly to a power plant.
[2] Flash wells produce a mixture of steam and water that must be separated.
[3] Binary wells produce hot water that is piped to a power plant, where heat is transferred to another fluid.
SOURCE: U.S. Geological Survey circular 1125, 1994.

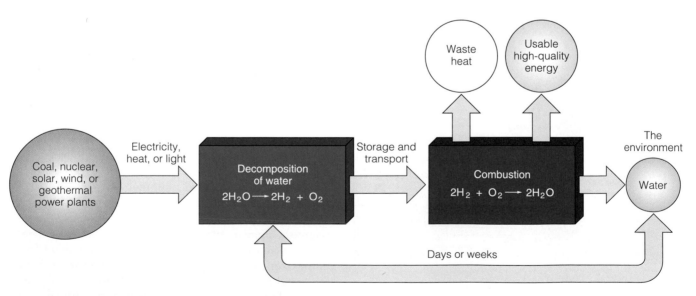

◆ **FIGURE 14.29** The hydrogen energy cycle. Electricity, heat, or solar energy is required for "splitting" (decomposing) water molecules into hydrogen gas and oxygen. Combustion of hydrogen gas yields energy and water. Using solar energy to produce hydrogen from water could eliminate most air pollution and greatly reduce the threat of global warming.

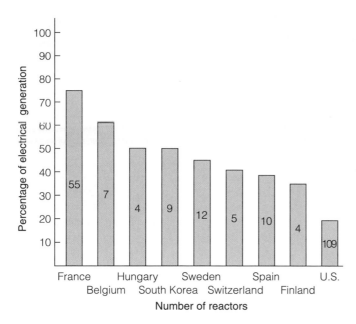

◆ **FIGURE 14.30** Percentage of total energy need supplied by nuclear energy and the number of operating reactors for selected countries, 1994.

Russia and Germany are working on a hydrogen-powered airliner, and Mercedes, BMW, and Mazda are road-testing prototype motor vehicles.

Nuclear Energy

Nuclear energy provides 16 percent of the world's electricity and 19 percent of the United States'. At one time it was considered *the* solution to the world's future energy needs—being clean, limitless, and "too cheap to meter." More than 400 nuclear power plants were built worldwide (◆ Figure 14.30). In December 1982, 42 nuclear power plants were operating in the United States, and by 1993 there were 109. Nevertheless, the Chornobyl accident in April 1986, after the 1979 Three Mile Island event in the United States, reinforced people's fears regarding nuclear power, resulting in the cancellation of all orders for new "nuke" plants in the United States (Case Study 14.5).

Nuclear reactors and fossil-fuel plants generate electricity by similar processes. Both processes heat a fluid, which then directly or indirectly makes steam used to spin turbine blades that drive an electrical generator. Uranium is the fuel of atomic reactors, because its nuclei are so packed with protons and neutrons that they are capable of sustained nuclear reactions. All uranium nuclei have 92 protons and between 142 and 146 neutrons. The common isotope uranium-238 (^{238}U)

has 146 neutrons and is barely stable. The next most common isotope, ^{235}U, is so unstable that a stray so-called slow neutron penetrating its nucleus can cause it to split apart completely, a process known as **fission**—a term particle physicists borrowed from the biological sciences. Nuclei that are split easily, such as those of ^{235}U, ^{233}U, and plutonium-239 (^{239}Pu), are called **fissile isotopes.** Fission occurs when a fissile nucleus absorbs a neutron and splits into lighter elements, called *fission products,* while at the same time emitting several "fast" neutrons and energy, 90 percent of which is heat (◆ Figure 14.31). The lighter fission products are elements that recoil from the split nucleus at high speeds. This energy of motion and subsequent collisions with other atoms and molecules creates heat, which raises the temperature of the surrounding medium.

Meanwhile, the stray fast neutrons collide with other nuclei, repeating the process in what is called a **chain reaction.** A controlled chain reaction occurs when one free neutron on average from each fission event goes on to split another nucleus. If more than one nucleus is split by each emitted neutron (on average), the rate of reaction increases rapidly, and the reaction eventually goes out of control. Because ^{235}U makes up only 0.7 percent of all naturally occurring uranium, the rest being ^{238}U, it must be "enriched" for use in commercial nuclear reactors.

In 1994, 109 commercial nuclear power plants were operating in the United States. Reactors that use slow neutrons to create controlled fission have a moderating material surrounding the "hot" fuel rods for slowing down the emitted fast neutrons and increasing the probability of nuclear collisions. Most U.S. reactors use water as the moderator and are known as *light-water reactors (LWR).* The components of an LWR reactor are shown in ◆ Figure 14.32. Note that a thick containment dome surrounds the fuel rods in the reactor core, the surrounding plumbing for cooling the core, and the plumbing for heat transfer to the turbines.

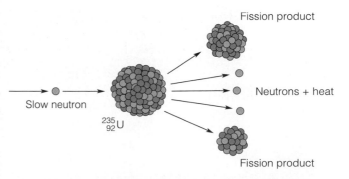

◆ **FIGURE 14.31** Nuclear fission. A slow neutron penetrates the nucleus of a U-235 atom, creating fission products (also called *daughter isotopes*), 2 or 3 fast neutrons, and heat.

Case Study | 14.5

Three Mile Island and Chornobyl—To Err Is Human; to Forgive Is Difficult

It started on 28 March 1979 with a valve that stuck and allowed core coolant to escape from the reactor vessel. No problem; the emergency core-cooling system (ECCS) automatically goes into action, adding "makeup" water to the core and keeping it cool. Enter the human factor. The operators misinterpreted the information fed to them by their instruments and *shut off* the ECCS; that is, they literally overrode the safety system that was designed and installed to mitigate just such an event. This human error at Three Mile Island Nuclear Station in Pennsylvania initiated the worst accident in the history of U.S. commercial power reactors.

With the ECCS shut down, the heat that built up in the core boiled off the cooling water surrounding the fuel rods. Before operators realized they had made a mistake, about a third of the fuel had melted and dropped into the cooling water in the bottom of the vessel. The vessel did not break, and the water quenched the molten rods, but the damage had been done. The material covering the fuel rods had deteriorated, and fission products had escaped into the steam to be carried into the containment building. Essentially all of the radioactivity remained in the building, but another human error allowed overflow water with low-level radioactivity to be pumped out of the leakproof containment structure to an adjacent building that was not leakproof.

Although "Three Mile Island" was a frightening incident, no adverse health effects have yet been detected. It is estimated that the gases that leaked from the containment structure gave the local residents a total radiation dosage equal to that of 4 chest X rays.

On 26 April 1986, Reactor Number 4 at Chornobyl, a nuclear facility 100 kilometers (60 mi) north of Kiev in the Ukraine, exploded (◁▷ Figure 1). The cause was similar to that at Three Mile Island, only here an experiment was being conducted in which the graphite-moderated reactor was running while the emergency core-cooling system was turned off. Miscalculation by operators allowed neutron buildup in part of the core, and the chain reaction went out of control—became *supercritical*, in the jargon of the nuclear scientist. Two explosions are known to have occurred, one of which blew the top off the reactor building. This latter explosion is believed to have been caused by a buildup of hydrogen when steam reacted with graphite, but both may have been gigantic steam explosions. With the core exposed, the graphite ignited and burned, furiously spewing radioactive materials directly into the atmosphere. Ten days were required to put out the fire, and when it was out, the reactor core was buried in concrete. Two workers at the plant were

◁▷ **FIGURE 1** Chornobyl Reactor Number 4 after the explosion.

killed instantly, and 29 others died later due to radiation exposure. In all, 500 people were hospitalized.

The significance of Chornobyl is that large quantities of radioactive materials were released directly into the atmosphere. In fact, scientists who were measuring atmospheric radiation in Sweden were the first to tell the world that an accident had occurred somewhere in the Soviet Union. Sadly, it took 36 hours to evacuate people within a 30-kilometer (18 mi) radius of the plant, and it is expected that within this group, there will be 500 more cancer-related deaths than normal in the next 70 years. Seventy-five million people lived within the fallout area in the western USSR and Eastern Europe, and the Chornobyl disaster may have increased the number of future cancer deaths in this group by as much as

24,000. All of the water, soil, and vegetation within the 30-kilometer fallout zone had to be decontaminated. Many areas are still "hot" and probably will never be cleaned up (◁▷ Figure 2). Gomel, 150 kilometers (90 mi) north of Chornobyl, has a child thyroid-cancer rate 80 times the world average. A reasonable medical explanation is that the Gomel residents' thyroid glands have absorbed and concentrated the radioactive iodine-131 that was released from the burning reactor. The prognosis is indeed grim for the long-term cancer toll from radiation exposure in the fallout area (◁▷ Figure 3).

The accidents at Three Mile Island and Chornobyl demonstrated differences in reactor design. The Russian reactor would not have been licensed in the United States, because it did not meet U.S. design and safety standards. Had Chornobyl Reactor Number 4 been enclosed by a containment building like the one at Three Mile Island, the escaping radioactivity would have been trapped and the disaster would have been averted, or at least minimized. Today, Number 4 is buried beneath concrete, but the other reactors at Chornobyl are back in operation. Six plant officials were convicted of negligence and are serving prison terms. Unit Number 3 is scheduled to be retired early in the twenty-first century.

◁▷ FIGURE 2 The Chornobyl fallout area.

LATVIA

Area of elevated radiation

LITHUANIA

RUSSIA

POLAND

Mensk

BELARUS

Gomel

Chernobyl Atomic
Energy Plant

Kiev

UKRAINE

0   100 mi

0   100 km

◁▷ FIGURE 3 A colt born in the Ukraine after the Chornobyl accident. Genetic damage is indicated by the colt's deformed legs and extra legs and hooves.

◆ **FIGURE 14.32** Main components of a light-water reactor. Note that the reactor is surrounded by a containment structure and that the steam supplied to the turbine is isolated from the reactor core.

This is to prevent the release of radioactivity to the environment in the event of an accident (see Case Study 14.5). If the reactor fuel vessel (the core) is not cooled properly, it will overheat, and it may reach the temperature where the fuel rods and the vessel melt. This is known as a **central-core meltdown.** When it occurs the molten material may eat its way through the floor of the containment structure and release radioactivity to the atmosphere.

The major risk associated with nuclear energy is the risk of contamination during processing, transportation, and disposal of ^{235}U and high-level nuclear waste products. The locations of these risks in the nuclear fuel cycle are apparent in ◆ Figure 14.33.

Geological Considerations

One of the many criteria in the siting of nuclear reactors is geologic stability. The selected site must be free from landsliding, tsunami, volcanic activity, flooding, and the like. The Nuclear Regulatory Commission (NRC), the federal government body that licenses nuclear reactors for public utilities, requires that all active and potentially active faults within a distance of 320 kilometers (200 mi) of a nuclear power plant be located and described. Reactors must be built a distance from an active fault determined by the fault's earthquake-generating potential. Further, the NRC defines an active fault as one that has moved once within the last 30,000 years or twice in 500,000 years. This is a very strict definition, and sites that meet these criteria are difficult to find along active continental margins. To provide additional safety and allow for geological uncertainty, reactors are programmed to shut down immediately at a seismic acceleration of 0.05 *g*. This is conservative indeed, and the shutdown would prevent overheating or meltdown should the earthquake be a damaging one.

Nuclear Waste Disposal

There is probably no more sensitive and emotional issue involved in geology today than the need to provide for safe disposal of the tons upon tons of radioactive materials that have accumulated over many decades. These wastes are the largest deterrent to further development of nuclear energy, as they present a legitimate hazard to humans, their offspring, and generations yet to come. Nuclear-waste products must be isolated, because they emit high-energy radiation that kills cells,

◆ **FIGURE 14.33** The nuclear fuel cycle. The steps include mining, enrichment, transportation, power generation, reclamation, and disposal.

causes cancer and genetic mutations, and causes death if individuals are exposed to large doses. The need to develop the means for safe, permanent disposal led Congress to pass the Nuclear Waste Policy Act in 1982 (NWPA). This law gave the Department of Energy (DOE) responsibility for locating a fail-safe underground disposal facility, a **geological repository,** for high-level nuclear waste and designated nine locations in six states as potential sites (◀▶ Figure 14.34). Based upon preliminary studies, President Reagan approved three of the sites for intensive scientific study (known as *site characterization*): Hanford, Washington; Deaf Smith County, Texas; and Yucca Mountain, Nevada. In 1987 Congress passed the Nuclear Waste Policy Amendments Act, which directed DOE to study only the Yucca Mountain site. This site's suitability is discussed later in this section.

Radioactive wastes are differentiated by the intensity of their radiation and by their physical form; that is, whether they are a solid, liquid, or gas. Their physical form determines how they need to be handled and their potential as a health and environmental hazard.

TYPES OF NUCLEAR WASTE

High-Level Wastes

High-level wastes, by-products of nuclear power generation and military uses, are the most intensely radioactive and dangerous. About once a year, a third of a reactor's fuel rods are removed and replaced with fresh rods. Replacement is necessary because, as the fissionable uranium in a reactor is consumed, the fission products capture more neutrons than the remaining uranium produces. This causes the chain reaction to slow,

and eventually it would stop. The radioactive rods, called *spent fuel,* are the major form of high-level nuclear waste. Currently they are stored in water-filled pools at the individual reactor sites, supposedly a temporary measure until a final grave is approved and prepared for them. More than 20,000 metric tons of spent fuel is stored across the country, and at many of the reactors, pool storage capacity is nearly filled (◀▶ Figure 14.35). Two major concerns about pool storage are that an unintended nuclear reaction might start in the pool and that the rods might deteriorate and release fuel pellets. It should be noted that although the fuel removed from a reactor every year weighs about 31,000 kilograms (65,000 lbs), because of its high density, it is only about the size of an automobile.

Alternatively, the rods are chopped up and *reprocessed* to recover unused uranium-235 and plutonium-239. However, the liquid waste remaining after reprocessing contains more than 50 radioactive isotopes, such as strontium (^{90}Sr) and iodine (^{131}I), that pose significant health hazards. In 1992, the Department of Energy (DOE) decided to phase out reprocessing of spent fuel for the purpose of recovering chemical stocks for nuclear weapons. Unfortunately, the volume of existing reprocessed waste in temporary storage would cover a football field to a depth of 60 meters (200 ft). These wastes are currently stored in metal tanks at four U.S. sites. The oldest site, the one at Hanford, Washington, originally stored wastes in single-walled steel tanks. In 1956 a tank leak was detected, and since then, 750,000 gallons of high-level waste have leaked underground from 60 of 149 storage tanks. It is said by some that "radioactive waste was invented here." Cleanup at Hanford may be the costliest environmental cleanup in world history; the cost has been estimated as at least $57 billion.

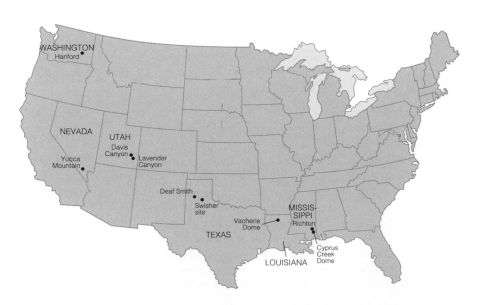

◀▶ **FIGURE 14.34** The nine potential repository sites for high-level nuclear waste established in 1982. A repository has yet to be built, and the Yucca Mountain location was the only one being studied in 2000. The Texas, Louisiana, and Mississippi sites are all salt domes or bedded salt deposits.

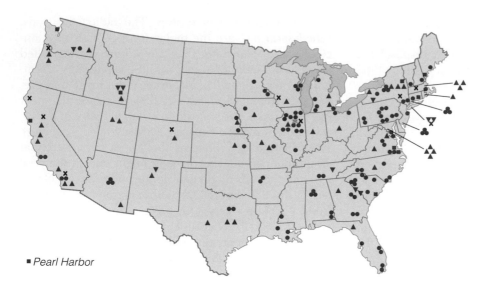

◆ **FIGURE 14.35** Location of temporarily stored spent nuclear fuel and other high-level nuclear waste awaiting permanent disposal, 2000. The symbols do not reflect precise locations.

■ *Pearl Harbor*

Transuranic Wastes

Most (96 percent) of the nuclear fuel in a reactor is composed of uranium-238, which is not involved in the chain reaction. Instead, ^{238}U absorbs neutrons and becomes a **transuranic element**—one whose atomic number is higher than that of uranium. Thus the ^{238}U in a fuel rod changes into one of the 11 unstable transuranic elements *(TRU),* of which plutonium, fermium, nobelium, einsteinium, and berkelium are examples. The DOE Waste Isolation Pilot Plant (WIPP) has been designed to store transuranic waste in vast salt deposits 660 meters (2,150 ft) beneath the desert surface 44 kilometers (26 mi) east of Carlsbad, New Mexico. These salt deposits are impermeable, easily mined, and capable of eventually self-sealing around voids, such as the waste-storage rooms that are created by mining. The site was certified by the EPA on 13 May 1998, and the first shipment of transuranic wastes was received from Los Alamos National Laboratory on 26 March 1999.

Low-Level Wastes

Low-level waste, by NCR definition, is anything that is not high-level waste or transuranic waste. Low-level wastes are created in research labs, hospitals, and industry as well as in reactors. Ocean dumping of these wastes in 55-gallon drums was a common practice between 1946 and 1970. Since then, shallow land burial has been the preferred disposal method, but many sites have been found to leak.

Mill Tailings

Uranium mill tailings, the finely ground residue that remains after uranium ore is processed, are typically found in mountainous piles outside the mill facility.

Large volumes of tailings exist that still contain low concentrations of radioactive materials, including thorium (^{230}Th) and radium (^{226}Ra), both of which produce radioactive radon gas (^{222}Rn), the so-called hidden killer (Case Study 14.6). Most of the abandoned tailings piles are in sparsely settled areas of the West (◆ Figure 14.36). A former Atomic Energy Commission policy permitted the use of uranium-mill tailings in the manufacture of building materials such as concrete blocks and cement. For almost twenty years, 30,000 citizens of Grand Junction, Colorado, lived in homes whose radon levels were up to seven times the maximum allowed for uranium miners. Uranium tailings were used in building materials in Durango, Rifle, and Riverton, Colorado; Lowman, Idaho; Shiprock, New Mexico; and Salt Lake City, Utah.

The Uranium Mill Tailings Radiation Control Act (1978) makes the DOE responsible for 24 inactive sites in 10 states left from uranium operations. The act specifies that the federal government is to pay 90 percent of the cleanup cost, with the respective state to pay the remainder. Without some form of management, uranium-mill tailings could become a serious problem, because they contain the largest volume of radioactive waste in the country.

ISOLATION OF NUCLEAR WASTE

Scientists agree that nuclear waste must be isolated from contact with all biological systems for at least 250,000 years in order for radioactive materials—particularly plutonium-239, whose half-life is 24,000 years—to decay to harmless levels. Should the ban on plutonium recovery be rescinded, the conditions for high-level-waste disposal may change. Several methods of storing wastes have been proposed, but any method must meet these conditions:

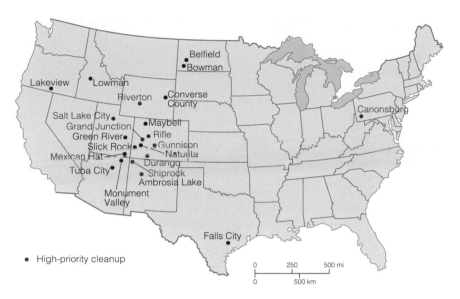

◆▶ **FIGURE 14.36** Inactive uranium ore mill sites. Locations indicated by red dots have been designated by the EPA as high-priority cleanup and stabilization sites.

● High-priority cleanup

- Safe isolation for at least 250,000 years,
- Safe from sabotage or accidental entry,
- Safe from natural disasters (hurricanes, landslides, floods, etc.),
- Noncontaminating to any nearby natural resources,
- Geologically stable site (no volcanic activity, earthquakes, etc.), and
- Fail-safe handling and transport mechanisms.

A number of disposal methods have been proposed. Six of the more serious ones are discussed here.

1. *Rocket the waste into the sun or deep space.* Rocket launchings are prohibitively expensive, and an explosion during launch could contaminate the atmosphere, ocean, and land.
2. *Continue the present tank storage.* This is not a permanent solution, because the tanks have limited lifetimes and leakage has already occurred.
3. *Dispose of waste in a convergent plate boundary (subduction zone).* The feasibility of this very imaginative proposal has not been demonstrated, because the rate of plate movement is very slow.
4. *Place containerized waste on the ice sheets of Antarctica or Greenland.* Radioactive heat would cause the containers to melt downward to the ice-bedrock contact surface, which could place the radioactive waste two miles beneath ice in either Greenland or Antarctica. This proposal assumes that long-term global warming will not occur in the 250,000-year period. This is unsound, because the last ice sheet of the Pleistocene melted back to Greenland in less than 10,000 years.
5. *Geologic repository: Salt mines or domes.* Salt is dry, flows readily, and is self-healing. Because it is a good conductor of heat, temperatures in the repository rocks would not become excessive. Salt is subject to underground solution, however, and some hydrated minerals in salt beds give up their water when heated. Salt beds and salt domes occur in geologically stable regions such as in Texas, Louisiana, and Mississippi.
6. *Geologic repository: Deep chambers in granite, volcanic rocks, or some other very competent, relatively dry rock formation.* This would require detailed site analysis to ensure that all geologic criteria are met.

Yucca Mountain Geologic Repository

The DOE's Civilian Radioactive Waste Management Program has been conducting a site characterization of Yucca Mountain, Nevada, as a high-level-nuclear-waste repository since the early 1980s (◆▶ Figure 14.37). The 7-square-kilometer (2.7-mi²) site is in welded ash-flow tuffs (see Chapter 5), and the waste would be placed more than 200 meters (650 ft) below the surface (◆▶ Figure 14.38). A favorable geologic condition at Yucca Mountain is the depth to the water table, 224–366 meters (800–1,200 ft) below the level of the proposed storage chambers. Important geologic questions in the site investigation include:

- What is the depth, thickness, and extent of the host rocks?
- How much water is in the zone of aeration, and how does it move through the rocks?
- What effect might nearby volcanic activity have on the repository?
- How might an earthquake affect the repository and the water table below it?

All geologic studies indicate that Yucca Mountain could be an acceptable geologic repository, but until it receives final certification, radioactive wastes are

Case Study | 14.6

Radon and Indoor Air Pollution

Radon (^{222}Rn) is a radioactive gas formed by the natural disintegration of uranium as it transmutes step-by-radioactive-step to form stable lead. The immediate parent element of radon is radium (^{226}Ra), which emits an alpha particle (^{4}He) to form ^{222}Rn. Radon-222 has a half-life of only 3.8 days, but it in turn breaks down into other elements that emit dangerous radionuclides. *Emanation* is sometimes used to describe the behavior of this element, and concern has been growing about the health hazard posed by radon emanating from rocks and soils and seeping into homes. Radon was first suggested as a danger to miners by Georgius Agricola (1494–1555). Known as "the father of mineralogy," Agricola proposed that the lung disease contracted by miners might have something to do with emanations inside the mines. The EPA estimates that between 5,000 and 20,000 people die every year of lung cancer because they have inhaled radon and its decay products.

Radon is only a problem in areas underlain by rocks whose uranium concentrations are greater than 10 ppm—three times the average amount found in granite. For comparison, uranium ore contains more than 1,000 ppm of uranium. As radon disintegrates in the top few meters of rock and soil, it seeps upward into the atmosphere, or into homes through cracks in concrete slabs or around openings in pipes. The measure of radon activity is *picocuries per liter (pCi/L)*—the number of nuclear decays per minute per liter of air. One pCi per liter represents 2.2 potentially cell-damaging disintegrations per minute. Inasmuch as the average human inhales between 7,000 and 12,000 liters (2,000–3,000 gallons) of air per day, high radon concentrations in household air can be a significant health hazard. In addition, radon can access the human body through well water if it is ingested within a few weeks of withdrawal from underground. Average uranium concentrations of some common rocks are

black shale	8.0 ppm
granite	3.0 ppm
schist	2.5 ppm
limestone	2.5 ppm
basalt	1.0 ppm

Bear in mind that these figures do not indicate the *range* of uranium content in the rocks, which for black shales and granites can be quite great.

Cancer due to bombardment of lung tissue by breathing radon-222 or its decay products is a major concern of health scientists. The EPA has established 4 pCi/L as the maximum allowable indoor radon level. This is equivalent to smoking a half-pack of cigarettes per day or having 300 chest X rays in a year. ◆ Figure 1, opposite, vividly illustrates the dangers of inhaling indoor radon (and of smoking). Such cancer is rarely apparent before 5–7 years of exposure, and its incidence is rare in individuals under the age of 40.

Widespread interest in radon pollution first appeared in the 1980s with the discovery of high radon levels in houses in Pennsylvania, New Jersey, and New York. The discovery was made accidentally when a nuclear-plant worker set off the radiation monitoring alarm when he arrived at work one day. Since his radioactivity was at a safe level, he did not worry about it. Returning to work after a weekend at home, he once again triggered the alarm. An investigation of his home revealed very high levels of radon. This was the first time radon was recognized as a public health hazard.

There are currently two types of radon "detectors." One is a charcoal canister that is placed in the household air for a week and then returned to the manufacturer for analysis. The other is a plastic-film alpha-track detector that is placed in the home for 2–4 weeks. It is recommended that the less-expensive canister type be used first, and that if a high radon level is indicated, a follow-up measurement with a plastic-film detector be made. Any reading above 4 pCi/L requires follow-up measurements. If the reading is between 20 and 200 pCi/L, one should consider taking measures to reduce radon levels in the home (◆ Figure 2). Common methods are increasing the natural ventilation below the house in the basement or crawlspace, using forced air to circulate sub-floor gases, and intercepting radon before it enters the house by installing gravel-packed plastic pipes below the floor slabs. For houses with marginal radon pollution, sealing all openings from beneath the house that go through the floor or the walls is usually effective.

piling up in temporary storage facilities at an alarming rate.

Completion of the Yucca Mountain repository was originally scheduled for 1998, but the year has been extended to 2010. In July 1999 the DOE issued a Preliminary Environmental Impact Statement with the conclusion that very little radiation would leak from the proposed Yucca Mountain repository. The report said that the repository would be as safe as and more eco-

nomical than storing wastes at the dozens of nuclear sites across the nation. The DOE also stated that several issues needed further study before the agency could recommend Yucca Mountain as the final geologic jail for nuclear waste. Approval and the start of construction are scheduled for 2005.

Radioactive waste disposal is a problem for all nations that generate nuclear power. The British are contemplating a site in the beautiful Lake District—

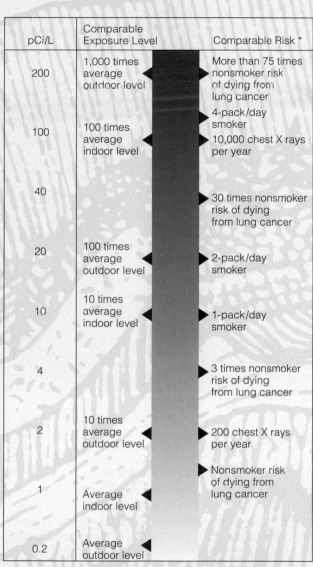

pCi/L	Comparable Exposure Level	Comparable Risk *
200	1,000 times average outdoor level	More than 75 times nonsmoker risk of dying from lung cancer
100	100 times average indoor level	4-pack/day smoker
		10,000 chest X rays per year
40		30 times nonsmoker risk of dying from lung cancer
20	100 times average outdoor level	2-pack/day smoker
10	10 times average indoor level	1-pack/day smoker
4		3 times nonsmoker risk of dying from lung cancer
2	10 times average outdoor level	200 chest X rays per year
		Nonsmoker risk of dying from lung cancer
1	Average indoor level	
0.2	Average outdoor level	

* Based on lifetime exposure

◇▷ FIGURE 1 Lung cancer risk due to radon exposure, smoking, and X-ray exposure. A few houses have been found with radon radiation levels greater than 100 pCi/L, and one in Pennsylvania had a level of 2,000 pCi/L.

◇▷ FIGURE 2 Routes of radon entry into a home. Plugging radon leaks is the most common method of reducing residential radon accumulations.

the backyard of Wordsworth, Constable, and Beatrix Potter—and the Swiss are considering drilling a tunnel into Mount Wellberg to sequester their waste. The problem in Sweden is being addressed by phasing out nuclear power until such a time as satisfactory disposal of high-level wastes is achieved. The Swedes have low- and intermediate-level-waste disposal sites that are to be envied—tunnels under the Baltic Sea.

HEALTH AND SAFETY STANDARDS FOR NUCLEAR WASTE

There is general agreement that any nuclear-waste repository will leak radiation over the course of the next 10,000 years. Nuclear waste emits three types of radiation:

1. Alpha (α) particles, the most energetic but least penetrating type of radiation (they can be stopped by a

(a)   (b)

◆▶ **FIGURE 14.37** (a) Location of the proposed Yucca Mountain nuclear repository. (b) Aerial view of the Yucca Mountain study area. The structures are at the site of the proposed repository.

piece of paper). If α emitters are inhaled or ingested, alpha radiation can attack lung and body tissue.

2. Beta (ß⁻) radiation, penetrating but most harmful when a ß⁻ emitter is inhaled or ingested. For example, strontium-90 is a ß⁻ emitter that behaves exactly like calcium (builds bone) in the food chain.

3. Gamma (γ) rays, nuclear X rays that can penetrate deeply into tissue and damage human organs (◆▶ Figure 14.39).

Nuclear waste policy limits the radioactivity that may be emitted from a repository, as measured by cancer deaths directly attributable to leaking radiation. The mortality limit is 10 deaths per 100 years, and for this reason the maximum allowable radiation dose per year per individual in the proximity of a repository is set at 25 millirems. The *rem* is the unit used for measuring the amounts of ionizing radiation that affect human tissue, and a millirem (mrem) is 0.001 rem. Doses on the order of 10 rems can cause weakness, redden the skin, and reduce blood-cell counts. A dose of 500 rems would kill half of the people exposed. Rem units take into account the type of radiation—alpha, beta, or

gamma—and the amount of radioactivity deposited in body tissue. ◆▶ Figure 14.40 shows how the average annual dose of 208 mrems is partitioned for U.S. citizens and common sources of radiation. (Case Study 14.6 explains the hazards of high levels of radon gas in the home and methods for testing and mitigating them.) It is estimated that a person living within 5 kilometers (3 mi) of the Yucca Mountain repository would receive less than 1 mrem per year as a result of nuclear waste being stored there (currently no one lives that close to the site). The 25-mrem standard for maximum annual dosage is equal to the radiation a person would receive in 2 or 3 medical X rays.

IS THERE A FUTURE FOR NUCLEAR ENERGY?

Nuclear reactors are very complex; a typical U.S. reactor may have 40,000 valves, whereas a coal-fired plant of the same size has only 4,000! This fact alone makes reactors relatively unforgiving of errors in operation, construction, or design. A newer generation of "standard" reactors are now under consideration that use natural forces such as gravity and convection instead of the network of pumps and valves currently re-

◈ **FIGURE 14.38** Paintbrush Tuff Formation at Yucca Mountain. Rock layers called *members* are distinguishable beds within the formation. The proposed position of the repository is shown.

◈ **FIGURE 14.39** Penetrating abilities of alpha, beta, and gamma radiation, the three types of radiant energy emitted from nuclear wastes.

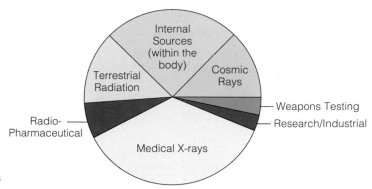

◈ **FIGURE 14.40** The typical American receives 208 millirems of radiation per year from the various sources. Some medical, research, and industrial workers' exposures may exceed the 208-millirem average.

quired. For example, if the core of one of these new reactors should overheat, a deluge of water stored in tanks above the reactor would inundate the core. In addition, design would be standardized, which would greatly simplify construction, maintenance, and repair. Standardization would also facilitate better operator training and would eliminate the possibility of some human errors.

The ultimate, clean nuclear-energy source is released by **fusion,** whereby two nuclei merge to form a heavier nucleus and in the process give off tremendous amounts of heat. Deuterium ($_1^2$H) is a heavy isotope of hydrogen that at very high temperatures fuses to form helium ($_2^4$He). This is the very same nuclear process that energizes our sun. The fuel is inexhaustible, the reaction cannot run out of control, and it is environmentally nonpolluting.

Energy from the Sea

OCEAN THERMAL ENERGY CONVERSION

Ocean thermal energy conversion (OTEC) utilizes the temperature difference between warm surface water

and colder deep waters, at least 22°C (72°F), to vaporize a low-boiling-point fluid and operate a generator. A potential OTEC site must be in a coastal area where water depth increases rapidly enough to obtain the needed temperature differential and close enough to shore for power transmission. The Island of Oahu in Hawaii has the population, need, and oceanographic conditions for this form of energy conversion. A mini-OTEC of about 50 kilowatts was operated in Hawaii in the 1980s. The electricity produced was not transmitted to the shore, but a continuously burning light bulb on a float near the village of Mokapu testified to OTEC capability (◈ Figure 14.41).

WAVE ENERGY

Capturing wave energy and converting it to electricity is not a new idea, and its practicality has improved very little since 1900. Wave energy is so diffuse—spread over such a large area—that concentrating and utilizing it is difficult. In order to convert a worthwhile amount of energy from wave motion, the energy must be focused by "horns," or funnels, extending seaward. In a 1971 experiment, a wave-powered generator hanging from a pier was used to power a string of lights of Pacifica, California. The Japanese currently market a wave-power machine for providing energy to lighthouses and buoys, and they have been working on wave-powered electrical-generating barges for offshore installation (Figure 14.41).

TIDAL CURRENTS

Tidal currents have been used by coastal dwellers to power mills for at least a thousand years. Restored, working tidal mills in New England and Europe are popular tourist attractions. Both flood (landward-moving) and ebb (seaward-moving) tidal currents are used. Modern designs utilize basins that fill with seawater at high tide and empty at low tide, generating electricity in the process. A potential site must have a large tidal range—the height difference between high and low tides—and adequate area for storing the elevated water. It is estimated that there are more than 100 suitable sites in the world that, together, are capable of generating 5 percent of the world's present generating capacity (Figure 14.41).

The first and best-known large-scale tidal power plant is on the Rance River near St. Malo on the English Channel in northwest France. The tidal range here

Tidal Power Plant

Wave Power Plant

Ocean Thermal Electric Plant

◆ **FIGURE 14.41** Energy from the sea.

(a) High tide

(b) Low tide

▷ **FIGURE 14.42** Proposed Passamaquoddy site tidal-power project. (a) Desired flow of water into Passamaquoddy Bay at high tide and (b) from Passamaquoddy Bay into Cobscook Bay and the Bay of Fundy at lowtide. A power plant at the narrows between Moose Island and the mainland would generate electricity as water drains from the high pool to the low pool.

In North America interest has focused on the Bay of Fundy in the Atlantic Ocean's Gulf of Maine. This bay has the greatest tidal range in the world, about 16 meters (52 ft), and a storage area in excess of 350 square kilometers (135 mi²). The bay is made up of 9 smaller bays with a total generation potential of 30 gigawatts (1 GW = 1,000 MW), equivalent to 30 conventional power plants. Under consideration by Canada and the United States for more than a half-century, the proposed project is known in both countries as the *Passamaquoddy Site*. Through a series of dams and locks, high-tide water would be stored in Passamaquoddy Bay and then drained at low tide to produce electricity (▷ Figure 14.42). It is a very ambitious and costly project, but as population and demand for nonpolluting energy increase, the project is likely to be built in the next century. With an anticipated generating capacity of 65–345 MW, its capacity as well as its size would dwarf the Rance power station.

varies from 9 to 14 meters (30–46 ft), and the power output is a reversible operation; that is, the plant generates power at both flood and ebb tides. Because the hydrostatic head (water drop) is only a few meters, 24 small 10-MW generators with a total capacity of 240 MW are used. The estuary of the Rance River serves as a reservoir for high water, and locks through the dam allow navigation upstream. The Rance project is nonpolluting, and it has been a boon to the tourist-dependent province of Brittany. On balance, the operation has been successful.

Consider This... ?
What nonfossil-fuel energy sources are likely to be expanded in the United States in the coming century? In other words, if you wanted to make a long-term investment in one of them, which one would you choose? ■

Gallery

Energy Is where You Find It

Humans devote a great deal of mental and physical energy to harnessing the earth's energy. Energy, like fresh water, is invaluable, and when it is in short supply a family, a tribe, or a country will do almost anything to obtain it. The photos in this gallery, ◆ Figures 14.43 to 14.46, illustrate some of the earth's nonconventional energy sources and some relics of fossil energy. Keep in mind that all these energy resources derive from our sun.

◆ FIGURE 14.43 A community gathers around its solar-cell-powered television set; the Republic of Niger. More than likely they are watching a cricket or soccer match, not "Wheel of Fortune."

◆ FIGURE 14.44 The Blue Lagoon, a great hot-water pool 40 kilometers (24 mi) from Reykjavík, Iceland. It uses geothermal power-plant outlet water, which is sally and grows algae, giving it its blue color. Open every day, the lagoon attracts about 100,000 visitors annually.

◆▶ **FIGURE 14.45** Peat mining in Northern Ireland. Cut into blocks and dried, the peat is used for heating homes in some parts of the British Isles.

◆▶ **FIGURE 14.46** A new windmill design, the Darrieus rotor, is utilized at Altamont Pass, California. The vertical-axis machine is nicknamed the "eggbeater" because of its appearance. The design is a major departure from horizontal-axis machines used to drive electrical generators. Darrieus rotors are more efficient and less expensive than propeller-driven types, and they do not need to face the wind, so they do not require pitch and pivoting mechanisms. Their principal drawback is their inability to self-start; a generator is needed to start the blades spinning.

Summary

ENERGY
DEFINED
Capacity or ability to do work. A society's quality of life depends in large part on the availability of energy.

TYPES
1. Renewable energy is replaced at least as fast as it is consumed; e.g., solar, wind, and hydrologic energy.
2. Nonrenewable energy is replenished much more slowly than it is utilized; e.g., energy derived from coal, oil, and natural gas (fossil fuels).

PETROLEUM
DEFINED
Volatile hydrocarbons composed mostly of hydrogen and carbon. Crude oil contains many different hydrocarbon compounds.

ORIGIN AND ACCUMULATION
The necessary conditions for an oil field:
1. Source rock—origin as organic matter;
2. Reservoir rock—porous and permeable;
3. Caprock—impermeable overlying stratum; and
4. Geologic trap—such as an anticline, faulted anticline, reef, or stratigraphic trap.

PRODUCTION
Drilled wells are either pumped or they flow if under gas and water pressure. Even with the best drilling and producing techniques, primary production may recover only half the oil.

SECONDARY RECOVERY
Oil remaining in pore spaces underground is stimulated to flow to a recovery well by injecting water, chemicals, steam, or CO_2. Much of the "stuck" oil can be stimulated to flow to the extraction well by these methods.

VALUE
Varies with weight. Light crude oil is the most valuable.

RESERVES
At the 1988 rate of consumption, known world reserves of crude oil will last until 2050 and natural gas until 2110. The latest forecast is that world oil reserves will last until 2100.

ENERGY GASES
May be the fuel of the first half of the twenty-first century.
1. Methane hydrates, ice balls of natural gas
2. Natural gas from unconventional sources
3. Hydrogen gas

COAL
DEFINED
Carbonaceous residue of plants that has been preserved and altered by heat and pressure.

RANK
As coal matures, it increases in heat content and decreases in the amount of water and volatile matter it contains. The sequence is plant matter to peat, to lignite, to subbituminous, to bituminous, to anthracite.

RESERVES
The United States has enough coal to last 200–300 years.

SYNFUELS
The manufacture of combustible gas, alcohol, and gasoline from coal has a promising future.

OTHER FOSSIL FUELS
TAR SANDS
Sands containing oil that is too thick to flow and that can be surface-mined. Oil is then washed from the sand. One place the sands are mined is the Athabasca Field in Alberta, Canada.

OIL SHALE
Eocene lake deposits containing light kerogen oil that can be removed by heating. Deposits are found in Wyoming, Utah, and Colorado.

PROBLEMS WITH FOSSIL-FUEL COMBUSTION
AIR POLLUTION
Products of combustion contribute to global warming, smog, and acid rain. Acid rain is a hazard in the U.S. East and Southeast and in eastern Canada, which receives some of the U.S. emissions. This is a major international air-pollution problem.

MITIGATION
Sulfur content must be reduced below 1 percent. Methods include fluidized-bed combustion, neutralizing sulfur dioxide in smokestacks with limestone scrubbers, cleaning coals by gravity separation, and chemical leaching methods.

SUBSIDENCE AND COLLAPSE
Areas underlain by shallow lignite are subject to subsidence.

ALTERNATIVE ENERGY SOURCES
DIRECT SOLAR
1. Solar heat collectors warm a fluid, which is then put to work.
2. Photovoltaic cells convert light directly into electricity.

INDIRECT SOLAR
Wind is produced by unequal heating of the earth's surface.

GEOTHERMAL
Facilities for generating electricity with geothermal energy currently are in place and may supply as much as 2 percent of U.S. electrical energy needs. Geothermal heat is also used worldwide for heating interior spaces, keeping sidewalks free of winter snow and ice, etc.

HYDROELECTRIC
Hydroelectric dams harness the power of water distributed over the earth by the hydrologic cycle.

OCEANS
The energy in waves, currents, and tides can be converted to do work for humans. Ocean thermal energy conversion (OTEC) shows some promise, but little has been done with waves and currents. Tidal power is being used successfully only at the Rance River, France, at present.

FUEL CELLS
Fuel cells generate electricity fueled by hydrogen. Many think this is the fuel–motor combination of the future.

NUCLEAR ENERGY
Provides 16 percent of world and 19 percent of U.S. energy needs. Heat generated by fission of uranium isotopes and their daughter-isotopes such as plutonium is converted to electricity. There are 109 commercial nuclear power plants operating in the United States.
1. *Problems*—Mining, transportation, and disposal of high-level nuclear materials are the most dangerous aspects of the nuclear energy cycle. If the chain reaction runs out of control, the heat generated may lead to a central-core meltdown such as occurred at Chornobyl.
2. *Geological considerations*—Very careful geological site studies and investigations are performed to assess hazards from earthquakes, mass-wasting, subsidence, and coastal and riverine floods.
3. *Nuclear waste considerations*—High-level nuclear waste is a major problem. Tons of these wastes sit in water-filled pools at reactor sites waiting for permanent storage. Yucca Mountain in Nevada is the only geological repository site being studied, and due to political and environmental delays, the earliest it can be ready to accept waste is 2010.

◆ Key Terms

acid rain	geological repository	OTEC	slant drilling
anticline	geothermal energy	photochemical smog	source rock
biomass fuel	high-level nuclear wastes	radon	stratigraphic trap
caprock	hydrocarbon compound	rank (coal)	synthetic fuels (synfuels)
central-core meltdown	kerogen	renewable resource	tar sands
chain reaction	low-level nuclear wastes	reserves	transuranic element
fissile isotopes	miscible	reservoir rock	wildcat well
fission	nonrenewable resource	salt dome	
fusion	OPEC	secondary recovery	

◆ Study Questions

1. What four conditions are required for an oil deposit to form?
2. What oceanographic conditions favor the formation of a good source rock for oil?
3. Draw a geologic cross section that shows an anticline and a faulted anticline. Assume your cross section is looking north. What are the directions of the dips and strikes of the two limbs of the anticline? Dip and strike determination is explained in Appendix 3.
4. What nonrenewable energy resources exist in the United States, and which one is most abundant?
5. Describe atomic fission. How does it create heat for electrical generation?
6. What are methane hydrates and what is their potential as fuel for the twenty-first century?
7. What are the environmental consequences of burning fossil fuels? What will be the impact upon the global environment of continued dependence upon fossil fuels?

8. How can solar energy be collected and used for heating, generating electricity, and producing fuels?

9. What sector of the U.S. economy uses the most fossil fuel, and what international problem is connected with this use?

10. Explain how photochemical smog is formed. (Show the chemical steps and the required conditions.)

11. Explain the meaning of renewable energy. Which renewable energy sources have been developed, and which undeveloped renewable sources show the most promise?

12. Solar–hydrogen is believed by many to be the fuel for the 21st century. What is solar–hydrogen, and how can it be put to work for us?

◁▷ For Further Information

BOOKS AND PERIODICALS

Abelson, Philip H. J. 1987. Energy futures. *American scientist* 75 (Nov./Dec.):584–593.

Carter, L. M. H., ed. 1995. *Energy and the environment: Application of geosciences to decision-making.* U.S. Geological Survey circular 1108.

Charlier, Roger. 1982. *Tidal energy.* New York: Van Nostrand Reinhold Co.

Churchill, Ron. 1997. Radon mapping. *California geology,* November–December:167–177.

Cooper, George A. 1994. Directional drilling. *Scientific American,* May:82–87.

Deutsch, Robert W. 1987. *Nuclear power: A rational approach.* 4th ed. Columbia, Md.: G. P. Courseware.

Duffield, W. A.; J. H. Sass; and M. Sorrey. 1994. *Tapping the earth's natural heat.* U.S. Geological Survey circular 1125.

Finch, Warren I. 1997. Uranium, its impact on the national and global energy mix. U.S. Geological Survey circular 1141.

Goodger, E. M. 1975. *Hydrocarbon fuels.* New York: John Wiley and Sons.

Hafele, Wolf. 1990. Energy from nuclear power. *Scientific American* 263, no. 3.

Haq, Bilal U. 1998. Gas hydrates: Greenhouse nightmare? Energy panacea or pipe dream? *GSA today* (Geological Society of America) 8, no. 11 (November).

Hirsch, Robert L. 1987. Impending United States energy crisis. *Science* 235:1467–1473.

Kvenvolden, Keith A. 1998. Potential effects of gas hydrate on human welfare. *National Academy of Sciences colloquium: Geology, mineralogy, and human welfare.* Irvine, California, November 7–9.

McCabe, Peter J., and others. 1993. *The future of energy gases.* U.S. Geological Survey circular 1115, Public issues in earth science series.

Magoon, Leslie B., ed. 1988. *Petroleum systems of the United States.* Washington, D.C.: U.S. Geological Survey professional paper 1870.

Mossop, Grant D. 1980. Geology of the Athabasca oil sands. *Science* 207, p. 145.

National Research Council. 1986. *Acid deposition: Long-term trends.* Washington, D.C.: National Academy Press.

Reese, April. 1999. Bad air days. *E: The environmental magazine,* November–December:28–35.

Robertson, John B. 1996. Why have earth scientists failed to find suitable nuclear waste disposal sites? *Geotimes,* August:16–19.

Ross, David. 1987. *Energy from waves,* 2d ed. New York: Pergamon Press.

Schmandt, Jurgen; J. Clarkson; and Hilliard Roderick, eds. 1988. *Acid rain and friendly neighbors: The policy dispute between Canada and the United States.* Durham, N.C.: Duke University Press.

Scientific American. 1990. Energy for planet earth. *Scientific American* special issue, September.

Sperling, Daniel. 1996. The case for electric vehicles. *Scientific American,* November:54–59.

Suess, Erwin; Gerhard Bohrmann; Jens Greinert; and Erwin Lausch. 1999. Flammable ice. *Scientific American,* November:76–83.

Sweet, William. 1988. *The nuclear age,* 2d ed. Washington: Congressional Quarterly, Inc.

U.S. Department of Energy. 1999. Draft environmental impact statement for a geologic repository for the disposal of spent nuclear fuel and high-level radioactive waste at Yucca Mountain, Nye County, Nevada—Summary.

U.S. Geological Survey. 1995. *National assessment of the United States oil and gas resources.* Circular 1118, to accompany CD-ROM, USGS digital data series 30, for Macintosh- and Windows-based PC systems. Available on Internet for downloading.

Weart, Wendell, and others. 1998. The waste isolation pilot plant. *Geotimes,* October:14–19.

Young, Robert, and Jan Young. 1971. *Gusher: The search for oil in America.* New York: Julian Messner, a Division of Simon and Schuster.

INTERNET SITES

Solar cars
  http://www.solarcar.arizona.edu/default.asp
Three Gorges Dam
  http://www.threegorges.org
U.S. Geological Survey
  http://www.usgs.gov

Waste Isolation Pilot Plant (WIPP)
   http://www.wipp.carlsbad.nm.us/wipp/
   _____. EPA home page
   http://www.epa.gov/radiation/wipp/

Wind energy
   http://www.eren.doe.gov/
   _____. California
   http://www.energy.ca.gov/wind/windfacts.html

◀▶ World Wide Web Activities

Did Methane Hydrates Cause the End-Cretaceous Extinction Event?
http://www.newscientist.com/ns/19991120/newsstory1.html

This article from *New Scientist* reports that scientists from the Naval Research Laboratory in Washington, D.C. have postulated a role for methane hydrates in the end-Cretaceous extinction event, which wiped out the dinosaurs and all other large land and sea animals.

1. What is the proposed relationship between the role of methane hydrates and asteroid impact in this scenario?
2. Do scientists have any evidence for explosive methane hydrate releases since the end-Cretaceous extinction event?

Experimental Approaches to Solar Energy Production
http://www.eren.doe.gov/sunlab/TechOverview.htm

SunLab is a joint project of Sandia National Laboratories and the National Renewable Energy Laboratory. This page introduces technological strategies of SunLab currently being explored to harness solar energy.

1. What are the three technological approaches illustrated here to trap the sun's energy? Pick one to explore further, using Web site links under the photos.
2. Is solar energy competitive today in general? In what specific instances might it be feasible as a power source?

Alternative Energies Display from the Smithsonian
http://americanhistory.si.edu/csr/powering/images/alter.htm

This two-page display from the Smithsonian Institution is a photo gallery of alternate energy sources. Peruse the two pages.

1. Name the sites shown here that use heat energy from Earth's interior.
2. Name the sites shown here that use solar energy either directly or indirectly.
3. Which of the sites uses gravity as an energy source, along with solar energy?

The Green River Formation: National Fossil Treasure and Petroleum Reserve
http://www.crcpress.com/index.htm?catalog/4615
http://www.ucmp.berkeley.edu/tertiary/eoc/greenriver.html

This pair of Web sites addresses the Green River Formation oil shales from very different perspectives. Both sites are technically excellent. Comparing them will provide insight into how people may view the same resource quite differently.

1. Peruse the abstract and table of contents of this authoritative text on the Green River site. Is there any mention of the Green River Formation as a fossil trove of global significance?
2. Read through the text for the second site (and view the glorious pictures). Would the reader have a clue regarding the potential petroleum resource of the Green River Formation from this material? Note that even the definition of oil shale here is uninformative.

Earth Systems Today CD Correlations

CHAPTER CONCEPT: Petroleum Deposits Require Organic Source Material, and Low-to-Moderate Temperature and Pressure to Mature

The rock cycle segment of the CD explains conditions of formation for igneous, metamorphic, and sedimentary rocks.

SECTION MODULE: The Rock Cycle and Rock Laboratory

Observe the physical conditions (pressure, temperature) and chemical changes that take place as Earth materials are cycled from one form to the next.

QUESTION: *Given the prerequisites for oil and gas formation and preservation, in rocks of which family are almost all petroleum resources found?*

CHAPTER CONCEPT: Oil Floats on Water, and Natural Gas is Less Dense than the Oil

The ground-water segment of the CD demonstrates the disposition of the water table and its relationship to the surface in a variety of terrain and rock-type situations.

SECTION MODULE: Ground-Water Contamination

Observe the dispersal of chemical versus petroleum contaminants in ground water.

QUESTION: *What is a common point source of petroleum contaminants in urban areas? Where is the petroleum pollutant located relative to the water table? What specific risk factor beyond toxicity would petroleum leaking to the surface present?*

chapter fifteen

Waste Management and Geology

Serious Trash Talk

A waste-disposal site's ability to isolate solid and liquid waste is determined almost entirely by its geologic and hydrologic conditions. The primary goal in site selection is to protect the local water supply. Surface and underground water become polluted when rain water percolates through solid wastes. Geologists assess a potential site's suitability through field observation, soil tests for percolation and strength, drill-hole sampling of underground rock materials, and evaluations of the region's underground water conditions based on tests of nearby wells. In addition, an environmental impact report must be submitted that addresses potential problems related to the local biology, air quality, traffic congestion, and visual blight.

Wrangell, Alaska, is a town of 2,000-some people on Wrangell Island in the state's beautiful southern panhandle. In 1999 the town's open trash dump was deemed unacceptable and was sealed (◈ Figure 1). It was determined that the island has no geologically suitable landfill site and that the town's solid waste must be transferred by ocean barge to acceptable sites in the lower 48 states, hundreds of kilometers away. Much of the town's small budget now goes to pay for this new means of waste disposal.

In sharp contrast with the picture at Wrangell is an engineered sanitary landfill being manufactured in Southern California (◈ Figure 2). Graded slopes are designed to provide stability, and a plastic liner has been placed beneath a compacted, impermeable clay layer. The clay-layer "seal" will prevent liquids from migrating out of the fill into the ground-water environment. Yes, geology is very important in the world of trash. ■

◈ **FIGURE 1** Wrangell's open trash dump before it was closed and sealed in 1999.

◈ **FIGURE 2** An engineered solid-waste landfill; Southern California.

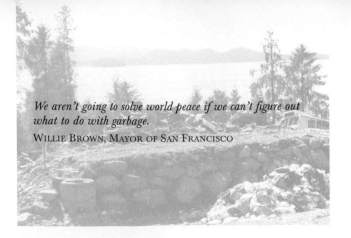

Getting rid of trash is a major environmental problem in industrialized nations. Growing populations of consumers have created an explosion of solid, liquid, and hazardous wastes, much of which requires special handling. At risk due to careless waste disposal are ground- and surface-water purity, air quality, public health, and less threatening but still important, scenic beauty and land-surface integrity. Municipal solid waste generated in the United States in 1996 amounted to 1.9 kilograms (4.3 lbs) per person per day. This category of waste includes yard cuttings, garbage, construction materials, paper products, metal cans, plastics, and glass (◆ Figure 15.1). It does not include wastes from agriculture, industry, utilities, or mining, which account for more than 96 percent of the solid waste generated in the United States (◆ Figure 15.2). After recovery for recycling and composting, 1.6 kilograms (3.6 lbs) per person per day went to a landfill or was incinerated. Just one day's total U.S. waste would cover 15 square kilometers (almost 6 mi²) to a depth of 3 meters (10 ft). If it were loaded into ten-ton trucks lined up bumper-to-

bumper, the trucks would stretch around the world 20 times. Those trucks containing only the day's *municipal solid waste* would circle the earth almost 3 times. The spectrum of solid-waste disposal problems ranges from the need to isolate highly dangerous nuclear waste to the challenge of dismantling and scrapping an astounding number of cars every year, estimated at 15 million in the United States alone. (One year 60,000 cars were abandoned on the streets of New York City.)

The amount of trash a country or governmental entity generates *per unit of land area* is just as important as the *total* amount it generates. If small countries with limited disposal sites generate large quantities of trash, major problems are created. Although the United States generates the highest total amount of waste in the world, the tiny city-state of Singapore generates the most trash per unit area of land, almost 2,500 tons per square kilometer per year in 1985. This is in comparison to less than 2 tons per square kilometer for Canada and 20 tons per square kilometer for the United States. Poland generates the most *industrial* waste per unit area, followed by Japan, and Hungary produces the most *toxic* waste per unit area, followed by the United States.

The impetus to develop environmentally safe trash disposal sites, known as *landfills,* was the passage of two important laws, the National Environmental Policy Act of 1969 (**NEPA**) and the Resource Conservation and Recovery Act of 1976 (**RCRA**). NEPA requires an in-depth field study and issuance of an Environmental Impact Statement on the consequences of all projects on federal land. The RCRA mandates federal regulation of waste products and encourages solid-waste planning by the states. The RCRA addresses the problems of so-called hazardous waste and its impact upon surface-

(a)

Construction/demolition 40%
Industrial 2%
Miscellaneous 8%
Commercial 20%
Residential 30%

(b)

Demolition debris, tires, rubber, textiles 14.6%
Yard trimmings 17.9%
Food 6.7%
Paper 37.5%
Plastic 8.3%
Glass 6.7%
Metal 8.3%

◆ **FIGURE 15.1** Components of urban solid waste (a) by sector of the economy and (b) by material. Note that paper products—packaging and printed paper—comprise the largest volume of waste and that residential trash accounts for only about a third of all urban solid waste.

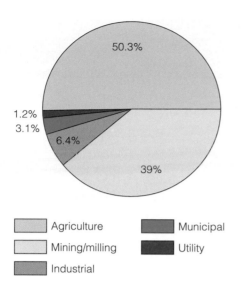

Agriculture

Mining/milling

Industrial

Municipal

Utility

◆ **FIGURE 15.2** Estimated relative contributions to total U.S. solid waste, 1992.

and ground-water quality, and it created the framework for regulating hazardous wastes. These laws sent the message that the then-prevailing "out of sight, out of mind" approach to disposing of wastes of all kinds, hazardous and benign, was no longer acceptable. The laws mandated "cradle to the grave" systems for impounding wastes and monitoring them to ensure their low potential for migrating into fresh-water supplies. The Environmental Protection Agency (**EPA**) is charged with monitoring and policing the provisions of these and subsequent laws and amendments intended to prevent pollution of natural systems (Case Study 15.1).

Disposing of solid and liquid wastes, regardless of how they are generated, falls into two main categories:

■ **isolation,** encapsulating, burying, or in some other way removing waste from the environment; and
■ **attenuation,** diluting, incinerating, or spreading trash or a pollutant so thinly that it has little impact.

Attenuation (dilution) is effective with sewage, some waste chemicals, and certain gaseous pollutants. About 10 percent of our waste, including sewage sludge, is incinerated in large, modern incinerators, usually found where land-disposal sites are not available. Historically, most of our solid waste has been isolated in dumps. Open dumps, the local "garbage dumps" of small towns such as the one pictured in the chapter opener, are no longer acceptable because of their attendant insect, vermin, odor, and air-pollution problems. In the mid 1990s, more than 70 percent of U.S. trash was being disposed of in 5,000 landfills, down from 7,900 landfills in 1988. In 1993 alone almost 1,000 landfills were closed, most because they had reached their ca-

pacity, but many because they could not meet the stringent EPA standards for protecting ground water. Deep wells are used successfully for isolating hazardous liquid wastes, although the tremendous volume of used motor oil remains a problem. Today, good **waste management** consists not only of waste disposal, but also of reducing the amount of waste at its sources and promoting waste recovery and recycling programs.

Municipal Waste Disposal

MUNICIPAL WASTE DISPOSAL METHODS

Sanitary Landfills

In 1912 the first solid-waste sanitary landfill was established in Great Britain. Known as "controlled tipping," this method of isolating wastes expanded rapidly and had been adopted by thousands of municipalities by the 1940s. The city of Fresno, California, claims to be the first city in the United States to employ a sanitary landfill for waste disposal. A **sanitary landfill** differs from an open dump in that each day's trash is covered with a layer of soil to isolate it from the rest of the environment. Although this sounds simple, the method involves spreading trash in thin layers, compacting it to the smallest practical volume with heavy machinery, and then covering the day's accumulation with at least 15 centimeters (6 in) of soil (◆ Figure 15.3). When finished, a landfill is sealed by 50 centimeters (20 in) of compacted soil and is graded so that water will drain off the finished surface. This prevents water infiltration and the production of potentially toxic fluids within the fill (◆ Figure 15.4). **Leachate** refers to the water that filters down through a landfill, acquiring (leaching out) dissolved chemical compounds and/or fine-grained solid and microbial contaminants as it goes. Protecting the local environment from leachate is a major consideration in designing and operating waste-disposal sites. The dump-and-cover procedure carried out at sanitary landfills is known as the "cell" method, because it isolates waste from the environment in rhomb-shaped compartments, or cells. A well-operated sanitary landfill is relatively free of odors, blowing dust, and debris (Case Study 15.2).

Landfills and individual disposal sites within a given landfill are classified on the basis of their underlying geology and the potential impact that leachate could have upon local ground or surface water. Each of the three classes is certified to accept specific wastes of differing degrees of reactivity (see ◉ Table 15.1 and ◆ Figure 15.5). Some landfill sites encompass all three classifications within their boundaries (◆ Figure 15.6).

Case Study | 15.1

The Sociology of Waste Disposal

The EPA is charged with the Herculean tasks of monitoring existing land and ocean dump sites, approving new hazardous-waste disposal sites, and supervising the cleanup of historic toxic sites and accidental spills and leaks. In carrying out these responsibilities, EPA personnel have encountered the spectrum of political and public attitudes about waste disposal. A wall poster at EPA headquarters in Washington, D.C., provides the following acronyms for those attitudes:

NIMBY	Not In My Back Yard
NIMFYE	Not In My Front Yard Either
PIITBY	Put It In Their Back Yard
NIMEY	Not In My Election Year
NIMTOO	Not In My Term Of Office
LULU	Locally Unavailable Land Use
NOPE	Not On Planet Earth

Incineration

Burning is the only proven way to significantly reduce the volume of garbage, and it can reduce the volume by as much as 90 percent. In 1993 about 10 percent of municipal solid waste was being burned at 162 modern municipal incinerators, up about a fifth from 136 in 1988. Use of the process is not expanding rapidly because of the high initial facility cost and the continuing costs of maintaining air-quality standards. In New York City, for example, burning costs are about $200 per ton, higher than the cost of landfilling waste in the mid-Atlantic states, the most expensive waste-disposal area in the United States. Heat energy is produced in the process, some of which is converted to electrical energy that can be sold to the local utility or used internally to run the plant. The use of incineration is highest in areas with high population density, little open land, or high water tables. Massachusetts, New Jersey, Connecticut, Maine, Delaware, and Maryland, for instance, burn more than 20 percent of their waste.

New York City is a leader in trash incineration, burning about 720,000 tons per year, or 11 percent of all its waste. Wastes must be burned at very high temperatures, and incinerator exhausts are fitted with sophisticated "scrubbers" that remove dioxins and other toxic air pollutants. Incinerator ash presents another problem, because it is in itself a hazardous waste, containing high ratios of heavy metals that are chemically active (see Case Study 15.2).

The City of Philadelphia operates trash incinerators but lacks landfills for isolating the ash. In desperation, in the mid 1980s 14,500 metric tons (16,000 tons)

◆ **FIGURE 15.3** (a) Method of dump and cover employed in a sanitary landfill. (b) An urban sanitary landfill in action; Palos Verdes, California.

(a)

(b)

Final cover
(20 inches
minimum)

Working cell

Visual barrier

Front face
(7 feet minimum)
1.75:1 slope

Drainage bench

Leachate
collection
system

Lift
(16–23
feet)

Daily cover
(6 inches minimum)

Compacted
trash cell

Impermeable seal or clay liner

Original ground

Leachate
collection
system

(a)

◆▶ FIGURE 15.4 (a) Local-government landfill specifications for cells, final cover, and leachate-collection system. Note the barrier for reducing the visual blight of the landfill. (b) A full, graded landfill before planting; Puente Hills, Los Angeles County, California. Final grades and slope benches for intercepting runoff are visible in the photo.

(b)

of the ash from two of its incinerators was loaded onto the steamship *Khian Sea* and sent to sea (◆▶ Figure 15.7). Like the sailor in Coleridge's "Rime of the Ancient Mariner," the *Khian Sea* sailed "alone, alone on a wide, wide sea" for two years, seeking a place to offload its unwanted cargo, with port after port refusing. In 1988 a miracle must have occurred, as the ship sailed into Singapore empty. No explanation has ever been given.

Ocean Dumping

Greek mythology relates how Herakles (Latin *Hercules*) was given the task of cleaning King Augeas's stables, which contained 30 years' accumulation of filth from 3,000 head of cattle. Inasmuch as one cow produces roughly 18 wet tons of manure per year, Herakles was up to his ears in 1.6 million tons of you-know-what, about the amount of sewage sludge that New York City generates in four months. Because Herakles was to per-

form the cleanup in a day, he ingeniously diverted the courses of two rivers to make them flow through the stables and wash the filth into an estuary. Had an environmental impact statement been required, it would have noted that a huge mass of solid sludge would cover a large area of the wetlands and bury bottom-dwelling organisms. In addition, phosphates and nitrates in the effluent would promote explosive algal blooms at the expense of other organisms, and dissolved oxygen in the water would decrease to the point of mass mortality of swimming and bottom-dwelling organisms.

Consider This... ? Incinerating all our waste would go a long way toward solving the landfill capacity problem, and the heat that would be generated could be put to use. Some negative aspects of burning trash must be considered, however. What are they? ■

Case Study | 15.2

The Highest Point between Maine and Florida

Fresh Kills landfill on Staten Island, New York, is the world's largest landfill, with 25 times the volume of the Great Pyramid of Giza, Egypt (◁▷ Figure 1). Fresh Kills (Dutch *kil*, "stream") serves New York City, its suburbs and parts of New Jersey. It accepts about 13,000 tons of trash daily. By 1998 it rose about 165 meters (500 ft) above sea level, the highest elevation on the eastern shore of the United States. The fill was established in 1948 on a salt marsh with no provisions for constraining leachate. For years, more than 4.2 million liters (1 million gal) of leachate leaked into the marsh and nearby waters each day. Recent cleanup has solved many of the pollution problems, but Fresh Kills remains a mountain of an eyesore to Staten Islanders.

What makes Fresh Kills unique, aside from being built on marshland, is that trash is taken to the fill by city-owned barges, each with a 600–700-ton capacity, that are loaded from garbage trucks at eight stations throughout the city and towed by tugboats. Specially built "skimmer" boats follow to pick up trash that falls from the barges. At the docks the refuse is unloaded by huge cranes into custom-made vehicles that carry it to a fill site. The refuse is dumped, sprayed with a deodorant, compacted by bulldozers, covered by clean earth, and eventually landscaped to "reclaim" the land.

New York City also has three active trash incinerators. In the past, ash residue from these was taken offshore and dumped into the ocean. Now the ash is deposited at Fresh Kills along with raw garbage. Because the ash contains toxic metals, specially designed ash-only fill areas are planned, each with a double liner and a double leachate-collection system to keep toxic substances within the fill. A clay slurry "wall" has been excavated around the fill to further confine leachate if there are leaks. A waste-to-energy incinerator is also planned; its ash will be dumped in Fresh Kills (◁▷ Figure 2).

Major efforts are underway to extend the landfill's life by recovering resources from the trash. Methane is extracted and used for heating and cooking in nearby homes, and construction debris is recycled. Yard waste from Staten Island and the Borough of Queens is composted and mixed with soil to promote healthy plant growth on the finished fill. Two more recycling facilities are planned, as is an intermediate processing center to sort and process recyclable glass, cans, paper, and plastics. At the present rate of delivery, Fresh Kills landfill will be filled to capacity early in this century.

Great Pyramid of Giza

Fresh Kills landfill

(a)

(b)

◁▷ **FIGURE 1** (a) Comparison of the volumes of the Great Pyramid of Giza and the Fresh Kills landfill. (b) The Great Pyramid (right) built by King Cheops at Giza on the west bank of the Nile River in the third century B.C. The length of each side at the base averages 756 feet, and its present height is 451 feet (original height, 481 feet). The smaller pyramid was built by Chephren, a later king. The automobiles give an idea of the immensity of these structures.

◁▷ **FIGURE 2** Fresh Kills landfill on Staten Island.

⊙ TABLE 15.1	**Classifications of Disposal Sites and Waste Groups**
	DISPOSAL SITES
Class I	No possibility of discharge of leachate to usable waters. Inundation and washout must not occur. The underlying lining material, whether soil or synthetic, must be essentially impermeable; that is, have a permeability less than 0.3 cm/year. All waste groups may be received (Figure 15.5, part a).
Class II	Site overlies or is adjacent to usable ground water. Artificial barriers may be used for both vertical and lateral leachate migration. Geologic formation or artificially constructed liners or barriers should have a permeability of less than 30 cm/year. Groups 2 and 3 waste may be accepted (Figure 15.5, part b).
Class III	Inadequate protection of underground- or surface-water quality. Includes filling of areas that contain water, such as marshy areas, pits, and quarries. Only inert Group 3 wastes can be accepted (Figure 15.5, part c).
	WASTE GROUPS
Group 1	Consists of but not limited to toxic substances that could impair water quality. Examples are saline fluids, toxic chemicals, toilet wastes, brines from food processing, pesticides, chemical fertilizers, toxic compounds of arsenic, and chemical-warfare agents.
Group 2	Household and commercial garbage, tin cans, metals, paper products, glass, cloth, wood, yard clippings, small dead animals, and hair, hide, and bones.
Group 3	Non-water-soluble, nondecomposable inert solids such as concrete, asphalt, plasterboard, rubber products, steel-mill slag, clay products, glass, and asbestos shingles.

Historically, all coastal countries have used the sea for waste disposal. This practice is based on the *assimilative capacity* approach to waste disposal; that is, the assumption that a body of seawater can hold a certain amount of material without adverse biological impact. This attitude is still held in many coastal areas, particularly in developing countries, where floatables in trash present a highly unsightly appearance. No U.S. municipality has dumped garbage in the sea since New York City stopped the practice in 1934.

Ocean pollution is governed by the Marine Protection Reserve and Sanctuary Act of 1972, better known as the *Ocean Dumping Act*. This act requires anyone dumping waste into the ocean to have an EPA permit and to provide proof that the dumped material will not degrade the marine environment or endanger human health. Hazardous substances entering the sea from the land (⊙ Table 15.2) are contained in sewage, sewage sludge (the solid part of sewage), dredged materials, industrial effluents, and natural runoff. Sludge and other particulate matter have been found to create a biological imbalance in the ocean in three ways:

(a) Class I landfill

(b) Class II landfill

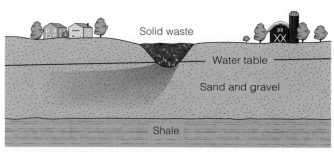

(c) Class III landfill

◆ **FIGURE 15.5** Geology of the three classes of landfills. Note that a Class I fill offers maximum protection to underground water and that a Class III fill offers no protection.

◆ **FIGURE 15.6** Oil-field brine is dumped in a Class I landfill. Other sections of this landfill have lower classifications because their bedrock is permeable.

- they are consumed by marine life as a food source,
- they introduce heavy-metal and chlorinated organic compounds when they are attached to the particles, and
- they inhibit light in the water column.

These findings led to the Ocean Dumping Ban of 1988 that required total cessation of ocean dumping by 1991. Coastal cities that once ocean-dumped were forced to begin dewatering sludge and exporting it as a soil additive or depositing it in landfills. In the west Texas town of Sierra Blanca, instead of "Home on the Range," they now sing "sludge on the range." Two hundred tons of processed toilet and sewer waste arrives daily by train from New York whereupon it is worked into the soil of a 200-square-mile sludge "ranch." The community's 700 people are divided on the issue of whether New York's waste benefits anyone in Texas, but the EPA says the spreading is environmentally safe. Time will tell.

PROBLEMS OF MUNICIPAL WASTE ISOLATION

Organic refuse in landfills ultimately decomposes. As it does, the ground surface settles; methane, carbon dioxide, and other gases are generated; and noxious or even toxic leachate from it may seep into the water table or bordering streams. Thus, a landfill's *pollution potential* depends upon the change in volume that accompanies decomposition, the waste's reactivity (which determines the chemistry of the leachate), the local geology, and the climate. Organic matter decomposes more slowly in cold, dry climates than in warm, wet ones.

Stabilization

Landfills continue to settle as decomposition goes on for many decades after burial. Typically, settlement is rapid at first and diminishes to a much slower rate after about ten years (◆ Figure 15.8). The rate of subsidence depends upon many factors, including the climate, the kind of waste, and the amount of compaction during filling. Thirty-ton tractors left on a fill over a weekend have been known to settle as much as a meter into the trash, and total settlement of as much as 30 percent has been measured in landfills (Figure 15.8b). Variation is great. A 6-meter-thick landfill in Seattle settled 23 percent in its first year, whereas a 23-meter-thick fill in Los Angeles settled only 1 percent per year in its first three years. Settlement is a problem, because it results in cracks and fissures, ponds, sags, and infestation by vermin. It is monitored to assure that water does not pond on the fill cover and percolate downward, creating a high volume of polluting leachate.

As microorganisms break down organic material *(biodegradation),* they release heat, methane (natural gas), and other gases including hydrogen sulfide, carbon dioxide, and oxygen. Research at the University of Arizona and elsewhere has shown that biodegradation can be surprisingly slow, as illustrated by exhumed garbage containing virtually unblemished 10–20-year-old newspapers (◆ Figure 15.9), a 14-year-old mound of recognizable guacamole, and positively ancient hot dogs (testimony to the effectiveness of sausage preservatives). The most efficient decomposers are the voracious oxygen-using aerobic bacteria, such as those that quickly turn grass cuttings into compost for our gardens. Most landfills are sealed off from the atmosphere, however, giving rise to anaerobic (no oxygen) microorganisms that manufacture methane and hydrogen sulfide in the slow process of decomposing cellulose. Pumping air into a fill, a procedure known as **composting,** stimulates aerobes, accelerates decomposition and settlement, and creates more space for trash.

Gas Generation

Methane (CH_4), a gaseous hydrocarbon, is the principal component of natural gas. Also known as *marsh gas,* it bubbles forth from stagnant ponds and swamps, the result of complex chemical fermentation of plant material by bacteria. It is the same gas that causes the explosions in coal mines that have taken innumerable lives. Methane is explosive when present in air at concentrations between 5 and 15 percent. Although enormous volumes of methane are generated by anaerobic microorganisms in a landfill, there is little danger of the fill exploding because no oxygen is present. Being lighter than air, methane migrates upward in a fill.

◆ **FIGURE 15.7** Greenpeace hung a protest banner on the loaded garbage barge in New York Harbor. In the past trash such as this was destined for sea burial. Disposal of solid waste is a problem in urban areas in all parts of the world.

⊙ TABLE 15.2	Examples of Recognized Seawater Pollutants from Human and Natural Sources*
ORGANIC POLLUTANTS	
PCBs	polychlorinated biphenyls
PCDFs	polychlorinated dibenzofurans
PCDDs	polychlorinated dioxins
DDT	chlorinated hydrocarbon pesticide
INORGANIC POLLUTANTS	
Cd	cadmium
Pb	lead
Co	cobalt
Hg	mercury
Zn	zinc

* Pollutants originating in sludge, materials dredged from harbors or estuaries, and sewage-treatment and industrial wastewater that are known to pose health hazards.
SOURCE: EPA (adapted).

Methane leaking through the cover of a landfill converted to a golf course near Santa Monica, California, was known to "pop" from ignition when cigarettes were dropped on its greens (◆ Figure 15.10).

A peculiar incident occurred at a soccer field planted on top of a sealed landfill in the Charlotte, North Carolina area. A ball was kicked into a small cavity created by settlement and a coach looked for the ball using a cigarette lighter as a light source. The outgassing methane exploded seriously injuring the individual. Methane can also migrate laterally, and it did so several hundred yards into the wall spaces of structures near a landfill south of Los Angeles. Fortunately, the potentially explosive situation was discovered, and wells were drilled to intercept and extract the gas (◆ Figure 15.11). One strategy for handling the large volumes of methane generated at large landfills is to extract it through perforated plastic pipes, remove the impurities (mostly CO_2), and use it as fuel—the so-called refuse-derived fuel, *RDF*. This can be accomplished either by piping the gas away via pipelines or by using it on the site to generate electricity. Because

(a)   (b)

◆ **FIGURE 15.8** (a) Graph of landfill settlement over time. The percentage of settlement is calculated by dividing the amount of settlement (ΔL) by the original thickness of the landfill (L) and multiplying by 100. The settlement values here are for illustration only and do not depict typical values. (b) Landfill weigh station on the Palos Verdes Peninsula, California, built on trash-filled land. The station eventually collapsed, requiring that a new structure and scale be established at a different location.

the gas may be as much as 50 percent CO_2, it is more expedient to use the fuel directly to generate electricity than to clean out the impurities and put it into pipelines. The Los Angeles County (California) Sanitation District extracts landfill methane and generates sufficient electricity to supply up to 45,000 homes.

Leachate

The compositions of leachates ("garbage juice") in and leaving landfills are highly variable, but they must be considered dangerous to the environment until established otherwise (◆ Figure 15.12). A leachate may be such that a receiving body of water can assimilate it without any impairment in water quality. Some purifi-

(a)

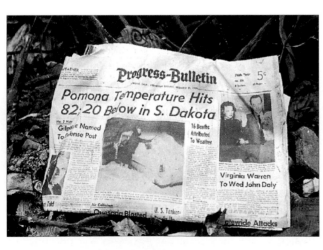

◆ **FIGURE 15.9** Underground newspaper—still readable after at least ten years' burial in a Pomona, California, landfill.

(b)

◆ **FIGURE 15.10** (a) Mission Canyon landfill near Santa Monica, California, as it was being finished. (b) The reclaimed landfill is now a golf course and upscale residential neighborhood.

◁▶ **FIGURE 15.11** Methane-extraction well and drill rig. The extracted gas is conveyed to a central location for conversion to electricity.

cation of a leachate occurs as it filters through clayey soils, and some purification also occurs when contained organic pollutants are oxidized. Ground-water-quality measurements taken in the vicinity of a landfill, at the fill, and downstream from the fill are shown in ◉ Table 15.3. Note that the leachate was causing deterioration of ground-water quality in the vicinity of the monitoring well. To correct this, the leachate must be intercepted by wells or a barrier must be installed between the leachate and the water table. Under existing regulations, a landfill that is in a geologic setting where leachate threatens local water bodies may accept only inert (Group 3) wastes.

The Completed Landfill

A completed state-of-the-art landfill is illustrated in ◆ Figure 15.13. It is designed to collect leachate above

the bottom liner and to recover methane for use in power generation. Note the inclusion of leachate- and gas-monitoring wells in the design. Double-lined landfills have been required in the United States since 1996. These landfills are designed to accept waste for 10–40 years, and owners are required to monitor the fills for at least 30 years after closure. If they begin to leak after that period, the contamination problem belongs to whomever is living in the area at the time.

RECYCLING

Resource recovery has become a component of waste management. It is the removal of certain materials from the waste "stream" for the purpose of recycling or composting them. It has been estimated that each year Americans throw out enough wood and paper to heat a billion homes for a year, and that Sunday newspapers alone—many of which are thrown away after 30 minutes—require cutting down 220 million trees a year. We throw away enough aluminum cans in a year to quadruple the size of our airline fleet (Case Study 15.3). Much of the material we toss out is recycled, but we must recycle more. The industrial nations are the "Saudi Arabias" of trash, and all have the same problems—dwindling resources and dwindling space for solid waste.

Until recently, **recycling** held little appeal; that is, it cost more to separate, sort, and recycle trash than to landfill it. It is now mandated in many states, however, because existing landfills are filling, and acceptable sites for new ones are difficult to find. An effective way to extend the lifetimes of landfills and natural resources is to recycle municipal wastes. The consuming public is "thinking green" these days, and manufacturers are realizing favorable marketing results by using recycled materials in packaging their products. President Clinton's 1993 mandate that the federal government use

◁▶ **FIGURE 15.12** The flow of leachate from a landfill can contaminate underground water, ponds, and streams. See also Table 15.3.

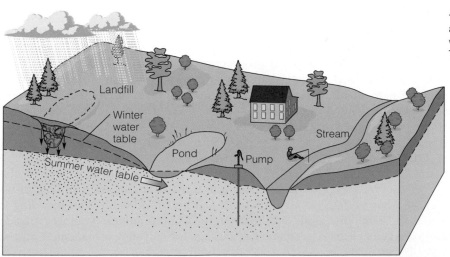

⊙ TABLE 15.3 Ground-Water Quality Measured Near One Landfill

	MEASUREMENT SITE		
WATER CHARACTERISTICS	*Local Ground Water*	*Fill Leachate**	*Monitoring Well***
Total dissolved solids, ppm	636	6,712	1,506
pH***	7.2	6.7	7.3
BOD, mg/L****	20	1,863	71
Hardness, ppm	570	4,960	820
Sodium content, ppm	30	806	316
Chloride content, ppm	18	1,710	248

* A saturated fill.
** A monitoring well about 150 feet downstream from the landfill at a depth of 11 feet in sandy material.
*** A solution with a pH of 7 is neutral, less than 7 is acidic, and greater than 7 is basic (Case Study 14.3).
**** Biological oxygen demand: the amount of oxygen per unit volume of water required for total aerobic decomposition of organic matter by microorganisms. This is discussed later in the chapter in the section on wastewater treatment.
SOURCE: D. R. Brunner and D. J. Keller, *Sanitary Landfill Design and Operation* (Washington, D.C.: U.S. Environmental Protection Agency, 1992).

more recycled paper helped to boost waste-paper prices. Because of helpful mandates and a growing awareness that mineral resources are finite, recycling increased from 10 percent to 25 percent of municipal waste between 1985 and 1996. (◆ Figure 15.14 summarizes resource recovery at landfills by category.) One indicator of the anticipated value of trash is that glass and high-grade plastics have joined corn and pork bellies as commodities bought and sold on the Chicago Board of Trade. Trash futures??

Curbside recycling programs grew enormously between 1988 and 1996, with more than 100 million people participating in 1993—almost 40 percent of the U.S. population. The "champion" city is Seattle, whose recycling rate is 45 percent. Many communities use a "pay as you throw" program, which bases collection charges on the amount of true waste that is added to a landfill; materials sorted out for recycling are collected free of charge. Many communities require separation of glass, metal, paper products, and plastics from other trash materials. Plastic recycling is difficult to handle and expensive, because there are seven kinds of plastic that must be separated by hand. Only 2 percent of waste plastic was recycled in 1995. Ingenuity pays, however. A small Massachusetts company grew into a $1.3 million business by recycling spent plastic ink-jet printer cartridges. It sells refurbished cartridges to consumers and the worn plastic and metal components to recyclers.

Not easily solvable is the problem of what to do with old automobile tires. An estimated three billion tires are stockpiled in the United States today, and each year 240 million more are added. We discard three times more tires than we recycle. Resting in dumps, they add to environmental problems because they collect water, promoting the growth of mosquitoes, vermin, and other disease-spreading pests. Also, accidental tire fires have been known to burn for months, giving off billows of irritating black smoke and fumes. No large-scale use for old tires has been found, although incidental small-scale uses are seen occasionally (◆ Figure 15.15). Chopped tires are mixed with asphalt to surface and repair highways and to fuel power plants in some places. In Modesto, California, for example, a specially built power plant burned five million tires each year to generate electricity. It was closed in 1996 for unexplained reasons. Because old tires are such an enormous environmental problem, many engineers and scientists are working on a technological "fix" that will rid us of this blight.

Besides extending the life of our resources, recycling makes sense because it saves energy. The energy used in processing many scrap materials is less than that required for processing the raw material. Producing a pound of aluminum from scrap takes only 5 percent of the energy required to produce a pound from bauxite; scrap iron requires one-fourth the energy needed to make iron from ore; and producing a ton of paper from recycled stock uses only 30 percent of the energy used to make paper from trees. However, the costs of collection, transportation, and handling of recycled materials can be considerable, and recycling is still subsidized in most communities. As more curbside separation of garbage for recycling is mandated, and as

Case Study | 15.3

Don't Mess With Texas

Cleaning up highway litter cost the Texas Department of Highways and Public Transportation $24 million in 1985. To reduce this unnecessary expenditure, Department snoops followed drivers to study their littering habits. The trash spies found that nearly 50 percent of the litterbugs were 18–34-year-old good old boys in pickups. Personal interviews disclosed that the slogans Texas had been using, such as "Don't Be A Litterbug" and "Pitch In," impressed these offenders about as much as drinking a Diet Pepsi on Friday nights.

A new litter-reduction campaign called "Don't Mess With Texas" was implemented in 1986. The thinking behind the slogan was that no-nonsense language would be remembered and make drivers think before discarding. It worked. A year later almost three quarters of all Texans were familiar with the slogan, and five years later highway litter had been reduced by 64 percent. The slogan now appears on bumper stickers, T-shirts, and trash cans, and Willie Nelson sings "Don't Mess with Texas" jingles on television and radio. Congratulations to the citizens of the Lone Star State!

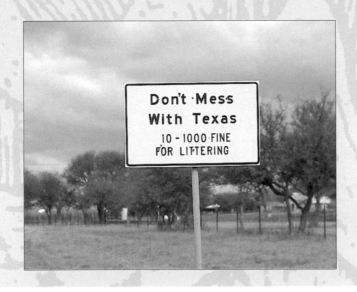

more states enact container-bill legislation encouraging the return of beverage containers, the flow of recyclables increases. The "champion" states (% recycled) in the United States are Minnesota (41%), New Jersey (39%), and Washington (35%). Taiwan is the world champion paper recycler; 98 percent of the paper used in that almost treeless country is recycled.

Composting

When decomposed by bacteria, biodegradable wastes, the so-called green garbage, can be used as a fertilizer and soil amendment. Yard cuttings, most kitchen wastes, and organic wastes from commercial operations are acceptable compost material. Because yard wastes are second only to paper in landfill tonnage (see Figure 15.14), many states and municipalities have mandated separate collection for them. As a result, the number of composting facilities in the United States grew dramatically from 651 facilities in 1988 to more than 3,000 in 1993.

A state-of-the-art composting method, borrowed from the Netherlands, is used in Columbia County, Wisconsin. Leaves, grass, food scraps, and even road kill are placed in one end of gently sloping, rotating, 3½-meter × 50–meter drums. A week later, the oxygen, bacteria, and constant churning have produced crumbly, black, high-quality compost, which is removed at the low end of each drum and sold to farmers and home gardeners. Heat generated by oxidation and bacterial action kill any pathogens present in the material. Mecklenburg County, North Carolina, earned $100,000 in 1991 from its yard-waste composting program called "Curb It."

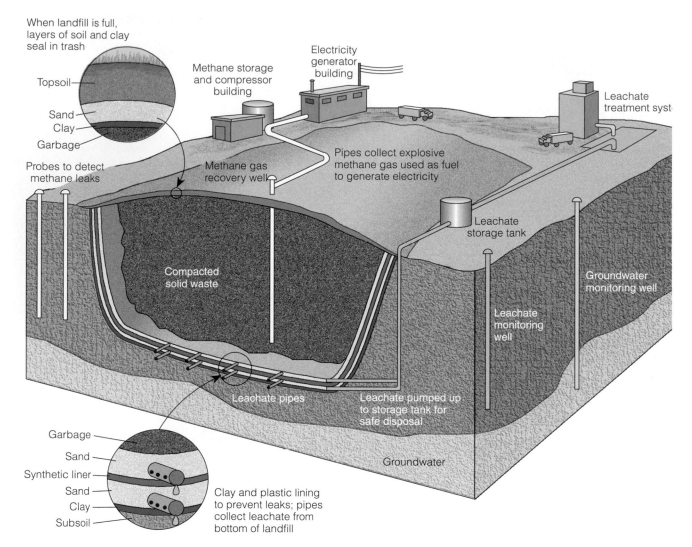

When landfill is full, layers of soil and clay seal in trash

Topsoil
Sand
Clay
Garbage

Probes to detect methane leaks

Methane storage and compressor building

Electricity generator building

Methane gas recovery well

Leachate treatment system

Pipes collect explosive methane gas used as fuel to generate electricity

Leachate storage tank

Compacted solid waste

Groundwater monitoring well

Leachate monitoring well

Leachate pipes

Leachate pumped up to storage tank for safe disposal

Groundwater

Garbage
Sand
Synthetic liner
Sand
Clay
Subsoil

Clay and plastic lining to prevent leaks; pipes collect leachate from bottom of landfill

◆▶ **FIGURE 15.13** State-of-the-art landfills are designed to eliminate the problems of toxic leachate and methane escaping into the environment. Note the double lining (clay and plastic layers) to prevent leakage. Unfortunately, 85 percent of U.S. sanitary landfills are unlined.

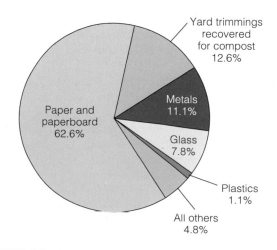

Yard trimmings recovered for compost 12.6%

Metals 11.1%

Glass 7.8%

Paper and paperboard 62.6%

Plastics 1.1%

All others 4.8%

◆▶ **FIGURE 15.14** Materials recovered at landfills as percentage of the total weight recovered, 1990.

Waste-Reduction Strategies

An integral component of waste management is reducing the amount of trash that is produced in the first place. Examples of waste-reduction strategies are now found in almost every office, manufacturing plant, school, and home in the United States. One company that now requires double-sided copying reduced its paper purchases by 77 million sheets a year. Some supermarkets sell their customers nylon-mesh grocery bags at cost. At least one furniture manufacturer ships its products in reusable padded quilts instead of heavy cardboard. Consumers can help by refusing to buy overpackaged items, patronizing firms that use some recycled materials, and returning beverage containers to the store to redeem the deposits.

◀▶ FIGURE 15.15 Three thousand tires were recycled as children's playground equipment and landscaping material at this park in Japan.

Consider This... ?
Composting is a "win-win" process, yet we have barely scratched the surface of its potential. What are the benefits of small-scale and large-scale composting to individuals, cities, and states? ■

(a)

(b)

◀▶ FIGURE 15.16 Palos Verdes Peninsula, California. (a) A diatomaceous-earth quarry that operated for several decades was converted into a landfill in the 1960s. (b) After filling was complete, the land was reclaimed again, as the South Coast Botanical Gardens, a tourist attraction and passive recreation area for local residents.

MULTIPLE LAND USE STRATEGIES

Land, like trash, can be recycled. Where the local geology is amenable to it, abandoned rock quarries and sand and gravel pits have been converted to sanitary landfills (◀▶ Figure 15.16). Population pressure and the need to conserve resources in the twenty-first century will require that we not only dispose of our waste efficiently, but also that we gain some benefit from the stored trash and the disposal site. The average new home requires about 65 cubic meters of concrete in its construction, and its inhabitants will generate about a ton of trash a year. Thus, it *makes sense* to place the household's trash in an opening created to build the house. After a landfill is completed, methane gas can be recovered from it and converted to energy, and the land may be reshaped into a golf course, athletic field, public garden (as in Figure 15.16), or the like. An ideal, beneficial cycle of multiple land uses is shown in ◀▶ Figure 15.17.

As explained earlier, structures placed directly on a landfill are subject to settlement and to methane emanations. A restaurant and shops built upon a thin landfill (<8 meters thick) south of Los Angeles were placed on piles driven through the fill to keep them from settling. The walkways around the buildings were not, however, and after a few years they settled and had to

◀▶ **FIGURE 15.17** Multiple land use, a cycle that can benefit everyone.

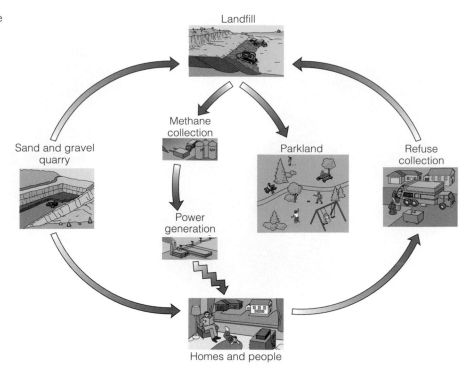

be built up to the base of the structures (◀▶ Figure 15.18). In the same commercial development, methane is extracted and used to illuminate exterior gas lanterns.

As cities expand and coalesce—such as in the New York-to-Boston corridor and the strip between Los Angeles and San Diego—waste-disposal-site selection must become very creative. Between 1940 and 1983 the Kaiser Corporation operated an open-pit iron mine at Eagle Mountain in the desert 115 kilometers (70 mi) east of Indio, California. The mining operation has now been abandoned, leaving an opening 3 kilometers (1.9 mi) long and more than 300 meters (about 1,000 ft) deep (◀▶ Figure 15.19). It is the proposed site for a landfill for all Southern California cities' household waste (no toxic or hazardous waste), which would come to it in closed railroad cars. The tracks that served the mine are still in place, and the site is known to be free from faults and potential for ground-water contamination. Far from any population center, the site could accommodate half of all the household trash generated in Southern California in the next 115 years! The open pit is in dense bedrock that would be lined with clay and high-density polyethylene plastic to protect ground water. The proposed site is acceptable geologically, and the only objections raised at public hearings regarding the proposal, from a technical point of view, seem to be entirely NIMBY, PIITBY, and NIMTOO (see Case Study 15.1).

◀▶ **FIGURE 15.18** Commercial development built on a landfill. Whereas the buildings are founded on piles through the fill to bedrock, the pavement was placed directly on the fill. The lanterns are fueled by methane from the fill.

Consider This... ?
During the past few decades in the United States, recycling waste has evolved from a concept to a major component of waste management for many communities. Why do some environmental groups object to waste-problem solutions that emphasize recycling? ▪

◁▷ **FIGURE 15.19** Eagle Mountain Mine, site of the proposed Eagle Mountain landfill; Riverside County, California. As designed, the fill could accommodate up to 20,000 tons of trash per day, which would make it the largest-capacity landfill in the world. It would accept only nonhazardous waste and would have a recycling center.

Hazardous Waste Disposal

Hazardous wastes are primarily such industrial products as sludges, solvents, acids, pesticides, and PCBs (polychlorinated biphenyls; highly toxic organic fluids used in plastics and electrical insulation). The threats that these and other pollutants pose to ground water are discussed in Chapter 8. The United States produces the largest total amount of hazardous waste and the second-largest amount per unit area. EPA estimates are that U.S. industries were generating 260 million metric tons of hazardous waste per year in the mid 1990s. A typical rate at which liquid hazardous waste was generated in the early 1980s was 150 million metric tons per year, which can be thought of as 40 billion gallons. If that amount could be placed in 55-gallon drums and the drums placed end to end, the drums would encircle the earth 16 times (◁▷ Figure 15.20). In the mid 1990s this would be almost 28 times. Examples of hazardous wastes individuals introduce to the environment include drain cleaners, paint products, used crankcase oil, and discarded fingernail polish (not trivial). The EPA has estimated that 90 percent of our hazardous waste is disposed of improperly. Fortunately, knowledge of how to dispose of these substances with a high degree of safety exists.

HAZARDOUS-WASTE DISPOSAL METHODS

Secure landfills are designed to totally isolate the received waste from the environment. Wastes are packaged and placed in an underground vault surrounded by a clay barrier, thick plastic liners, and a leachate re-

moval system (see, for example, Figure 15.13). Unfortunately, few of the nation's existing 75,000 industrial landfills have plastic liners, clay barriers, or leachate drains. Many older fills will either have to be retrofitted with modern environmental-protection systems or shut down. Additional protection is provided by existing regulations that mandate ground-water monitoring at disposal sites. An ideal secure landfill for hazardous waste would be a site in a dry climate with a very low water table. In humid climates where water tables are high, alternative means of safe disposal must be found.

Deep-well injection into permeable rocks far below fresh-water aquifers is another method of isolating toxic and other hazardous liquid substances (◁▷ Figure 15.21). Before such a well is approved, thorough studies of subsurface geology must be made in order to determine the location of faults, the state of rock stresses at depth, and the impact of fluid pressures on the receiving formations (Case Study 15.4). The injection boreholes are lined with steel casing to prevent the hazardous fluids from leaking into fresh-water zones above the injection depth.

Disposal of hazardous wastes, both liquid and solid, is regulated by the Resource Conservation and Recovery Act of 1976 (RCRA). Exempt from the requirements of this act are businesses that generate less than a metric ton of waste per month. According to EPA estimates, about 700,000 of these small firms generate 90 percent of our hazardous waste. Consequently, most of these wastes end up in sanitary landfills.

Discharge into sealed pits is the least expensive way to dispose of large amounts of water containing relatively small amounts of hazardous substances. If the pit is well sealed and evaporation of water equals or exceeds the input of contaminated water, the pit may

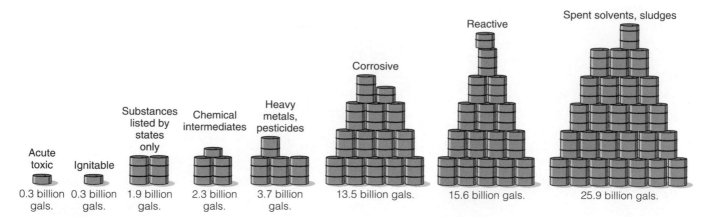

Acute
toxic

0.3 billion
gals.

Ignitable

0.3 billion
gals.

Substances
listed by
states
only

1.9 billion
gals.

Chemical
intermediates

2.3 billion
gals.

Heavy
metals,
pesticides

3.7 billion
gals.

Corrosive

13.5 billion gals.

Reactive

15.6 billion gals.

Spent solvents, sludges

25.9 billion gals.

◁▶ **FIGURE 15.20** Breakdown by category of the 40 billion gallons of toxic waste U.S. industries generated in 1981. The total of all categories is greater than 40 billion due to overlap; for example, a fluid may be both corrosive and reactive.

◁▶ **FIGURE 15.21** Deep-well injection of hazardous waste. This method presumes that wastes injected into strata of porous rock deep within the earth are isolated from the environment forever. What could go wrong? (1) Toxic spills may occur at the ground surface. (2) Corrosion of the casing may allow injected waste to leak into an aquifer. (3) Waste may migrate upward through rock fractures into the ground water.

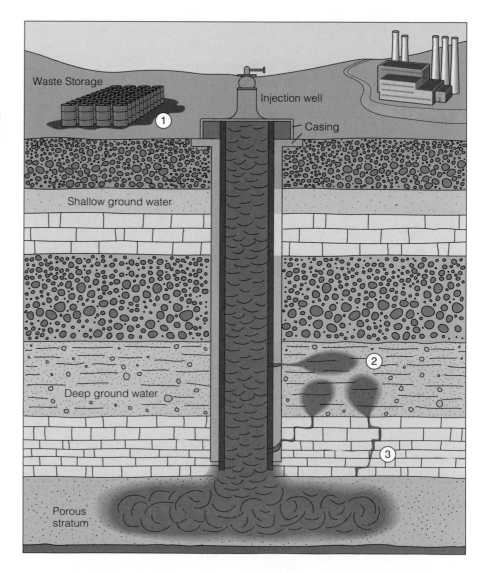

Case Study | 15.4

Deep-Well Injection and Earthquakes

The U.S. Army's Rocky Mountain Arsenal in suburban Denver drilled a 3,764-meter (2⅓-mi) deep well for disposing of chemical-warfare wastes in 1961. The Tertiary and Cretaceous sedimentary rocks into which the well was drilled were cased off to prevent any pollution of these formations, and the lower 23 meters (75 ft) were open to Precambrian granite. About 40 kilometers (25 mi) to the west of the arsenal is the frontal fault system of the Rocky Mountains (◁▶ Figure 1).

Injection began in 1962 at a rate of about 700 cubic meters per day (176,000 g/day). Soon after injection was started, numerous earthquakes were felt in the Denver area. The earthquake epicenters lay within 8 kilometers (5 mi) of the arsenal and plotted along a line northwest of the well. A local geologist attributed the quakes to the fluid injection at the arsenal. Although the army denied any cause-and-effect relationship, a graph of injection rates versus the number of earthquakes showed an almost direct correlation—especially during the quiet period when fluid injection was stopped (◁▶ Figure 2).

The fractured granite into which the fluid was being injected was in a stressed condition, and the high fluid pressures at the bottom of the hole decreased the frictional resistance on the fracture surfaces. Earthquakes were generated along a fracture propagating northwest of the well as the rocks adjusted to relieve internal stresses.

Thus earthquakes *can* be "triggered" by human activities, and the Denver earthquakes initiated research into the possibility of modifying fault behavior by fluid injection; that is, relieving fault stresses by producing a large number of small earthquakes. Seismic studies conducted during a water-injection (see Chapter 14) program at the nearby Rangely Oil Field confirmed that earthquakes *could* be turned on or off at will, so to speak, in areas of geologic stress. To date no group or government entity has assumed responsibility for triggering earthquakes by fluid injection, for obvious reasons.

◁▶ FIGURE 1 Relationship between the Rocky Mountain Arsenal and the Rocky Mountain frontal fault system.

◁▶ FIGURE 2 Correlation of fluid waste injection and the number of earthquakes per month in the Denver area, 1962–1965.

receive and hold hazardous waste almost indefinitely. Problems can arise from leaky seals and overflow of holding ponds during heavy storms or floods.

SUPERFUND

What about the thousands of old and mostly unrecorded sites where hazardous waste has been dumped improperly in the past? Hazardous substances have been disposed of indiscriminately since the eighteenth century in the United States. (⊙ Table 15.4 gives some examples.) **Superfund,** the Comprehensive Environmental Response, Compensation, and Liability Act (CERCLA), was passed in 1980 to rectify past and present abuses from toxic-waste dumping. It authorizes the EPA to clean up a spill and send the bill to the responsible party. CERCLA also empowered the EPA to establish a National Priorities List (NPL) for cleaning up the worst of the 40,000 identified abandoned toxic

and hazardous waste sites now known as *Superfund sites,* some of which date back many decades (Case Study 15.5). By 1999 1,400 NPL sites had been designated or were pending. Of those, 427 sites had been cleaned up, and work was under way or had been completed on 474 others. New Jersey has the dubious distinction of harboring 6 of the 15 worst Superfund sites as ranked by impact on the environment—5 landfills and 1 industrial site. ⊙ Table 15.5 reports the number of Superfund sites in each state in 1990 and ranks the states by number of sites. (How does your state rank?)

Although the costs of cleaning them up are staggering, toxic dump sites do not just "go away" on their own. They need to be cleaned up. Some problems with Superfund also need to be cleaned up. The Superfund concept was well intended, but the program has been poorly executed. As a result, 40 percent of its funding has gone to litigation expenses. Superfund has become a "lawyers' money pit," as one journalist put it. A case

⊙ TABLE 15.4 **Chronology of Hazardous-Substance Identification in the United States**

DATE	SUBSTANCE	NATURE OF HAZARD	SOURCE
1860	Copper acetoarsenite (Paris Green)	serious health problems or death with chronic exposure	used against the Colorado potato beetle
1925	Ethylene, ethylene oxide, ethylene glycol	carcinogenic, liver disease, mutations, nervous disorders	antifreeze
1927	Polychlorinated biphenyls (PCBs)	carcinogenic, mutations, birth defects, liver disease	used for heat transfer in electrical transformers, motors, appliances
1940s	Chlorinated hydrocarbons (chlordane, toxaphene, dieldrin, aldrin, DDT) and organophosphate (parathion)	highly toxic or carcinogenic; persistent for years, low biodegradation, bioaccumulative in humans	insecticides
1955	Halogenated hydrocarbons	toxic	chemical warfare agents, pesticides, refrigerants; some are ozone scavengers
1968	Lead	human health hazard	paint and gasoline
1980s	Cyanide, pesticides, acids	toxic	aquifers at Swartz Creek, Michigan,* contaminated by illegal burial of drums containing these substances
1980s	Dioxin-containing oil	carcinogenic, birth defects	road surfaces at Times Beach, Missouri* (EPA bought the town and relocated 2,200 residents at cost of $33 million in 1985.)
1985	Paint thinner, pesticides, TV sets, chemical bleaches, batteries, fingernail polish	various	dumped with ordinary wastes at many landfills
1990s	Cyanide, mercury, lead, sulfur, hydrocarbons, etc.	various	abandoned mines and oil wells

* Superfund site.
SOURCE: Adapted from Allen Hatheway, "Pre-RCRA History in Industrial Waste Management in Southern California," B. W. Pipkin and R. J. Proctor, eds., *Engineering Geology Practice in Southern California,* Association of Engineering Geologists Special Publication 4 (Belmont, Calif.: Star Publishing Company, 1992).

Case Study | 15.5

Love Canal—No Love Lost Here

Love Canal, in the honeymoon city of Niagara Falls, New York, may sound very romantic, but it's not. Excavated by William T. Love in the 1890s but never finished, the canal is 1,000 meters long and 25 meters wide (3,300 ft × 80 ft). Unused and drained, Love Canal was bought by Hooker Chemical and Plastics Corporation in 1942 for use as a waste dump. In the next 11 years Hooker dumped 21,800 tons of toxic waste into the canal, which was believed to be impermeable and thus an ideal "grave" for hazardous substances.

After the canal was filled and covered with a clay cap, Hooker sold it to the Niagara Falls School Board for one dollar. The company inserted in the deed a disclaimer denying legal liability for any injury caused by the wastes. In 1957 Hooker warned the school board not to disturb the clay cap because of the possible danger from leakage of toxic chemicals. A school was built on the landfill, and hundreds of homes and all the infrastructure needed to support a suburban community were built nearby.

In the early 1970s, after several years of heavy rainfall, water leaked through the "impermeable" clay cap and into basements and yards in the Love Canal area, and also into the local sewer system. Soon the new community was experiencing high rates of miscarriages, birth defects, liver cancer, and seizure-inducing diseases among children. Residents complained that toxic chemicals in the water were causing these adverse health effects.

In 1978 the New York State Health Commissioner requested the EPA's assistance in investigating the chemistry of fluids that were leaking into a few houses around the canal. The study revealed the presence of 82 toxic chemicals including benzene, chlorinated hydrocarbons, and dioxin, and the state commissioner declared Love Canal a health "emergency." Five days later President Carter declared it a federal disaster area, the first human-caused environmental problem to be so designated in the United States. The recommendation that pregnant women and children under the age of two be evacuated from the area followed, and New York State appropriated $22 million to buy homes and repair the leaks. In all, a thousand families were relocated (Figure 1).

The results of an EPA monitoring study were released to the public in 1982. An effective ground-water barrier and drain system were installed to intercept the toxic chemicals. The presence of relatively impermeable clays in the area suggests that future long-distance migration of pollutants from Love Canal is unlikely. Because the shallow and deep aquifers in the area are not hydraulically connected, there is little danger of contaminants migrating into deeper ground-water zones.

Love Canal was a precedent-setting case, because it spurred Congress to enact CERCLA, which provides federal money for toxic cleanup without long appeals. It also author-

izes the EPA to sue polluters for the costs of cleanup and relocation of victims. Although Love Canal has been essentially contained, some chemicals still infect a nearby stream and schoolyard, and maintenance costs amount to a half-million dollars annually. The EPA no longer ranks Love Canal high among the nation's most dangerous waste sites, and the New York State Health Commissioner has announced that recent federal-state studies found four of the seven polluted Love Canal areas to be habitable. In spite of the commissioner's statement, there has been no stampede to reinhabit the area.

After 16 years of litigation, the legal battles over Love Canal ended in 1995. Occidental Chemical Corporation, which purchased Hooker Chemical in 1968, agreed to pay Superfund $102 million, and the Federal Emergency Management Agency (FEMA) agreed to pay $27 million. The State of New York had made a $98 million settlement with EPA in 1994. The cleanup is estimated to have cost $275 million.

(a)

(b)

 FIGURE 1 Where the honeymoon ended and the nightmares began; Love Canal, Niagara Falls, New York.

○ TABLE 15.5	Hazardous Waste Sites on the National Priority List (Superfund) by State, 1990*						
STATE	SITES	RANK	PERCENT DISTRIBUTION	STATE	SITES	RANK	PERCENT DISTRIBUTION
Alabama	12	27	1.0	Nebraska	6	43	0.5
Alaska	6	43	0.5	Nevada	1	50	0.1
Arizona	11	29	0.9	New Hampshire	16	22	1.3
Arkansas	10	34	0.8	New Jersey	109	1	9.1
California	88	3	7.4	New Mexico	10	34	0.8
Colorado	16	22	1.3	New York	83	4	6.9
Connecticut	15	24	1.3	North Carolina	22	17	1.8
Delaware	20	19	1.7	North Dakota	2	48	0.2
District of Columbia	0			Ohio	33	12	2.8
Florida	51	6	4.3	Oklahoma	11	29	0.9
Georgia	13	26	1.1	Oregon	8	40	0.7
Hawaii	7	42	0.6	Pennsylvania	95	2	7.9
Idaho	9	38	0.8	Rhode Island	11	29	0.9
Illinois	37	10	3.1	South Carolina	23	16	1.9
Indiana	35	11	2.9	South Dakota	3	46	0.3
Iowa	21	18	1.8	Tennessee	14	25	1.2
Kansas	11	29	0.9	Texas	28	13	2.3
Kentucky	17	21	1.4	Utah	12	27	1.0
Louisiana	11	29	0.9	Vermont	8	40	0.7
Maine	9	38	0.8	Virginia	20	19	1.7
Maryland	10	34	0.8	Washington	45	7	3.8
Massachusetts	25	14	2.1	West Virginia	5	45	0.4
Michigan	78	5	6.5	Wisconsin	39	9	3.3
Minnesota	42	8	3.5	Wyoming	3	46	0.3
Mississippi	2	48	0.2	United States	1,197		100.0
Missouri	24	15	2.0	Guam	1		
Montana	10	34	0.8	Puerto Rico	9		
				TOTAL	1,207		

* Includes both proposed and final sites listed on the National Priorities List for the Superfund program as authorized by the Comprehensive Environmental Response, Compensations, and Liability Act of 1980 and the Superfund Amendments and Reauthorization Act of 1986.
SOURCE: *Statistical Abstract of United States, 1991,* p. 211.

in point is the Brio Superfund Site 18 miles south of Houston, Texas. Two waste-disposal plants once occupied the area that was developed as a subdivision of 2,800 people. Subsequently, water wells and soils there were discovered to be contaminated with chloroform, xylene, tar, TCE, and other health-endangering chemicals. High incidence of birth defects, heart disease, and other problems followed, and most residents fled their homes. By early 1996, more than a decade had passed since Brio was declared a Superfund site, and nearly a billion dollars of taxpayers' money had been paid in litigation fees and monetary damages to residents, but nothing had been done to remediate the site. In 1995

about $4 billion of the $20.4 billion spent on Superfund to date had been consumed by lawyers and filing fees, and legislation was under way to reduce litigation and settle liability issues more quickly.

Consider This... ? Perhaps no other recent environmental legislation has received as much public attention as the act with the catchy nickname "Superfund." What is Superfund? Who manages it, and what legislation is behind it? From an environmental standpoint, what is its main problem? ■

Wastewater Treatment

Sewage is required to be purified to a predetermined standard before it is delivered to a receiving stream or aquifer. An important measure of the purity of sewage effluent is **biological oxygen demand (BOD),** the amount of oxygen per unit volume of water—milligrams of O_2 per liter of water—required for the total aerobic decomposition of organic matter by microorganisms. Also referred to as *biochemical oxygen demand,* it indicates the degree of a water's organic pollution. BOD is determined by measuring the amount of oxygen in a closed liter bottle of treated water at intervals until uptake ceases. The total oxygen used is the BOD. Water that is totally free of organic matter and microorganisms has a BOD of zero.

TREATMENT LEVELS

As many as three levels of sewage treatment may be required to bring BOD, water clarity, and chemical quality to within acceptable limits for a particular receiving water body (⦾ Table 15.6). The levels are explained here.

1. *Primary treatment* consists mostly of removing solids by gravity settling, screening, and then aerating the fluid portion to remove some organic material. There is considerable debris in raw sewage, including wood, rags, paper, and sediment.
2. *Secondary treatment* reduces the BOD of the wastewater to a specified level. It introduces bacteria to digest organic matter and facilitates further sludge settling and clarification of the effluent (◆ Figure 15.22). Secondary treatment requires the presence of microorganisms and a mechanism for replenishing oxygen in the wastewater. One method is to spray wastewater onto a trickling filter, where microbial films develop on the filter material. Another method uses *activated sludge,* whereby a microbial growth that consumes organic matter is suspended in the wastewater and settles to the floor of the reaction tank. Some of the settled sludge is recycled to the incoming wastewater to maintain the biomass required for decomposition. Regardless of the level of wastewater treatment, a huge volume of solids remains that requires disposal. The alternative of dumping sludge at sea will soon be history. In landlocked areas and coastal areas where ocean dumping is prohibited, sludge must be dried and hauled to a landfill. Some is used commercially as a soil amendment for holding moisture and for fertilizer, although dried sludge has low nitrogen and little plant-nutrient value.
3. *Tertiary treatment* removes nitrogen and phosphorus compounds, reducing their levels to below those that cause eutrophication of the receiving waters. **Eutrophication** (from the Greek word, "to thrive") is a water problem that arises from nutrient enrichments and their associated explosive algal blooms. Sewage effluent is high in ammonia (NH_3), which is converted to nitrate $(NO_3)^-$, a plant fertilizer, by nitrifying bacteria. Tertiary treatment is expensive, but it is essential where the discharge of partially treated sewage effluent would jeopardize the health of the receiving water body, such as at Lake Tahoe (Case Study 15.6). Tertiary-treated effluent is pure enough for human consumption.

PRIVATE SEWAGE DISPOSAL

Modern household sewage-disposal systems utilize a septic tank, which separates solids from the effluent, and leach lines, which disseminate the liquids into the surrounding soil or rock (◆ Figure 15.23). A septic tank is constructed in such a way that all solid material stays within the tank and only liquid passes on into the leach field. A leach field consists of rows of tile pipe surrounded by a gravel filter through which the effluent

⦾ TABLE 15.6	What's Left after Primary and Secondary Sewage Treatment	
	PERCENTAGE REMAINING AFTER TREATMENT	
WASTEWATER COMPONENT	*Primary*	*Secondary*
Biological oxygen demand (BOD) Dissolved oxygen necessary to decompose organic matter in wastewater	65	15
Nitrogen Nitrogen becomes a plant nutrient when it combines with hydrogen or oxygen to form ammonia, nitrate, or nitrite.	80–85	70
Phosphorus Phosphorus becomes a plant nutrient when it combines with oxygen to form phosphate.	80–85	70
Suspended solids Undissolved material suspended in water (sludge)	40	15

SOURCE: Data from T. M. Hawley, "Herculean Labors to Clean Wastewater," *Oceanus* 33 (1990), no. 2, p. 73.

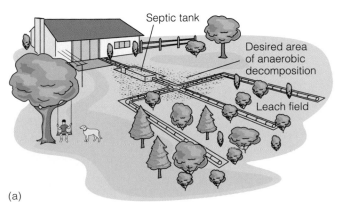

FIGURE 15.22 Wastewater samples, left to right: raw sewage, water after primary treatment (referred to as "mixed liquor"), and after secondary treatment (labeled FINAL EFF, for "final effluent"). After secondary treatment the water is clear and free of harmful chemicals and bacteria. It is suitable for agricultural and landscape watering uses and for discharge to the ocean.

percolates prior to percolating into the soil or underlying rock. Sewage is broken down by *anaerobic* bacteria in the septic tank and the first few feet of the leach lines. As the effluent percolates into the soil or bedrock, it is filtered and oxidized, and aerobic conditions are established. Studies have shown that the movement of effluent through a few feet of unsaturated fine-grained soil under aerobic conditions reduces bacterial counts to almost nil. Fine-grained soils have been found to be the best filters of harmful pathogens, whereas coarse-grained soils allow pathogenic organisms and viruses to disperse over greater distances.

Under the best of conditions, leach lines have a life expectancy of 10–12 years. Factors that reduce infiltration rates and thus limit the life of a septic system are dispersion of clay particles by transfer of sodium ions (Na^+) in sewage to clay particles in the soil, plugging by solids, physical breakdown of the soil by wetting and drying, and biological "plugging." This plugging occurs when the leach system becomes overloaded with

FIGURE 15.23 (a) A septic tank near the house drains into a leach field of drain tiles that are laid in trenches and covered with filter gravel and soil. (b) A common but not recommended method of private sewage disposal.

(a)

(b)

Case Study | 15.6

The Cinder Cone Disposal Site—Getting Your Sewage Together

Prior to 1965 sewage disposal in the north Lake Tahoe area was not much of a problem. Most households used septic tanks to dispose of their waste. With the area's population growth in the 1960s, however, it became clear that too much phosphate and nitrate were seeping from private septic-systems into the lake. Nutrients in the effluent were stimulating algal growth, which was clogging the shore zone and causing problems. It was decided to build a sewage disposal site outside the Tahoe basin as an interim solution until a permanent tertiary-treatment facility could be built.

Geologic studies showed a volcanic cinder cone near the Truckee River northwest of Tahoe City to be an ideal temporary disposal site (◆ Figure 1). With no evidence of recent activity, its clinkery cinders and aa-type lava flows would be good percolation media for sewage effluent (◆ Figure 2). Also, springs on the cone's flanks would help in monitoring the effluent to determine its purity after percolation through the volcano. An interim primary-treatment plant was built in Tahoe City to separate the solid waste from sewage. Sludge from that plant was dried and trucked to a disposal site some 30 kilometers (19 mi) away, and the ef-

fluent was collected in ponds and pumped to several miles of percolation trenches in the central crater (◆ Figure 3). Winter freezing was not a problem, and chemical analyses indicated that the water emerging from the cone was in a relatively pure state.

The Cinder Cone disposal site took 2.8 million gallons of sewage each day during its operation, adequately serving the area's needs until the permanent facility with a 30-million-gallon daily capacity was completed. Although the volume of effluent handled at Cinder Cone is impressive, it is small in comparison to large cities' sewage plants, which handle several hundred million gallons of sewage per day. Nevertheless, this imaginative enterprise is notable in enlisting a natural geologic feature into the service of humankind.

◆ **FIGURE 2** Geologic cross section of Cinder Cone disposal site (vertical scale exaggerated). The risk of polluting the Truckee River was a major consideration in geological studies of the proposed interim solution.

◆ **FIGURE 3** Solids were separated from the sewage in a clarifier and trucked to a Class II landfill near the California–Nevada border. Only the remaining liquid waste was pumped up to Cinder Cone disposal site.

◆ **FIGURE 1** Temporary Cinder Cone disposal site and new water treatment plant; Tahoe City, California.

sewage and anaerobic conditions invade the leach lines, resulting in a growth of organisms that plug the filter materials.

After passing through the leach field, the effluent is still foul, smelly, and highly toxic—a harmful, pathogen-carrying fluid. It must be kept underground until oxidation and bacteria can purify it, making it indistinguishable from local ground water. The geologic condition that most commonly causes a private sewage-disposal system to fail is a high water table that intersects or is immediately beneath the leach lines (◆ Figure 15.24a). This condition initiates the anaerobic growth that plugs filter materials and tile drains. Water tables may rise due to prolonged heavy precipitation or due to the addition of a number of septic tanks in a small area. One way to avoid exceeding the soil's ability to accept effluent is to design leach fields on the basis of percolation tests performed during wet periods. If necessary, a system's lifetime can be extended by constructing multiple leach fields whose use can be rotated regularly by switching a valve. Another problem

is the improper placement of leach lines. This results in effluent seepage on hill slopes or stream valley walls (◆ Figure 15.24b; see also Chapter 8). The proper horizontal distance of leach lines from a hill face is a function of *vertical permeability* and thus is best determined by a qualified professional. Although typical plumbing codes specify a minimum of 5 meters (16 ft) horizontal distance from a hill face, this may not be sufficient if the vertical permeability is low and the hill slope is steep.

Nitrates in sewage effluent pose a eutrophication problem if households are built near a lake or pond. Eutrophication is a natural process in the life cycle of a lake, but cultural (human-caused) eutrophication accelerates the process and can convert an attractive, "real-estate" lake into a weed-choked bog in a short time. Nutrients from sewage systems and fertilized lawns must not be allowed to flow unchecked into bodies of water. Solving this problem requires the help of such experts in water flow and quality as aquatic biologists and hydrogeologists.

(a)   (b)

◆ **FIGURE 15.24** Conditions under which raw effluent can seep into the ground. (a) A drain tile and filter material intersect a high water table. When this occurs sewage rises, and the growth of anaerobic bacteria is promoted in the leach field. This leads to biological plugging of the filter material. (b) A leach line is underlain by low-permeability layers adjacent to a steep slope.

Gallery

Waste Disposal Then and Now

All humans generate solid waste, and disposing of it has always been a societal challenge. Safely disposing of containers of all kinds, old tires, large and small appliances, and just plain trash presents major problems today. As individuals we must recycle, demand recycled and recyclable containers, and actively assist in reducing the volume of waste (◁▷ Figure 15.25). Historians point out that Ancient Rome was a filthy city; its cobblestones were noisy, its air was polluted from thousands of wood fires, and municipal sewage disposal was nonexistent. Medieval castles, monasteries, and convents had primitive waste-disposal systems, most of which were unhealthy and emanated noxious odors (◁▷ Figure 15.26).

More recently, environmental paranoia has demanded the removal of all asbestos from all public buildings, and this has created another problem—how to get rid of asbestos-containing debris safely (◁▷ Figure 15.27). Finally, certain activities at sanitary landfills appear to prove that some people have fun while "down in the dumps" (◁▷ Figure 15.28).

◁▷ **FIGURE 15.25** Norris McDonald organized a major cleanup of the Anacostia River in Washington, D.C., and adjacent Maryland.

◆ **FIGURE 15.26** An adjacent stream was diverted to this medieval abbey in Scotland to create what was called a "great drain." All of the abbey's wastes went into the drain, which carried them to the stream, thus polluting its water for downstream users. Little wonder that lifespans were so short in those days; diseases such as cholera, typhoid fever, and dysentery and pestilence caused by the lack of sanitation took their toll. The correlation between stream pollution and disease was not widely recognized until the 1800s.

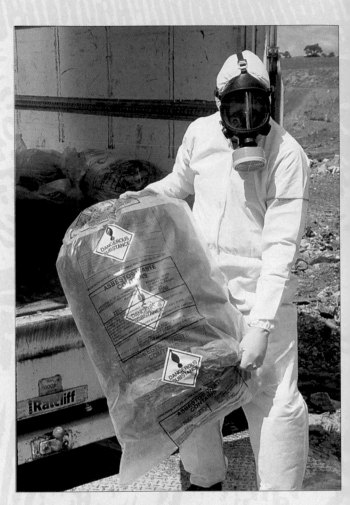

◆ **FIGURE 15.27** Not an astronaut; a sanitation worker. Protective suits and masks are required for asbestos handlers at sanitary landfills.

◆ **FIGURE 15.28** Caterpillar tractor operators play "king of the mountain" at a landfill. The 25-ton "Cats" are used to compact and flatten the debris to reduce its volume.

Summary

WASTE MANAGEMENT

MANAGEMENT COMPONENTS
Geology must be considered in all components:
1. Waste disposal
2. Source reduction
3. Waste recovery
4. Recycling

RATIONALE
Extending the lifetimes of landfills and natural resources.

CATEGORIES OF WASTE-DISPOSAL METHODS
Isolation and attenuation.

MUNICIPAL WASTE DISPOSAL

SANITARY LANDFILL
Trash is covered with clean soil each day.

OCEAN DUMPING
Favored in the past for disposal of sewage, sludge, dredgings, and canistered nuclear wastes. Now regulated by Marine Protection Reserve and Sanctuary Act (Ocean Dumping Act).

LANDFILL PROBLEMS
Leachate, settlement, gas generation, visual blight.

RESOURCE RECOVERY
Amount of waste delivered to landfills can be reduced through recycling programs, composting, and implementing waste-reduction strategies.

MULTIPLE LAND USE CONCEPT
A mine or quarry becomes a landfill, which when full may be converted to a park or other recreational area. Although some shallow landfills have been built upon, settlement is always a problem.

HAZARDOUS-WASTE DISPOSAL

SECURE LANDFILL
Lined with plastic and impermeable clay seal. Built with leachate drains.

DEEP-WELL INJECTION
Into dry permeable rock body sealed from aquifers above and below. Has triggered earthquakes in Denver area.

SUPERFUND
Comprehensive Environmental Response, Compensation, and Liability Act (CERCLA) created in 1980 to correct and stop abuses from irresponsible toxic-waste dumping.

WASTEWATER TREATMENT

PURPOSE
To clarify and purify water. The measure of purification is biological oxygen demand (BOD).

LEVELS
1. *Primary*—Solids are separated by settling and some aeration to remove organic materials.
2. *Secondary*—BOD is reduced by bacterial action and aeration.
3. *Tertiary*—Inorganic nutrients such as nitrates and phosphates are removed.

PRIVATE SEWAGE DISPOSAL
Requires permeability and porosity and can lead to local pollution.

◆ Key Terms

attenuation (waste disposal)

biological oxygen demand (BOD)

composting

deep-well injection

EPA

eutrophication

isolation (waste disposal)

leachate

NEPA

RCRA

recycling

resource recovery

sanitary landfill

secure landfill

Superfund

waste management

◆ Study Questions

1. Explain the federal legislation known as NEPA and RCRA and describe their impacts on the environmental safety of landfills.
2. What is leachate, and what methods are used to prevent landfill leachates from entering bodies of water?
3. Explain the "cell" method of placing trash in a landfill.
4. How much solid waste does the average American generate each day in kilograms? In pounds?
5. What problems are associated with operating a sanitary landfill?
6. What resources are found in a landfill, and how can they be recovered? What can individuals do to assist in recovering and extending the lifetimes of certain limited natural resources?
7. Describe two common methods of isolating hazardous wastes from the environment. Explain the steps that are taken to keep these wastes from polluting underground water.
8. What is eutrophication, and what can be done in sewage treatment to minimize this threat to lakes and rivers?

◆ For Further Information

BOOKS AND PERIODICALS

Alexander, J. H. 1993. *In defense of garbage.* New York: Praeger.

Brown, Lester, and others. 1999. *State of the world.* New York: Worldwatch Institute and W. W. Norton.

Brunner, D. R., and D. K. Keller. 1972. *Sanitary landfill design and operations.* Washington, D.C.: U.S. Environmental Protection Agency.

Grove, Noel. 1994. Recycling. *National geographic* 186, no. 1 (July).

Lewin, David. 1991. Solid waste: Bury, burn, or reuse. *Mechanical engineering,* August:75.

McKinney, Michael, and Robert Schoch. 1996. *Environmental science.* St. Paul, Minn.: West Educational Publishing.

Miller, G. Tyler. 1999. *Environmental science.* Belmont, California: Wadsworth Publishing Company.

Rathje, William L. 1991. Once and future landfills. *National geographic,* May:114–134.

Rathje, William L., and Cullen Murphy. 1992. *Rubbish! The archeology of garbage.* New York: Harper Collins.

U.S. Environmental Protection Agency. 1992. *Characterization of municipal solid waste in the United States: Final report.*

White, Peter T. 1983. The wonderful world of trash. *National geographic,* April.

World Resources Institute. 1992. *World resources, 1991–1992.* New York: Oxford University Press.

INTERNET SITES

Environmental Protection Agency (EPA). EPA sites and links are the best sources of information about waste disposal and problems of toxic waste in the United States.
_____. Home site.
http://www.epa.gov
_____. Citizen's guide to pollution and waste disposal.
http://www.epa.gov/epahome/Citizen.html
_____. Governing laws. More than a dozen major statutes or laws form the legal basis for the programs of the EPA. Full text of these laws may be found at:
http://www.epa.gov/epahome/laws.html
_____. Recycling handbook.
http://www.epa.gov/OSWRCRA/non-hw/reduce/catbook
_____. Superfund.
http://www.epa.gov/superfund/

◆ World Wide Web Activities

A Biological Approach to Solid Waste Reduction
http://www.ecochem.com/t_landfill.html
This commercial site offers an approach to solid waste volume and toxicity reduction. Link to the **Comparative Analysis** page for a summary of the advantages to this approach.

1. What is the specific approach to waste management being offered by this company? Summarize the advantages to this approach, as stated in the Comparative Analysis.
2. What possible problems do you see with this approach?

Love Canal Summary
http://ublib.buffalo.edu/libraries/units/sel/exhibits/lovecanal.html
This Web site summarizes the history of Love Canal, a notorious case of toxic waste release from a landfill in New York State.

1. What was the original source of toxic chemicals on the site?
2. What action brought the site into the public domain before its hazards were recognized?

3. What natural factors mobilized the waste, endangering residents?

Recycling Disposables: Beyond Paper, Metals, and Glass
http://www.obviously.com/recycle/guides/hard.html
This intriguing Web site contains information on recycling the more obscure disposables of our culture–smoke detectors, eyeglasses, holiday greeting cards, and so on.

1. Why should we be concerned about recycling smoke detectors?
2. What *does* Nike do with the sneakers they accept for recycling?

Deep-Well Injection for Disposal of Toxic Fluids
http://www.frtr.gov/matrix2/section4/4_57.html
Deep-well injection, one method for dealing with toxic fluid waste, is summarily examined on this Web page.

1. What do you think is the key warning phrase, or "caveat," in section 4.57?
2. Under **Limitations,** what specific aspects of an injection site might rule it out as a candidate for deep-injection disposal?

Earth Systems Today CD Correlations

CHAPTER CONCEPT: Class I Disposal Sites Must Have No Possibility of Discharge to Usable Waters

This part of the CD presents the conditions of formation, and resulting physical properties, of the three rock families and their major subgroup rocks.

SECTION MODULE: The Rock Cycle and Rock Laboratory

Observe the physical characteristics of the common rock categories.

QUESTION: *Since Class I disposal sites must be located in essentially impermeable materials, what specific rock type(s) would be most suitable for siting of these very hazardous waste disposal dumps? Give reasons for your selection(s).*

CHAPTER CONCEPT: Point Source Leakages of Petroleum Products Are a Common Hazard in Urban Areas

The **Ground-Water Contamination** segment of the CD explains sources of pollution.

SECTION MODULE: Ground Water Contamination

Observe the dispersal of chemical contaminants in ground water below a leaking landfill.

QUESTION: *What is a common point source of petroleum contaminants in urban areas? Where is the petroleum pollutant located relative to the water table? What specific risk factor beyond toxicity would petroleum leaking to the surface present?*

Periodic Table of the Elements

Period IA

	1																		Noble Gases

Atomic Number — 1
Symbol of Element — H
Atomic Mass — 1.008
(rounded to four significant figures)

1
| 1 | | | | | | | | | | | | | | | | | 2 |
| **H** 1.008 | IIA | | | | | | | | | | | IIΛ | IVA | VA | VIA | VIIA | **He** 4.003 |

2
| 3 | 4 | | | | | | | | | | | 5 | 6 | 7 | 8 | 9 | 10 |
| **Li** 6.941 | **Be** 9.012 | | | | | | | | | | | **B** 10.81 | **C** 12.01 | **N** 14.01 | **O** 16.00 | **F** 19.00 | **Ne** 20.18 |

3
| 11 | 12 | | | | | | | | | | | 13 | 14 | 15 | 16 | 17 | 18 |
| **Na** 22.99 | **Mg** 24.31 | IIIB | IVB | VB | VIB | VIIB | ——VIII—— | | | IB | IIB | **Al** 26.98 | **Si** 28.09 | **P** 30.97 | **S** 32.06 | **Cl** 35.45 | **Ar** 39.95 |

4
| 19 | 20 | 21 | 22 | 23 | 24 | 25 | 26 | 27 | 28 | 29 | 30 | 31 | 32 | 33 | 34 | 35 | 36 |
| **K** 39.10 | **Ca** 40.08 | **Sc** 44.96 | **Ti** 47.90 | **V** 50.94 | **Cr** 52.00 | **Mn** 54.94 | **Fe** 55.85 | **Co** 58.93 | **Ni** 58.71 | **Cu** 63.85 | **Zn** 65.37 | **Ga** 69.72 | **Ge** 72.59 | **As** 74.92 | **Se** 78.96 | **Br** 79.90 | **Kr** 83.80 |

5
| 37 | 38 | 39 | 40 | 41 | 42 | 43 | 44 | 45 | 46 | 47 | 48 | 49 | 50 | 51 | 52 | 53 | 54 |
| **Rb** 85.47 | **Sr** 87.62 | **Y** 88.91 | **Zr** 91.22 | **Nb** 92.91 | **Mo** 95.94 | **Tc**[x] 98.91 | **Ru** 101.1 | **Rh** 102.9 | **Pd** 106.4 | **Ag** 107.9 | **Cd** 112.4 | **In** 114.8 | **Sn** 118.7 | **Sb** 121.8 | **Te** 127.6 | **I** 126.9 | **Xe** 131.3 |

6
| 55 | 56 | 57 | 72 | 73 | 74 | 75 | 76 | 77 | 78 | 79 | 80 | 81 | 82 | 83 | 84 | 85 | 86 |
| **Cs** 132.9 | **Ba** 137.3 | **La** 138.9 | **Hf** 178.5 | **Ta** 180.9 | **W** 183.9 | **Re** 186.2 | **Os** 190.2 | **Ir** 192.2 | **Pt** 195.1 | **Au** 197.0 | **Hg** 200.6 | **Tl** 204.4 | **Pb** 207.2 | **Bi** 209.0 | **Po**[x] (210) | **At**[x] (210) | **Rn**[x] (222) |

7
| 87 | 88 | 89 | 104 | 105 | 106 | 107 | 108 | 109 |
| **Fr**[x] (223) | **Ra**[x] 226.0 | **Ac**[x] (227) | **Unq**[x] (261) | **Unp**[x] (262) | **Unh**[x] (263) | **Uns**[x] (262) | **Uno**[x] (265) | **Une**[x] (266) |

● Lanthanide series

| 58 | 59 | 60 | 61 | 62 | 63 | 64 | 65 | 66 | 67 | 68 | 69 | 70 | 71 |
| **Ce** 140.1 | **Pr** 140.9 | **Nd** 144.2 | **Pm**[x] (147) | **Sm** 150.4 | **Eu** 152.0 | **Gd** 157.3 | **Tb** 158.9 | **Dy** 162.5 | **Ho** 164.9 | **Er** 167.3 | **Tm** 168.9 | **Yb** 173.0 | **Lu** 175.0 |

■ Actinide series

| 90 | 91 | 92 | 93 | 94 | 95 | 96 | 97 | 98 | 99 | 100 | 101 | 102 | 103 |
| **Th**[x] 232.0 | **Pa**[x] 231.0 | **U**[x] 238.0 | **Np**[x] 237.0 | **Pu**[x] (244) | **Am**[x] (243) | **Cm**[x] (247) | **Bk**[x] (247) | **Cf**[x] (251) | **Es**[x] (254) | **Fm**[x] (257) | **Md**[x] (258) | **No**[x] (255) | **Lr**[x] (256) |

x: All isotopes are radioactive.

() indicates mass number of isotope with longest known half-life.

The Elements

ELEMENT	SYMBOL	ATOMIC NUMBER	ELEMENT	SYMBOL	ATOMIC NUMBER
Actinium	Ac	89	Mercury	Hg	80
Aluminum	Al	13	Molybdenum	Mo	42
Americium	Am	95	Neodymium	Nd	60
Antimony	Sb	51	Neon	Ne	10
Argon	Ar	18	Neptunium	Np	93
Arsenic	As	33	Nickel	Ni	28
Astatine	At	85	Niobium	Nb	41
Barium	Ba	56	Nitrogen	N	7
Berkelium	Bk	97	Nobelium	No	102
Beryllium	Be	4	Osmium	Os	76
Bismuth	Bi	83	Oxygen	O	8
Boron	B	5	Palladium	Pd	46
Bromine	Br	35	Phosphorus	P	15
Cadmium	Cd	48	Platinum	Pt	78
Calcium	Ca	20	Plutonium	Pu	94
Californium	Cf	98	Polonium	Po	84
Carbon	C	6	Potassium	K	19
Cerium	Ce	58	Praseodymium	Pr	59
Cesium	Cs	55	Promethium	Pm	61
Chlorine	Cl	17	Protactinium	Pa	91
Chromium	Cr	24	Radium	Ra	88
Cobalt	Co	27	Radon	Rn	86
Copper	Cu	29	Rhenium	Re	75
Curium	Cm	96	Rhodium	Rh	45
Dysprosium	Dy	66	Rubidium	Rb	37
Einsteinium	Es	99	Ruthenium	Ru	44
Erbium	Er	68	Samarium	Sm	62
Europium	Eu	63	Scandium	Sc	21
Fermium	Fm	100	Selenium	Se	34
Fluorine	F	9	Silicon	Si	14
Francium	Fr	87	Silver	Ag	47
Gadolinium	Gd	64	Sodium	Na	11
Gallium	Ga	31	Strontium	Sr	38
Germanium	Ge	32	Sulfur	S	16
Gold	Au	79	Tantalum	Ta	73
Hafnium	Hf	72	Technetium	Tc	43
Helium	He	2	Tellurium	Te	52
Holmium	Ho	67	Terbium	Tb	65
Hydrogen	H	1	Thallium	Tl	81
Indium	In	49	Thorium	Th	90
Iodine	I	53	Thulium	Tm	69
Iridium	Ir	77	Tin	Sn	50
Iron	Fe	26	Titanium	Ti	22
Krypton	Kr	36	Tungsten	W	74
Lanthanum	La	57	Uranium	U	92
Lawrencium	Lr	103	Vanadium	V	23
Lead	Pb	82	Xenon	Xe	54
Lithium	Li	3	Ytterbium	Yb	70
Lutetium	Lu	71	Yttrium	Y	39
Magnesium	Mg	12	Zinc	Zn	30
Manganese	Mn	25	Zirconium	Zr	40
Medelevium	Md	101			

Some Important Minerals

MINERAL	CHEMICAL COMPOSITION	COLOR	HARDNESS	OTHER CHARACTERISTICS	PRIMARY ROCK OCCURRENCE
1. amphibole (e.g., hornblende) (hydrous)	Na, Ca, Mg, Fe, Al silicate	green, blue brown, black	5 to 6	often forms needlelike crystals; two good cleavages forming 60°–120° angle crystals hexagonal in cross section	igneous and metamorphic minor amounts in all types
2. apatite	$Ca_5(PO_4)_3$ (F, Cl, OH)	blue-green	5	crystals hexagonal in cross section	minor amounts, all types
3. biotite (a mica)	hydrous K, Mg, Fe silicate	brown to black	2½–3	excellent cleavage into thin sheets	all types
4. calcite	$CaCO_3$	variable; colorless if pure	3	effervesces in weak acid	sedimentary and metamorphic
5. dolomite	$CaMg(CO_3)_2$	white or pink	3½ to 4	powdered mineral effervesces in acid	sedimentary
6. galena	PbS	silver-gray	2½	metallic luster; cubic cleavage; heavy	sulfide veins
7. garnet	Ca, Mg, Fe, Al silicate	variable; often dark red	7	glassy luster	metamorphic
8. graphite	C	dark gray	1 to 2	streaks like pencil lead	metamorphic
9. gypsum	$CaSO_4 \cdot 2H_2O$	colorless	2	2 directions of cleavage	sedimentary
10. halite	NaCl	colorless	2½	salty taste; cleaves into cubes	sedimentary
11. hematite	Fe_2O_3	red or dark gray	5½ to 6½	red-brown streak regardless of color	oxidation of iron-bearing minerals
12. limonite	$Fe_2O_3 \cdot 3H_2O$	yellow-brown	2 to 3	earthy luster; yellow-brown streak	hydration of iron oxides
13. magnetite	Fe_3O_4	black	6	strongly magnetic	minor amounts, all types
14. muscovite (mica)	K, Al, Si, silicate	colorless to light brown	2 to 2½	excellent basal cleavage	all types
15. olivine	Fe, Mg silicate	yellow-green	6½ to 7	glassy luster	igneous and metamorphic
16. plagioclase feldspar	Na, Ca, Al silicate	white to gray	6	2 cleavages at 90°, striations on cleavage surfaces	all types
17. orthoclase feldspar	K, Al silicate	white; or colored pink or aqua	6	two good cleavages forming 90° angle; no striations	all types
18. pyrite	FeS_2	brassy yellow	6 to 6½	metallic luster; black streak	all types
19. pyroxene (e.g., augite)	Na, Ca, Mg, Fe, Al silicate	usually green or black	5 to 7	two good cleavages forming 90° angle	igneous and metamorphic
20. quartz	SiO_2	variable; commonly colorless or white	7	glassy luster; conchoidal fracture; hexagonal crystals	all types

Planes in Space: Dip and Strike

Geologists describe the orientation of a plane surface by the plane's "strike" and "dip." Geologists measure such tangible planes as those made by sedimentary strata, foliation, and faults, because it allows them to anticipate hazards during excavations for such things as tunnels, dam foundations, and roadcuts. Accurate surface measurements of planes also enable geologists to predict the underground locations of valuable resources such as oil reservoirs, ore deposits, and water-bearing strata.

Strike is the direction of the trace of an inclined plane on the horizontal plane, and the direction is referenced to north. In the illustrated example, the strike is 30° west of north. In geologic notation, this is written *N30° W.*

Dip is the vertical angle that the same inclined plane makes with the horizontal plane. Dip is assessed perpendicular to the strike, because any other angle will give an "apparent dip" that is less than the true dip. Because a plane that strikes northwest may be inclined to either the southwest or the northeast, the general direction of dip must be indicated. In our example, the dip is 60° to the northeast, which is notated *60° NE.*

Thus, a geologist's notes for the bedding planes in this example would describe their orientation as "strike N30° W, dip 60° NE." All other geologists would understand that the plane has a strike that is 30° west of north and a dip that is 60° to the northeast.

Block diagram showing sedimentary rock that is striking N30° W and dipping 60° northeast. The apparent dip is shown on the north–south cross section at the lower front of the diagram.

Modified Mercalli Scale

The Rossi–Forel intensity scale was the first scale developed for evaluating the effects of an earthquake on structures and humans. It was developed in Europe in 1878 and assigned Roman numerals from *I* (for *barely felt*) to *X* (for *total destruction*) for varying levels of intensity. This scale was first modified by Father Giuseppi Mercalli in Italy and later in 1931 by Frank Neumann and Harry O. Wood in the United States. The Modified Mercalli scale is the scale most widely used today for evaluating the effects of earthquakes in the field. An abbreviated version is given here.

I. Not felt except by a very few persons under especially favorable circumstances.

II. Felt by a few persons at rest, especially by persons on upper floors of multistory buildings, and by nervous or sensitive persons on lower floors.

III. Felt quite noticeably indoors, especially on upper floors, but many people do not recognize it as an earthquake; vibrations resemble those made by a passing truck.

IV. Felt indoors by many persons and outdoors by only a few. Dishes, windows, and doors are disturbed, and walls make creaking sounds as though a heavy truck had struck the building. Standing cars are rocked noticeably.

V. Felt by nearly everyone. Many sleeping people are awakened. Some dishes and windows are broken, and some plaster is cracked. Disturbances of trees, poles, and other tall objects may be noticed. Some persons run outdoors.

VI. Felt by all. Many people are frightened and run outdoors. Some heavy furniture is moved, and some plaster and chimneys fall. Damage is slight, but humans are disturbed.

VII. General fright and alarm. Everyone runs outdoors. Damage is negligible in buildings of good design and construction, considerable in those that are poorly built. Noticeable in moving cars.

VIII. General fright approaching panic. Damage is slight in specially designed structures; considerable in ordinary buildings, with partial collapse; great in older or poorly built structures. Chimneys, smokestacks, columns, and walls fall. Sand and mud are ejected from ground openings (liquefaction).

IX. General panic. Damage is considerable in specially designed structures, well-designed frame structures being deformed; great in substantial buildings, with partial collapse. Ground is cracked; underground pipes are broken.

X. General panic. Some well-built wooden structures and most masonry and frame structures are destroyed. Ground is badly cracked; railway rails are bent; underground pipes are torn apart. Considerable landsliding on riverbanks and steep slopes.

XI. General panic. Ground is greatly disturbed, with cracks and landslides common. Sea waves of significant height may be seen. Few if any structures remain standing; bridges are destroyed.

XII. Total panic. Total damage to human engineering works. Waves seen on the ground surface. Lines of sight and level are distorted. Objects are thrown upward into the air.

five

Fluid Flow in Porous Rocks

Porosity is the ratio of the volume of pore spaces in a rock or other solid to the material's total volume. Expressed as a percentage, it is

$$P = \frac{V_p}{V_t} \times 100$$

where P = porosity

V_p = pore volume

V_t = total volume

If an aquifer has a porosity of 33 percent, a cubic meter of the aquifer is composed of ⅔ cubic meter of solid material and ⅓ cubic meter of pore space, which may or may not be filled with water. Porosities of 20 to 40 percent are not uncommon in clastic sedimentary rocks.

Permeability is a measure of how fast fluids can move through a porous medium and is thus expressed in a unit of velocity, such as meters per day. Knowing the permeability of an aquifer allows us to determine in advance how much water a well will produce and how far apart wells should be spaced. To determine permeability we use **Darcy's Law,** which states that the rate of flow (Q in cubic meters) is the product of hydraulic conductivity (K), the **hydraulic gradient,** or the slope of the water table (I in meters per meter), and the cross-sectional area (A in square meters). Thus,

Q	$=$	K	$\times$	I	$\times$	A
rate of flow		hydraulic conductivity		hydraulic gradient		cross-sectional area

Hydraulic conductivity, K, is a measure of the permeability of a rock or a sediment, its water-transmitting characteristics. It is defined as the quantity of water that will flow in a unit of time under a unit hydraulic gradient through a unit of area measured perpendicular to the flow direction. K is expressed in distance of flow with time, usually meters per day.

If we use a unit hydraulic gradient, then Darcy's Law reduces to:

$$Q = K \times A$$

Unit hydraulic gradient is the slope of 1-unit vertical to 1-unit horizontal ($I = 1/1 = 45°$), and it allows us to compare the water-transmitting ability of aquifers composed of materials with different permeabilities. Flow rates may be less than a millimeter per day for clays, many meters per day for well-sorted sands and clean gravels. In general, ground water moves at a snail's pace.

Pumice provides an interesting example. It is a porous but almost impermeable rock. It has porosities as high as 90 percent but few connected pore spaces. Pumice will float for a time because it is light-weight and its low permeability prevents immediate saturation. A cubic meter of pumice may weigh as little as 500 pounds, a fifth of the weight of solid rock of the same composition.

The unit hydraulic gradient (I) and unit prism (1 m²) of a sandstone aquifer. When the hydraulic conductivity (K) is known, then the discharge (Q) in liters or cubic meters per day per square meter of aquifer can be calculated.

glossary

aa A basaltic lava that has a rough, blocky surface.

ablation All processes by which snow or ice is lost from a glacier, floating ice, or snow cover. These processes include melting, evaporation, and calving.

ablation zone The part of a glacier or a snowfield in which ablation exceeds accumulation over a year's time; the region below the firn line.

abrasion The mechanical grinding, scraping, wearing, or rubbing away of rock surfaces by friction or impact. Abrasion agents include wind, running water, waves, and ice.

absolute age dating Assigning units of time, usually years, to fossils, rocks, and geologic features or events. The method utilizes mostly the decay of radioactive elements, but ages may also be obtained by counting tree rings or annual sedimentary layers.

accretion The mechanism by which the earth is believed to have formed from a small nucleus by additions of solid bodies, such as meteorites, asteroids, or planetesimals. Also used in plate tectonics to indicate the addition of terranes to a continent.

accumulation zone The part of a glacier or snowfield in which accumulation exceeds ablation over a year's time; the region above the firn line. Also the B-horizon of a soil.

acid mine drainage (AMD) The acid runoff from a mine or mine waste.

acid rain Corrosive precipitation due to the solution of atmospheric sulfuric acid in rain, which makes it acidic. Although it is generally considered human-made, it can also be natural, as from volcanic activity.

active layer The zone in permafrost that is subject to seasonal freezing and thawing.

active sea cliff A cliff whose formation is dominated by wave erosion.

adsorption The attraction of ions or molecules to the surface of particles, in contrast to *absorption,* the process by which a substance is taken into a clay-mineral structure.

aerosol A suspension of very fine, liquid and solid particles in the atmosphere.

aftershocks A series of earthquakes following the first or main shock. They are usually of lower magnitude than the main shock.

alluvial fan A gently sloping, fan-shaped mass of alluvium deposited in arid or semiarid regions where a stream flows from a narrow canyon onto a plain or valley floor.

alluvial placer A placer deposit that is concentrated in a streambed.

alluvium Stream-deposited sediment.

alpha decay The transformation that occurs when the nucleus of an atom spontaneously emits a helium atom, reducing its atomic mass by 4 units and its atomic number by 2; for example, $^{238}_{92}U \rightarrow {}^{234}_{90}Th + {}^{4}_{2}a$.

amorphous Descriptive term applied to glasses and other substances that lack an orderly internal crystal structure.

andesite line The petrographic boundary between basalts and andesites. It runs roughly around the Pacific basin margin. It also marks the boundary between continental (explosive) and oceanic (Hawaiian-type) volcanoes in the Pacific Ocean.

angle of repose The maximum stable slope angle that granular, cohesionless material can assume. For dry sand this is about 34°.

anomaly A deviation from the average or expected value. In paleomagnetism, a "positive" anomaly is a stronger-than-average magnetic field at the earth's surface.

anticline A generally convex-upward fold that has older rocks in its core. (*Contrast with* syncline.)

aphanitic texture A fine-grained igneous-rock texture resulting from rapid cooling. Aphanites are fine-grained igneous rocks.

A.P.I. gravity (American Petroleum Institute) The weight of crude oil arbitrarily compared to that of water at 60° F, which is assigned a value of 10°. Most crude oils range from 5° to 30° A.P.I.

aquiclude A layer of low-permeability rock that hampers ground-water movement. Often applied to the confining bed in a confined aquifer.

aquifer A water-bearing body of rock or sediment that will yield water to a well or spring in usable quantities.

area mining The mining of coal in flat terrain (strip mining).

artesian pressure surface An imaginary surface defined by the level to which water will rise in an artesian well.

artificial levee An embankment constructed to contain a stream at flood stage. *See* levee.

ash Very fine, gritty volcanic dust: particles are less than 4 mm in diameter.

ash fall A "rain" of airborne volcanic ash.

asperity Rough spot or "knot" along a fault that stops movement along that section for a period of time. The bigger the asperity, the greater the stress buildup and the larger the earthquake that results. (*See also* seismic gap.)

asthenosphere The *plastic* layer of the earth, at whose upper boundary the brittle overlying lithosphere is detached from the underlying mantle.

atom The smallest particle of an element that can exist either alone or in combination with similar particles of the same or a different element.

atomic number The number of protons (positively-charged particles) in an atomic nucleus. Atoms with the same atomic number belong to the same chemical element.

atomic mass The sum of the number of protons and neutrons (uncharged particles) in the nucleus of an atom.

attenuation (waste disposal) To dilute or spread out the waste material.

average crustal abundance The percentage of a particular element in the composition of the earth's crust.

a-zone The upper soil horizon (topsoil) capped by a layer of organic matter. It is the zone of leaching of soluble soil salts.

back slope The gentler slope of a dune, scarp, or fault block.

ball mill A large rotating drum containing steel balls that grinds ore to a fine powder. Batteries of ball mills are used in modern milling operations.

banded-iron formation An iron deposit consisting essentially of iron oxides and chert occurring in prominent layers or bands of brown or red and black.

barchan dune An isolated, moving, crescent-shaped sand dune lying transverse to the direction of the prevailing wind. A gently sloping, convex side faces the wind, and the horns of the crescent point downwind.

barrier beach (barrier island) A narrow sand ridge rising slightly above hightide level that is oriented generally parallel with the coast and separated from the coast by a lagoon or a tidal marsh.

basalt A dark-colored, fine-grained volcanic rock that is commonly called *lava*. Basaltic lavas are quite fluid.

base level The level below which a stream cannot erode its bed. The ultimate base level is sea level.

base shear Due to a rapid horizontal ground acceleration during an earthquake. It can cause a building to part from its foundation, thus the term.

batholith A large, discordant intrusive body greater than 100 km² (40 mi²) in surface area with no known bottom. Batholiths are granitic in composition.

bauxite An off-white, grayish, brown, or reddish brown rock composed of a mixture of various hydrous aluminum oxides and aluminum hydroxides.

beach cusps A series of regularly spaced points of beach sand or gravel separated by crescent-shaped troughs.

bedding *See* stratification.

bedding planes The divisions or surfaces that separate successive layers of sedimentary rocks from the layers above and below them.

bench placer Remnant of an ancient alluvial placer deposit that was concentrated beneath a former stream and that is now preserved as a bench on a hillside.

Benioff zone A zone of earthquake epicenters beneath oceanic trenches that extends from near the surface to a depth of 700 km (430 mi). The zone is named after Hugo Benioff and is thought to indicate active subduction.

bentonite A soft, plastic, light-colored clay that swells to many times its volume when placed in water. Named for outcrops near Fort Benton, Wyoming, bentonite is formed by chemical alteration of volcanic ash.

beta decay The radioactive emission of a nuclear electron (beta ray, β^-) that converts a neutron to a proton. The element is changed to the element of next-highest atomic number; for example, $^{14}_{6}C \rightarrow {}^{14}_{7}N + \beta^-$.

biogenic sedimentary rock Rocks resulting directly from the activities of living organisms, such as a coral reef, shell limestone, or coal.

biological oxygen demand (B.O.D.) Indicator of organic pollutants (microbial) in an effluent measured as the amount of oxygen required to support them. The greater the B.O.D. (ppm or mg/liter) the greater the pollution and less oxygen available for higher aquatic organisms.

biomass fuels Fuels that are manufactured from plant materials; for example, grain alcohol (ethanol) is made from corn.

bioremediation The use of microbes to clean hydrocarbon residues from contaminated soils.

black smoker A submarine hydrothermal vent associated with a mid-oceanic spreading center that has emitted or is emitting a "smoke" of metallic sulfide particles.

block (volcanology) A mass of cold country rock greater than 32 mm (1¼ in) in size that is ripped from the vent and ejected from a volcano.

block glide (translational slide) A landslide in which movement occurs along a well-defined plane surface or surfaces, such as bedding planes, foliation planes, or faults.

body wave Earthquake waves that travel through the body of the earth.

bolide A large meteor that explodes as it passes through the earth's atmosphere.

bomb (volcanology) A molten mass greater than 32 mm (1¼ in) in diameter that is ejected from a volcano and cools sufficiently before striking the ground to retain a rounded, streamlined shape.

boreal The northern and mountainous regions of the Northern Hemisphere in which the average temperature of the hottest season does not exceed 18° C (64.4° F).

breakwater An offshore rock or concrete structure, attached at one end or parallel to the shoreline, intended to create quiet water for boat anchorage purposes.

breeder reactor A reactor that produces more fissile (splittable) nuclei than it consumes.

British thermal unit (Btu) The amount of heat, in calories, needed to raise the temperature of 1 pound of water 1° F.

caldera Large crater caused by explosive eruption and/or subsidence of the cone into the magma chamber, usually more than 1.5 km (1 mi) in diameter.

caliche In arid regions, a sediment or soil that is cemented by calcium carbonate ($CaCO_3$).

calving The breaking away of a block of ice, usually from the front of a tidewater glacier, that produces an iceberg.

caprock An impermeable stratum that caps an oil reservoir and prevents oil and gas from escaping to the ground surface.

capillary fringe The moist zone in an aquifer above the water table. The water is held in it by capillary action.

capillary water Water that is held in tiny openings in rock or soil by capillary forces.

carrying capacity The number of creatures a given tract of land can adequately support with food, water, and other necessities of life.

central-core meltdown The melting of an atomic-reactor fuel containment vessel as a result of runaway atomic fission that generates extreme heat.

CERCLA The Comprehensive Environmental Response, Compensation and Liability Act of 1980. *See* Superfund.

chain reaction A self-sustaining nuclear reaction that occurs when atomic nuclei undergo fission. Free neutrons are released that split other nuclei, causing more fissions and the release of more neutrons, which split other nuclei, and so on.

chemical sedimentary rocks Rock resulting from precipitation of chemical compounds from a water solution. *See* evaporite.

chlorofluorocarbons (CFCs) A group of compounds that, on escape to the atmosphere, break down, releasing chlorine atoms that destroy ozone molecules; for example, Freon, which is widely used in refrigeration, air-conditioning, and formerly was a propellant in aerosol-spray cans.

cinder cone A small, straight-sided volcanic cone consisting mostly of pyroclastic material (usually basaltic).

cinders Glassy and vesicular (containing gas holes) volcanic fragments, only a few centimeters across, that are thrown from a volcano.

clastic Describing rock or sediment composed primarily of detritus of preexisting rocks or minerals.

clay minerals Hydrous aluminum silicates that have a layered atomic structure. They are very fine grained and become plastic (moldable) when wet. Most belong to one of three clay groups: kaolinite, illite, and smectite, the last of which are expansive when they absorb water between layers.

claypan A dense, relatively impervious subsoil layer that owes its character to a high clay-mineral content due to concentration by downward-percolating waters.

cleavage The breaking of minerals along certain crystallographic planes of weakness that reflects their internal structure.

climatic controls The relatively constant factors that dictate a region's climate.

coastal desert A desert on the western edge of continents in tropical latitudes; for example, near the Tropic of Capricorn or Cancer. The daily and annual temperature fluctuations are much less than in inland tropical deserts.

columnar jointing Vertically oriented polygonal columns in a solid lava flow or shallow intrusion formed by contraction upon cooling. They are usually formed in basaltic lavas.

compaction The reduction in bulk volume of fine-grained sediments due to increased overburden weight or to tighter packing of grains. Clay is more compressible than sand and therefore compacts more.

complex landslide When both rotational and translational sliding are found within one landslide mass.

composting Bacterial recycling of organic material.

concentrate Enriched ore, often obtained by flotation, produced at a mill.

concentration factor The enrichment of a deposit of an element, expressed as the ratio of the element's abundance in the deposit to its average crustal abundance.

cone of depression The cone shape formed by the water table around a pumping well when pumping exceeds the flow of water to the well.

confined aquifer A water-bearing formation bounded above and below by impermeable beds or beds of distinctly lower permeability.

contact metamorphism A change in the mineral composition of rocks in contact with an invading magma from which fluid constituents are carried out to combine with some of the country-rock constituents to form a new suite of minerals.

continental divide A drainage divide that separates streams flowing toward opposite sides of a continent.

continental drift The term applied by American geologists to Wegener's hypothesis of long-distance horizontal movement of continents he called "continental displacement."

contour mining The mining of coal by cutting into and following the coal seam around the perimeter of the hill.

convection The circulatory movement within a body of nonuniformly heated fluid. Warmer material rises in the center, and cooler material sinks at the outer boundaries. It has been proposed as the mechanism of plate motion in plate tectonic theory.

convergent boundary Tectonic boundary where two plates collide. Where a tectonic plate sinks beneath another plate and is destroyed, this boundary is known as a *subduction zone*. Subduction zones are marked by deep-focus seismic activity, strong earthquakes, and violent volcanic eruptions.

Coriolis effect A condition peculiar to motion on a rotating body such as the Earth. It appears to deflect moving objects in the Northern Hemisphere to the right and objects in the Southern Hemisphere to the left, to an observer on the rotating body.

craton A part of the earth's continental crust that has attained stability—that has not been deformed for a long period of time. The term is restricted to continents and includes their most stable areas, the continental shields.

creep (soil) The imperceptibly slow downslope movement of rock and soil particles by gravity.

cross-bedding Arrangement of strata, greater than 1 cm thick, inclined at an angle to the main stratification.

crystal settling In a magma, the sinking of crystals because of their greater density, sometimes aided by magmatic convection. (*See* gravity separation).

cubic feet per second (cfs) Unit of water flow or stream discharge. One cubic foot per second is 7.48 gallons passing a given point in one second.

cutoff A new channel that is formed when a stream cuts through a very tight meander. (*See* oxbow lake).

cyanide heap leaching A process of using cyanide to dissolve and recover gold and silver from ore.

c-zone Soil zone of weathered parent material leading downward to fresh bedrock.

Darcy's Law A derived formula for the flow of fluids in a porous medium.

debris avalanche A sudden, rapid movement of a water–soil–rock mixture down a steep slope.

debris flow A moving mass of a water, soil, and rock mixture. More than half of the soil and rock particles are coarser than sand, and the mass has the consistency of wet concrete.

deflation The lifting and removal of loose, dry, fine-grained particles by the action of wind.

delta The nearly flat land formed where a stream empties into a body of standing water, such as a lake or the ocean.

desert An arid region of low rainfall, usually less than 25 cm (10 in) annually, and of high evaporation or extreme cold. Deserts are generally unsuited for human occupation under natural conditions.

desert pavement An interlocking cobble mosaic that remains on the earth surface in a dry region after winds have removed all fine-grained sediments.

desertification The process by which semiarid grasslands are converted to desert.

detrital *See* clastic.

detrital sedimentary rocks Rocks composed of fragments of minerals or other rock materials.

dip The angle that an inclined geological planar surface (sedimentary bedding plane, fault, joint) makes with the horizontal plane.

discharge (Q) The volume of water, usually expressed in cfs, that passes a given point within a given period of time.

discontinuity In solid earth geophysics this means a chemical of phase change in materials with depth. At discontinuities the earth materials become denser and earthquake waves change velocity.

disseminated deposit A mineral deposit, especially of a metal, in which the minerals occur as scattered particles in the rock, but in sufficient quantity to make the deposit a commercially worthwhile ore.

divergent boundary Found at mid-ocean ridge or spreading center where new crust is created as plates move apart. For this reason they are sometime referred to as *constructive* boundaries and are the sites of weak shallow-focus earthquakes and volcanic action.

doldrums The area of warm rising air, calm winds, and heavy precipitation between 5° north latitude and 5° south latitude. It is where the northeast and southeast tradewinds meet.

doubling time The number of years required for a population to double in size.

drainage basin The tract of land that contributes water to a particular stream, lake, reservoir, or other body of surface water.

drainage divide The boundary between two drainage basins (*see* continental divide).

drawdown The lowering of the water table immediately adjacent to a pumping well.

dredging The excavation of earth material from the bottom of a body of water by a floating barge or raft equipped to scoop up, discharge by conveyors, and process or transport materials.

drift A general term for all rock material that is transported and deposited by a glacier or by running water emanating from a glacier.

dripstone A general term for calcite or other minerals deposited in caves by dripping water.

driving force (landslide) Gravity.

dune A mound, bank, ridge, or hill of loose, windblown granular material (generally sand), either bare or covered with vegetation. It is capable of movement, but maintains its characteristic shape.

Dust Veil Index (DVI) A measure of the impact of a volcanic eruption on incoming solar radiation and climate.

eccentricity The deviation of the earth's path around the sun from a perfect circle.

effective stress The average normal force per unit area (stress) that is transmitted directly across grain-to-grain boundaries in a sediment or rock mass.

effluent stream A stream that receives water from the zone of saturation.

elastic Said of a body in which strains are totally recoverable, as in a rubber band. (*Contrast with* plastic).

Elastic Rebound Theory The theory that movement along a fault and the resulting seismicity is the result of an abrupt release of stored elastic strain energy between two rock masses on either side of the fault.

element A substance that cannot be separated into different substances by usual chemical means.

end moraine (terminal moraine) A moraine that is produced at the front of an actively flowing glacier; a moraine that has been deposited at the lower end of a glacier.

engineering geology The application of the science of geology to human engineering works.

ENSO Abbreviation for the periodic meteorological event known as El Niño–Southern Oscillation.

environmental geology The relationship between humans and their geological environment.

eolian Of, produced by, or carried by wind.

epicenter The point at the surface of earth directly above the focus of an earthquake.

equatorial low The belt of low-pressure air masses near the equator. (*See* doldrums.)

erosion The weathering and transportation of the materials of the earth's surface.

erratic A rock fragment that has been carried by glacial ice or floating ice and deposited when the ice melted at some distance from the outcrop from which it was derived.

estuary A semienclosed coastal body where outflowing river water meets tidal sea water.

eutrophication The increase in nitrogen, phosphorous, and other plant nutrients in the aging of an aquatic system. Blooms of algae develop, preventing light penetration and causing reduction of oxygen needed in a healthy system.

evaporite A nonclastic sedimentary rock composed primarily of minerals produced when a saline solution becomes concentrated by evaporation of the water; especially a deposit of salt that precipitated from a restricted or enclosed body of seawater or from the water of a salt lake.

evapotranspiration That portion of precipitation that is returned to the air through evaporation and transpiration, the latter being the escape of water from the leaves of plants.

exfoliation The process whereby slabs of rock bounded by sheet joints peel off the host rock, usually a granite or sandstone.

exfoliation dome A large rounded dome resulting from exfoliation; e.g., Half Dome in Yosemite National Park.

exotic terranes Fault-bounded bodies of rock that have been transported some distance from their place of origin and that are unrelated to adjacent rock bodies or terranes.

expansive soils Clayey soils that expand when they absorb water and shrink when they dry out.

extrusive rocks Igneous rocks that formed at or near the surface of the earth. Because they cooled rapidly, they generally have an *aphanitic* texture.

fault Fracture in the earth along which there has been displacement.

fault creep (tectonic) The gradual slip or motion along a fault without an earthquake.

felsic Descriptive of magma or rock with abundant light-colored minerals and a high silica content. The term is derived from *fel*dspar + *si*lica + *c*.

fetch The unobstructed stretch of sea over which the wind blows to create wind waves.

fiord (fjord) A narrow, steep-walled inlet of the sea between cliffs or steep slopes, formed by the passage of a glacier. A drowned glacial valley.

firn A transitional material between snow and glacial ice, being older and denser than snow, but not yet transformed into glacial ice. Snow becomes firn after surviving one summer melt season; firn becomes ice when its permeability to liquid water becomes zero.

fissile isotopes Isotopes with nuclei that are capable of being split into other elements (*see* fission).

fission The rupture of the nucleus of an element into lighter elements (fission products) and free neutrons spontaneously or by absorption of a neutron.

fjord *See* fiord.

flood frequency The average time interval between floods that are equal to or greater than a specified discharge.

floodplain The portion of a river valley adjacent to the channel that is built of sediments deposited during times when the river overflows its banks at flood stage.

floodwall A heavily reinforced wall designed to contain a stream at flood stage.

floodway Floodplain area under federal regulation for flood insurance purposes.

flotation The process of concentrating minerals with distinct non-wettable properties by floating them in liquids containing soapy frothing agents such as pine oil.

fluidization The process whereby granular solids (corn, wheat, volcanic ash and lapilli) under high gas pressures become fluid-like and flow downslope or can be pumped.

fluvial Pertaining to rivers.

focus (hypocenter) Point within the earth where an earthquake originates.

foliation The planar or wavy structure that results from the flattened growth of minerals in a metamorphic rock.

foreshore slope The zone of a beach that is regularly covered and uncovered by the tide.

fossil fuels Coal, oil, natural gas, and all other solid or liquid hydrocarbon fuels.

fracture Break in a rock caused by tensional, compressional or shearing forces (*see also* joints).

frost wedging The opening of joints and cracks by the freezing and thawing of water.

fusion The combination of two nuclei to form a single heavier nucleus accompanied by a loss of some mass that is converted to heat.

Gaia The controversial hypothesis that proposes that life has a controlling influence on the oceans and atmosphere. It states that the earth is a giant self-regulating body with close connections between living and non-living components.

geologic agents All geologic processes—for example, wind, running water, glaciers, waves, mass wasting—that erode, move, and deposit earth materials.

geological repository (nuclear waste) An underground vault in terrane area free from geologic hazards. Repositories in salt mines, granite, and welded tuff have been studied.

geothermal energy Energy derived from circulating hot water and steam; usually associated with cooling magma or hot rocks.

gigawatt (GW) A thousand million (a billion) watts; 1,000 megawatts.

glacial outwash Deposits of stratified sand, gravel, and silt that have been removed from a glacier by meltwater streams.

glacier A large mass of ice formed, at least in part, on land by the compaction and recrystallization of snow. It moves slowly downslope by creep or outward in all directions due to the stress of its weight, and it survives from year to year.

global warming The general term for the steady rise in the average global temperatures over the last 100 years.

Gondwana, Gondwanaland The southern part of the Permo-Triassic drift landmass of Pangaea.

graben An elongate flat-floored valley bounded by faults on each side.

graded stream A stream that is in equilibrium, showing a balance of erosion, transporting capacity, and material supplied to it. Graded streams have a smooth profile.

gradient (stream) The slope of a streambed usually expressed as the amount of drop per horizontal distance in meters per kilometer (m/km) or in feet per mile (ft/mi).

gravity separation *See* crystal settling.

greenhouse effect The warming of the atmosphere by an increase in the amount of certain gases in the atmosphere, primarily CO_2 and CH_4, that increase the retention of heat that has been radiated by the earth.

groin A structure of rock, wood, or concrete built roughly perpendicular to a beach to trap sand.

ground water That part of subsurface water that is in the zone of saturation (below the water table).

gullying The cutting of channels into the landscape by running water. When extreme, it renders farmland useless.

half life The period of time during which one-half of a given number of atoms of a radioactive element or isotope will disintegrate.

hard rock (a) A term loosely used for an igneous or metamorphic rock, as distinguished from a sedimentary rock. (b) In mining, a rock that requires drilling and blasting for economical removal.

hardpan Impervious layer just below the land surface produced by calcium carbonate ($CaCO_3$) in the B horizon (*see* claypan).

high-grade deposit A mineral deposit with a high concentration of a desired element.

high-level nuclear waste Mostly radioactive elements in spent fuel with long half lives.

horse latitudes Oceanic regions in the *subtropical highs* about 30° north and south of the equator that are characterized by light winds or calms, heat, and dryness.

hot spot A point (or area) on the lithosphere over a plume of lava rising from the mantle.

hydration A process whereby anhydrous minerals combine with water.

hydraulic conductivity A ground-water unit expressed as the volume of water that will move in a unit of time through a unit of area measured perpendicular to the flow direction. Commonly called permeability, it is expressed as the particular aquifer's $m^3/day/m^2$ (or m/day) or $ft^3/day/ft^2$ (or ft/day).

hydraulic gradient The slope or vertical change (ft/ft or m/m) in water-pressure head with horizontal distance in an aquifer.

hydraulic mining, hydraulicking A mining technique by which high-pressure jets of water are used to dislodge unconsolidated rock or sediment so that it can be processed.

hydrocarbon One of the many chemical compounds solely of hydrogen and carbon atoms; may be solid, liquid, or gaseous.

hydrocompaction The compaction of dry, low-density soils due to the heavy application of water.

hydrogeologist A person who studies the geology and management of underground and related aspects of surface waters.

hydrograph A graph of the stage (height), discharge, velocity, or some other characteristic of a body of water over time.

hydrologic cycle (water cycle) The constant circulation of water from the sea to the atmosphere, to the land, and eventually back to the sea. The cycle is driven by solar energy.

hydrology The study of liquid and solid water on, under, and above the earth's surface, including economic and environmental aspects.

hydrolysis The chemical reaction between hydrogen ions in water with a mineral, commonly a silicate, usually forming clay minerals.

hydrothermal Of or pertaining to heated water, the action of heated water, or a product of the action of heated water—such as a mineral deposit that precipitated from a hot aqueous solution.

hypocenter See focus.

igneous rocks Rocks that crystallize from molten material at the surface of the earth (volcanic) or within the earth (plutonic).

inactive sea cliff A cliff whose erosion is dominated by sub-aerial processes rather than wave action.

indicated reserves The hydrocarbons that can only be recovered by secondary recovery techniques such as fluid injection. These reserves are in addition to the reserves directly recoverable from a known reservoir.

influent stream A stream or reach of a stream that contributes water to the zone of saturation.

intensity scale An earthquake rating scale (I–XII) based upon subjective reports of human reactions to ground shaking and upon the damage caused by an earthquake.

intrusive rocks Igneous rocks that have "intruded" into the crust, hence they are slowly cooled and generally have phaneritic texture.

ion An atom with a positive or negative electrical charge because it has gained or lost one or more electrons.

isolation (waste disposal) Buried or otherwise sequestered waste.

isoseismals Lines on a map that enclose areas of equal earthquake shaking based upon an intensity scale.

isostasy The condition of equilibrium, comparable to floating, of large units of lithosphere above the asthenosphere. Isostatic compensation demands that as mountain ranges are eroded, they rise—just as a block of wood floating in water would rise if a part of it above the waterline were removed.

isostatic rebound Uplift of the crust of the earth that results from unloading such as results from the melting of ice sheets.

isotope One of two or more forms of an element that have the same atomic number but different atomic masses.

J-curve The shape of a graphing of geometric (exponential) growth. Growth is slow at first, and the rate of growth increases with time.

jetties Structures built perpendicular to the shoreline to improve harbor inlets or river outlets.

joint Separation or parting in a rock that has not been displaced. Joints usually occur in groups ("sets"), the members of a set having a common orientation.

karst topography, karst terrane Area underlain by soluble limestone or dolomite and riddled with caves, caverns, sinkholes, lakes, and disappearing streams.

kerogen A carbonaceous residue in sediments that has survived bacterial metabolism. It consists of large molecules from which hydrocarbons and other compounds are released on heating.

kettle A bowl-shaped depression without surface drainage in glacial-drift deposits, often containing a lake, believed to have formed by the melting of a large detached block of stagnant ice left behind by a retreating glacier.

lahar A debris flow or mudflow consisting of volcanic material.

landslide The downslope movement of rock and/or soil as a semicoherent mass on a discrete slide surface or plane (see also mass wasting).

lapilli (cinders) Fragments only a few centimeters in size that are ejected from a volcano.

lateral spreading The horizontal movement on nearly level slopes of soil and mineral particles due to liquefaction of quick clays.

laterite A highly weathered brick-red soil characteristic of tropical and subtropical rainy climates. Laterites are rich in oxides or iron and aluminum and have some clay minerals and silica. Bauxite is an Al-rich deposit of a similar origin.

Laurasia The northern part of the Permo-Triassic landmass of Pangaea. (See Gondwana)

lava Molten material at the surface of the earth.

lava dome (volcanic dome) Steep-sided protrusion of viscous, glassy lava, sometimes within the crater of a larger volcano (e.g., Mount St. Helens) and sometimes free-standing (e.g., Mono Craters).

lava plateau A broad elevated tableland, thousands of square kilometers in extent, underlain by a thick sequence of lava flows.

leachate The noxious fluids that percolate through sanitary landfills. If released to the surface or underground environment, leachate can pollute surface and underground water.

levees (natural) Embankments of sand or silt built by a stream along both banks of its channel. They are deposited during floods, when waters overflowing the stream banks are forced to deposit sediment.

liquefaction See spontaneous liquefaction.

lithification The conversion of sediment into solid rock through processes of compaction, cementation, and crystallization.

lithified To change into stone as in the transformation of loose sand to sandstone.

lithosphere The rigid outer layer from the earth's surface down to the asthenosphere. It includes the crust and part of the upper mantle and is of the order of 100 km (60 mi) thick. The "plates" in plate tectonic theory.

littoral drift Sediment (sand, gravel, silt) that is moved parallel to the shore by longshore currents.

lode A mineral deposit consisting of a zone of veins in consolidated rock, as opposed to a placer deposit.

loess A blanket deposit of buff-colored calcareous silt that is porous and shows little or no stratification. It covers wide areas in Europe, eastern China, and the

Mississippi valley. It is generally considered to be windblown dust of Pleistocene age.

longitudinal wave (P-wave) A type of seismic wave involving particle motion that is alternating expansion and compression in the direction of wave propagation. Also known as *compressional waves,* they resemble sound waves in their motion.

longshore current (littoral current) The current adjacent and parallel to a shoreline that is generated by waves striking the shoreline at an angle.

low-grade deposit A mineral deposit in which the mineral content is minimal but still exploitable at a profit.

low-level nuclear wastes Wastes generated in research labs, hospitals, and industry. Shallow burial in containers is sufficient to sequester this waste.

luster The manner in which a mineral reflects light using such terms as metallic, resinous, silky, and glassy.

M-discontinuity (Moho) The contact between the earth's crust and the mantle. There is a sharp increase in earthquake P-wave velocity across this boundary.

maar A low-relief, broad volcanic crater formed by multiple shallow explosive eruptions. Maars commonly contain a lake and are surrounded by a low rampart or ring of ejected material. They typically form from *phreatic eruptions.*

mafic Descriptive of a magma or rock rich in iron and magnesium. The mnemonic term is derived from *ma*gnesium + *f*erric + *ic.*

magma Molten material within the earth that is capable of intrusion or extrusion and from which igneous rocks form.

magnitude (earthquake) A measure of the strength or the strain energy released by an earthquake at its source. (*See* moment magnitude, Richter magnitude.)

manganese nodule A small irregular black to brown, laminated concretionary mass consisting primarily of manganese minerals with some iron oxides and traces of other metallic minerals, abundant on the floors of the world's oceans.

mass wasting A general term for all downslope movements of soil and rock material under the direct influence of gravity.

massive sulfide deposit Usually volcanogenic, often rich in zinc and sometimes in lead.

meander One of a series of sinuous curves in the course of a stream.

megawatt (MW) A million watts; 1,000 kilowatts.

meltdown *See* central-core meltdown.

metamorphic rocks Preexisting rocks that have been altered by heat, pressure, or chemically active fluids.

mid-latitude desert A desert area occurring within latitudes 30°–50° north and south of the equator in the deep interior of a continent, usually on the lee side of a mountain range that blocks the path of prevailing winds, and commonly characterized by a cold, dry climate (*see* rainshadow desert).

millirem A measure of radioactive emissions in the environment.

mineral Naturally occurring crystalline substance with well-defined physical properties and a definite range of chemical composition.

mineral deposit A localized concentration of naturally occurring mineral material (e.g., a metallic ore or a nonmetallic mineral), usually of economic value, without regard to its mode of origin.

mineral reserves Known mineral deposits that are recoverable under present conditions but as yet undeveloped. The term excludes *potential ore.*

mineral resources The valuable minerals of an area that are presently legally recoverable or that may be so in the future; include both the known ore bodies (*mineral reserves*) and the *potential ores* of a region.

miscible Soluble; capable of mixing.

Modified Mercalli Scale Earthquake intensity scale from I (not felt) to XII (total destruction) based upon damage and reports of human reactions.

Mohorovičić discontinuity *See* M-discontinuity.

Moh's hardness scale A scale that indicates the relative hardness of minerals on a scale of 1 (talc, very soft) to 10 (diamond, very hard).

moment magnitude (M_w or M) A scale of seismic energy released by an earthquake based on the product of the rock rigidity along the fault, the area of rupture on the fault plane, and the amount of slip.

moraine A mound, ridge, hill, or other distinct accumulation of unsorted, unstratified glacial drift, predominantly till, deposited chiefly by direct action of glacial ice in a variety of landforms.

mother lode (a) A main mineralized unit that may not be economically valuable in itself but to which workable deposits are related; e.g., the Mother Lode of California. (b) An ore deposit from which a placer is derived; the *mother rock* of a placer.

mountain-top mining The mining of coal that underlies the tops of mountains.

natural levee *See* levee.

neutron An electronically neutral particle (zero charge) of an atomic nucleus with an atomic mass of approximately 1.

nonrenewable resource An energy or material resource that once used is not available for reuse in human time spans. Coal, oil, and metallic minerals are examples.

nucleus The positively charged central core of an atom containing protons and neutrons that provide its mass.

nuée ardenté (pyroclastic flow) French for "glowing cloud," it is a highly heated, almost incandescent cloud of volcanic gases and pyroclastic material that travels with great velocity down the slopes of a volcano. Produced by the explosive disintegration of viscous lava in a vent.

NWPA Nuclear Waste Policy Act of 1982 mandating selection of a repository for nuclear waste.

obliquity The tilt of the earth's axis.

OPEC The Organization of Petroleum Exporting Countries.

open-pit mining Mining from open excavations, most commonly for low-grade copper and iron deposits, and coal.

ore deposit The same as a *mineral reserve* except that it refers only to a metal-bearing deposit.

ore mineral The part of an ore, usually metallic, that is economically desirable.

OTEC Ocean thermal energy conversion.

outgassing The release of gases and water vapor from molten rocks, leading to the formation of the earth's atmosphere and oceans.

oxbow lake A crescent-shaped lake formed along a stream course when a tight meander is cut off and abandoned.

pahoehoe Basaltic lava typified by smooth, billowy, or ropy surfaces.

paleomagnetism Study of the remanent magnetism in rocks, which indicates the strength and direction of the earth's magnetic field in ancient landmasses.

paleoseismicity The rock record of past earthquake events in displaced beds and liquefaction features in trenches or natural outcrops.

Pangaea A supercontinent that existed from about 300–200 Ma, which included all the continents we know today.

pedalfer Soil of humid regions characterized by an organic-rich A horizon and clays and iron oxides in the B horizon.

pedocal Soil of arid or semiarid regions that is rich in calcium carbonate.

pedologist (Greek *pedo*, "soil," and *logos*, "knowledge.") A person who studies soils.

pedology The study of soils.

pegmatite An exceptionally coarse-grained igneous rock with interlocking crystals, usually found as irregular dikes, lenses, or veins, especially at the margins of batholiths.

perched water table The upper surface of a body of ground water held up by a discontinuous impermeable layer above the static water table.

permafrost Permanently frozen ground, with or without water, occurring in arctic, subarctic, and alpine regions.

permafrost table The upper limit of permafrost.

permeability The degree of ease with which fluids flow through a porous medium (*see* hydraulic conductivity).

phaneritic texture A coarse-grained igneous-rock texture resulting from slow cooling. Phanerites are coarse-grained igneous rocks.

photochemical smog Atmospheric haze that forms when automobile-exhaust emissions are activated by ultraviolet radiation from the sun to produce highly reactive oxidants known to be health hazards.

phreatic eruption Volcanic eruption, mostly steam, caused by the interaction of hot magma with underground water, lakes, or seawater. Where significant amounts of new (magmatic) material are ejected in addition to steam, the eruptions are described as *phreatomagmatic*.

phytoremediation The use of plants to remove soil contaminants, particularly excess salt.

piping Subsurface erosion in sandy materials caused by the percolation of water under pressure. A somewhat similar effect is obtained by placing a garden-hose nozzle in a sandbox and jetting an opening through the sand.

placer deposit (pronounced "plass-er") A surficial mineral deposit formed by settling from streams of mineral particles from weathered debris. (*See also* lode.)

plastic Capable of being deformed continuously and permanently in any direction without rupture.

Pleistocene The geologic Epoch of the Quaternary Period known as the Ice Age, falling by definition between 1.6 million and 12,000 years ago. Extensive glaciation is known as far back as the Pliocene Epoch 2.5 million years ago. The exact boundaries shift in time as new data are revealed.

Plinian eruption (volcanology) An explosive eruption in which a steady, turbulent, nearly vertical column of ash and steam is released from a vent at high velocity.

plutonic Pertaining to rock formed by any process at depth, usually with a phaneritic texture.

point bar A deposit of sand and gravel found on the inside curve of a *meander*.

polar desert The bitterly cold, arid region centered at the south pole that receives very little precipitation.

polar front The boundary in the atmosphere where cold polar surface air and mild-temperature surface air converge.

polar high The region of cold, high-density air that exists in the polar regions.

polarity The magnetic positive (north) or negative (south) character of a magnetic pole.

population dynamics The study of how populations grow.

population growth rate The number of live births less deaths per 1,000 people, usually expressed as a percentage; a growth rate of 40 would represent 4.0% growth.

pore pressure The stress exerted by the fluids that fill the voids between particles of rock or soil.

porosity A material's ability to contain fluid. The ratio of the volume of pore spaces in a rock or sediment to its total volume, usually expressed as a percentage.

porphyry copper A copper deposit, usually of low grade, in which the copper-bearing minerals occur in disseminated grains and/or in veinlets through a large volume of rock.

potable (water) Drinkable without ill effects.

potential ore As yet undiscovered mineral deposits and also known mineral deposits for which recovery is not yet economically feasible.

potentiometric surface The hydrostatic head of ground water and the level to which the water will rise in a well. The water table is the potentiometric surface in an unconfined aquifer. Also called *piezometric surface*.

precession The slow gyration of the Earth's axis, analogous to that of a spinning top, which causes the slow change in the orientation of the Earth's axis relative to the sun that accounts for the reversal of the seasons of winter and summer in the Northern and Southern Hemispheres every 11,500 years.

precipitation The separation of a solid substance from a solution by a chemical reaction.

precursors (of earthquakes) Observable phenomena that occur before an earthquake and indicate that an event is soon to occur.

pregnant pond A catchment basin that holds a gold- and silver-bearing cyanide solution.

proglacial lake An ice-marginal lake formed just beyond the frontal moraine of a retreating glacier and generally in direct contact with the ice.

proton A particle in an atomic nucleus with a positive electrical charge and an atomic mass of approximately one.

protostar A star in the process of formation that has entered the slow gravitational contraction phase.

pumice Frothy-appearing rock composed of natural glass ejected from high-silica, gas-charged magmas. It is the only rock that floats.

P-wave *See* longitudinal wave.

pyroclastic ("fire broken") Descriptive of the fragmental material, ash, cinders, blocks, and bombs ejected from a volcano. (*See also* tephra.)

quad A quadrillion British thermal units (10^{15} Btu).

quick clay (sensitive clay) A clay possessing a "house-of-cards" sedimentary structure that collapses when it is disturbed by an earthquake or other shock.

radioactivity The property possessed by certain elements that spontaneously emit energy as the nucleus of the atom changes its proton-to-neutron ratio.

radiometric dating *See* absolute age dating.

radon A radioactive gas emitted in the breakdown of uranium.

rain-shadow desert A desert on the lee side of a mountain or a mountain range (*see* mid-latitude desert).

rank (coal) A coal's carbon content depending upon its degree of metamorphism.

recessional moraine One of a series of nested end moraines that record the stepwise retreat of a glacier at the end of an ice age.

recrystallization The formation in the solid state of new mineral grains in a metamorphic rock. The new grains are generally larger than the original mineral grains.

recurrence interval The return period of an event, such as a flood or earthquake, of a given magnitude. For flooding, it is the average interval of time within which a given flood will be equalled or exceeded by the annual maximum discharge.

regional metamorphism A general term for metamorphism that affects an extensive area.

regolith Unconsolidated rock and mineral fragments at the surface of the earth.

Regulatory Floodplain The part of a floodplain that is subject to federal government regulation for insurance purposes.

relative age dating The chronologic ordering of geologic strata, features, fossils, or events with respect to the geological time scale without reference to their absolute age.

renewable resource An energy or material resource that continually renews itself as it is being consumed. Timber, solar energy, wind, and water are examples.

replacement rate The birth rate at which just enough offsprings are born to replace their parents.

reserves Those portions of an identified resource that can be recovered economically.

reservoir rock A permeable, porous geologic formation that will yield oil or natural gas.

residence time The average length of time a substance (atom, ion, molecule) remains in a given reservoir.

residual soil Soil formed in place by decomposition of the rocks upon which it lies.

resisting force (landslide) Friction along the slide plane and normal forces across the plane.

resonance The tendency of a system (a structure) to vibrate with maximum amplitude when the frequency of the applied force (seismic waves) is the same as the vibrating body's natural frequency.

revetment A rock or concrete structure built landward of a beach to protect coastal property.

Richter Magnitude Scale (earthquakes) The logarithm of the maximum trace amplitude of a particular seismic wave on a seismogram, corrected for distance to epicenter and type of seismometer. A measure of the energy released by an earthquake.

rift zone (volcanology) A linear zone of weakness on the flank of a shield volcano that is the site of frequent flank eruptions.

rill erosion The carving of small channels, up to 25 cm (10 in) deep, in soil by running water.

riparian Pertaining to or situated on the bank of a body of water, especially of a river.

rock cycle The cycle or sequence of events involving the formation, alteration, destruction, and reformation of rocks as a result of processes such as erosion, transportation, lithification, magmatism, and metamorphism.

rocks Aggregates of minerals or rock fragments.

rotational landslide (slump) A slope failure in which sliding occurs on a well-defined, concave-upward, curved surface, producing a backward rotation of the slide mass.

runoff That part of precipitation falling on land that runs into surface streams.

runup The elevation to which a wind wave or tsunami advances onto the land.

salt dome A column or plug of rock salt that rises from depth because of its low density and pierces overlying sediments.

sanitary landfill A method of burying waste where each day's dumping is covered with soil to protect the surrounding area. This is also monitored to ensure protection of local underground water.

sea-floor spreading A hypothesis that oceanic crust forms by convective upwelling of lava at mid-ocean ridges and moves laterally to trenches where it is destroyed.

seawall A rock or concrete structure built landward of a beach to protect the land from wave action. (*See also* revetment.)

secondary enrichment Processes of near-surface mineral deposition, in which oxidation produces acidic solutions that leach metals, carry them downward, and reprecipitate them, thus enriching sulfide minerals already present.

secondary recovery Production of oil or gas by artificially stimulating a depleted reservoir with water, steam, or other fluids in order to mobilize the remaining hydrocarbons.

sedimentary rock A layer of rock deposited from water, ice, or air and subsequently lithified to form a coherent rock.

seif (longitudinal) dune A very large, sharp-crested, tapering chain of sand dunes; commonly found in the Arabian Desert.

seismic gap The segment along an active fault that has seismically been quiet relative to segments at either end. It is the part of the fault most likely to rupture and generate the next earthquake. (*See also* asperity.)

seismogram The recording from a seismograph.

seismograph An instrument used to measure and record seismic waves from earthquakes.

sensitive clay *See* quick clay.

settlement This occurs when the applied load of a structure's foundation is greater than the bearing strength of the foundation material (soil or rock).

sheet erosion The removal of thin layers of surface rock or soil from an area of gently sloping land by broad continuous sheets of running water (sheet flow), rather than by channelized streams.

sheet joints (*See* joint) Cracks more or less parallel to the ground surface that result from expansion due to deep erosion and the unloading of overburden pressure.

shield volcano (cone) The largest type of volcanoes, composed of piles of lava in a convex-upward slope. Found mainly in Hawaii and Iceland.

sinkhole Circular depression formed by the collapse of a shallow cavern in limestone.

slant-drilling (whipstocking) Purposely deflecting a drill hole from the vertical in order to tap a reservoir not directly below the drill site. The deflecting tool is a whipstock.

slide plane (slip plane) A curved or planar surface along which a landslide moves.

slip face The steeper, lee side of a dune, standing at or near the angle of repose of loose sand and advancing downwind by a succession of slides whenever that angle is exceeded.

slump See rotational landslide.

smelter An industrial plant that mechanically and chemically produces metals from their ores.

SMCRA The Surface Mining Control and Reclamation Act of 1977.

soil (a) Loose material at the surface of the earth that supports the growth of plants (pedological definition). (b) All loose surficial earth material resting on coherent bedrock (engineering definition).

soil horizon A layer of soil that is distinguishable from adjacent layers by properties such as color, texture, structure, or chemical composition.

soil profile A vertical section of soil that exposes all of its horizons or zones.

solution (weathering) The dissolving of rocks and minerals by natural acids or water.

source rock The geologic formation in which oil and/or gas originate.

specific yield The ratio of the volume of water drained from an aquifer by gravity to the total volume considered.

speleology The study of caves and caverns.

spheroidal weathering Weathering of rock surfaces that creates a rounded or spherical shape as the corners of the rock mass are weathered faster than its flat faces.

spontaneous liquefaction Process whereby water-saturated sands, clays, or artificial fill suddenly become fluid upon shaking, as in an earthquake.

spreading center See divergent boundary.

stack A pillarlike rocky island that has become detached from a sea cliff or headland by wave erosion.

stage The height of flood waters in feet or meters above an established datum plane.

stalactite Conical deposit of mineral matter, usually calcite, that has developed downward from the ceiling of a limestone cave.

stalagmite Conical deposit of mineral matter, usually calcite, developed upward from the floor of a cavern or cave.

steppe An extensive treeless grassland in the semiarid midlatitudes of Asia and southeastern Europe. It is drier than the prairie that develops in the mid-latitudes of the United States.

storm surge The sudden rise of sea level on an open coast due to a storm. Storm surge is caused by strong onshore winds and results in water being piled up against the shore. It may be enhanced by high tides and low atmospheric pressure.

strain Deformation resulting from an applied stress (=force per unit area). It may be elastic (recoverable) or ductile (nonreversible).

stratification (bedding) The arrangement of sedimentary rocks in strata or layers.

stratigraphic trap An accumulation of oil that results from a change in the character (permeability) of the reservoir rock, rather than from structural deformation.

stratovolcano A large volcano that is shaped slightly concave upward and is composed of layers of ash and lava.

stress The force applied on an object per unit area, expressed in pounds per square foot (lb/ft^2) or kilograms per square meter (kg/m^2).

strike The direction west or east of north of the trace of the inclined plane (sedimentary bedding plane, fault, or joint surface) on the horizontal plane.

strip-mining Surficial mining, in which the resource is exposed by removing the overburden.

subduction The sinking of one lithospheric plate beneath another at a convergent plate margin. (See convergent boundary.)

subsidence Sinking or downward settling of the earth's surface due to solution, compaction, withdrawal of underground fluids, or to cooling of the hot lithosphere or to loading with sediment or ice.

subtropical desert A hot, dry desert occurring between latitudes 15°–30° north and south of the equator where subtropical air masses prevail. (See trade-wind desert.)

subtropical highs The climatic belts at latitudes of approximately 30°–35° north and south of the equator characterized by semipermanent high-pressure air masses, moderate seasonal rainfall, summer heat, dryness, and generally clear skies.

Superfund A spinoff of CERCLA insuring financial accountability for environmental polluters.

surcharge Weight added to the top of a natural or artificial slope by sedimentation or by artificial fill.

surface wave Seismic waves that travel the surface of the earth. They are generated by the unreflected energy of P- and S-waves striking the earth's surface. They are the most damaging of earthquake waves.

surging glacier A glacier that alternates periodically between brief periods (usually 1–4 years) of very rapid flow, called *surges,* and longer periods (usually 10–100 years) of near stagnation.

suspension (a) A mode of sediment transport in which the upward currents in eddies of turbulent flow are capable of supporting the weight of sediment particles and keeping them held indefinitely in the surrounding fluid (as silt in water or dust in air). (b) The state of a substance in such a mode of transport.

sustainable society A population capable of maintaining itself while leaving sufficient resources for future generations.

sustained yield (hydrology) The amount of water an aquifer can yield on a daily basis over a long period of time.

S-wave See transverse wave.

syncline A concave-upward fold that contains younger rocks in its core. (Contrast with anticline.)

synthetic fuels (synfuels) A manufactured fuel; for example, a fuel derived by liquefaction or gasification of coal.

tailings The worthless rock material discarded from mining operations.

talus Coarse angular rock fragments lying at the base of a cliff or steep slope from which they were derived.

tar sand A sand body that is large enough to hold a commercial reserve of asphalt or other thick oil. It is usually surface excavated to remove the thick hydrocarbons.

tectonism (adj. tectonic) Deformation of the earth's crust by natural processes leading to the formation of ocean basins, continents, mountain systems, and other earth features.

tephra A collective term for all *pyroclastic* material ejected from a volcano.

terranes (*See* exotic terannes.)

texture The size, shape, and arrangement of the component particles or crystals in a rock.

till Unsorted and unstratified drift, generally unconsolidated, that is deposited directly by and beneath a glacier without subsequent reworking by water from the glacier.

tombolo A sandbar that connects an island with the mainland or another island.

total fertility rate Average number of children born to women of child-bearing age.

trace element An element that occurs in minute quantities in rocks or plant or animal tissue. Some are essential for human health.

trade winds Steady winds blowing from areas of high pressure at 30° north and south latitudes toward the area of lower pressure at the equator. This pressure differential produces the northeast and southeast trade winds in the Northern and Southern Hemispheres, respectively.

trade wind desert *See* subtropical desert.

transform boundary *See* transform fault.

transform faults Plate boundaries with mostly horizontal (lateral) movement that connect spreading centers to each other or to subduction zones.

translational slide *See* block glide.

transported soils Soil developed on *regolith* transported and deposited by a geological agent. Hence the term *glacial* soils, alluvial soils, *eolian* soils, *volcanic* soils, and so forth.

transuranic elements (TRU) Unstable (radioactive) elements of atomic numbers greater than 92.

transverse dune A strongly asymmetrical sand dune that is elongated perpendicularly to the direction of the prevailing winds. It has a gentle windward slope and a steep leeward slope that stands at or near the angle of repose of sand.

transverse wave (S-wave) A seismic wave propagated by a shearing motion that involves oscillation perpendicular to the direction of travel. It travels only in solids.

troposphere That part of the atmosphere next to the earth's surface, in which temperature generally decreases rapidly with altitude and clouds and convection are active. Where the weather occurs.

tsunami Large sea wave produced by a submarine earthquake, a volcanic eruption, or a landslide. Incorrectly known as a tidal wave.

tuff A coherent rock formed from volcanic ash, cinders, or other *pyroclastic* material.

tundra (soil) A treeless plain characteristic of arctic regions with organic rich, poorly drained tundra soils and permanently frozen ground.

unconfined aquifer Underground body of water that has a free (static) water table; that is, water that is not confined under pressure beneath an aquiclude.

upwelling A process whereby cold, nutrient-rich water is brought to the surface. Coastal upwelling occurs mostly in trade-wind belts.

vapor extraction The use of vacuum pumps placed in shallow bore holes to remove vaporized organic contaminants in the soil, such as gasoline or TCE.

vein deposit A thin, sheetlike igneous intrusion into a crevice.

ventifact Any stone or pebble that has been shaped, worn, faceted, cut, or polished by the abrasive or sandblast action of windblown sand, generally under arid conditions.

Volcanic Explosivity Index (VEI) A scale for rating volcanic eruptions according to the volume of the ejecta, the height to which the ejecta rise, and the duration of the eruption.

volcanogenic deposit Of volcanic origin; for example, volcanogenic sediments or ore deposits.

water cycle *See* hydrologic cycle.

water table The contact between the zone of aeration and the zone of saturation.

wave base The depth at which water waves no longer move sediment, equal to approximately half the wavelength.

wavelength The distance between two equivalent points on two consecutive waves; for example, crest to crest or trough to trough.

wave of oscillation A water wave in which water particles follow a circular path with little forward motion.

wave of translation A water wave in which the individual particles of water move in the direction of wave travel, as in broken waves (surf).

wave period (T) The time (in seconds) between passage of equivalent points on two consecutive waveforms.

wave refraction The bending of wave crests as they move into shallow water.

weathering The physical and chemical breakdown of materials of the earth's crust by interaction with the atmosphere and biosphere.

welded tuff A glassy pyroclastic rock that has been made hard by the welding together of its particles under the action of retained heat.

wetted perimeter (stream) The length of stream bed and bank that are in contact with running water in the stream in a given cross section.

wind The movement of air from a region of high pressure to a region of low pressure.

zero population growth A population growth rate of 0; that is, the number of deaths in a population is equal to the number of live births on average.

zone of aeration (vadose zone) The zone of soil or rock through which water infiltrates by gravity to the water table. Pore spaces between rock and mineral grains in this zone are filled with air; hence the name.

zone of flowage The lower part of a glacier that deforms due to the glacier's own weight.

zone of saturation The ground-water zone where voids in rock or sediment are filled with water.

credits

This page constitutes an extension of the copyright page. We have made every effort to trace the ownership of all copyrighted material and to secure permission from copyright holders. In the event of any question arising as to the use of any material, we will be pleased to make the necessary corrections in future printings. Thanks are due to the following authors, publishers, and agents for permission to use the material indicated.

Chapter 1 1, Fig. 1: USGS; Fig. 2: Jim Dolan, Dept. of Earth Sciences, Univ. of Southern Calif. 2, Fig. 1.1: Blue Planet Group. 3, Fig. 1.2(a): B. Pipkin; (b): Courtesy Alastair Black. 3, 4, Figs. 1.3, 1.4: From "Why the United States is Becoming More Vulnerable to Natural Disasters," by G. van der Vink, et al., *EOS: Transactions of the American Geophysical Union,* 1998, vol. 79, no. 44. 5, Fig. 1.6: D. D. Trent; Fig. 1.7: USGS. 6, Fig. 1.7: USGS; Fig. 1.8: Jim Mori, USGS. 11, Fig. 1.10: NOAA data. 11, Fig. 1.11: Blue Planet Group. 16, Fig. 1.12: B. Pipkin; Fig. 1.13: Cathlyn Melloan/Tony Stone Images. 17, Fig. 1.14(a)–(c): Owen Frankyn/Sygma. 18, Fig. 1.16: United Nations Food and Agriculture Organization data.

Chapter 2 22, Fig. 1: NASA. 24, Fig. 2.1: D. D. Trent. 27, Fig. 2.3: After NASA, "Earth System Science Report, 1986," in NRC, *Solid Earth Sciences and Society* (Washington, D.C.: National Academy Press, 1993). 30, Fig. 2.7(a): USGS; (b)–(d): Paul Silverman/Fundamental Photographs; Fig. 2.8: top left, Michael Dalton/Fundamental Photographs; top right and bottom left: courtesy Sue Monroe; bottom right: Paul Silverman/Fundamental Photographs. 31, Fig. 2.9(a)–(d): Courtesy Sue Monroe. 32, 33, Figs. 1, 2: Janet Blabaum, Highland Geotechnical, courtesy AIPG. 36, Fig. 2.13: B. Pipkin; Fig. 2.14(a)–(c): Heather Dutton. 37, Fig. 2.15: Courtesy Sue Monroe. 38, Fig. 2.16(a)–(c): B. Pipkin. 39, Fig. 2.18(a)–(e): B. Pipkin. 40, Figs. 2.19, 2.20: D. D. Trent. 41, Fig. 2.22(a)–(c): B. Pipkin. 42, Fig. 1(a): B. Pipkin; (b): Martin Land, Science Photo Library/Photo Researchers; (c): Carl Frank/ Photo Researchers. 43, Fig. 2: After C. B. Gunn, "A Description Catalog of the Drift Diamonds of the Great Lakes Region, North America," *Gems and Gemology* 12, nos. 10 and 11 (1968), pp. 297–303, 333–334, © Gemological Inst. of America; Fig. 3(a): Courtesy Smithsonian Inst.; (b)–(e): courtesy Sue Monroe. 44, Fig. 1: © 1994 Margaret Miller/The National Audubon Society Collection/Photo Researchers. 45, Fig. 2.23: Courtesy Sue Monroe; Fig. 2.24: D. D. Trent. 47, Fig. 2.26: Adapted with permission from "The Decade of North American Geology, 1983 Geologic Time Scale," by A. R. Palmer, *Geology,* 1983, 504. 48, Table 2.6: From *Earth Science Today,* by B. Murphy and D. Nance, p. 74, Brooks/Cole, 1999. All rights reserved. 51, 52, Figs. 2.31(b), 2.32: From *The Changing Earth,* by J. Monroe and R. Wicander, Brooks/Cole, 1997. All rights reserved; Fig. 2.33: Theodore Roosevelt Collection, Harvard Coll. Library; Fig. 2.34: B. Pipkin.

Chapter 3 56, Fig. 1: Earth Satellite Corp. 58, Fig. 3.1: After A. Wegener, *Continents and Ocean Basins* (Dover, 1996). 59,

Fig. 3.2(a), (b): Patricia G. Gensel, Univ. of N. Carolina. 60, Fig. 3.3(c): Scott Katz. 61, 62, Figs. 3.5, 3.6: From *Earth Science Today,* by B. Murphy and D. Nance, p. 105, Brooks/Cole, 1999. All rights reserved. 64, Fig. 3.8: NGDC. 67, 68, Figs. 3.13, 13.14(a): From *Essentials of Geology,* 2nd ed., by R. Wicander and J. Monroe, p. 35, Brooks/Cole, 1999. All rights reserved; (b) Photo Researchers; (c): USGS. 71, Fig. 2: After I. W. D. Dalziel, A. Cox and D. Engebretson, GSA special paper 207, 1987. 72, Fig. 3.16(b): From *Earth Science Today,* by B. Murphy and D. Nance, p. 133. Brooks/Cole, 1999. All rights reserved. 74, 75, Fig. 3.18: Bildarchiv Preussischer Kulturbesitz; Fig. 3.19: Courtesy Rodney Supple; Fig. 3.20(a)–(c): Woods Hole Oceanographic.

Chapter 4 80, Fig. 1: Yi-Ben Tsai, National Central Univ., Taiwan. 81, Fig. 2: From "The Chi-Chi, Taiwan Earthquake: Large Surface Displacements on an Inland Thrust Fault," by M. Kuo-Fong, et al., *EOS,* December 1999, vol. 80, no. 50. 83, Fig. 4.1(b): D. D. Trent; (d): B. Bradley, Univ. of Colo.; (f): NOAA. 84, Fig. 4.2(e): USGS; Fig. 4.3: From *Essentials of Geology,* 2nd ed., by R. Wicander and J. Monroe, p. 228, Brooks/Cole, 1999. All rights reserved. 86, Fig. 4.5: B. Pipkin; Fig. 4.6: Trustees of the Science Museum, London. 87, 89, Figs. 4.7, 4.10: SCEC data. 91, Fig. 4.12: B. Pipkin. 92, Fig. 4.13(b): Dorothy Stout, Cypress Coll.; (d): Dennis Fox. 93, Fig. 4.14: James C. Anderson, USC; Fig. 4.15: C. J. Langer, USGS. 94, Fig. 1: Kerry E. Sieh, Seismological Laboratory, CIT; Fig. 2: Redrawn courtesy E. Harp and R. Jibson, USGS. 95, Fig. 4.16(c): B. Pipkin; Fig. 4.17(b): B. Bradley, Univ. of Colo. 96, Fig. 4.18(a): NOAA; (b): courtesy Richard Hilton; Fig. 4.19: NGDC 97, Fig. 4.21(a): Courtesy Dave Johnson; (b): USGS, Menlo Park. 98, Fig. 4.22: Martin E. Klimek, *Marin Independent Journal;* Fig. 4.23: Katsushika Hokusai, *The Great Wave off Kanagawa.* Metropolitan Museum of Art, bequest of Mrs. H. O. Havemeyer, 1929, H. O. Havemeyer Collection (JP1847). 100, Fig. 4.25: USGS data. 100, 101, Fig. 4.26: NOAA; Figs. 4.27, 4.28: From "Program Aims to Reduce Impact of Tsunamis on Pacific States," by E. Bernard, *EOS:Transactions of the American Geophysical Union,* June 2, 1998, vol. 79, no. 2. 103, Fig. 2: USGS Circular 187; Fig. 3: B. Pipkin; Fig. 4: USN. 105, Fig. 4.29(a): USGS data. 106, Fig. 1: SCEC; Fig. 2: From M. Peterson, The January 17, 1994, Northridge Earthquake, *California Geology,* March/April 1994, pp. 40–45. 107, Fig. 4.31: Dennis Fox; Fig. 4.32: Frederick Larson, *San Francisco Chronicle.* 108, Fig. 4.33: © American Geophysical Union, EOS. 109, Fig. 4.35: USGS data. 110, 111, Figs. 1, 2: After *The Los Angeles Times.* 112, Fig. 4.36(a), (b): Gregory A. Davis, USC; (c): Peter Weigand, Calif. State Univ.–Northridge. 113, Fig. 4.37: Grant Marshall, USGS. 113, Fig. 4.38(a), (b): From Univ. of Calif., San Diego, after C. Arnold, *Earthquake Disaster Prevention Planning in Japan* BSD (Palo Alto, Calif.: Buildings System Development, 1982); planning density example by R. Eisner and H. Shapiro. 114, Fig. 4.39(a): Carol Prentice, USGS, courtesy GSA; (b): C. Rojahn, Applied Technology council, courtesy GSA; (c): Thomas L. Holzer, USGS. 114, Fig. 4.40: After T. Fujiwara, Y. Suzuki, M. Iwai Nakashima, A. S. Kitahara, and M. Bruneau, "Overview of Building Damage from the 1995 Great Hanshin Earthquake," *Newsletter of the Disaster Prevention Institute,* Kyoto Univ., Preliminary Report on the Great Hanshin Earthquake of January 17, 1995, February 1995 special issue, pp. 13–20. 115, Fig. 4.41: USGS data. 116, Fig. 4.42(a)–(c): Courtesy James Dolan, Dept. of Earth Sciences, Univ. of Southern Calif., Los Angeles. 117, Fig. 4.43: From M. L. Blair and W. W. Spangle, USGS prof. paper 941-B, 1979. 118, Fig. 4.44: USGS Fact Sheet 168-95. 119, 122, Figs. 4.45, 4.48: USGS data. 120, Fig. 4.46: Oreg. Dept. of Geology and Mineral Industries. 123, Fig. 4.49: After Sykes and Seeber, "Great Earthquakes and Great Asperities, San Andreas fault, California, *Geology,* 13 (Dec. 1985), pp. 835–838; Fig. 4.50: Kerry E. Sieh, Seismological Laboratory, CIT. 124, Fig. 1: B. Pipkin. 125, Fig. 4.51(a), (b): FEMA. 126, 127, Fig. 4.52(a): Joseph and Helga Lehemen; (b)–(d): Jose Martinez, all courtesy of Jose Borerro, Univ. of Southern Calif.; Fig. 4.53: Yi-Ben Tsai, National Central Univ., Taiwan; Fig. 4.54: CDMG; Fig. 4.55: Courtesy Ted Reeves, Chaffey High School, Upland, Calif.

Chapter 5 132, Fig. 1: Courtesy Joyce M. Warren and USGS. 133, Fig. 2(b): NASA; Fig. 3(a): USGS; (b): Map courtesy David Schneider, Mich. Technological. 135, Fig. 5.1: After R. I. Tilling, C. Heliker, and T. L. Wright, *Eruptions of Hawaiian Volcanoes: Past, Present, and Future* (USGS,1987). 135, Fig. 5.2: After Hays, USGS Prof. paper 1240-B, 1981; data compiled by Donald Mullineaux, USGS. 137, 138, Fig. 1(a), (c): Karen Williams, *Volcanoes of the Southwind: A Field Guide to the Volcanoes and Landscape of Tongariro National Park* (Tongariro National History Society); (b): David M. Bailey, Grant Co. Airport, Moses Lake, Wash.; Fig. 2: NOAA. 140, Fig. 5.4: After T. Simpkin and L. Siebert, *Volcanoes of the World,* 2nd ed. (Tucson, Ariz.: Geoscience Press, 1994). 140, Fig. 5.5: Data from T. Simpkin, et al., *Volcanoes of the World* (Dowden, Hutchinson, and Ross). 142, Fig. 5.7: B. Pipkin; Fig. 5.8(a): Eros Data Center; (b): Modified from Moore and Clague, 1992. Volcano World Web site http://volcano.und.nodak.edu/. 143, Fig. 5.9: USGS; Fig. 5.10(a): USGS; (b): J. D. Griggs, Hawaiian Volcano Observatory; (c): D. D. Trent. 144, Fig. 5.11: J. B. Stokes, USGS; Fig. 5.12(a): *Pele, Goddess of Hawaii's Volcanoes,* by artist Herb Kawainui Kane; (b): B. Pipkin. 145, Fig. 5.13(b): B. Pipkin; Fig. 5.14(a): B. Pipkin; (b): Arturo Abuta. 146, 147, Figs. 5.15, 5.16, 5.17(a): B. Pipkin; (b): Tau Rho Alpha 148, Fig. 5.18: USGS; Fig. 5.19: B. Pipkin. 149, Fig. 5.20: USGS image, Fact Sheet 002-97, "What are Volcano Hazards?" 150, Fig. 5.21(a): B. Pipkin; (b): NOAA, National Geophysical Data Center, Boulder, Colo.; Fig. 5.22: NOAA. 151, Fig. 5.23: Jim Mori, USGS; Fig. 5.24: USGS Photo Library. 152, Fig. 1: Dr. Simon R. Young, Montserrat Volcano Observatory; Fig. 2: Montserrat Volcano Observatory Web site map, January 1998, British Geological Survey. 153, Fig. 5.25(a): Keith Ronnholm; (b): B. Pipkin; Fig. 5.26(a), (b): Smithsonian Inst. 154, Fig. 5.27(a)–(c): B. Pipkin. 155, Fig. 5.28: B. Pipkin. 156, Fig. 5.29: USGS data. 157, Fig. 5.30: B. Pipkin; Fig. 5.31: Modified from USGS fact sheet 169-97, 1997. 158, 159, Fig. 1: USGS data; Fig. 2(a), (b): John Rogie © 1999, with permission; Fig. 3: B. Pipkin. 160, Fig. 1(a): Mike Tuttle, USGS; (b): T. Orban/Sygma. 161, Fig. 5.32: USGS. 162, 163, Figs. 5.33–5.35: USGS data. 164, 165, Fig. 5.36: John Mead, Science Photo Library/Photo Researchers; Fig. 5.37: Jack Green, Calif. State Univ., Long Beach; Figs. 5.38, 5.41: J. D. Griggs, Hawaiian Volcano Observatory, USGS; Fig. 5.39: Jim Mori, USGS; Fig. 5.40: B. Pipkin.

Chapter 6 170, Fig. 1: UPI/Bettmann. 171, Fig. 3: Bettmann Archive. 173, Fig. 6.2: From *The Changing Earth,* by J. Monroe and R. Wicander, Brooks/Cole, 1997. All rights reserved. 174, Fig. 6.3: John Shelton. 176, 177, Figs. 6.5, 6.6: B. Pipkin. 177, Fig. 6.7: © Mark E. Gibson/Visuals Unlimited. 178, Fig. 6.9: B. Pipkin. 180, Fig. 6.11: Rodman Snead/JLM Visuals; Fig. 6.12(a): USDA data; (b): Data from USDA modified from *Physical Geography,* 2nd ed., by R. Scott, p. 299, Brooks/Cole. All rights reserved. 182, Fig. 1: Visuals Unlimited. 183, Fig. 6.14(a): Walt Anderson/Visuals Unlimited; (b): courtesy Sue Monroe; Fig. 6.15: B. Pipkin. 184, Fig. 6.17: Stew Monroe; Fig. 6.18: USDA. 186, Fig. 6.20: Howard G. Wilshire, USDS; Fig. 6.21: Modified from *Natural Resources Inventory,* 1997 State of the Land Update, USDA-NRCS. 187, Fig. 6.22: USDA Soil Conservation Service, 1994; Fig. 6.23: Reprinted by permission of the American Geological Inst.; Fig. 6.24: Doug Burbank, USC. 188, Fig. 6.25: Courtesy Edward B. Nuhfer and AIPG. 190, Figs. 6.28, 6.29: Troy L. Pewe, USGS; Fig. 6.30(a): R. A. Combellick, Alaska Div. of Geological and Geophysical Sciences; (b): Troy L. Pewe, Ariz. State Univ. 191, Fig. 6.31: B. Pipkin. 193, Fig. 6.32: ARCO. 193, Fig. 6.34: Douglas J. Lathwell, Cornell Univ. 194, 195, Fig. 1: J. M. Spencer, *Texas Journal of Science,* 21(1970); Fig. 2: Verle Bohrman, Dept. of Animal Sciences, Univ. of Nev.–Reno. 197, Fig. 1: Al Robb III. 198, 199, Fig. 6.35: Alain Nogues/Sygma; Fig. 6.36: Robert Thayer/Projections; Fig. 6.37: B. Pipkin; Fig. 6.38: J. P. Laffont/Sygma.

Chapter 7 203, Fig. 1: B. Pipkin. 204, Fig. 7.1: J. P. Krohn and J. E. Slosson, "Landslide Potential in the United States," *California Geology* 29, no. 10 (1976), pp. 224–231. 206, Figs. 1, 2: B. Pipkin. 207, Fig. 7.3(b): B. Pipkin. 208, Fig. 7.4: USGS; Fig. 7.5(b): Randall W. Jibson, USGS. 209, Fig. 7.6: *California Geology* data from CDMG 209, Fig. 7.7(a): National Weather Service and NOAA; (b): Robert L. Schuster, USGS. 210, Fig. 7.8: B. Pipkin. 212, Fig. 7.10(c): B. Pipkin; (d): D. D. Trent. 213, Fig. 7.11(a): B. Pipkin; (b): Reed Wicander. 214, 215, Fig. 1: Robert Schuster, USGS; Fig. 2: Redrawn courtesy Lynn Highland, *Earthquakes and Volcanoes* 24, no. 3 (1993); Fig. 3: Redrawn courtesy Lynn Highland, USGS. 216, Fig. 7.12: Randall W. Jibson, USGS. 218, Fig. 7.14: W. R. Hansen, USGS. 219, Fig. 7.15: USGS; Fig. 7.17: NGDC. 221, Fig. 7.19: USGS. 222, Fig. 7.21: D. D. Trent; Fig. 7.22: Pam Irvine, CDMG. 223, Fig. 7.23: Siang Tan, CDMG; Fig. 7.24(b): B. Pipkin. 225, Fig. 7.26: B. Pipkin. 226, Fig. 7.28(a): Bob Hollingsworth; (b): D. D. Trent; Fig. 7.29: Rick Giardino, Texas A&M Univ.; Fig. 7.30: Dedman's Photo Shop, Skagway, Alaska. 227, Fig. 7.31: USGS data. 228, 229, Fig. 1: Ronny Jacques/Photo Researchers; Fig. 2: Adapted from P. Gatto and L. Carbognin, "The Lagoon of Venice," *Hydrological Sciences Bulletin* 26, no. 4 (1981). 231, Fig. 7.33: City of Long Beach, Dept. of Oil Properties; Fig. 7.34: Courtesy T. L. Holzer, USGS. 232, Fig. 7.35: USGS Photo Library; Fig. 7.36: R. V. Dietrich. 233, Fig. 7.37(a): State of Calif. data; (b): Courtesy James L. Ruhle & Associates. 235, 236, Figs. 7.38, 7.40: Paul Bauer, Cuesta Coll.; Fig. 7.39: USGS; Fig. 7.41: Robert Sydnor, Calif. Div. of Mines and Geology.

Chapter 8 241, Jim Richardson. 242, Fig. 8.1: From *Environmental Science,* 7th ed., by G. T. Miller, p. 300,

Brooks/Cole. All rights reserved. 243, Fig. 8.2: From *Essentials of Geology,* 2nd ed., by R. Wicander and J. Monroe, p. 27, Brooks/Cole, 1999. All rights reserved. 244, Fig. 8.3: USGS data. 246, Fig. 8.8: From *Essentials of Geology,* 2nd ed., by R. Wicander and J. Monroe, p. 27, Brooks/Cole, 1999. All rights reserved. 244, 245, Figs. 8.4, 8.6: USGS. 248, Fig. 1: AIPG, *Ground Water: Issues and Answers,* 1985. 249, Fig. 8.11: USGS data. 251, Fig. 8.15(c): D. D. Trent. 253, Fig. 2: Texas Geological Survey. 255, Fig. 8.19: Kevin McGillicuddy. 257–259, Figs. 1–3: After B. L. Foxworthy, USGS prof. paper 950, 1978; Fig. 4: Courtesy Nassau Co. Dept. of Public Works. 260, Fig. 8.21: USGS. 261, Fig. 8.23(a): USGS; (b): Frank Kujawa, Univ. of Central Florida, GeoPhoto Pub. Co.; (c): B. Pipkin. 262, Fig. 8.24: USGS. 266, Fig. 1: Courtesy K. W. Brown, Texas A&M Univ.; Fig. 2: After S. Testa in Pipkin and Proctor, *Engineering Geology Practice in Southern California* (Belmont, Calif.: Star, 1993). 268, 269, Fig. 8.26(a): James Farrell; (b): From *Essentials of Geology,* 2nd ed., by R. Wicander and J. Monroe, p. 27, Brooks/Cole, 1999. All rights reserved; Fig. 8.27: Scott Fee, American Speleological Society; Fig. 8.28: Visuals Unlimited; Fig. 8.29: Martin Land/Science Photo Library/Photo Researchers.

Chapter 9 274, B. Pipkin. 277, Fig. 9.2: USGS data 281, Fig. 9.5(c): W. L. Fisher et al., *Delta Systems in the Exploration for Oil and Gas —A Research Colloquium,* 1969. 279, Fig. 9.3(b): 280, Fig. 9.4(a): From *Physical Geology,* 2nd ed., by J. Monroe and R. Wicander, Brooks/Cole. All rights reserved. 282, Fig. 9.6: USN, courtesy Robert Stevenson; Fig. 9.7: NASA/JPL. 283, Fig. 1(a): After R. T. Saucier and J. Snowden, "Engineering Geology of the New Orleans Area," 1995 GSA annual meeting *Field Trip Guide Book No. 6;* (b): EROS; (c): B. Pipkin. 284, Fig. 2: B. Pipkin; Fig. 3: Redrawn courtesy Jesse O. Snowden, Dept. of Geosciences, Southeast Mo. State Univ., from Snowden, Simmons, Traughber, and Stephens, *Transactions, Gulf Coast Association of Geological Societies* 27. 285, Fig. 4: B. Pipkin. 286, Fig. 9.8(b): Victor R. Baker, Univ. of Ariz. 287, Fig. 9.9(e): From R.C. Scott, Introduction to *Physical Geography,* 2nd ed., Brooks/Cole. All rights reserved. 288, Fig. 9.11: Chris Stewart/Black Star; Fig. 9.12: B. Pipkin. 291, Fig. 9.16: Modified from W. D. Hays, USGS prof. paper 1240-B, 1981 293, Fig. 9.18(a): NRC data. 293, Fig. 9.18(b): U.S. Geological Survey Fact Sheet FS-097-97. 294, Fig. 9.19: USGS. 295, Fig. 9.20(a): Data from North Dakota Geological Survey miscellaneous series 35. 295, Fig. 9.20b: USGS data. 296, Fig. 1: U.S. Geological Survey circular 112. 296, Fig. 9.21: Redrawn from USGS 1993 Yearbook. 297, Fig. 9.22: Landsat imagery courtesy Earth Observation Satellite Co., Lanham, Md. 299, Fig. 2: M. Carlowicz (staff writer), "Controlled Flood of Colorado River Creates Stream of Data," 1996, *EOS,* vol. 77, no. 24, (11 June). 300, Fig. 9.23(a), (b): B. Pipkin; (c): Courtesy Ed Nuhfer and AIPG. 301, Fig. 9.24: A. O. Waananen, et al., USGS; Fig. 9.25: San Diego Co. Engineers Office prof. paper 942, 1977. 302–305, Fig. 9.27: B. Pipkin; Fig. 9.28(a), (b): *Los Angeles Times;* (c): B. Pipkin; Fig. 9.29(a), (b): Security Pacific National Bank Historical Collection, Los Angeles Co. Library; Fig. 9.30: B. Pipkin.

Chapter 10 310, NASA. 311, © Mike Booher. 313, Fig. 10.2(b): From *Earth Science Today,* by B. Murphy and D.

Nance, p. 233, Brooks/Cole, 1999. All rights reserved. 314, Figs. 10.3, 10.4: Tom Servais, *Surfer* magazine; Fig. 10.5(b): B. Pipkin. 316, Fig. 10.7(b): B. Pipkin; Fig. 10.8: John S. Shelton. 317, 318, Figs. 10.10, 10.11: B. Pipkin. 319, Fig. 10.12(a), (b): B. Pipkin; (c): Craig Everts; (d): USAF. Fig. 10.13: From *Essentials of Geology,* 2nd ed., by R. Wicander and J. Monroe, p. 348, Brooks/Cole, 1999. All rights reserved. 320, Fig. 10.14: NOAA. 321, Fig. 10.15: Data from USGS. 322, Fig. 10.16: B. Pipkin. Fig. 10.17: After R. Dolan, et al., 1985, "Coastal Erosion and Accretion," *USGS National Atlas.* 323, Fig. 10.18: USGS data. 324, Fig. 10.19(a): J. S. Williams et al., *Coasts in Crisis,* USGS circular 1075, 1009; (b): Craig H. Everts; (c): From *Environmental Science,* 7th ed., by G. T. Miller, Brooks/Cole. All rights reserved. 324, 325, Figs. 10.20, 10.21, 10.22: B. Pipkin. 327, Fig. 10.24(a): Courtesy Jim Falls, Calif. Div. of Mines and Geology; (b): GeoPhoto Pub. Co.; Fig. 10.25: Courtesy Orrin Pilkey, Duke Univ. 328, Figs. 10.26, 10.27: B. Pipkin; Fig. 10.28(a): B. Pipkin; (b): U. S. Army Corps of Engineers. 329, Fig. 10.29: Ruth Flanagan, © 1993, Kalmbach Pub. Co., *Earth* magazine; Fig. 10.30: From *Essentials of Geology,* 2nd ed., by R. Wicander and J. Monroe, Brooks/Cole. All rights reserved. 330, 331, Figs. 1, 3: Courtesy Orrin Pilkey, Duke Univ.; Fig. 2: After O. H. Pilkey, et al., "The Concept of Shoreface Profile of Equilibrium: A Critical Review," *Journal of Coastal Research* 9(1993), no. 1, p. 255. 332, Fig. 10.31: USGS. 332, Fig. 10.32: NOAA. 333, Fig. 10.33(a): From *Meteorology Today,* 6th ed., by C. D. Ahrens, p. 422, Brooks/Cole, 2000. All rights reserved; (b): NOAA. 334, 335, Fig. 1: B. Pipkin; Fig. 2: From *Environmental Science,* 7th ed., by G. T. Miller, Brooks/Cole; Fig. 3: After Peierls, et al., 1991. 336, Fig. 10.34: From *Meteorology Today,* 6th ed., by C. D. Ahrens, p. 426, Brooks/Cole, 2000. All rights reserved. 338, Fig. 10.36: Coastal Weather Research Station, Univ. of South Alabama. 339, Fig. 10.37(a), (b) National Weather Service (NOAA), National Hurricane Center, courtesy Ed Nuhfer and AIPG. 340, Fig. 10.38: NOAA. 341, Fig. 10.39: From *Meteorology Today,* 6th ed., by C. D. Ahrens, p. 295, Brooks/Cole, 2000. All rights reserved; Fig. 10.40: NOAA. 343, Fig. 10.42: Suzanne and Nick Geary/Tony Stone Worldwide; Figs. 10.43, 10.45: B. Pipkin; Fig. 10.44: National Weather Service (NOAA), Hurricane Information Center, courtesy Ed Nuhfer and AIPG.

Chapter 11 347, (a) From J. Monroe and R. Wicander, *Physical Geology,* 3rd ed., Brooks/Cole. All rights reserved; (b): D. D. Trent. 350, Fig. 11.3: Wilfried Haeberli. 351, Fig. 11.4: NPS data; Fig. 11.5: John S. Shelton. 352, Fig. 1: Jurg Alean. 354, Fig. 11.6: Richard P. Jacobs/JLM Visuals. 355, Figs. 11.8, 11.9: D. D. Trent; Fig. 11.10: B. Pipkin. 356, 357, Figs. 2, 3(a): D. D. Trent. 358, Fig. 11.11: C. Hillaire-Marcel; Fig. 11.12: After Dott and Batten, *Evolution on the Earth,* 3rd ed., p. 514, McGraw-Hill, 1981. 359, Fig. 11.13: Foundation for Glacier and Environmental Research data. 360, Fig. 11.14: Rolf A. Larsson. 361, Fig. 11.15: After Weiss and Newman, "The Channeled Scablands of Eastern Washington," USGS pamphlet, 1973; and USGS prof. paper 424-B, 1961. 362, Figs. 11.16, 11.17: D. D. Trent; Fig. 11.18: John S. Shelton. 364, Fig. 11.20: From *Earth Science Today,* by B. Murphy and D. Nance, p. 331, Brooks/Cole, 1999. All rights reserved. 369, Fig. 2: D. D. Trent. 370, Fig. 11.24: After Nansen and Lebedeff, "Global

Trends of Measured Surface-Air Temperatures," *Journal of Geophysical Research,* 29, no. D11 (20 Nov. 1987), fig. 15, p. 13370. 372, Fig. 11.25: Data from W. S. Broecker, "Chaotic Climate," *Scientific American* 273, no. 5 (Nov. 1995), pp. 62–68. 373, Fig. 1: Data from Greenland Ice Core Project (GRIP), "Climate Instability During the Last Interglacial Period Recorded in the GRIP Ice Core," *Nature* 364, no. 6434 (15 July 1993), pp. 203–207. 375, Fig. 11.27: Data from Baggeroer and Munk, Heard Island feasibility test, *Physics Today,* 45, no. 9 (Sept. 1992), fig. 2, p. 24. 378, Fig. 11.28: © Dan Guravich/National Audubon Society Collection/Photo Researchers. 380, Figs. 11.29, 11.30: D. D. Trent. 381, Fig. 11.31: After R. Byerly, Jr. "The Policy Dynamics of Global Change," *EarthQuest,* Spring 1989 (Univ. Corp for Atmospheric Research, Office of Interdisciplinary Earth Sciences, Boulder, Colo.) pp. 11–24. 384, 385, Figs. 11.32–11.35: D. D. Trent; Fig. 11.36: Austin Post, USGS.

Chapter 12 390, Fig. 1: British War Museum. 395, Fig. 12.5: Soil Conservation Service. 396, 397, Fig. 1: Steve McCurry/Magnum; Fig. 2: From *Meteorology Today,* 6th ed., by C. D. Ahrens, Brooks/Cole, 1999. All rights reserved. 398, Fig. 12.6: Howard G. Wilshire, USGS; Figs. 12.7, 12.8: D. D. Trent. 399, Fig. 12.9: Georg Gerster/Photo Researchers; Figs. 12.10, 12.11: D. D. Trent. 400, Fig. 12.12(a), (b), (d): D. D. Trent; (c): Joseph Allen.

Chapter 13 403, Barrick Gold Corp. 409, Fig. 13.3: B. Pipkin; Fig. 13.4: Wendell E. Wilson, 1985. 410, Fig. 13.5: Clifford Taylor, USGS bulletin 2156. Fig. 13.6: After "Alaska's Oil/Gas and Minerals," *Alaska Geographic,* Vol. 9, No. 4, 1982. 411, Fig. 13.7: Courtesy Kennecott Corp. 412, Fig. 13.8: John E. Gray, USGS bulletin 2156; Fig. 13.9: John Delaney, Univ. of Wash.; Fig. 13.10: Stew Monroe. 413, Fig. 13.11: Bruce Heezen, courtesy Scripps Inst. of Oceanography, Univ. of Calif.–San Diego. 414, 415, Fig. 2: Courtesy Calif. State Library; Fig. 3: Donald B. Sayner collection; Fig. 4: D. D. Trent collection. 416, Fig. 13.12: After "Alaska's Oil/Gas and Minerals," *Alaska Geographic,* Vol. 9, No. 4, 1982. 417, Fig. 13.13: T. K. Buntzen, Silverado Mines U.S., Inc. Fig. 13.15: After "Alaska's Oil/Gas and Minerals," *Alaska Geographic,* Vol. 9, No. 4, 1982. 418, Fig. 13.16: D. D. Trent. 421, Fig. 13.17: Data from Mineral Commodity Summaries, U.S. Bureau of Mines, 1995. 424, 425, Fig. 1: TVA data; Fig. 2: A. Keith, USGS; Figs. 3, 4: Jack A. Muncey, TVA. 426, Fig. 13.18: Courtesy Homestake Mining Co. 428, Fig. 1: Edward Nuhfer, The Citizen's Guide to Geologic Hazards, AIPG, 1993); Figs. 2, 3: Edward Nuhfer, AIPG, 1993; Fig. 4: Chuck Meyers, Office of Surface Mining, Reclamation, and Enforcement. 429, Fig. 13.21: D. D. Trent. 430, Fig. 13.22: Based on USGS data. 431, Fig. 13.23: W. Virginia Geological and Economic Survey; Figs. 13.24, 13.25: D. D. Trent; Fig. 13.26: T. K. Buntzen. 432, 433, Figs. 1, 3: D. D. Trent; Fig. 2: Marion Galant, Colo. Dept of Public Health and Environment. 434, Fig. 13.27: Marc Jay Plumley. 435, Fig. 13.28: Courtesy IMC Fertilizer, Inc. 436, 437, Figs. 13.30, 13.31: D. D. Trent. 439, Fig. 13.32: John E. Gray, USGS bulletin 2156. 440, Fig. 13.33: E. Frank Schnitzer, Oreg. Dept. of Geology and Mineral Industries;

Figs. 13.34, 13.35: ARCO. 444, 445, Figs. 13.36–13.40: D. D. Trent; Fig. 13.41: Ruhrkole Niederrhein AG.

Chapter 14 450, 451, Fig. 1: USGS; Fig. 2: McCabe, Peter, et al., 1993. USGS Circular 115, p. 46; Fig. 3: Keith A. Kenvolden, USGS; 4: Dillon, USGS. 452, Fig. 14.1: USGS data. 453, Fig. 14.2: From *Environmental Science,* 7th ed., by G. T. Miller, p. 512, Brooks/Cole. All rights reserved. 458, Fig. 1: Drake Well Museum, Titusville, Penn. 461, Fig. 14.10: API data; Fig. 14.11: Department of Energy. 462, Fig. 1: From *Environmental Science,* 7th ed., by G. T. Miller, p. 329, Brooks/Cole. All rights reserved. 463, Fig. 2: Bill Nation/Sygma; Fig. 3: Data from J. Chapman, Kenai Fjords National Park. 464, Fig. 14.12: After AAPG *Explorer,* March 1991, p. 39. 466, Fig. 14.13: Energy Information Administration. 467, Fig. 14.14(b): Visuals Unlimited. 468, Fig. 14.15(a): DOE; (b): From *Earth Science Today,* by B. Murphy and D. Nance, p. 555. Brooks/Cole, 1999. All rights reserved. 469, Fig. 14.16: Peggy Foster. 471, Fig. 14.17: Courtesy Hobart King, Mansfield Univ.; Fig. 14.18: After A. C. Stern, ed., *Air Pollution,* Vol. I (San Diego: Academic Press). 473, Fig. 14.19(a), (b): Robert Finkelman, USGS. 475, Fig 2: Data from P. Vrone, *Air Pollution,* Vol. I (San Diego: Academic Press). 476, Fig. 14.20: USGS. 477, Fig. 14.21: Richard Keeler, Southern Calif. Edison Co. 478, Fig. 14.22: D. D. Trent; Fig. 14.24: From "Energy from the Sun," by C. J. Weinberg and R. Williams, *Scientific American,* 1990. Copyright © 1990 Scientific American. All rights reserved. 480, Fig. 14.26: B. Pipkin. 481, Fig. 14.27: Data from W. Duffield, et al., USGS circular 1125, 1994; Fig. 14.28: B.. Pipkin. 482, Fig. 14.29: From *Environmental Science,* 7th ed., by G. T. Miller, p. 534, Brooks/Cole. All rights reserved. 484, 485, Fig. 1: Sovfoto/Eastfoto; Fig. 3: Igor Kostin/Sygma. 487, Fig. 14.34: DOE data 488, Fig. 14.35: DOE data, Office of Environmental Management, August, 1994. 492, Fig. 14.37(a): DOE data; (b): DOE. 493, Fig. 14.40: National Academy of Sciences, NRC. 494, Fig. 14.41: From *Environmental Science,* 7th ed., by G. T. Miller, p. 530, Brooks/Cole. All rights reserved. 496, 497, Fig. 14.43: John Chiasson/Gamma Liaison; Fig. 14.44: D. D. Trent; Fig. 14.45: Angela O'Malley, Cypress Coll., courtesy NAGT; Fig. 14.46: B. Pipkin.

Chapter 15 503, Figs. 1, 2: B. Pipkin. 505, Fig. 15.2: EPA 518, Fig. 15.13: From *Environmental Science,* 7th ed., by G. T. Miller, p. 384, Brooks/Cole. All rights reserved. 506, 507, Figs. 15.3(b), 15.4(b): B. Pipkin. 508, Fig. 1(b): B. Pipkin. 510, Fig. 15.6: B. Pipkin. 511, Fig. 15.7: Dennis Capolongo, Greenpeace. 512, Figs. 15.8(b), 15.9, 15.10: B. Pipkin. 513, Fig. 15.11: B. Pipkin. 515, D. D. Trent photo. 516, Fig. 15.14: EPA data. 517, Fig. 15.15: Mitsuhiro Wada/Gamma Liaison; Fig. 15.16: B. Pipkin. 518, Fig. 15.18: B. Pipkin. 519, Fig. 15.19: Courtesy Richard Proctor. 520, Fig. 15.21: Data from AIPG. 523, Fig. 1(a), (b): Michael J. Philippot/Sygma. 526, Fig. 15.22: B. Pipkin; Fig. 15.23(b): Joe Halbig, Univ. of Hawaii–Hilo. 529, 530, Fig. 15.25: © Robert Rathe; Figs. 15.26, 15.28: B. Pipkin; Fig. 15.27: Adam Hart-Davis/Science Source, Photo Researchers.